Handbook of ELECTROMAGNETIC MATERIALS

Monolithic and Composite Versions and Their Applications

Perambur S. Neelakanta, Ph.D., C. Eng.
Professor of Electrical Engineering
Florida Atlantic University
Boca Raton, Florida

CRC Press
Boca Raton New York London Tokyo

Library of Congress Cataloging-in-Publication Data

Neelakanta, Perambur S.
Handbook of electromagnetic materials: monolithic and composite versions and their applications/Perambur S. Neelakanta.
p. cm.
Includes bibliographical references and index.
ISBN 0-8493-2500-5 (acid-free paper)
1. Electronics–Materials–Electric properties–Handbooks, manuals, etc. 2. Electronics—Materials–Magnetic properties–Handbooks, manuals, etc. 3. Composite matertials–Electric properties–Handbooks, manuals, etc. 4. Composite materials–Magnetic properties–Handbooks, manuals, etc. I. Title.
TK7871.N44 1995
620.1′.1297—dc20 95-2747
CIP

Direct all inquiries to CRC Press, Inc., 2000 Corporate Blvd., N.W., Boca Raton, Florida 33431.

International Standard Book Number 0-8493-2500-5
Library of Congress Card Number 95-2747
Printed in the United States of America 2 3 4 5 6 7 8 9 0
Printed on acid-free paper

Dedicated to my
beloved and respected teachers
Professor (Late) S. K. Chatterjee
and
Professor (Mrs.) R. Chatterjee
of
Indian Institute of Science
Bangalore, India.

". . . The *teacher who walks in the shadow of the temple among his followers gives not of his wisdom, but rather of his faith and his lovingness. If he is indeed wise he does not bid you enter the house of his wisdom, but rather leads you to the threshold of your own mind.. . "*

Kahlil Gibran

FOREWORD I

The growing appreciation among practicing scientists and engineers that the properties of monolithic materials have their origin in the interaction of the internal structure with external fields has led to widespread interest in tailoring materials to meet specific applications. Composite materials, those formed by combining metals, ceramics, polymers and semiconductors in various configurations, offer possibilities for exploiting selected properties of each constituent to optimize overall performance in specific applications. Electromagnetic materials offer a particularly wide range of opportunities for materials design. In order to exploit these opportunities it is essential for the materials engineer and the designers of components to have information on relevant properties of candidate materials in order to optimize materials selection in the design.

As the use of composites in electromagnetic applications grows, this handbook will become increasingly valuable not only to designers of electromagnetic components but also to developers of materilas for this purpose. This work provides a single source for information on the fundamentals of electromagnetic theory, relevant properties of a wide range of materials and models for determining the electromagnetic response of composites composed of different materials. All involved in the development, design and construction of electromagnetic components will welcome the convenience of having such a readily accessible reference for these subjects.

January 1995

Craig S. Hartley, Ph. D., P. E.
Fellow ASM, Fellow AAAS.
Dean, College of Engineering
Florida Atlantic University
Boca Raton, Florida

FOREWORD II

As we enter the 21st century, the roles played by materials in our society become more demanding requiring materials with specific properties. The understanding of the theoretical basis for specific behaviors and knowledge of what materials are available is becoming increasingly critical and complicated. Such knowledge is critical because of the increasingly specific materials properties that are required for specific applications. Such knowledge is complicated because of the wide variety of materials that are available. Electronics is a watershed, critical area for the sustained development of our complex, technology dependant society and the understanding and continued development of electromagnetic materials associated with electronics is a major critical aspect. Perambur Neelakanta's book is a valuable contribution to this effort. It presents the necessary theoretical background laying the foundation for future advances and for the materials presented in the book. The foundational material is presented at a level that allows those not directly involved in the area to "come up to speed". It also gives a "state-of-the-art" presentation of materials that are currently available arranged in a sound systematic fashion and of the applications of these materials. Finally, potential applications are given.

This book also describes other critical uses of electromagnetic materials in such diverse areas as shields, smart materials, bioelectromagnetic phantoms and ferrites.

While the book is not intended as a textbook, it would be a quite acceptable text for an advanced course dealing with electromagnetic materials.

I fully recommend this book to scientists and engineers interested in the broad field of electromagnetic materials. This book is a valuable contribution that will have a major impact in the ongoing evolution of electromagnetic applications.

Charles Carraher, Jr., Ph. D., Fellow AIC.
Dean, College of Science
Florida Atlantic University
Boca Raton, Florida
and
Co-chair, Polymer Education Committee
American Chemical Society

January 1995

PREFACE

The increasing importance of electromagnetic applications of monolithic and composite materials in modern electrical and electronic engineering has made it highly imperative that engineers and scientists be prepared to comprehend and utilize these materials effectively in relevant engineering designs and scientific applications. This handbook is intended to explain the basic concepts underlying the electromagnetic properties of materials, address the ways of deploying them in modern applications, and supply pertinent data not heretofore compiled in one volume.

This handbook has the objective of providing a comprehensive record based on the foundation of established material science considerations relevant to electromagnetic material properties at micro- and macrolevels. Its main purpose is to extend rapid information on the properties observed and techniques and ideas developed by the scientists and technologists on the existing as well as futuristic versions of electromagnetic materials.

It attempts to cover the topics in logical sequence with necessary mathematics and concepts of physics in a self-contained manner conducive for a projected audience comprised of students, material scientists, technologists, and practicing engineers.

The introductory sections are written to acquaint the reader with electromagnetic fundamentals supplemented by the details on the interaction of electromagnetic force fields with matter on a microscopic scale. This is done primarily to provide a foundation for the topics that follow and to establish a background for more advanced and macroscopic considerations.

Subsequent chapters extend the microscopic electromagnetic properties of materials to the macroscopic level so as to depict the electric and/or magnetic behavior of a given material by a set of macroscopic measurable quantities. Hence, the basic aspects of monolithic materials as well as the way to perceive a realization of multicomponent composites with a prescribed extent of electromagnetic properties at the macrolevel are explained. Emphasis is placed to elaborate the theoretical aspects of electromagnetic interaction with monolithic as well as multiconstituent materials; and engineering potentials of using these materials are explained in detail. Throughout this handbook, examples (*via* tables, charts, graphs, etc.) are furnished from the viewpoint of practical applications of electromagnetic materials in various fields.

In recent times, use of electromagnetic materials in high technological and sophisticated engineering items has grown stupendously. Electromagnetic shields, radar absorbing materials, bioelectromagnetic phantoms, smart materials, electromagnetically active surfaces, exotic magnets, application-specific electrodes, ferrites, electrets, etc. are a few to mention. Beneficiaries of wide-scale applications of electromagnetic materials include aerospace, electrical, chemical, and medical industries.

Readers of this handbook will get an overview of the advances made in electromagnetic materials and appreciate state-of-the-art potentials that exist in the relevant research, development, and manufacturing areas. Also, the yet-to-come strategies, new concepts, and translation of existing ideas from laboratory level to manufacturing stage are presented to portray the futuristic scope of the subject matter.

This handbook will help practicing engineers to acquire information on the topics of interest. It will be a companion to researchers in material science and can aptly serve as a reference text for a college-level elective course in an electromagnetic and/or material science discipline.

In short, spurred on by imminent aspects of electromagnetic materials, this handbook would be an immensely practical and compact encyclopedia of such materials with the citation of vast references from the world archival literature.

Boca Raton 1995 P.S. Neelakanta

Contents

CHAPTER 1

Introduction

1.1 General

Recent technological advancements are largely indebted to the development of better materials. It is well known, for example, that the prolific growth in engineering realms of industries engaged in aerospace projects, electrical/electronic products, nuclear engineering, and similar activities is strictly due to the recent breakthroughs made in material science and technology. The demand for unique electromagnetic properties warranted in the spheres of civilian, military, and aerospace technology has set the emergence of a new class of monolithic materials and composites which break down remarkably the performance barriers encountered with conventional materials used in electromagnetic applications.

Materials are the basis of the produce in any industry and a great deal has been known for many years about the properties of simple materials commonly used in the manufacturing of engineering items. Metals, alloys, ceramics, chemicals, plastics, etc., for example, have been well comprehended in their engineering applications and their technology is a long-established subject with a sound understanding of the associated physical and chemical attributions.

While technology decides how a material can be utilized appropriately, the science of materials attempts to decipher why materials behave as they do. The success of development of materials and their applications is governed by the abundance of results perceived in material science *vis-a-vis* ingenious methods of harnessing the material behavior *via* technological strategies. In this cohesive blend of scientific and technological approaches, the deeper understanding and utilization of materials have come from realizing that the bulk properties of matter depend largely on the way microscopic constituents of the matter (at atomic and/or molecular level) are arranged under compaction. In the modern era, scientific investigations at such microscopic levels have been feasible owing to the progress made in realizing advanced instruments like the electron microscope.

With the advent of studies made on simple materials, state-of-the-art technological advancements have facilitated the formation of newer monolithic materials as well as the synthesis of composite materials constituted by combining two or more materials, so that the combination has the most desirable features of the constituents. Again, modern instruments have given access to seeing the morsels of the materials at their microscopic level and enabled the assessment of their behavior to a fair degree of accuracy. As a result, exotic versions of monolithic and composite materials have become a new generation of electromagnetic materials, widely accepted and used in a variety of disciplines, such as aerospace, medical, nuclear, and electrical/electronic technology.

Inasmuch as material science itself is a rapidly advancing subject, the science of realizing new and better monolithic and/or composite materials has become an even more challenging subject with exciting prospects for the future. It has the potential to facilitate the new materials for specific, and in many cases, unique applications which cannot be otherwise achieved with simple (conventional) materials. The making of application-specific monolithic materials and the conjecture of multicomponents in composites provide greater flexibility to develop an end product with optimal material performance.

Being a subset of material science, the technological art of emulating special purpose monolithic and composite materials has been a topic of study in the last few decades. As a result, several versions of such materials have been developed and beneficially utilized in recent times.

1.2 Interaction of Electrical and Magnetic Fields with Materials

The electromagnetic properties of materials at the macroscopic level are decided by the mode(s) of interactions involved between the material constituents at the microscopic (atomic and/or molecular) level and the electric and/or magnetic force fields. Classification of materials on an electrical engineering perspective is based on these interaction mechanisms and the corresponding responses perceived at the macroscopic level.

The electromagnetic force field is dictated basically by two phenomenological situations: (i) Force field due to a system of static (electric) charges and (ii) force field due to the dynamics of charges in motion. The electric charge is an attribution to the fundamental particles of the matter, namely, electrons and protons. The characteristics of charges of electrons and protons are that they are equal in magnitude, but opposite in sign. Designating the electrons as *negative charges* and protons as *positive charges*, the following general characteristics of electric charges can be stated :

- Electric charge is quantized and appears as integral multiples of the charge of the electron (law of charge quantization).
- Charges cannot be created or destroyed (conservation principle of charges).
- Charges exist in two distinct classes specified as positive and negative charges.
- Charges of same polarity repel each other and those of opposite attract each other.

To understand the mode(s) of interaction between electric charges present in a material and the applied electromagnetic force field, it is necessary to briefly discuss the underlying microscopic aspects of materials, in general, and of solids, in particular, specific to the static and dynamic status and availability of charges which participate in the interaction process with the applied (external) force fields.

1.3 Microscopic Structure of Matter

Solids, in comparison to most liquids and gases, have their molecules or atoms arranged much closer to each other in an almost regularly arranged pattern which is called a *crystalline state*. In some materials, this arrangement could, however, be less orderly, but in all solids, the atoms are tightly compacted. Individual (isolated) atoms are characterized by discrete *energy levels* surrounding it, in which the electrons form orbital paths around the parent nuclei (*Bohr's model*). The energy level indicated refers to the potential energy attributed to the electrons in a particular orbit as decided by the force of interaction between them and the protons in the parent nuclei. However, when compacted as in solids, the interaction between the atoms leads to the formation of *bands of energy levels* as illustrated in Figure 1.1. That is, when atoms are brought together to constitute a solid, the discrete atomic energy levels change under the influence of neighboring nuclei. The total number of energy levels, however, remains constant but the discrete levels that correspond to a given isolated atomic level become closely packed in energy forming a *band* (which can be viewed as *a continuum of energy states*).

The occupancy of electrons in the energy levels within a band is dictated by what is known as *Fermi-Dirac statistics*. At absolute zero temperature individual atoms are in the ground state. It implies that the electrons reside and fill all the energy levels below a reference energy level known as the *Fermi level*. However, at higher temperatures (T °K), the electrons acquire thermal energy (proportional to the temperature and given by $k_B T$, where k_B is a constant of proportionality known as the *Boltzmann constant*) and may shift to the band of energy levels higher than the Fermi level. Thus, at 0 °K, the Fermi level specifies the probabilistic dividing line between the fully occupied energy levels below it and fully empty energy levels above it.

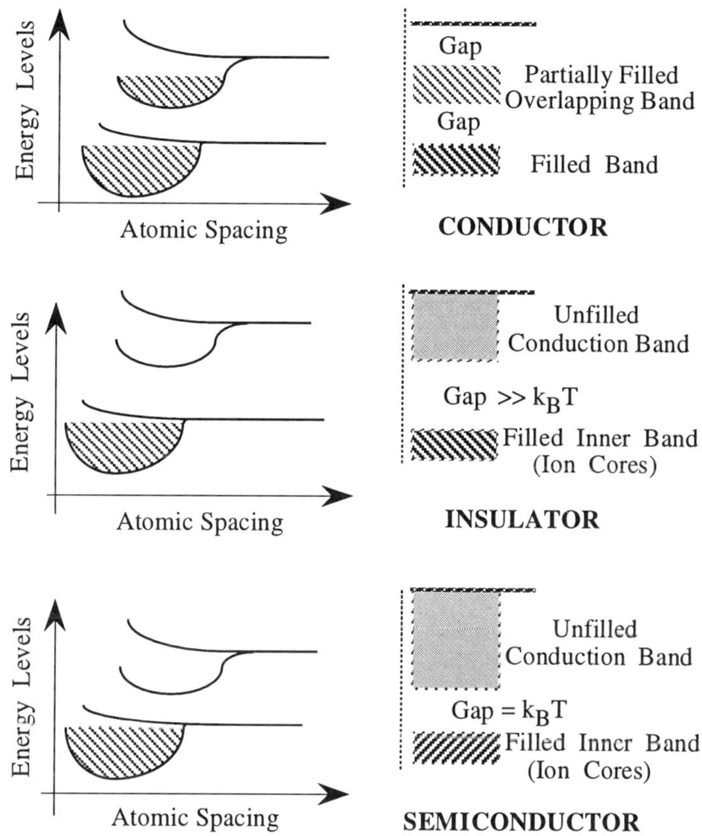

Figure 1.1 Energy bands of conductors, dielectrics, and semiconductors.

The band of energy levels closest to the nuclei which are totally filled at 0 °K is known as the *valence band*. Above the uppermost occupied shells of this band, there is a continuum of disallowed states which is designated as the *forbidden band* and an (unoccupied) empty band (at 0 °K) lies above these disallowed states of energy levels. At temperatures higher than 0 °K, the thermal energy imparted to the electrons in the valence band could allow the electrons to "jump" across the forbidden gap and reach the unoccupied stretch of energy levels above the forbidden band. The electrons which reach these energy levels are less influenced by the (coulombic) force of attraction by the nuclei and hence are *free* to move. They are called *free electrons* and the band in which the free electrons are present is known as the *conduction band*. On the contrary, the electrons which reside in the valence band are tightly bound to the parent nuclei and are hence known as *bound electrons*.

The extent of forbidden gap permits the classification of a material as a *dielectric (insulator)*, a *semiconductor*, or *a conductor*. The simple *band theory* suggests that the availability of a number of free electrons in the conduction band (at a given temperature) depends on the forbidden gap energy. Where the forbidden gap is very large, it is possible that no free electrons are available (even at room temperature) for conduction. Commonly, this refers to the electrical insulation property of a material which is then designated as a *dielectric*.

With a small forbidden gap energy, the availability of electrons in the conduction band though limited, permits an electrical conduction. Such materials are classified as *semiconductors*. In the so-called conducting materials (such as metals), the band structure is such that the conduction and valence bands may overlap, permitting a copious amount of free electrons available at ambient temperatures.

More detailed microscopic and macroscopic aspects of dielectrics, conductors, and semiconducting materials are presented in Chapters 2, 9, and 10, respectively.

1.4 Dielectric Characterization of Materials

The force of attraction or repulsion between two systems of charges perceived in a medium depicts a force field (or *electric field*) and its magnitude is specified by a quantity called the *electric field intensity* $\boldsymbol{E}$, defined as the net force (of attraction or repulsion) exerted on a unit charge due to the presence of other charges. This is determined by Coulomb's law (an experimental postulation) and is given by:

$$\boldsymbol{E} = q\boldsymbol{a_r}/4\pi\varepsilon_o r^2 \qquad \text{newton/coulomb} \qquad (1.1)$$

where $\boldsymbol{a_r}$ is the unit vector along $\boldsymbol{r}$, r being the distance between the charge q and the point of observation where $\boldsymbol{E}$ is being measured; and, ε_o, the *absolute permittivity of free space*, is a proportionality constant (of magnitude $1/36\pi \times 10^{-9}$ farad/meter). In the elucidation of $\boldsymbol{E}$ as per Equation 1.1, it is assumed that the locales of the charge q and the unit charge are in free space (vacuum). Should these charges reside in a dielectric region (medium) other than free space, ε_o is replaced by $\varepsilon_o\varepsilon_r$ where ε_r is referred to as the *relative permittivity* or *dielectric constant* of the medium. For the free space, therefore, $\varepsilon_r = 1$.

The presence of electric field in a dielectric medium could be visualized by a system of *flux lines* (of forces). The density of these flux lines (per meter^2) known as the *electric flux density* or *electric displacement* ($\boldsymbol{D}$) is given by:

$$\boldsymbol{D} = \varepsilon\boldsymbol{E} \qquad (1.2)$$

where $\varepsilon = \varepsilon_o\varepsilon_r$. Applying Equation 1.1 to a closed surface S surrounding a charge, it can be shown that

$$|\boldsymbol{D}| = |\varepsilon \int_S \boldsymbol{E}\bullet d\boldsymbol{s}| = q \qquad (1.3)$$

which is known as *Gauss' law*.

Due to the existence of a force field in a region containing a system of charges, work is necessary to be exerted on a charge (q) in moving it from one point to another in the field domain. The corresponding work done per unit charge is called the *electrostatic potential difference* (V) between the points given by:

$$V = -\int_{\ell} \boldsymbol{E}\bullet d\boldsymbol{\ell} \qquad \text{joule/coulomb (volt)} \qquad (1.4)$$

or alternatively in the differential form as a *gradient*, $\boldsymbol{E}$ and V are specified by:

$$E = -\boldsymbol{\nabla} V \qquad (1.5)$$

The unit of V, namely, joule/coulomb is called a *volt* and $\boldsymbol{\nabla}$ is the *gradient operator* known as the *del operator*. Further, the work required to bring a test charge from infinity to the point of observation is called the *absolute potential* at that point.

The net flux emanating from a unit volume is defined by the *divergence operator** as follows:

$$\varepsilon\iiint \boldsymbol{\nabla}\bullet\boldsymbol{E}\, d_{vol} = \varepsilon\iint \boldsymbol{E}\bullet ds = \iiint q_{vol}\, d_{vol} \tag{1.6}$$

where q_{vol} is the charge contained in the elemental volume d_{vol}. It follows from the above relation that:

$$\varepsilon\boldsymbol{\nabla}\bullet\boldsymbol{E} = \boldsymbol{\nabla}\bullet\boldsymbol{D} = q_{vol} \tag{1.7}$$

or in terms of the potential function V,

$$\boldsymbol{\nabla}\bullet(\boldsymbol{\nabla}V) = \boldsymbol{\nabla}^2 V = -\,q_{vol}/\varepsilon \tag{1.8}$$

which is known as *Poisson's equation.* If $q_{vol} = 0$, $\nabla^2 V = 0$ refers to *Laplace's equation.*

The electric property of a material with $\varepsilon_r > 1$ is decided by the polarizable atoms or molecules in the material. A detailed description of the relevant considerations of dielectric properties is presented in Chapter 2. When a material is characterized by the parameter ε_r, it represents a *dielectric material.* Further a dielectric material is *homogeneous* if its dielectric property (as specified by ε_r) does not vary from point-to-point in the medium, and it is *linear* when ε_r remains constant when the magnitude of the charges inducing an electric field force in it is varied. Also, a dielectric material could be classified as *isotropic*, if the dielectric characteristics are independent of direction.

1.5 Chemical Bonding and Covalent Structure of Dielectrics

The interatomic forces keep the atoms in a solid as conglomerated clusters. The binding forces between them are called *chemical bonds.* These bonds are made up of attractive and repulsive forces that tend to hold atoms (ions and molecules as well) at a different spacing such that the opposing forces just balance out. The process of holding the atomic unit field together is known as *bonding.* The *bonding strength* classifies the chemical bonds as *primary* or *secondary bonds.* The interatomic bonds constitute the *primary bonds* whereas the intermolecular bonds are the *secondary bonds.*

The type of bonding is dependent on the atomic structure. Considering the permissible discrete electronic orbits around the nuclei (each associated with a definite potential energy or *stationary state*), the maximum number of electrons permitted in each orbit is decided by $2n^2$ law, where n denotes the orbital number. Thus, the first orbit (closest to the nucleus) holds a maximum of 2 electrons, the second 8 electrons and so on. When a particular orbit has the maximum number of electrons (as per $2n^2$ *law*), it is said to be *saturated.* When the outer-most orbit or the shell has less than the maximum, it is ready to accept electrons to attain the saturation. Likewise, when the outermost orbit has more electrons than required for saturation, the excess electrons are ready to be given up. The electrons in the outermost orbit or shell are known as *valence electrons.* They essentially determine the chemical activity and binding patterns at atomic and/or molecular levels. Implicitly, they decide the bulk material properties for engineering applications.

Electrons can break free of atoms with excess valence electrons producing *positive ions* and unite with atoms having an incomplete outer shell to form *negative ions.* The positive and negative ions so formed constitute a strong bond known as *ionic bonding.* A typical

---

* *Divergence operator*: This refers to the *dot product* of the del operator and a vector.

example is: $Na^+ + Cl^- \Rightarrow NaCl$ which is a totally neutral chemical compound. Generally, ionic compounds have electrical insulating properties.

In certain cases, the linked atoms find it possible to produce saturated shells by *electron transfer*, thereby achieving a stable electronic structure. For example, a tetravalent atom (such as silicon) can share its four valence electrons with neighboring silicon atoms constituting a *covalent bonding* (Figure 1.2). Electrical conductivity of such covalent-bonded materials depends on the strength of bonding which, in turn, is decided by the forbidden gap energy. Weak bond strength may render the materials as partial electrical conductors, which is the case with semiconducting materials (like silicon, germanium, etc.). However, when the forbidden gap energy is excessive, as in dielectric materials, the associated bonding strength is too overwhelming to permit the disassociation of free electrons available for electrical conduction. In other words, the material behaves as an insulator. Covalent-bonded engineering materials are ceramic or polymeric in nature.

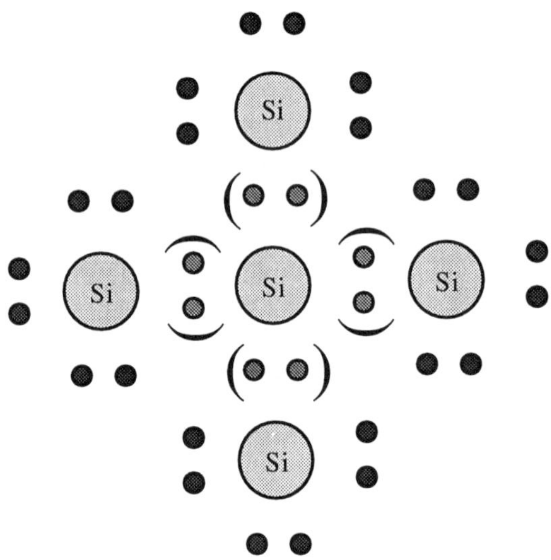

Figure 1.2 Covalent bonding.

Another class of *primary bonding* refers to *metallic bond*. In certain elements having a small number of valence electrons which are loosely held (due to the absence of forbidden gap) they become available as free electrons for electronic conduction. Typically as in metals, the bonding structure is such that every atom contributes its valence electrons to the formation of a free-electron population, which are highly mobile. Therefore, metals have extremely high electrical conductivity.

The *secondary bonds* known as weak bonds are due to *van der Waals forces* which link (in certain materials) the molecules having asymmetric charge distribution so as to form a stretch of linked dipoles as shown in Figure 1.3. The linked dipole pattern is known as *polarization*.

A second version of secondary bonding results from the momentary polarization caused by random spatial dispersion of electrons. The bonding so formed has, however, weak strength.

The third kind of weak bond refers to a *hydrogen bridge* caused by a hydrogen nucleus held in bonding by two electrons of two different atoms forming a triangular linkage. This is perceived in biological materials.

Mostly, secondary bonds are insignificant in engineering aspects of bulk materials.

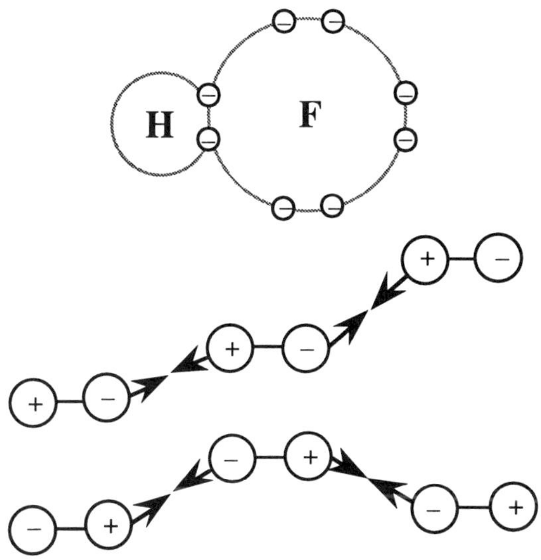

Figure 1.3 Dipole bond.

1.6 Displacement Current in a Dielectric Material

When the system of charges present in a dielectric medium varies with time, the displacement vector (or the electric flux lines) alternates in direction at a rate specified by the time dependency of the charges. The rate of change of electric flux across the dielectric material would force a corresponding movement of charges in the closed path (circuit), of which the dielectric medium is a constituent part. This time-varying current (or flow of charges) so induced in the circuit is specified by :

$$\boldsymbol{J}_D = \varepsilon \partial \boldsymbol{D}/\partial t \qquad \text{ampere/meter}^2 \qquad (1.9)$$

where J_D represents the displacement current per unit cross-sectional area (or the *displacement current density*).

1.7 Flow of Charges in Conductors

As indicated before, the "free electrons" available in the conduction band at temperatures greater than 0 °K could constitute a flow, the rate of which is defined as the *conduction current.* That is, electric conduction current, a macroscopic quantity, is the result of motion of many microscopic charges. Such a flow in a conducting material requires an application of an electric field to force the charges to move (in the direction of the electric vector field). This flow of charges constitutes a *conduction current density (J_c)* given by:

$$\boldsymbol{J}_c = ne\boldsymbol{v} \qquad \text{ampere/meter}^2 \qquad (1.10)$$

where n is the number of charges, e is the electronic charge, and $\boldsymbol{v}$ is the velocity vector along the applied electric force field, $\boldsymbol{E}$.

Since the associated force field in the movement of an electron is $e\boldsymbol{E}$ newton, the corresponding acceleration can be specified by Newton's second law as:

$$d\boldsymbol{v}/dt = e\boldsymbol{E}/m \qquad \text{meter/sec}^2 \qquad (1.11)$$

where m is the electronic mass. Upon integration, the above equation yields:

$$v = (eE/m)t \qquad \text{meter/sec} \qquad (1.12)$$

assuming that the initial velocity is zero.

Movement of electrons in a solid material after being accelerated may result in collisions with the lattice. Denoting the *mean-collision time* between successive collisions as τ, it represents the time required for the electron velocity to be randomized by the electron-lattice collisions when electrons starting from rest acquire a *drift velocity*. The average drift velocity (v_d) due to the applied electric field can then be written as:

$$\langle v \rangle = v_d = (eE/m)\tau \qquad \text{meter/sec} \qquad (1.13)$$

The lattice collision represents energy loss and a damping mechanism. With the inclusion of this mechanism, the equation of motion of electrons in a conductor is given by:

$$m(dv_d/dt) + (mv_d/\tau) = eE \qquad (1.14)$$

where m/τ represents the *coefficient of friction.* Further, in terms of the drift velocity, the conduction current density becomes:

$$J_c = (ne^2\tau/m)E \qquad \text{ampere/meter}^2 \qquad (1.15)$$

Here, the constant of proportionality between J_c and E, namely, $ne^2\tau/m$, is called the *electrical conductivity* (σ) of the material. Considering an elemental volume of material of side $d\ell$ as illustrated in Figure 1.4, the conduction current (I_c) flow can be specified as:

$$I_c = J_c(d\ell \times d\ell) = \sigma(Ed\ell)d\ell \qquad (1.16a)$$

That is,

$$I_c = J_c dA = \sigma V\, dA/d\ell \qquad (1.16b)$$

where V $(= E \bullet d\ell)$ is the electrical potential across the cube of cross-sectional area dA (Figure 1.4). Hence, it follows that:

$$\begin{aligned} V/I_c &= (1/\sigma)\,(d\ell/dA) \\ &= R \qquad \text{volt/ampere (ohm)} \qquad (1.17) \end{aligned}$$

where R is called the macroscopic or bulk *resistance* of the material. The above relation is the well-known *Ohm's law*. Further, $(1/\sigma)$ is called the *resistivity* of the medium and denoted by ρ. The resistance R has the dimension volt/ampere which is called *ohm*. The corresponding unit for the resistivity is *ohm-meter* and that of the conductivity is $(\text{ohm-meter})^{-1}$ or *siemen/meter.*

The collision loss mechanism manifests as thermal (heat) energy dissipation and constitutes a power loss due to electric conduction. Using *Ohm's law*, this power loss $P = I_c^2R$ or V^2/R watts. The corresponding thermal energy dissipated in the conducting material over a time t second is $W = I^2Rt$ joule.

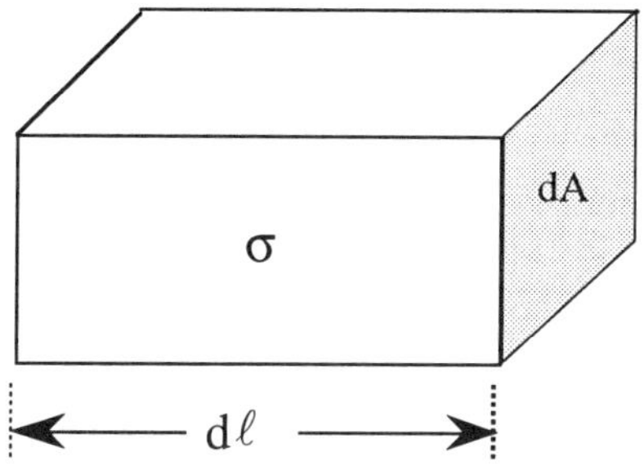

Figure 1.4 An elemental volume of a conducting material.

The conductivity σ is temperature dependent inasmuch as the collision mechanism (or the number of collisions per unit time) is influenced by the thermal energy that the material may acquire as a result of increased temperature. Therefore, the mean collision time τ is a function of temperature and so is the conductivity of the material.

In terms of this conductivity parameter, an ideal dielectric can be characterized with σ equal to zero and a semiconductor is a material with a conductivity lower than that of a metallic conductor. More details on electrical conductivity are furnished in Chapter 9.

1.8 Magnetic Characterization of Materials

While the presence of charges causes an electric field force, the movement of such charges induces a *magnetic field force*. That is, the electric current (either due to finite rate of change of charges or due to time rate of change of electric flux) in a material is associated with a force field known as the *magnetic field*. There are three experimental postulations which describe the characteristics of the magnetic field *vis-a-vis* the dynamics of charge flow or electric-flux change. These experimental laws are as follows:

1.8.1 *Biot-Savart's law*

The flow of a current I through an elemental length $d\boldsymbol{\ell}$ will produce a magnetic field $\boldsymbol{B}$ which at a distance r from the element is given by:

$$d\boldsymbol{B} = (\mu/4\pi)\, I \times \boldsymbol{a}_r\, d\boldsymbol{\ell}/r^2 \tag{1.18a}$$

or,

$$d\boldsymbol{H} = d\boldsymbol{B}/\mu = I \times \boldsymbol{a}_r\, d\boldsymbol{\ell}/4\pi r^2 \tag{1.18b}$$

where $\boldsymbol{a}_r$ is a unit vector along r and μ is a material constant of the medium wherein the magnetic field is perceived. In free space $\mu = \mu_o$ with $\mu_o = 4\pi \times 10^{-7}$ henry/meter, known as the *absolute permeability* of free space. In other media, $\mu = \mu_o\, \mu_r$ where μ_r is called the *relative permeability* of the medium concerned. The permeability is a parameter that depicts the magnetic property of the medium and $\boldsymbol{H}$ refers to a force field (analogous to the electric force field $\boldsymbol{E}$). The magnetic field $\boldsymbol{B}$ represents the flux lines of force per unit area. Therefore, it is termed as the *magnetic flux density*. It is directly proportional to the *magnetic field intensity* $\boldsymbol{H}$ by the relation $\boldsymbol{B} = \mu\boldsymbol{H}$. Inasmuch as $\boldsymbol{B}$ depicts a force field, if another current element is brought into $\boldsymbol{B}$, a force of interaction (attraction or repulsion) would result. This force is quantified by *Ampere's force law*, as detailed below.

1.8.2 *Ampere's force law*

When a current element $I d\boldsymbol{\ell}$ is immersed in a magnetic field $\boldsymbol{B}$, it experiences a force $d\boldsymbol{F}$ given by the following *Ampere's force law*:

$$dF = \boldsymbol{I} \times \boldsymbol{B} d\boldsymbol{\ell} \qquad \text{newton} \qquad (1.19)$$

Alternatively, two current elements $\boldsymbol{I}_1 d\boldsymbol{\ell}_1$ and $\boldsymbol{I}_2 d\boldsymbol{\ell}_2$ will experience a mutual magnetic force (F_{12}) of interaction between them as per the relation:

$$F_{12} = (\mu_o/4\pi)\,(\boldsymbol{I}_1 d\boldsymbol{\ell}_1) \times (\boldsymbol{I}_2 d\boldsymbol{\ell}_2) \times \boldsymbol{a}_{r12}/r^2_{12} \qquad (1.20)$$

where r_{12} is the distance between the current elements and $\boldsymbol{a}_{r12}$ is the unit vector along r_{12}. The above expression is similar to Coulomb's law concerning mutual force of interaction between two static charges kept separated by a distance.

From Ampere's force law, it can be seen that the unit for B is newton/ampere-meter, which is written simply as a *tesla or weber/meter*^{2}. (In CGS unit, 1 *tesla* = 10^4 *gauss*.) The corresponding unit for H is *ampere-turn/meter*.

In view of the above considerations, magnetic fields result from the charges in motion; that is, all magnetic effects are attributable to the time-varying system of electric charges which governs the microscopic aspects of *magnetism*. The material response to magnetism is decided by its extent of interaction with the magnetic field force. The parameter that quantifies such an interaction is the magnetic permeability.

The orbital and spin motion of electrons in materials constitute currents, the vector sum of which when finite results in *ferromagnetism* (as in iron), which when "retained" causes *permanent magnetism*. The ferromagnetic materials are more easily influenced by external magnetizing force (or *magnetization*). On the contrary, materials (like plastic, copper etc.) are not susceptible to magnetization and are termed as d*iamagnetic materials*. Certain materials (like aluminum) are feebly magnetizable and are designated as *paramagnetic materials*. The ferromagnetic property is identified by the large value of μ_r whereas for diamagnetic materials $\mu_r = 1$. More discussions on magnetic materials are presented in Chapter 14.

1.8.3 *Ampere's circuital law*

Considering the nondivergence (solenoidal) property of magnetic field $\boldsymbol{B}$ (with neither its source nor sink specified), the closed-path integration of $\boldsymbol{B}$ can be related to the electric current that sets up this magnetic field. Again, the extent of such a relation is dictated by the magnetic property of the medium, namely, the permeability μ. Hence, *Ampere's circuital law* postulates that over a closed path,

$$\oint_{\text{closed path}} \boldsymbol{B} \bullet d\boldsymbol{\ell} = \mu I \qquad \text{weber-meter} \qquad (1.21)$$

where I is the electric current. Using the constitutive relation $\boldsymbol{B} = \mu \boldsymbol{H}$, it follows that over a closed path,

$$\oint \boldsymbol{H} \bullet d\boldsymbol{\ell} = I \qquad \text{ampere} \qquad (1.22)$$

In other words the *magnetic force* (or *the magnetomotive force, mmf*) integrated over the closed path (wherein the magnetism is perceived) is equal to the electric current enclosed. When the current is distributed through the medium with a density $\boldsymbol{J}$, Ampere's circuital law can be written modified as:

$$\oint \boldsymbol{H} \bullet d\boldsymbol{\ell} = \iint_A \boldsymbol{J} \bullet d\boldsymbol{A} \qquad (1.23)$$

where A refers to the area bounded by the closed path of B. Thus, only current that flows through the area A contributes to the line integral of $\boldsymbol{H}$.

1.9 Electromagnetic Induction

A time-varying magnetic field (or a time-dependent interaction of magnetic field) on a conductor will induce a potential difference across the conductor terminals. The voltage so induced is given by:

$$V = - d\phi/dt \tag{1.24}$$

where ϕ is the *total magnetic flux linkage* with the conductor and the negative sign depicts that the current flow through the conductor (taken as a closed path) due to the induced voltage produces a magnetic field which will oppose the change in flux. The above relation is known as *Faraday-Lenz's law.*

At the microscopic level Faraday-Lenz's law is written as:

$$\oint \boldsymbol{E} \bullet d\boldsymbol{\ell} = -(\partial/\partial t) \iint_A \boldsymbol{B} \bullet d\boldsymbol{A} \tag{1.25}$$

The above equation (1.25) describes the work done per unit charge to its motion under the force of $\boldsymbol{E}$ which when added around the circuit length must be equal to the *electromotive force* (potential) induced in the circuit. Again, A represents the area enclosed by the closed path of $\boldsymbol{B}$.

1.10 Macroscopic Effects in Electromagnetic Materials

As a result of the various fundamental electric and magnetic phenomena at the microscopic level as described above, materials at the macroscopic level exhibit corresponding characteristic electric and magnetic behavior. That is, in bulk form, the response of the material while interacting with electric and magnetic field forces is decided by a set of fundamental relations known as *Maxwell's equations* detailed below:

1.10.1 *Maxwell's equations*

$$\iint \boldsymbol{D} \bullet dA = \iiint q_v \, d_{vol} \quad \Leftrightarrow \quad \nabla \bullet \boldsymbol{D} = q_v$$

$$\iint \boldsymbol{B} \bullet dA = 0 \quad \Leftrightarrow \quad \nabla \bullet \boldsymbol{B} = 0$$

$$\oint \boldsymbol{H} \bullet d\boldsymbol{\ell} = \iint (\boldsymbol{J}_c + \boldsymbol{J}_d) \bullet dA \quad \Leftrightarrow \quad \nabla \times \boldsymbol{H} = \partial D/\partial t + \boldsymbol{J}_c$$

$$\oint \boldsymbol{E} \bullet d\boldsymbol{\ell} = -\partial/\partial t \iint \boldsymbol{B} \bullet dA \quad \Leftrightarrow \quad \nabla \times \boldsymbol{E} = - \partial B/\partial t \tag{1.26}$$

1.10.2 *Constitutive relations*

$$\begin{aligned} \boldsymbol{D} &= \varepsilon E \\ \boldsymbol{B} &= \mu H \\ \boldsymbol{J} &= \sigma E + \varepsilon \, \partial E/\partial t \end{aligned} \tag{1.27}$$

The interaction of electromagnetic energy with macroscopic solids can be described by the *Helmoltz wave equation* derived from the aforesaid Maxwell's equations and the constitutive relations. The electromagnetic wave equation describes the electric and magnetic field forces as functions of time and position in a given material. Written explicitly, the wave equation is:

$$\nabla^2 \boldsymbol{E} = \mu\varepsilon\,(\partial^2 E/\partial t^2) + \sigma\mu\,(\partial E/\partial t) + \nabla(q/\varepsilon) \tag{1.28a}$$

or,

$$\nabla^2 \boldsymbol{H} = \mu\varepsilon\,(\partial^2 \boldsymbol{H}/\partial t^2) - \nabla \times \boldsymbol{J} \tag{1.28b}$$

The set of electric and magnetic fields $(\boldsymbol{E},\boldsymbol{H})$ perceived in a material is therefore dictated by the material parameters, namely, permittivity (ε), permeability (μ), and conductivity (σ); and by the presence of charges (q), if any. As characterized by these parameters, the electromagnetic materials can be classified macroscopically as follows:

- Lossless dielectrics (closest approximation of perfect insulators)
- Lossy dielectrics
- Perfect conductors
- Good conductors
- Semiconductors
- Lossless magnetic materials (nonconductors)
- Lossy magnetic materials (conductors)
- Materials with free charges

Not only do (ε,μ,σ) parameters decide the extent of macroscopic behavior of materials *vis-a-vis* applied electromagnetic fields, but also they characterize a bulk material as:

- A linear or a nonlinear medium;
- A homogeneous or a heterogeneous medium; and/or,
- An isotropic or an anisotropic medium.

In view of the above considerations the subsets of generic electromagnetic materials are as listed in Table 1.1.

The attributions of linearity (or nonlinearity), isotropicity (or anisotropicity), and homogeneity (or inhomogeneity) can be observed in any of the above subsets. The characteristics of homogeneity render the material as *monolithic*; on the contrary, the heterogeneous property that portrays the distinct spatial variations of the parametrers (ε,μ,σ) is invariably due to the presence of multiconstituents in the material concerned. In otherwords, heterogeneity in the material property implicitly refers to the *composite* nature of the material.

Another interesting macroscopic aspect of electromagnetic materials is that the material behavior or response to electromagnetic forces is *frequency dependent.* That is, the material parameters ε, μ, and σ are in general complex, namely, ε^*, μ^*, and σ^*.

Such complex electromagnetic parameters decide the spatial extent (or *depth*) to which the electromagnetic energy can couple to the material at a given frequency. In other words, considering the bulk of electromagnetic material, it is possible that only a part of it would be macroscopically interacting with the applied electromagnetic energy under dynamic conditions. Then, the depth of penetration (*skin depth*) of the electromagnetic energy in the material could be so small that the interaction may just be limited to the surface only. This happens at very high frequencies. In such cases, the engineering considerations on the utility and behavior/response of the material therefore refer not only to the *bulk characteristics* but also to the *surface properties* of the material.

Table 1.1 Subclassifications of Dielectrics, Conductors, and Magnetic Materials

Dielectric materials

Perfect dielectrics ($\sigma = 0$)

Imperfect (lossy) dielectrics ($\sigma \neq 0$)
- Low-loss dielectrics
- High-loss dielectrics

Conducting materials

Perfect conductors ($\sigma = \infty$)

Conductors
- Low resistivity conductors
- High resistivity conductors

Semiconductors

Intrinsic types
N-type
P-type

Superconductors

Magnetic materials

Diamagnetics ($\mu_r = 1$)

Paramagnetics

Ferromagnetics ($\mu_r >> 1$)
- Nonconductors (lossless)
- Conductors (lossy)

1.11 Concluding Remarks

Electromagnetic materials (whether monolithic or composite) are characterized essentially by three material parameters, namely, μ, ε, and σ. These parameters represent implicitly the extent of interaction between the material *versus* electromagnetic force(s) to which the material is subjected. Such interactions perceived at the microscopic level decide the global (macroscopic) properties of the materials.

Engineering utility of a monolithic electromagnetic material or a composite version (constituted by multicomponent monolithic materials) depends on effective harnessing of the aforesaid material parameters so that the bulk medium represents a usable constituent in the engineering pragmatics. Modern technology strives to understand the properties of these materials with the aim of manufacturing conceivable products. The chapters that follow present the details on a wide selection of materials of such practical interest.

References (General Reading)

[1] P. S. Neelakanta and K. Subramaniam: Controlling the properties of electromagnetic composites. Adv. Mater. Process., vol. 141(3), 1992: 20-25.

[2] L. Solymar and D. Walsh: *Lectures on the Electrical Properties of Materials.* (Oxford University Press, Oxford, 1993).

[3] D. D. Pollock: *Physics of Engineering Materials.* (Prentice-Hall Inc., Englewood Cliffs, NJ: 1990).

[4] A. J. Dekker: *Electrical Engineering Materials.* (Prentice-Hall of India Pvt. Ltd., New Delhi: 1988).

[5] M. A. Plonus: *Applied Electromagnetics.* (McGraw-Hill Book Co., New York: 1978).

[6] S. Ramo, J. R. Whinnery and T. Van Duzer: *Fields and Waves in Communication Electronics.* (John Wiley and Sons Inc., New York: 1994).

[7] D. J. Griffiths: *Introduction to Electrodynamics.* (Prentice-Hall of India Pvt. Ltd., New Delhi: 1989).

[8] K. K. Chawla: *Composite Materials.* (Springer-Verlag, New York: 1987).

Defining Terms

Composite materials: A multiconstituent material formed by two or more discrete (monolithic) materials.

Electric field: State of a region in which charged bodies are subjected to forces of interaction (by virtue of their charges) with charges in their vicinity.

Electromagnetic field: A co-existing electric and magnetic fields under time-varying conditions.

Electromagnetic material: A material which responds to the applications of electric or magnetic (or both) force field(s) and exhibits corresponding characteristic behavior.

Magnetic field: State of interactive force caused by flow of charges (or movement of charges) exerted on a moving charge in the field.

Monolithic material: A single constituent discrete version of material.

APPENDIX 1.1

Electromagnetic (EM) Fields And Waves: A Summary

Electrostatics

A. Electrical Charge

• A physical attribution to electrons and protons of their electrical property. Electronic charge is negative $(-e)$ and the protonic charge is positive $(+e)$. $|e| = 1.6 \times 10^{-19}$ coulomb.

• Electric charge (Q) is quantitized: i.e. $Q = \pm ne$, $n = 1,2,3...$, (law of charge quantization). Like-charges repel each other and unlike-charges attract each other. Charges cannot be created or destroyed (law of conservation of charge).

B. Electrostatic Concepts

• Coulomb's law, an experimental postulation: Given two-point charges separated in a medium by a distance r, the force (of attraction or repulsion) between them, $\boldsymbol{F} = (1/4\pi\varepsilon) \times [Q_1 Q_2 \boldsymbol{a}_r/r^2]$ newton where $\boldsymbol{a}_r$ is the unit vector along r and ε is the permittivity, a property of the medium. $\varepsilon = \varepsilon_o \varepsilon_r$ with $\varepsilon_o = (1/36\pi) \times 10^{-9}$ farad/meter being the absolute permittivity of free space and ε_r is the relative permittivity (or dielectric constant) of the medium.

• Electric field: Electrostatic force field perceived in a medium due to the presence of a charge or a set of charges. Electric field intensity $(\boldsymbol{E})$ is the force exerted on a unit charge at a point due to other charge(s), Q. By Coulomb's law, $\boldsymbol{E} = Q\boldsymbol{a}_{r_n}/4\pi\varepsilon r^2$ newton/coulomb or volt/meter. Due to a system of N point charges $(Q_1, Q_2, ..., Q_N)$ located at vector distances $|\boldsymbol{r} - \boldsymbol{r}_n|$, $n = 1, 2, ..., N$, $\boldsymbol{E}(\boldsymbol{r})$ is equal to $(1/4\pi\varepsilon) \sum_n Q_n \boldsymbol{a}_r/|\boldsymbol{r} - \boldsymbol{r}_n|^2$ where $\boldsymbol{a}_{r_n}$ is the unit vector along r_n.

• Electric flux and electric displacement: Force field due to electric charges can be depicted by a set of flux lines (per unit area) or flux density given by $\boldsymbol{D} = \varepsilon\boldsymbol{E}$ known as electric displacement. Electric flux $(d\psi)$ streaming through an elemental area dA in a medium is $\boldsymbol{D} \bullet d\boldsymbol{A}$ and total flux $\psi = \iint \boldsymbol{D} \bullet d\boldsymbol{A}$. Total flux out of a surface enclosing a volume is equal to the total charge enclosed; or $\iint \boldsymbol{D} \bullet d\boldsymbol{A} = Q_{enclosed} = \iiint \rho_v \, d_{vol}$: (Gauss's law), where ρ_v = volume charge density and d_{vol} is an elemental volume.

• Electrical potential: In an electric force $(\boldsymbol{E})$ field, work done in moving a unit charge from infinity to the point of observation $= -\int_\infty^r \boldsymbol{E} \bullet d\boldsymbol{\ell}$, known as the absolute potential (in volt) at that point. When moving a charge from r_1 to r_2, the potential difference is: $-\int_{r_2}^{r_1} \boldsymbol{E} \bullet d\boldsymbol{\ell}$. Work field $V = -\int \boldsymbol{E} \bullet d\boldsymbol{\ell} \Longleftrightarrow \boldsymbol{E} = -\boldsymbol{\nabla} V$ where $\boldsymbol{\nabla}$ is the differential gradient operator specifying a maximum spatial derivative $(d/d\ell)_{max}$. The integral $-\int \boldsymbol{\nabla} V \bullet d\boldsymbol{\ell}$ over a closed path = 0 specifies conservation of energy.

• Energy stored in an electric field (E): $W = (1/2)\, \varepsilon \iiint |E|^2 \, d_{vol}$.

• Charge distribution: Electric charges can exist as (i) Discrete, point charges; (ii) a cloud of bulk (volume) charges of density ρ_v coulomb/meter^3; (iii) charges distributed along a line (linear charges of density ρ_ℓ coulomb/meter); and (iv) charges distributed on an area (surface charges of density ρ_s coulomb/meter^2).

• Boundary conditions for electric field at dielectric-dielectric interface:
a. Tangential components of electric field across the interface of two dielectric media are continuous regardless of presence or absence of surface charges; that is, $E_{t1} = E_{t2}$ in Figure 1.5.

b. Normal components of electric flux density across a charge-free interface of two dielectric media are continuous, that is, $(D_{n1} = \varepsilon_1 E_{n1}) = (D_{n2} = \varepsilon_2 E_{n2})$ in Figure 1.5. Should a surface charge of density ρ_s coulomb/meter^2 exist at the interface, then $D_{n1} - D_{n2} = \rho_s$ (Figure 1.6).

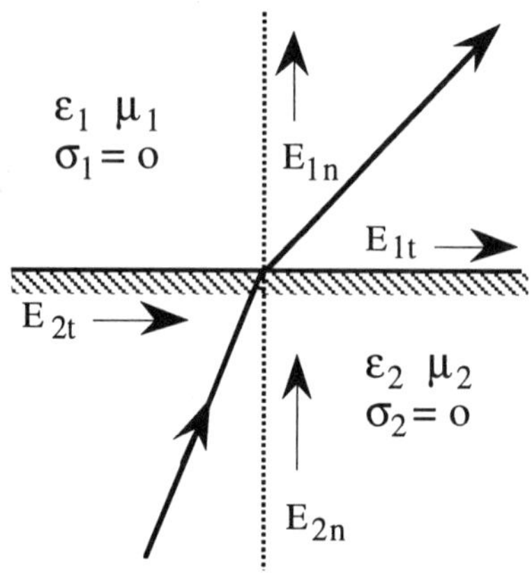

Figure 1.5 Electric field components at the charge-free boundary of two dielectric media.

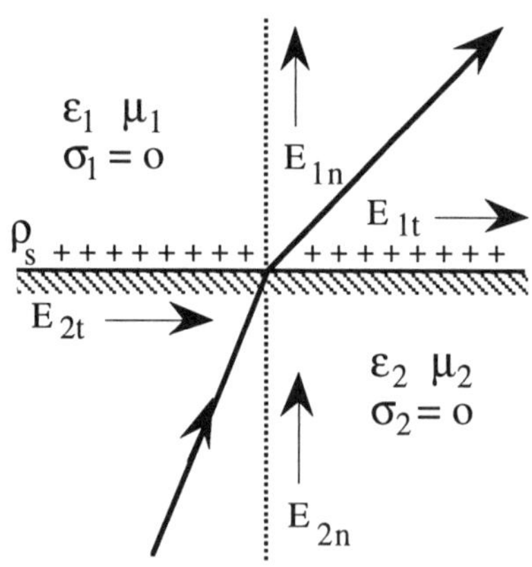

Figure 1.6 Electric field components at the boundary of two dielectric media with a surface charge distribution.

• Divergence of electric flux: Net electric flux through a charge-free, closed volume is zero; that is, the total flux entering and emerging a charge-free medium remains the same (solenoidal property) => divergence operation on $\boldsymbol{D}$, $\nabla \bullet \boldsymbol{D} = 0$. With ρ_v enclosed, $\nabla \bullet \boldsymbol{D}$ (net flux through the volume) is equal ρ_v (differential form of Gauss's law).

• Poisson's and Laplace's equations: With $\nabla \bullet D = \rho_v$, $D = \varepsilon E$ and $E = \nabla V$, $\nabla \bullet \nabla V = -\rho_v/\varepsilon$. $\nabla \bullet \nabla$ => divergence of a gradient known as the Laplacian operator, ∇^2 $(= \nabla \bullet \nabla)$. Hence, $\nabla^2 V = -\rho_v/\varepsilon$ => Poisson's equation. If $\rho_v = 0$, $\nabla^2 V = 0$ => Laplace's equation. Poisson's and Laplace's equations with appropriate boundary conditions lead to explicit determination of the potential V at any point in the medium.

Electric Current

A. Conduction Current (I_c)

• I_c = Time rate of change of charges; that is, $I_c = dQ/dt$ coulomb/sec ⇒ ampere. It is the conduction current due to flow of free electrons in a metal, free electrons and holes in a semiconductor or ions in a medium like an electrolyte or plasma.

• Conduction current density $J_c \Rightarrow I_c$ /area (ampere/meter^2). It is directly proportional to the electric field that causes the charges to flow. That is, $J_c = \sigma E$ where σ (siemen/meter) is the electrical conductivity (a material property). Resistivity, ρ (ohm-meter) = $1/\sigma$.

B. Ohm's Law

• Microscopic Ohm's law: $J_c = \sigma E$ amperes/meter^2. For a bulk material of length ℓ meter, and area of cross-section A meter^2, from the microscopic Ohm's law it follows that $I_c = J_c A = E\ell/(\rho\ell/A)$ = voltage/resistance. $(\rho\ell/A)$ => Resistance (R ohm) of the bulk material. Ohm's law ⇒ $V/I_c = R$: Current is directly proportional to the applied voltage with R as the constant of proportionality (at a given temperature).

• Conductivity *versus* temperature: σ is decided by $ne^2\tau/m$ where n = number of electrons under movement, e = electronic charge, m = electronic mass, and τ (relaxation time) = the mean collision time of moving electrons. When temperature changes, τ is affected due to the thermal energy acquired. Hence σ, ρ and R depend on temperature.

• Perfect conductors, metals, semiconductors, and insulators: $\sigma = \infty$ ⇒ perfect conductor. For metals say, Cu, σ is very high (5.7×10^7 siemen/meter) and for pure semiconductors like Ge, σ is low (2.3 siemen/meter). For ideal insulators (dielectrics), $\sigma - 0$. For practical insulators like Bakelite™ $\sigma \cong 10^{-9}$ siemen/meter. Ions also contribute conductivity. For example, σ (salt water) ≅ 4 siemen/meter.

Conductor Media

A. Boundary Conditions at a Conductor-Dielectric Interface

• Perfect conductor: E within a perfect conductor is equal to 0; or $\nabla V = 0$. That is, V is constant in a perfect conductor making it an equipotential region.

• Electric charges reside on the surface of a conductor; that is, $|D|_{normal} = \rho_s$ on the surface where ρ_s is the surface charge density.

• A conductor being an equipotential surface, the voltage drop tangential to its surface is zero. Therefore, $|E|_{tangential} = 0$.

• Electric flux lines at the conductor terminate normal to the surface on the surface charges.

• Charges located above a conducting surface, induce opposite image charges.

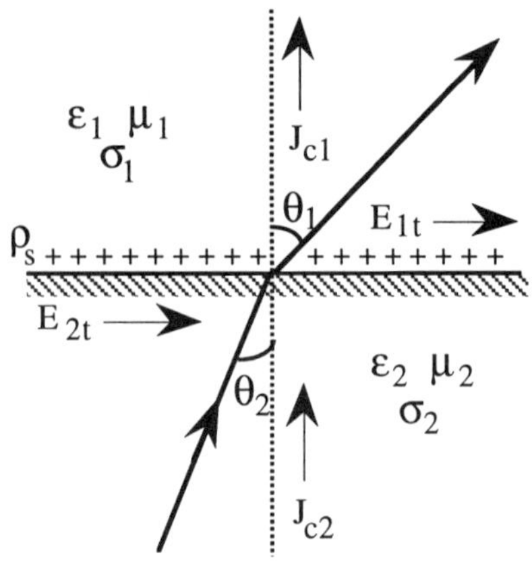

Figure 1.7 Refraction of current at the interface of two lossy dielectric media.

• Continuity equation in the conductor interior: Since $\boldsymbol{E} = 0$, $\nabla \bullet \boldsymbol{D} = 0$ => Interior of a metallic body is point-wise neutral. $\nabla \bullet \boldsymbol{J}_c = 0$ inside a conductor refers to the solenoidal property denoting I_c has no source nor sink.

• Refraction of current at the boundary of two imperfect conductors: Since $\boldsymbol{J}_c = \sigma \boldsymbol{E}$, normal and tangential currents at the boundary are: $(E_{t1} = E_{t2}) \Rightarrow (J_{ct1}/\sigma_1 = J_{ct2}/\sigma_2)$, and $(J_{cn1} = J_{cn2}) \Rightarrow (\sigma_1 E_{n1} = \sigma_2 E_{n2})$. Therefore, in Figure 1.7 $tan\theta_1/tan\theta_2 = \sigma_1/\sigma_2 \Rightarrow$ current refraction condition. Since $(\varepsilon_1 E_{n1} = \varepsilon_2 E_{n2})$, $(E_{n1} = J_{cn1}/\sigma_1$ and $E_{n2} = J_{cn2}/\sigma_2)$ and $(J_{cn1} = J_{cn2} = J_n)$, $J_n\,(\varepsilon_1/\sigma_1 - \varepsilon_2/\sigma_2)$ is finite, equal to a surface charge ρ_s at the interface.

B. Laplace's Equation for Conductor Media

• Since $\nabla \bullet \boldsymbol{J}_c = 0$ and $\nabla \bullet V = 0$, $\nabla \bullet \sigma \boldsymbol{E} = \nabla \bullet \nabla V = 0$ *and* $\nabla^2 V = 0 \Rightarrow$ (II order partial differential equation) determines V at any point in a given coordinate system.

• To solve $\nabla^2 V = 0$: (a) Specify the region with appropriate (one-, two-, or three-dimensional boundaries) in a coordinate system. (b) Stipulate the following boundary conditions: (i) Potential all over the boundaries (Dirichlets' condition); or (ii) gradient of the potential all over the boundaries (Neumann condition); or (iii) mixed Dirichlet and Neumann conditions in sections over the boundaries. (c) Stipulate $V \rightarrow 0$ at infinity. (d) Solve $\nabla^2 V = 0$ *via* variable separable method or numerically.

Dielectric Media

A. Microscopic Characterization of Dielectric Materials

Neutral molecules have a set of equal positive and negative charges separated by a distance forming electric dipoles which align or orient along $\boldsymbol{E}$-field applied on the material. This state of alignment is known as dielectric polarization.

• Dipole pairs of $+q$ and $-q$ charges separated by a small distance, d experience a dielectric moment $\boldsymbol{p} = q\boldsymbol{d}$ coulomb-meter in aligning along the E field experiencing a torque, $\boldsymbol{T} = \boldsymbol{p} \times \boldsymbol{E}$.

• Dielectrics consist of two types of molecules: Nonpolar molecules in which the $+q$ and $-q$ are located at the same locale with no dipole moment in the absence of electric field, but under E field they get stretched and experience a torque. In polar type, $+q$ and $-q$ are separated by d with a permanent dipole moment.

• Polarization $(\boldsymbol{P})$: This is defined as the dielectric moment per unit volume.

• Rotating force due to the torque on a dipole, $\boldsymbol{F} = \nabla(\boldsymbol{p} \bullet \boldsymbol{E})$; and work done when a dipole orients itself by rotation. Potential energy $W = -\boldsymbol{p} \bullet \boldsymbol{E}$ joules.

• Under polarization $(\boldsymbol{P})$, electric flux density $(\boldsymbol{D}_d)$ inside a dielectric $= (\varepsilon_o \boldsymbol{E}_d + \boldsymbol{P})$. Denoting $\boldsymbol{D}_d = \varepsilon_d \boldsymbol{E}_d$, where ε_d is the permittivity of the dielectric, $\varepsilon_d = (\varepsilon_o + \boldsymbol{P}/\boldsymbol{E}_d)$. $\chi = \boldsymbol{P}/\boldsymbol{E}_d \Rightarrow$ dielectric susceptibility and $\varepsilon_d = \varepsilon_o(1 + \chi)$; or $(\varepsilon_{rd} - 1) = \chi$ with $\varepsilon_d = \varepsilon_o \varepsilon_{rd}$, ($\varepsilon_{rd} =$ dielectric constant).

• Polarized dipoles in a dielectric is equivalent to a collection of volume charges of density, $-\nabla \bullet \boldsymbol{P}$ plus a system of charges of density, $\boldsymbol{P} \bullet \boldsymbol{n}$ on the surface (S) enclosed by the dielectric volume. Here, $\boldsymbol{n}$ is the normal vector to the surface.

• Electric field exterior to the dielectric under polarized state at a vector distance $\boldsymbol{R}$ from an elemental volume d_{vol} is:

$$\boldsymbol{E}(\boldsymbol{r}) = -\nabla V(\boldsymbol{r}) = (\boldsymbol{a}_R/4\pi\varepsilon_o) \iint (\boldsymbol{P} \bullet \boldsymbol{n})\, dS/R^2 + (\boldsymbol{a}_R/4\pi\varepsilon_o) \iiint (\nabla \bullet \boldsymbol{P})\, d_{vol} / R^2$$

where $\boldsymbol{a}_R$ = unit vector along R; $\boldsymbol{r}$ = vector distance of the point of observation from the origin.

• Polarization current: Polarization $(\boldsymbol{P})$ in a dielectric changing with time corresponds to time-varying bound charges constituting a polarization current, $\boldsymbol{J}_p = dP/dt$ amperes.

B. Capacitance

• Capacitance: State of acquiring electric charges by a particular geometry of isolated conductors at different potential levels. With charges on a pair of conductors as $+Q$ and $-Q$ and the potential difference as V, capacitance (C) is specified by a linear relation between Q and V. That is, $Q = CV$.

• Capacitance depends on: Surface (area) of the conductors (A); distance (d) between the conductors and permittivity (ε) of the medium interposed. $C = \varepsilon A/d$ farads.

• Capacitor is an electric passive component formed by a pair of conductors with an interposed dielectric to yield a specified value of capacitance.

• Electrostatic energy stored in a capacitor: $W = (1/2)CV^2$ joules $= Q^2/2C$ joules.

• Net capacitance (C_p) of parallel-connected capacitors $(C_1, C_2, \ldots, C_n)$: $Cp = C_1 + C_2 + \ldots + C_n$; net capacitance (C_s) of series-connected capacitors $(C_1, C_2, \ldots, C_n)$: $1/C_s = 1/C_1 + 1/C_2 + \ldots + 1/C_n$.

Magnetostatics

A. Magnetic Field

• Magnetic field is a force field perceived in a region where electric charges are time varying.

• Magnetic flux: Magnetic field depicted by a set of flux lines has a density per unit area *($\boldsymbol{B}$)* $\Rightarrow$ Magnetic flux density ($\boldsymbol{B}$ in tesla). (1 tesla = 1 weber/meter^2 = 10^4 gauss in C.G.S. units.)

• Driving force causing $\boldsymbol{B} \Rightarrow$ magnetic flux intensity $\boldsymbol{H}$ (ampere-turn per meter).

• ***B** versus **H***: $\Rightarrow$ Constitutive relation, $\boldsymbol{B} = \mu\boldsymbol{H}$ where μ is the permeability of the medium $= \mu_o\mu_r$. μ_o: Absolute permeability of free space equal to $4\pi \times 10^{-7}$ henry per meter representing magnetic property of free space or the extent of magnetic flux setup in free space due to a magnetizing force (caused by time-varying charges). μ_r is the relative permeability of a medium. It denotes the magnetic characteristics of a medium relative to free space.

B. Magnetic Field Postulations

• Ampere's force law: A current element $\boldsymbol{I}\, d\boldsymbol{\ell}$ (or equivalently a flow of q charges with velocity, $\boldsymbol{v}$) placed in a magnetic field $\boldsymbol{B}$ will experience a force $d\boldsymbol{F}$ equal to $\boldsymbol{I} \times \boldsymbol{B} d\boldsymbol{\ell}$ or $q\boldsymbol{v} \times \boldsymbol{B}$ newtons. The vector cross-product indicates $d\boldsymbol{F}$, $\boldsymbol{I}$ and $\boldsymbol{B}$ are mutually perpendicular to each other as given by left-hand thumb rule.

• Biot-Savart's law: A current element $\boldsymbol{I}\, d\boldsymbol{\ell}$ produces a magnetic field $\boldsymbol{B}$ which at a distance R from the element given by $d\boldsymbol{B} = (\mu_o/4\pi)\, \boldsymbol{I} \times \boldsymbol{a}_R d\boldsymbol{\ell}/R^2$ where $\boldsymbol{a}_R$ is the unit vector along R.

$|d\boldsymbol{B}| = (\mu_o/4\pi)\, I\ell \sin\theta/R^2 \Rightarrow$ This is similar to Coulomb's law.

• Lorentz force: Force *($\boldsymbol{F}$)* on a moving charge q with velocity $\boldsymbol{v}$ in a magnetic field $\boldsymbol{B}$ and electric field $\boldsymbol{E}$ equal to $q\,(\boldsymbol{E} + \boldsymbol{v} \times \boldsymbol{B})$ newtons.

• Torque *($\boldsymbol{T}$)* on a current element $\boldsymbol{I}d\boldsymbol{\ell}$ in a magnetic field $\boldsymbol{B}$, due to Ampere's force: $\boldsymbol{T} = \boldsymbol{m} \times \boldsymbol{B}$ newton-meter. $\boldsymbol{m}$: Magnetic moment of the current carrying loop with area A and $|\boldsymbol{m}| = IA$.

• Setting $IA = q_m d$, where q_m is a fictitious magnetic isolated charge (analogous to electric charge) separated from an opposite magnetic charge $(-q_m)$ by a distance d. $(+q_m$ *and* $-q_m)$ constitute a magnetic dipole. The poles of a bar magnet (of length ℓ) can be equated to a dipole with $(+q_m, -q_m)$.

• Torque experienced by a magnetic dipole in a magnetic field forms the basis of electric motors.

C. Magnetic Properties of Matter

• Material magnetism: In materials, movement of electrons in the orbit or due to spin refers to time-varying charges inducing a magnetic field. The net vector motion may in total cancel to zero; that is, current loop does not exist or the magnetic field induced is zero. Such materials are diamagnetic. In so-called ferromagnetics, the atomic/molecular arrangement permits a finite set of current loops (or dipoles). Random arrangement of such magnetic dipoles, however, cancel with each other producing null magnetism. Dipole magnets permanently aligned by external lead to "permanent" magnetism.

• Relative permeability: In diamagnetics $\mu_r = 1$ (for example, water, Cu); for ferromagnetics $\mu_r >> 1$ (for example, iron). In paramagnetics, μ_r is slightly larger than one (for example, air, aluminum).

• Hysteresis: Under cyclic magnetization, ***B versus H*** forms a hysteresis loop as shown in Figure 1.8.

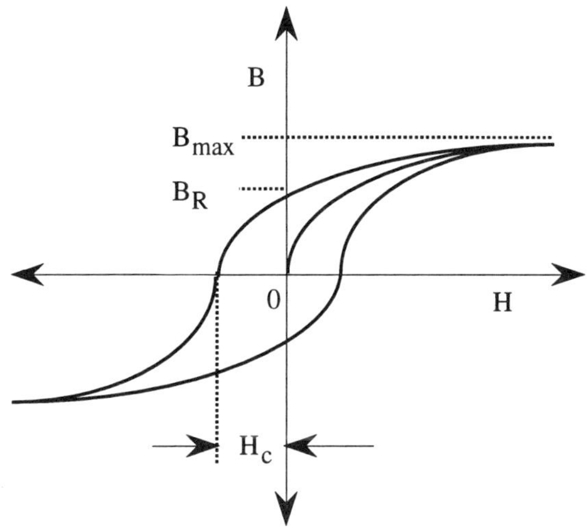

Figure 1.8 Magnetic hysteresis curve.
B_R: Retentive magnetic field; H_c: Coersive magnetic field intensity, B_{max}: Saturation field.

D. Ampere's Current Law

• Line integral of a magnetic field $\boldsymbol{B}$ around any closed path is equal to μI. That is, $\oint \boldsymbol{B} \bullet d\boldsymbol{\ell} = \mu \mathbf{I}$, if the closed path of integration encloses I; otherwise zero. $\oint \boldsymbol{H} \bullet d\boldsymbol{\ell} = \oint \boldsymbol{B} \bullet d\boldsymbol{\ell} = \mu \iint \boldsymbol{J} \bullet d\boldsymbol{A}$ with dA = Area bounded by the closed path $d\boldsymbol{\ell}$.

• Differential form of Ampere's law: By Stoke's theorem, $o\int \boldsymbol{B} \bullet d\boldsymbol{\ell} = \iint (\nabla \times \boldsymbol{B}) \bullet d\boldsymbol{A}$. Therefore, $\nabla \times \boldsymbol{B} \rightarrow$ curl of $\boldsymbol{B} = \mu \boldsymbol{J}$.

• Magnetomotive force *(mmf)*: $\Rightarrow o\int \boldsymbol{H} \bullet d\boldsymbol{\ell} = NI$ with N being the number of turns and representing the driving influence in the induction of $\boldsymbol{B}$ (or $\mu \boldsymbol{H}$) due to a current through a loop of N turns.

E. Inductance

• Self-inductance *(L)*: Measure of magnetic flux which links a circuit when a current I flows in the circuit. It (like capacitance) depends on physical arrangement of conductors. For a flux linkage $\boldsymbol{\Phi}$, $L \triangleq \Phi/I = N \iint \boldsymbol{B}.d\boldsymbol{A}/I$ henries.

• Energy stored in an inductor: $W = (1/2)LI^2 \Rightarrow 1/2\mu|\boldsymbol{H}|^2$ joules.

• Mutual Inductance: It relates the flux linkage produced in circuit 2 as caused by current in circuit 1 or *vice versa* . That is, $M_{21} = d\Phi_{21}/dI_1$; or $M_{12} = d\Phi_{12}/dI_2$.

F. Faraday-Lenz's Law

• An electromotive force *(emf)* is induced in a conductor when magnetic flux linking the conductor is time varying. Magnitude of emf is proportional to the time rate of change of flux linkage with the conductor. Direction of induced emf tends to oppose the cause (Lenz's law). $emf = - d\Phi/dt$.

• Electric generators and transformers work on the principles based on Faraday-Lenz's law.

G. Boundary Conditions for Magnetic Fields
• Normal component of **B** and tangential component of **H** are continuous across the interface of two nonconducting materials.

• Above conditions are also valid across the interface of conducting media for static fields.

• For time-varying fields, however, tangential compoment of **H** is discontinuous by the extent of surface current density (J_s) in the conductor; but B_{normal} is continuous.

Maxwell's Equations

• Magnetomotive force *(mmf)* around a closed path is equal to the conduction current plus the displacement current through any surface bounded by the path:

$$\oint \mathbf{H} \bullet d\boldsymbol{\ell} = \iint(\mathbf{J}_c + \mathbf{J}_d) \bullet d\mathbf{A} \Leftrightarrow \nabla \times \mathbf{H} = (\mathbf{J}_c + \mathbf{J}_D); \; (\mathbf{J}_c = \sigma \mathbf{E}, \; \mathbf{J}_D = \varepsilon \, d\mathbf{D}/dt)$$

• Electromotive force around a closed path is equal to the time derivative of magnetic flux density through any surface bounded by the path:

$$\oint \mathbf{D} \bullet d\mathbf{A} = -\iint(d\mathbf{B}/dt) \bullet d\mathbf{A} \Leftrightarrow \nabla \times \mathbf{E} = -d\mathbf{B}/dt, \; (\mathbf{B} = \mu \mathbf{H}, \; \mathbf{D} = \varepsilon \mathbf{E})$$

• Total electric flux density through a surface enclosing a volume is equal to the total charge within the volume: $\iint \mathbf{D} \bullet d\mathbf{A} = \iiint q_{vol} \, d_{vol} \Leftrightarrow (\nabla . \mathbf{D} = q_{vol})$.

• The net magnetic flux emanating from a closed surface is zero: $\iiint \mathbf{B} \bullet d\mathbf{A} = 0 \Leftrightarrow (\nabla \bullet \mathbf{B} = 0)$.

Wave Equation

• For an isotropic, homogeneous, linear medium (with ε, μ, σ):

$$\nabla^2 \mathbf{E} - \mu\varepsilon \, d^2\mathbf{E}/dt^2 - \mu\sigma \, d\mathbf{E}/dt - \nabla \rho_v/\varepsilon = 0; \quad \nabla^2 \mathbf{H} - \mu\varepsilon \, d^2\mathbf{H}/dt^2 + \nabla \times \mathbf{J} = 0$$

with $1/(\mu\varepsilon)^{1/2}$ = velocity of propagation of EM wave (= $c = 3 \times 10^8$ meter/second, in free space).
ρ_v : volume charge-density in the medium and σ : conductivity of the medium.
Charge-free medium: $\rho_v = 0$; lossless medium: $\sigma = 0$.

• In a charge-free medium, propagation constant $\gamma = (\alpha + j\beta)$; ω: $2\pi \times frequency$

α: Attenuation constant = $\omega \, [(\mu\varepsilon/2) \, \{(1 + \sigma^2/\omega^2\varepsilon^2)^{1/2} - 1\}]^{1/2}$

β : Phase constant = $\omega \, [(\mu\varepsilon/2) \, \{(1 + \sigma^2/\omega^2\varepsilon^2)^{1/2} + 1\}]^{1/2}$

• Z_o: Characteristic impedance of the medium = $[j\omega\mu/\sigma + j\omega\varepsilon]^{1/2}$

• Perfect dielectric $(\sigma = 0)$: $\alpha = 0$; β: $\omega(\mu\varepsilon)^{1/2}$

• Z_o: $(\mu/\varepsilon)^{1/2}$

Good dielectric $(\sigma/\omega\varepsilon << 1)$: $\alpha = (\sigma/2) \, (\mu/\varepsilon)^{1/2}$; $\beta = \omega(\mu\varepsilon)^{1/2} \, (1 + \sigma^2/8\omega^2\varepsilon^2)$

$Z_o = (\mu/\varepsilon)^{1/2}\ (1 + j\sigma/2\omega\varepsilon)$

Good conductor $(\sigma/\omega\varepsilon >> 1)$: $\gamma = \omega\mu\sigma\ \angle 45^o$; $\alpha = \beta = (\omega\mu\sigma/2)^{1/2}$; $Z_o = (\omega\mu/\sigma)^{1/2}\ \angle 45^o$

• Poynting vector $(\boldsymbol{P})$: Electromagnetic power flow per unit area: $\boldsymbol{P} = \boldsymbol{E} \times \boldsymbol{H}$ watts/meter.

• Plane wave (TEM wave): Electric and magnetic field components are transverse to the direction of power flow ⇔ plane polarized along the $\boldsymbol{E}$ direction.

• Velocity of wave propagation: $v = \omega/\beta$ meters/second.

Wavelength (λ): Frequency (f in Hz) = v (in meters/second)/λ (meters)

Perfect dielectric ⇒ $v = 1/(\mu\varepsilon)^{1/2}$ meters/second

Good dielectric ⇒ $v = [(\mu\varepsilon)^{1/2}\ (1 + \sigma^2/8\omega^2\varepsilon^2)]^{-1}$ meters/second.

Good conductors ⇒ $v = (2\omega/\mu\sigma)^{1/2}$ meters/second.

Reflection and Refraction of Plane Waves

• At a dielectric-dielectric interface, for normal incidence, the (complex) reflection coefficient is $\Gamma = (\eta_2 - \eta_1)/\eta_2 + \eta_1)$ and the complex transmission coefficient $T = \eta_2/(\eta_2 + \eta_1)$. η_1: Intrinsic (complex) impedance of medium 1; η_2: Intrinsic (complex) impedance of medium 2. Standing Wave Ratio $(SWR) = (1 + |\Gamma|)/(1 - |\Gamma|)$.

• For oblique incidence:

Case (i): E in the plane of incidence (Parallel polarization) (Figure 1.9).

$\Gamma_{||} = (\eta_2 cos\theta_t - \eta_1 cos\theta_i)/(\eta_2 cos\theta_t + \eta_1 cos\theta_i)$.

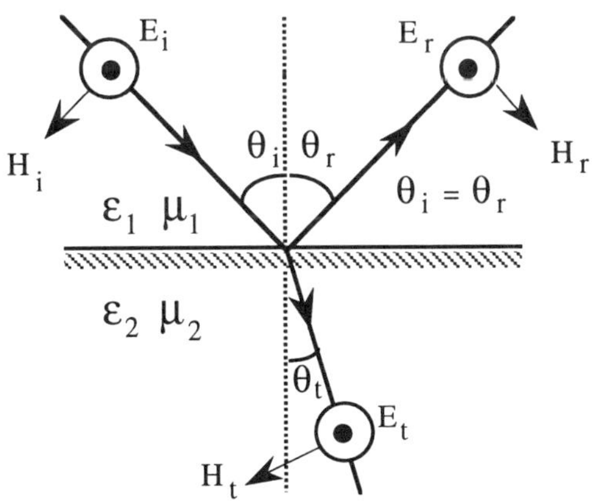

Figure 1.9 Reflection of a plane wave obliquely incident on a dielectric surface (parallel polarization).

Case (ii): E normal to the plane of incidence (perpendicular polarization) (Figure 1.10).

$\Gamma_{\perp} = (\eta_2 cos\theta - \eta_1 cos\theta_t)/(\eta_2 cos\theta + \eta_2 cos\theta_t)$.

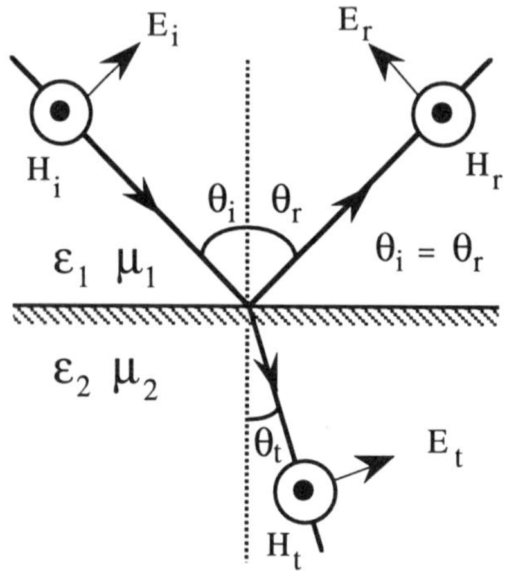

Figure 1.10 Reflection of a plane wave obliquely incident on a dielectric surface (perpendicular polarization).

• Brewster angle (polarizing angle): No reflection condition with parallel polarization, $(\Gamma_{\parallel} = 0) \Rightarrow \theta_B = \arcsin [\varepsilon_2/(\varepsilon_1 + \varepsilon_2)]^{1/2}$.

• Total reflection: In both polarizations, total internal reflection occurs (with no transmission to other medium) when the permittivity of the incident medium is larger than the other medium (ε_2). That is, with $\varepsilon_1 > \varepsilon_2$, the critical angle for total reflection is $\theta_c = \sin^{-1} [(\varepsilon_2/\varepsilon_1)]^{1/2}$.

• Plane polarized (TEM) wave (Figure 1.11) and circularly polarized wave (Figure 1.12).

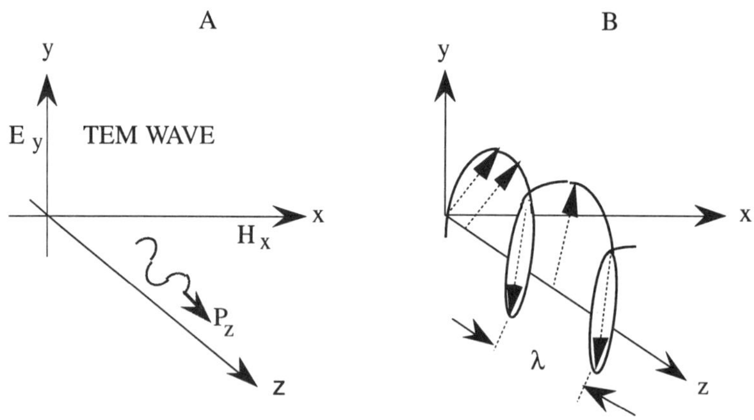

Figure 1.11 (A) Plane polarized wave and (B) circularly/elliptically polarized wave.

APPENDIX 1.2

International System (SI) of Units

[Adopted by the 11th General Conference on Weights and Measures (CGPM) in 1960]

Meter— The meter is the length of path traveled by light in free space (vaccum) during a time interval of 1/299, 792, 458 of a second (17th CGPM, 1983).

Kilogram— The kilogram is the unit of mass; it is equal to the mass of the international prototype of the kilogram (3rd CGPM, 1901).

Second— The second is the duration of 9, 192, 631, 770 periods of the radiation corresponding to the transition between the two hyperfine levels of the ground state of the cesium-133 atom (13th CGPM, 1967).

Ampere— The ampere is that constant current which, if maintained in two straight parallel conductors of infinite length, and negligible circular cross-section, and placed 1 meter apart in vaccum, would produce between these conductors a force equal to 2×10^{-7} newton per meter of length (9th CGPM, 1948).

Kelvin— The kelvin, a unit of thermodynamic temperature, is the fraction 1/273.16 of the thermodynamic temperature of the triple point of water (13th CGPM, 1967).

Mole— The mole is the amount of substance of a system which contains as many elementary entities as there are atoms in 0.012 kilogram of carbon-12. When the mole is used, the elementary entities must be specified and may be atoms, molecules, ions, electrons, or other particles, or specified groups of such particles (14th CGPM, 1971).

Examples of the use of the mole:

1 mol of H_2 contains about 6.022×10^{23} H_2 molecules, or 12.044×10^{23} H atoms.
1 mol of HgCl has a mass of 236.04 g.
1 mol of Hg_2Cl_2 has a mass of 472.08 g.
1 mol of Hg_2^{2+} has a mass of 401.18 g and a charge of 192.97 kC.
1 mol of $Fe_{0.91}S$ has a mass of 82.88 g.
1 mol of e^- has a mass of 548.60 μg and a charge of –96.49 kC.
1 mol of photons whose frequency is 10^{14} Hz has energy of about 39.90 kJ.

Candela— The candela is the luminous intensity, in a given direction, of a source that emits monochromatic radiation of frequency 540×10^{12} hertz and that has a radiant intensity in that direction of (1/683) watt per steradian (16th CGPM, 1979).

Names and Symbols for the SI Base Units

Physical quantity	Name of SI unit	Symbol for SI unit
length	meter	m
mass	kilogram	kg
time	second	s
electric current	ampere	A
thermodynamic temperature	kelvin	K
amount of substance	mole	mol
luminous intensity	candela	cd

Derived SI Units and Their Symbols

Physical quantity	Name of SI unit	Symbol for SI unit	Expression in terms of SI base units	
frequency[1]	hertz	Hz	s^{-1}	
force	newton	N	M kg s^{-2}	
pressure, stress	pascal	Pa	N m^{-2}	= m^{-1} kg s^{-2}
energy, work, heat	joule	J	N m	= m^2 kg s^{-2}
power, radiant flux	watt	W	J s^{-1}	= m^2 kg s^{-3}
electric charge	coulomb	C	A s	
electric potential, electromotive force	volt	V	J C^{-1}	= m^2 kg s^{-3} A^{-1}
electric resistance	ohm	Ω	V A^{-1}	= m^2 kg s^{-3} A^{-2}
electric conductance	siemens	S	Ω^{-1}	= m^{-2} kg^{-1} s^3 A^2
electric capacitance	farad	F	C V^{-1}	= m^{-2} kg^{-1} s^4 A^2
magnetic flux density	tesla	T	V s m^{-2}	= kg s^{-2} A^{-1}
magnetic flux	weber	Wb	V s	= m^2 kg s^{-2} A^{-1}
inductance	henry	H	V A^{-1} s	= m^2 kg s^{-2} A^{-2}
Celsius temperature[2]	degree Celsius	^oC	K	
luminous flux	lumina	lm	cd	
illuminance	lux	lx	sr	
activity (radioactive)	becquerel	Bq	cd sr m^{-2}	
absorbed dose (of radiation)	gray	Gy	s^{-1}	
dose equivalent (dose equivalent index)	sievert	Sv	J kg^{-1}	= m^2 s^{-2}
plane angle	radian	rad	1	= m m^{-1}
solid angle	steradian	sr	1	= m^2 m^{-2}

1 For radial (circular) frequency and for angular velocity the unit rad s^{-1}, or simply s^{-1}, should be used, and this may not be simplified to Hz. The unit Hz should be used only for frequency in the sense of cycles per second.

2 The Celsius temperature θ is definied by the equation:

$$\theta/^oC = T/K - 273.15$$

The SI unit of Celsius temperature interval is the degree Celsius, ^oC, which is equal to the kelvin, K. ^oC should be treated as a single symbol, with no space between the ^{o} sign and the letter C. (The symbol ^oK, and the symbol ^{o}, should no longer be used.)

SI Units and Symbols of Electromagnetic Entities

Name	Symbol	Definition	SI unit
		Electricity and Magnetism	
quantity of electricity, electric charge	Q		C
charge density	ρ	$\rho = Q/V$	C m^{-3}
surface charge density	σ	$\sigma = Q/A$	C m^{-2}
electric potential	V, ϕ	$V = dW/dQ$	V, J C^{-1}
electric potential difference	U, ΔV, $\Delta\phi$	$U = V_2 - V_2$	V
electromotive force	**F**	$F = \int(\mathbf{F}/Q)\bullet dS$	V
electric field strength	**E**	$E = \mathbf{F}/Q = -\text{grad } V$	V m^{-1}
electric flux	ψ	$\psi = \int \mathbf{D}\bullet d\mathbf{A}$	C
electric displacement	**D**	$\mathbf{D} = \varepsilon\mathbf{E}$	C m^{-2}
capacitance	C	$C = Q/U$	F, C V^{-1}
permittivity	ε	$\mathbf{D} = \varepsilon\mathbf{E}$	F m^{-1}
permittivity of vaccum	ε_o	$\varepsilon_o = \mu_o^{-1} c_o^{-2}$	F m^{-1}
relative permittivity	ε_r	$\varepsilon_r = \varepsilon/\varepsilon_o$	1
dielectric polarization (dipole moment per volume)	**P**	$\mathbf{P} = \mathbf{D} - \varepsilon_o\mathbf{E}$	C m^{-2}
electric susceptibility	χ_e	$\chi_e = \varepsilon_r - 1$	1
electric dipole moment	**p**, μ	$p = Qr$	C m
electric current	**I**	$I = dQ/dt$	A
electric current density	**j**, **J**	$I = \int \mathbf{j}\bullet d\mathbf{A}$	A m^{-2}
magnetic flux density, magnetic induction	**B**	$\mathbf{F} = Qv \times \mathbf{B}$	T
magnetic flux	Φ	$\Phi = \int \mathbf{B}\bullet d\mathbf{A}$	Wb
magnetic field strength	**H**	$\mathbf{B} = \mu\mathbf{H}$	A M^{-1}
permeability	μ	$\mathbf{B} = \mu\mathbf{H}$	N A^{-2}, H m^{-1}
permeability of vaccum	μ_o		H m^{-1}
relative permeability	μ_r	$\mu_r = \mu/\mu_o$	1
magnetization (magnetic dipole moment per volume)	**M**	$\mathbf{M} = \mathbf{B}/\mu_o - \mathbf{H}$	A m^{-1}
magnetic susceptibility	χ, κ, (χ_m)	$\chi = \mu_r - 1$	1
molar magnetic susceptibility	χ_m	$\chi_m = V_m\chi$	m^3 mol^{-1}
magnetic dipole moment	**m**, μ	$E_p = -\mathbf{m}\bullet B$	A m^2, J T^{-1}
electrical resistance	R	$R = U/I$	Ω

conductance	G	$G = 1/R$	S
loss angle	δ	$\delta = (\pi/2) + f_I - f_U$	1, rad
reactance	X	$X = (U/I)\sin\delta$	Ω
impedance (complex impedance)	Z	$Z = R + iX$	Ω
admittance (complex admittance)	Y	$Y = 1/Z$	S
susceptance	B	$Y = G + iB$	S
resistivity	ρ	$\rho = E/j$	Ω m
conductivity	κ, γ, σ	$\kappa = 1/\rho$	S m^{-1}
self-inductance	L	$E = -L(dI/dt)$	H
mutual inductance	M, L_{12}	$E_l = L_{12}(dI_{12}/dt)$	H
magnetic vector potential	**A**	$\mathbf{B} = \Delta \times \mathbf{A}$	Wb m^{-1}
poynting vector	**S**, **P**	$\mathbf{P}, \mathbf{S} = \mathbf{E} \times \mathbf{H}$	W m^{-2}

Electromagnetic Radiation

wavelength	λ		m
speed of light	c_o		m s^{-1}
in vaccum	c	$c = c_o/n$	m s^{-1}
in a medium			
wave number in vaccum	$\tilde{v}$	$\tilde{v} = v/c_o = 1/n\lambda$	m^{-1}
wave number (in a medium)	σ	$\sigma = 1/\lambda$	m^{-1}
frequency	v	$v = c/\lambda$	Hz
circular frequency, pulsatance	ω	$\omega = 2\pi v$	s^{-1}, rad s^{-1}
refractive index	n	$n = c_o/c$	1
Planck's constant	h		J s
Planck's constant/2p	$\hbar$	$\hbar = h/2p$	J s
radiant energy	Q, W		J
radiant energy density	ρ, w	$\rho = Q/V$	J m^{-3}
spectral radiant energy density			
in terms of frequency	ρ_v, w_v	$\rho_v = d\rho/dv$	J m^{-3} Hz^{-1}
in terms of wavenumber	$\rho_{\tilde{v}}$, $w_{\tilde{v}}$	$\rho_{\tilde{v}} = dr/d\tilde{v}$	J m^{-2}
in terms of wavelength	ρ_λ, w_λ	$\rho_\lambda = d\rho/d\lambda$	J m^{-4}

Einstein transition probabilities spontaneous emission	A_{nm}	$dN_n/dt = -A_{nm}N_n$	s^{-1}
stimulated emission	B_{nm}	$dN_n/dt = -\rho_{\tilde{\nu}}(\tilde{\nu}_{nm}) \times B_{nm}N_n$	$s\ kg^{-1}$
stimulated absorption	B_{mn}	$dN_n/dt = \rho_{\tilde{\nu}}(\tilde{\nu}_{nm}) \times B_{mn}N_m$	$s\ kg^{-1}$
radiant power, radiant energy per time	Φ, P	$\Phi = dQ/dt$	W
radiant intensity	I	$I = d\Phi/d\Omega$	$W\ sr^{-1}$
radiant excitance (emitted radiant flux)	M	$M = d\Phi/dA_{source}$	$W\ m^{-2}$
irradiance (radiant flux received)	E, (I)	$E = d\Phi/dA$	$W\ m^{-2}$
emittance	ε	$\varepsilon = M/M_{bb}$	1
Stefan-Boltzmann constant	σ	$M_{bb} = \sigma T^4$	$W\ m^{-2}\ K^{-4}$

Conversion of Other Units Commonly Used in Engineering Practice

To obtain	Multiply	By
Atmospheres	Feet of water @ 4°C	2.950×10^{-2}
Atmospheres	Inches of mercury @ 0°C	3.342×10^{-2}
Atmospheres	Pounds per square inch	6.804×10^{-2}
BTU	Foot-pounds	1.285×10^{-3}
BTU	Joules	9.480×10^{-4}
Cubic feet	Cords	128
Degree (angle)	Radians	57.2958
Ergs	Foot-pounds	1.356×10^{7}
Feet	Miles	5280
Feet of water @ 4°C	Atmospheres	33.90
Foot-pounds	Horsepower-hours	1.98×10^{6}
Foot-pounds	Kilowatt hours	2.655×10^{6}
Foot-pounds per min	Horsepower	3.3×10^{4}
Horsepower	Foot-pounds per sec	1.818×10^{-3}
Inches of mercury @ 0°C	Pounds per square inch	2.036
Joules	BTU	1054.8
Joules	Foot-pounds	1.35582
Kilowatts	BTU per min	1.758×10^{-2}
Kilowatts	Foot-pounds per min	2.26×10^{-5}
Kilowatts	Horsepower	0.745712
Knots	Miles per hour	0.86897624
Miles	Feet	1.894×10^{-4}
Nautical miles	Miles	0.86897624
Radians	Degrees	1.745×10^{-2}
Square feet	Acres	43560
Watts	BTU per min	17.5796

Physical Constants

Equatorial radius of the earth = 6378.388 miles (statute).
Polar radius of the earth, 6356.912 km = 3949.99 miles (statute).
1 degree of latitude at 40° = 69 miles.
1 international nautical mile = 1.15078 miles (statute) = 1852 m = 6076.115 ft.
Mean density of the earth = 5.522 $g.cm^3$ = 344.7 lb/ft^3.
Constant of gravitation $(6.673 \pm 0.003) \times 10^{-8}$ cm^3 gm^{-1} s^{-2}.
Acceleration due to gravity at sea level, latitude 45° = 980.6194 cm/s^2 = 32.1726 ft/s^2.
Length of seconds pendulum at sea level, latitude 45° = 99.3575 cm = 39.1171 in.
1 knot (international) = 101.269 ft/min = 1.6878 ft/s = 1.1508 miles (statute)/h.
1 micron = 10^{-4} cm.
1 ångstrom = 10^{-8} cm.
Mass of hydrogen atom = $(1.67339 \pm 0.0031) \times 10^{-24}$ g.
Density of mercury at $0^{o}C$ = 13.5955 g/ml.
Density of water at $3.98^{o}C$ = 1.000000 g/ml.
Density, maximum, of water, at $3.98^{o}C$ = 0.999973 g/cm^3.
Density of dry air at $0^{o}C$, 760 mm = 1.2929 g/l.
Velocity of sound in dry air at $0^{o}C$ = 331.36 m/s – 1087.1 ft/s.
Velocity of light in vaccum = $(2.997925 \pm 0.000002) \times 10^{10}$ cm/s.
Heat of fusion of water $0^{o}C$ = 79.71 cal/g.
Heat of vaporization of water $100^{o}C$ = 539.55 cal/g.
Electrochemical equivalent of silver 0.001118 g/s international amp.
Absolute wavelength of orange-red line of krypton 86 = 6057.802 Å.

CHAPTER 2

Dielectric Materials

2.1 Introduction

Dielectric materials refer to those having the basic electric property of being polarized in the presence of an electric field and having an electrostatic field within them under the state of polarization. (*Polarization* here refers to the molecular alignment along the direction of the applied electric field.) Also synonymously referred to as an *electrical insulation material*, the dielectric has the ability to prevent the leakage of electrical charges from the body on which it is deployed. In general, electrical insulation materials which possess dielectric properties offer a very high resistance to the passage of electric current under the action of an applied direct-current voltage. Hence, such materials differ distinctly in basic electric properties from those classified as *electric conductors,* such as metals.

Naturally occurring materials like wood, vegetable fibers, cotton, silk, oils, rubber, resin, rocks, etc. are well-known as electric insulators and are characteristically dielectric. Added to these, is a gamut of man-made materials such as plastics, polymers, epoxies, etc., which are also used widely as insulating/dielectric materials.

The dielectric properties can be perceived in the materials in all four *states of matter*, namely, solid, liquid, gas, and plasma. A detailed interpretation of the dielectric properties in solids and liquids, however, is considerably more involved than for the dilute phase states, namely, gases and plasma.

Nevertheless, the basis of dielectric properties stems from the interaction of matter at the microscopic level with an external electric field force and its manifestation as the macroscopic electrical insulation (dielectric) property is rather common in all states of matter. This basic microscopic property dictates the extent of electric field force of interaction between electric charges in the medium as quantified by the well-known *Coulomb's law.* That is, the extent of dielectric properties of a material is stipulated by a constant proportionality ε in Coulomb's law given by:

$$\boldsymbol{F} = (1/4\pi\varepsilon)q_1 q_2 \boldsymbol{a}_r/r^2 \qquad \textit{newton} \qquad (2.1)$$

where $\boldsymbol{F}$ is the coulombic force of interaction between two charges q_1 and q_2 (each expressed in coulombs) separated by a distance of r meters in the dielectric medium; and $\boldsymbol{a}_r$ is the unit vector along the direction of r.

Further, as indicated in Chapter 1, ε is normally expressed as $\varepsilon_o\varepsilon_r$ where ε_o is the absolute dielectric permittivity of free space and ε_r is the dimensionless permittivity of the medium, relative to the free space. ε_o has a value of $(1/36\pi)10^{-9}$ farad/meter in SI units. Tables (2.A-2.C) list various, commonly known dielectric materials (both natural and man-made) and their relative permittivities which are also known popularly as the *dielectric constants*.

Microscopically, dielectric permittivity of a material can be defined as the ratio of the electric field $\boldsymbol{E}$ in free space (vacuum) to that in the material for the same distribution of charges in the media conceived. This quantity is perceived in the so-called *capacitors*, the most simple form of which consists of two parallel conductor plates of area A meter^2 separated by a distance d meter when a difference of potential ϕ between the plates is applied (Figure 2.1). Also, spacing d is much smaller than the plate dimensions. When the charges developed on the plates placed in free space are taken as $+Q$ and $-Q$, the electric field $\boldsymbol{E}$ inside the capacitor will be perpendicular to the plates and its intensity in SI units is given by:

$$|E_o| = \rho/\varepsilon_o \qquad \textit{volt/meter} \qquad (2.2)$$

where ρ is the surface charge density (in coulomb/meter^2) on each plate equal to Q/A.

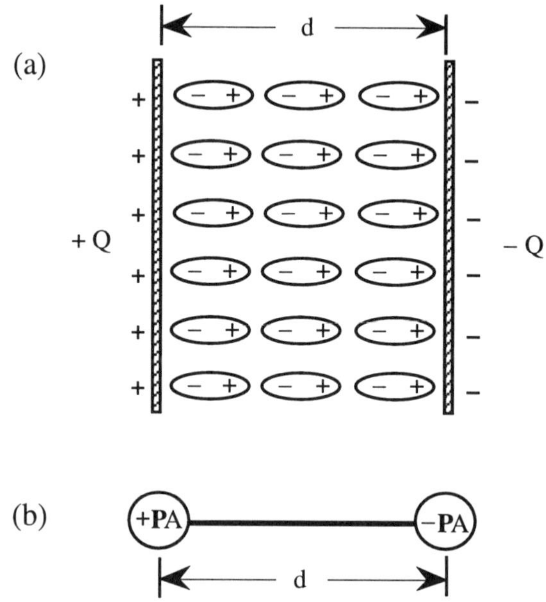

Figure 2.1 (a) Dielectric polarization in a parallel plate capacitor with a plate area A meter^2; (b) Equivalent electric dipole.

The corresponding capacitance in free space is given by:

$$C_o = \varepsilon_o A/d \qquad \textit{farad} \qquad (2.3)$$

Similarly $\boldsymbol{E}$ and C pertinent to a dielectric material (having a relative permittivity ε_r) interspaced between the plates can be expressed as:

$$|\boldsymbol{E}| = \rho/\varepsilon_o\varepsilon_r \qquad \textit{volt/meter} \qquad (2.4)$$

and

$$C = \varepsilon_o\varepsilon_r\ A/d \qquad \textit{farad} \qquad (2.5)$$

From Equations 2.2 and 2.5, the decrease in electric field strength due to the effect of the dielectric is found to be:

$$|\boldsymbol{E}_o - \boldsymbol{E}| = \rho(\varepsilon_r - 1)\varepsilon_o\varepsilon_r \qquad \textit{volt/meter} \qquad (2.6)$$

which implies that the dielectric reduces the surface electric density by an amount

$$|P| = [\rho(\varepsilon_r - 1)/\varepsilon_r] \qquad \textit{coulomb/meter}^2 \qquad (2.7)$$

The entity $\boldsymbol{P}$ can be considered as the polarization vector with a magnitude equal to the charge density on the surface of the dielectric; or the dielectric can be regarded as a *dipole* with a total charge $+PA$ on one face and $-PA$ on the other separated by a distance d. The *dielectric moment* $\boldsymbol{m}_e$ of this dipole is PAd. Hence,

$$\boldsymbol{P} = |\boldsymbol{P}|\,(Ad)/(Ad) \Rightarrow \boldsymbol{m}_e \qquad \textit{(per unit volume)} \tag{2.8}$$

The dielectric polarization can also be regarded as an average dipole moment per unit volume. Therefore, $\boldsymbol{P}$ is related to the dipole moment of individual molecules. It follows from the above relations that the *electric susceptibility* (χ) of the material is:

$$\chi = |\boldsymbol{P}/\boldsymbol{E}| = (\varepsilon_r - 1) \tag{2.9}$$

A related quantity concerning dielectrics is the *electric displacement* across the material defined as $|\boldsymbol{D}| = \rho$. It is related to $\boldsymbol{E}$ and $\boldsymbol{P}$ as follows:

$$\boldsymbol{D} = \varepsilon \boldsymbol{E} + \boldsymbol{P} \tag{2.10}$$

and as explained in Chapter 1, D depicts the electric flux density or the electric flux per unit area across the cross-section of the dielectric medium.

In Chapter 1, the difference between conductors and dielectrics was briefly discussed. It was pointed out that atoms of dielectric materials have their outermost electron shell almost completely filled. A characteristic of complex shells is that it is relatively difficult to dislodge an electron from the shell — it is a bound electron. The result is that dielectric materials have only a few electrons available for the conduction of electric current (and hence are classified as insulators in contrast to metals which have an abundance of free electrons). When a set of charges is placed in a dielectric, they rearrange over an extremely large period of time and for many practical purposes one can consider a placed charge in a dielectric to remain at the point of placement indefinitely. (On the contrary, the rearrangement time for conductors is extremely small, on the order of 10^{-9} seconds, and most of the charges will vanish from the interior of the conducting body and will rearrange to appear on the surface.)

2.2 Polar and Nonpolar Dielectrics

Dielectric materials can be classified into the following versions on the basis of electrical, physical, and chemical properties of the materials:

- *Polar* (dipole) dielectrics.
- *Nonpolar* (neutral) dielectrics.

Depending upon the spatial arrangement of charges in a molecule of a material, a positive-negative pair of charges could form a lumped entity with a common center of gravity; or they could be separated by a distance with their positional centers not coinciding in space. The first case refers to a nonpolar molecule and the second type is designated as a polar molecule inasmuch as, even in the absence of an external electric field, the polar molecule exhibits a *permanent or rigid dipole moment* (Figure 2.2).

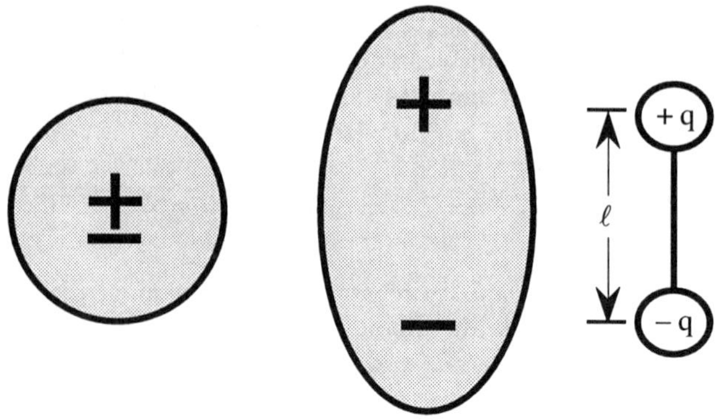

Figure 2.2: (a) Nonpolar molecule;
(b) Polar molecule represented by an equivalent electric dipole.

In nonpolar dielectrics, an external electric field causes an elastic displacement of electronic shells relative to the nucleus of the atoms of a dielectric. This *electronic polarization* occurs in all dielectrics without exception. Nonpolar dielectrics displaying pure electronic polarization have low relative permittivity of the order 1 to 3. On the other hand, the relative permittivity of polar dielectrics could be higher.

Dielectrics of an ionic crystal structure (such as mica, electrical ceramics etc.) exhibit *ionic polarization* adjunct to electronic polarization. This ionic polarization results from the displacement of elastically bound ions and the corresponding relative permittivity is invariably large, being 8 or more.

Spontaneous polarization is another dielectric phenomenon perceived in, special category of dielectrics known as *ferroelectrics*. Details on ferroelectrics are presented in Chapter 12.

Referring to Figure 2.2, the separated positive and negative charges in a polar molecule constitute an *electric dipole*. When subjected to an external electric field, a dipole rotates to align itself along the field. The associated electric moment is given by:

$$|\boldsymbol{p}| = |\boldsymbol{m}_e| = q\ell \qquad \textit{coulomb-meter} \qquad (2.11)$$

where q is the magnitude of the charges constituting the dipole and ℓ is the length of separation between the positive and the negative charges in the dipole. In a nonpolar molecule $\ell = 0$ and therefore $\boldsymbol{m}_e = 0$. In most cases, the electric moment is on the order of 10^{-30} coulomb-meter. It is dependent on the molecular (chemical) bond of the material. The values of the moment for certain groups and bonds of typical organic dielectrics are given in Table 2.1.

2.3 Physical Significance of Dielectric Polarization

Polarization refers to some ordering in space of the electrically charged particles with possible displacement (rotation) in a dielectric under the influence of an external electric field. This results in the formation of an electric moment in the bulk of the material constituted by polarizing particles (atoms, ions and molecules). In linear dielectrics, the induced dielectric moment $\boldsymbol{m}_e$ setup in the polarization process is directly proportional to the applied electric field intensity. That is:

$$m_e = \alpha E \tag{2.12}$$

where α is the constant of proportionality and is known as the *polarizability* of a given particle. It has the unit farad/meter^2 and it reflects the dielectric properties of individual particles of matter and therefore is a microscopic parameter. The sum total of such polarization in the entire medium, however, depicts the bulk dielectric property. The displacement of charged particles during polarization may be regarded as the *elastic shift* of charges. That is, when the applied electric field is removed, the displaced charges return to their initial positions. (In ferroelectric materials (see Chapter 12), however, such a total reversal may not take place.)

Table 2.1 Dipole Moments for Typical Groups and Bonds of Certain Organic Dielectrics

Group or Bond	Organic Compound	$\lvert m_e \rvert \times 10^{30}$ (coulomb-meter)	Direction of Moment Vector
C - H, C - CH_3	Saturated compounds	1.33	H → C
O H	Alcohols	5.27	H → C
N - H	-	5.53	H → N
NO_2	-	13.20	N → O_2
C - N	-	2.03	C - N
C ≡ N	-	13.10	C - N
C - O	Ethers	3.73	C → O
C - Cl	Saturated compounds	6.83	C → Cl
C - Cl	Saturated benzenes	5.17	C → Cl
C - F	-	6.10	C - F

As a measure of intensity of polarization, a polarization vector $\boldsymbol{P}$ as defined earlier is:

$$\boldsymbol{P} = p\ (\text{or } \boldsymbol{m}_e) \qquad \text{per unit volume} \tag{2.13}$$

That is, $\boldsymbol{P}$ refers to total dipole moment per unit volume.

In the absence of an orienting field, the thermal agitation of polar molecules randomizes their direction (Figure 2.3). On the other hand, nonpolar dielectrics which have no dipole moment in the absence of an electric field $(\boldsymbol{E})$ as shown in Figure 2.4, may get stretched and polarize along the applied external field $\boldsymbol{E}$.

Despite random thermal agitation tending to randomize the alignment, each dipole, however, experiences a *torque* given by:

$$T = P \times E \tag{2.14}$$

which realigns the dipoles along the applied electric field.

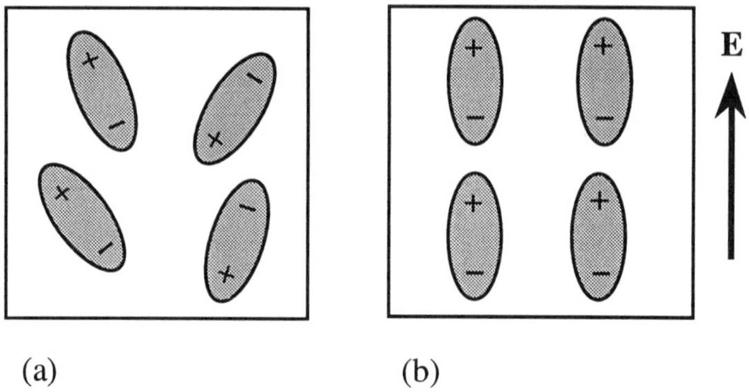

Figure 2.3 (a) Randomly oriented unpolarized set of polar molecules;
(b) Polarized set of polar molecules under the influence of an electric field **E**.

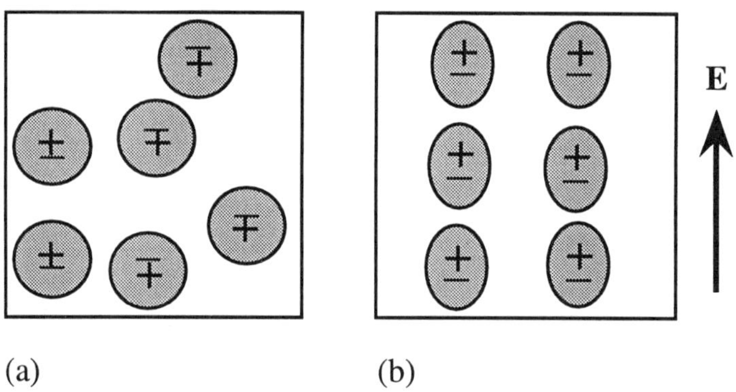

Figure 2.4 (a) Unpolarized, randomly dispositioned set of nonpolar molecules;
(b) Polarized and stretched set of nonpolar molecules under the influence of an electric field **E**.

(In the materials specified as ferroelectrics, there is a permanent or spontaneous polarization present even in the absence of an applied field. This is similar to permanent magnets exhibiting permanent *magnetic polarization.* This permanency exists only below a certain critical temperature known as the *Curie point.*) It should be noted that the polarization in a dielectric does not give rise to a net charge inside the material. That is, even under the state of polarization, the net charge inside the dielectric is zero; or the bulk dielectric is neutral and holds an equal amount of positive and negative charges. In the event of polarization, the layer of dipole charges nearest to the boundary appears as a distribution of surface polarization charges. The magnitude of this polarization charge on the surface is proportional to the applied electric field. The dielectric polarization is a property common to all states of dielectric matter. It can be observed in gases, liquids, solids, and, plasma. The following sections summarize the polarization phenomenon in such states of matter.

2.4 Dielectric Polarization in a Gaseous Material

The application of an electric field *(**E**)* to a dielectric gas will produce a partial alignment of the randomly oriented permanent dipole in the direction of the field together with the separation of the residual charges of the molecules. The polarization vector $\boldsymbol{P}$ of a gas is given by:

$$\boldsymbol{P} = N(\alpha_o + \alpha_d)\boldsymbol{E} \tag{2.15}$$

where N is the number of molecules per unit volume, α_o is the *polarizability* of the molecule and α_d is the *orientational polarizability*. The polarizability of the molecule α_d refers to the constant of proportionality that linearly relates the applied electric field and the permanent dipole moment $\boldsymbol{m}_{ed}$ of the molecules. That is:

$$\boldsymbol{m}_{ed} = \alpha_d \boldsymbol{E} \tag{2.16}$$

and the orientational polarizability is given by $\boldsymbol{m}_{ep}^2/3k_BT$ where $\boldsymbol{m}_{ep}$ is the component of moment due to the initial orientation of the polar molecule and the total moment $(\boldsymbol{m}_e)$ is the vector sum of $\boldsymbol{m}_{ed}$ and $\boldsymbol{m}_{ep}$. The orientation polarizability depends on the thermal state of the system and therefore is dictated by the *Boltzmann energy* k_BT (where k_B is the Boltzmann constant and T is the temperature).

In terms of α_o and α_d, the *molar polarization* of the gas $\boldsymbol{P}$ can be written as:

$$\boldsymbol{P} = (4/3)\pi N_A(\alpha_o + \alpha_d) \tag{2.17}$$

where N_A is *Avogadro's number*. The corresponding relation that specifies the dielectric constant of the gas is:

$$(\varepsilon - 1)(\varepsilon + 2) = (4\rho/3M)\pi N_A(\alpha_o + \alpha_d) \tag{2.18}$$

where ρ and M are the density and molecular weight of the gas, respectively. The above relation (Equation 2.18) is popularly known as *Debye's equation*.

Clausius-Mossotti Equation

In terms of the dielectric constant ε_∞ of the gaseous medium specified at very high frequencies (or at optical wavelengths), the Debye relation is written modified as the *Mossotti - Clausius equation* given by:

$$[(\varepsilon - 1)/(\varepsilon + 2)] - [(\varepsilon_\infty - 1)/(\varepsilon_\infty + 2)] = 4\pi N_A \boldsymbol{m}_{ep}^2/3mk_BT \tag{2.19}$$

The parameter ε_∞ can also be written in terms of the *refractive index* of the medium n by the relation $\varepsilon_\infty = n^2$. The Debye equation is found to hold for a wide variety of gases and vapors at ordinary pressures. It is possible to calculate the approximate values of molecular dipole moment from the dielectric constants and densities at dilute solutions of the polar molecules in nonpolar solvents. However, it fails for pure liquids, gases, and vapors in which association, dissociation, and changing intermolecular energy occur invariably.

2.5 Dielectric Polarization in Liquids

Rigorous analysis pertinent to liquids leads to the following equation (which is similar to the Debye equation of gases):

$$[(\varepsilon - 1)/(\varepsilon + 2)][M/\rho] - [(\varepsilon_\infty - 1)/(\varepsilon_\infty + 2)][M/\rho] = [3\varepsilon(\varepsilon_\infty + 2)/(2\varepsilon + \varepsilon_\infty)(\varepsilon + 2)][4\pi N_A m_{ep}^2/9k_BT] \quad (2.20)$$

Here again, ε_∞ is the dielectric constant at the frequency tending to infinity or at the optical wavelength limits, and $\varepsilon_\infty = n^2$ where n represents the refractive index of the medium. The above relation (Equation 2.20) is known as the *Onsager equation*. It converges to the Debye equation when $\varepsilon \rightarrow \varepsilon_\infty$ as occurs with gases at atmospheric or lower pressures. The Onsager relation presumes a uniformity of dielectric constant throughout the medium. However, if m_{ep}^2 is replaced by gm_{ep}^2 in Equation 2.20, with the hindrance to molecular orientation due to molecular interaction specified by a coefficient g, Equation (2.20) then would represent the *modified Onsager equation* due to Kirkwood and Fröhlich. A summary of such local field corrections is presented in Table 2.2

Table 2.2 Local Field Corrections

Model	Correction Factor Δ in the Expression $(\varepsilon_s - \varepsilon_\infty) = Nm_{ep}^2\Delta/3k_BT$
Dilute phase in which the local field is equal to the applied field	1
Clausius-Mossotti-Lorentz	$(\varepsilon_s + 2)(\varepsilon_\infty + 2)/9$
Onsager	$\varepsilon_s(\varepsilon_\infty + 2)^2/[3(2\varepsilon_s + 2)]$
Fröhlich	$g\,\varepsilon_s(\varepsilon_\infty + 2)^2/[3(2\varepsilon_s + \varepsilon_\infty)]$

2.6 Dielectric Response: The Relaxation Process

The dielectric response of homogeneous materials to applied electric fields has been studied extensively in the past. Most of the relevant work sought to explain the macroscopic behavior of the dielectrics stipulated by the permittivity in terms of microscopic quantities (namely, the average dipole moment and the number of dipoles per unit volume). For example, the Debye model describes the materials in terms of individual dipole moments containing variable charge separation in a viscous medium. A dipole subjected to an electric field would tend to align itself and elongate in the direction of the applied field in a finite time. When the externally applied field is removed, the dipole "relaxes" to the initial state as an exponential decay process. The time constant for this exponential decay is referred to as the *relaxation time* τ_r. The relaxation behavior for a single dipole, or a volume filled with identical non-interacting dipoles having the same orientation would yield the classical *Debye response* given by:

$$\varepsilon^* = \varepsilon_\infty + (\varepsilon_s - \varepsilon_\infty)/(1 + j\omega\tau_r) \quad (2.21)$$

where ε^* is the complex permittivity, ε_∞ is the permittivity at infinite frequency, ε_s is the static or d.c. permittivity, ω is the applied radian frequency, and τ_r is the relaxation time for the particular dipole structure.

The following relations refer to commonly used terminology adopted in the literature concerning the permittivity parameters and are largely adopted in this handbook. (More notations are presented in Appendix 2A.)

$$\varepsilon^* = (\varepsilon' - j\varepsilon'') = \varepsilon_o \varepsilon_r^* \quad \text{farad/meter} \tag{2.22a}$$

$$= \varepsilon_o(\varepsilon_r' - j\varepsilon_r'') \quad \text{farad/meter}$$

$$= \varepsilon_o(\varepsilon_r' - j\varepsilon_r' \tan\delta) \quad \text{farad/meter}$$

$$= \varepsilon_o(\varepsilon_r' - j\, s/\omega\varepsilon_o) \quad \text{farad/meter} \tag{2.22b}$$

$$(\varepsilon - \varepsilon_\infty) = (\varepsilon_s - \varepsilon_\infty)/(1 + \omega^2 \tau_r^2) \tag{2.23}$$

$$\varepsilon'' = (\varepsilon_s - \varepsilon_\infty)\omega\tau_r/(1 + \omega^2 \tau_r^2)$$

ε' = Real part of the complex permittivity (farad/meter).
ε'' = Imaginary part of the complex permittivity (farad/meter).
ε_o = Permittivity of free space (8.8542×10^{-12} (farad/meter).
ε_r^* = Complex relative permittivity, (complex dielectric constant).
ε'_r = Real part of complex dielectric constant.
ε''_r = Loss factor, (imaginary part of complex dielectric constant).
$\tan\delta$ = Loss tangent ($\tan\delta = \varepsilon''/\varepsilon' = \varepsilon_r''/\varepsilon_r' = \sigma/\omega\varepsilon_o\varepsilon_r$).
σ = Conductivity (siemen/meter).
ε_s = Static (d.c.) dielectric constant.
ε_∞ = Dielectric constant at infinite frequency (at optical wavelengths).
n = $(\varepsilon_\infty)^{1/2}$ = Refractive index of the medium.

The conductivity term in the expression for loss tangent makes no distinction between long-range, free-charge carrier motion (d.c. conductivity) and short-range dipolar or confined charge motion (a.c. conductivity minus d.c. conductivity). Both short- and long-range charge motions are accounted for in the following expression which is only valid at non-zero frequencies:

A detailed study on dielectric relaxation is presented in the following section.

2.7 Polarization in Lossy Dielectrics: Concept of Complex Polarization

The molecular polarization process involves displacement of dipoles. Therefore, when subjected to an alternating electric field force, there is a time-dependent polarization response referred to as the *relaxation process*. The rate of response of the molecular polarization therefore influences the polarizability of the medium and hence the bulk dielectric property. As a consequence, both polarizability and the permittivity become complex numbers which can be expressed as:

$$\alpha^* = (\alpha' - j\alpha'') \tag{2.24a}$$

$$\varepsilon^* = (\varepsilon' - j\varepsilon'') \tag{2.24b}$$

When the frequency of applied field force is very high, the permanent dipole moments may be unable to "relax" along or reorient themselves with the electric field and their contribution to the polarization of the medium will decrease with increasing frequency.

Debye analyzed the dependence of the complex polarizability (α_o^*) of a dielectric sphere of radius a, immersed in a fluid of viscosity η and obtained the following relation:

$$\alpha^* = \boldsymbol{m}_{ep}^2/3k_BT\ [1/(1 + j\omega\tau)] \tag{2.24c}$$

where τ is the corresponding *relaxation time* which is given by:

$$\tau = 4\pi\eta a^3/k_BT \tag{2.25}$$

Extending Debye's theory to electrons and atoms, the corresponding complex polarizabilities α_e and α_a can be written in terms of the natural resonant frequencies of electrons and atoms and the associated absorptions.

The global effect of complex polarizability or a dielectric material can be portrayed as the *complex permittivity* of the medium. That is, in the most general form, the permittivity of a medium is complex and written as:

$$\varepsilon^* = (\varepsilon' - j\varepsilon'') \tag{2.26}$$

where ε' and ε'' are frequency dependent and ε'' represents the lossy nature of the dielectric.

Written in terms of the constitutive relation depicting the current density $(\boldsymbol{J})$ in the dielectric *versus* the applied electric field $\boldsymbol{E}$, namely,

$$\boldsymbol{J} = (\sigma + j\omega\varepsilon)\boldsymbol{E} \tag{2.27}$$

the complex permittivity can be specified in terms of the conductivity parameter (σ) as follows:

$$\varepsilon^* = (\varepsilon - j\ \sigma/\omega) = (\varepsilon' - j\ \varepsilon'') \tag{2.28}$$

The real part of the complex permittivity depicts the capacitive term and the imaginary part represents the energy dissipative term. A ratio of these two terms is known as the *loss tangent* given by:

$$\tan\delta = \sigma/\omega\varepsilon \tag{2.29}$$

In a dielectric material, the power loss or dissipation is not exclusively due to any free charge that may present, but also is due to bound charges.

2.8 Dielectric Dispersion

Considering an electromagnetic wave propagation through a lossy dielectric the *complex propagation constant* (γ^*) is given by

$$\gamma^* = (\alpha + j\ \beta) = (j\omega/c\ n^*) \tag{2.30}$$

where c is velocity of propagation of the electromagnetic wave and n^* is the *complex refractive index* of the lossy dielectric medium. Written explicitly:

$$n^* = [\varepsilon\ -\ j\varepsilon''/\varepsilon_o]^{1/2} = [\varepsilon'_r - j\varepsilon''_r]^{1/2} \tag{2.31}$$

where ε_o is the permittivity of free space and $(\varepsilon_r - j\varepsilon''_r) = \varepsilon_r^*$ represents the complex relative permittivity (or *complex dielectric constant* of the medium). Denoting $n^* = (n - jk)$, the following relations can be derived:

$$\alpha = \omega/c\ k \tag{2.32a}$$
$$\beta = \omega/c\ n \tag{2.32b}$$

or in terms of complex dielectric constant*,

$$n = 1/[\sqrt{2}\{\sqrt{(\varepsilon'_r)^2 + (\varepsilon''_r)^2} + \varepsilon'_r\}^{1/2}] \tag{2.33a}$$

$$k = 1/[\sqrt{2}\{\sqrt{(\varepsilon'_r)^2 + (\varepsilon''_r)^2} - \varepsilon'_r\}^{1/2}] \tag{2.33b}$$

and

$$\varepsilon_r^* = (\beta^2 - 2j\alpha\beta)(c/\omega)^2 \tag{2.33c}$$

2.9 Relaxation and Resonances

In any material the various types of charges and charge associations or groups lead to corresponding interactions with the applied electromagnetic field resulting in either relaxation or *resonance* phenomenon, as briefed below:

Inner bound electrons: These are tightly bound to the nuclei and are little influenced by the external field. They *resonate* only with a high energy ($\cong 10^4$ ev), extremely short wavelength ($\lambda \cong 10^{-10}$ meter) electromagnetic field such as due to X-rays.

Outer bound electrons: These correspond to valence electrons of outer electronic shells which contribute *atomic and/or molecular polarizabilities.*

Free electrons: These refer to those in the conduction band and contribute electric current through free movement under an electric field with a velocity characterized by a *mobility factor.*

Bound ions: These represent ionic dipoles (for example, Cl^-H^+) formed by a set of negative and positive ions or by a vacancy plus a substitutional cation, such as $[Li]^-Mg^+$ where $[Li]^-$ represents a vacancy. These are permanent dipoles which experience *orientational polarization* under an applied field.

Free ions: These are dissociated ions as in electrolytes or excess ions (nonstoichiometric ions) in crystals which move in the applied field with a low mobility.

Multipoles: These are mainly *quadrupoles* or an antiparallel disposition of a pair of dipoles which undergo a configurational strain under a uniform field and experience a torque in a divergent field.

Depending on the type of charges as enumerated above pertinent to a material, the electromagnetic field *versus* material interaction may result in either a relaxation or a resonance as depicted in Figure 2.5. Each category of the charges listed above has its own critical frequency above which the interaction with the field becomes vanishingly small.

As could be inferred from the above deliberations, the complex dielectric characterization of materials refers to the relaxation process involved. The subsets of the relaxation phenomenon are as follows:

---

* For very small dissipation involved, $n \simeq \sqrt{\varepsilon'_r}$ and $k \simeq \delta/2\sqrt{\varepsilon''_r} = \varepsilon''_r/2\sqrt{\varepsilon'_r}$.

- *Dipole orientational relaxation*: This is time-dependent polarization due to orientation of dipoles.
- *Interfacial relaxation*: In heterogeneous materials, the polarization corresponds to an evolution of short-time (high frequency) capacitive effects and long-time (low frequency) resistive effect. This is known as the *Maxwell-Wagner relaxation.*
- *Space-charge relaxation*: This occurs in materials wherein carriers do not recombine at the electrodes and therefore behave in a low frequency a.c. field as macroscopic dipoles which reverse their direction as the polarity of the applied field alternates.

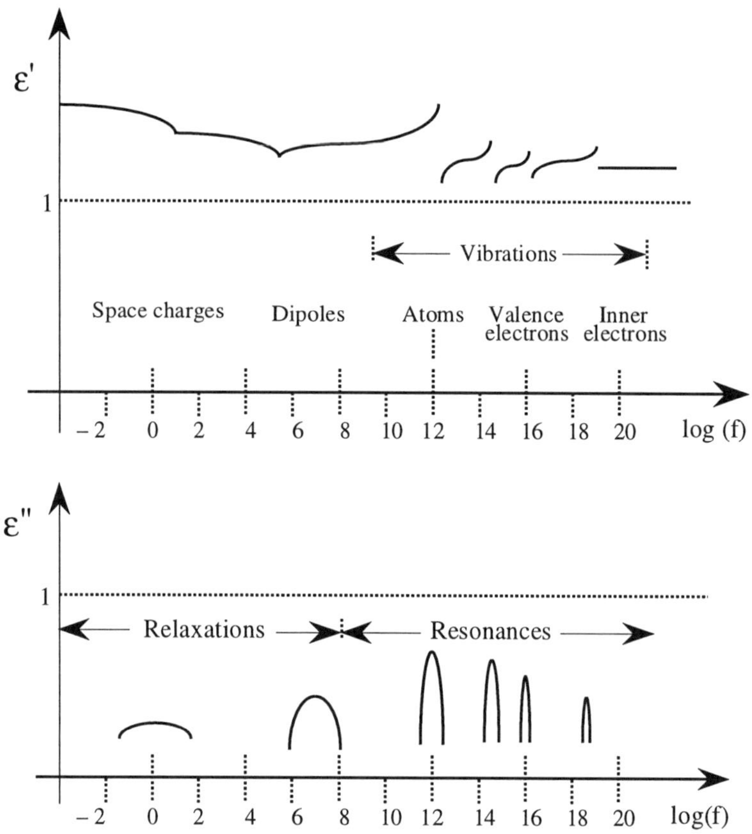

Figure 2.5 Response of a dielectric material interacting with an electromagnetic wave: Relaxation and resonance effects.

2.10 Dielectric Response in Time Domain

In a linear, isotropic homogeneous dielectric, subjected to a time-varying electric field $\boldsymbol{E(t)}$, the corresponding dielectric displacement $\boldsymbol{D(t)}$ can be specified by the following relation:

$$\boldsymbol{D(t)} = \varepsilon_\infty \boldsymbol{E(t)} + (\varepsilon_s - \varepsilon_\infty)[\boldsymbol{E(t)} * \phi(t)] \tag{2.34}$$

where * indicates the *convolution operation* and $\Phi(t)$ is known as the *decay function.* It is the derivative of a function that describes the time-dependent relaxational effect in the dielectric that causes a sluggish growth of polarization (instead of instantaneous response) when subjected to a step-functional electric field. The sluggishness or transient growth of polarization is dictated by the noninstantaneous (or delayed) molecular dipole orientations and other frictional processes.

In the case of harmonic field excitation, $\Phi(t)$ can be derived explicitly from which the following relations (known as *Kramers-Kröning equations*) have been obtained:

$$\varepsilon'(\omega) - \varepsilon_\infty = 2/\pi \int_0^\infty x\varepsilon''(x)/(x^2 - \omega^2)\, dx \tag{2.35a}$$

$$\varepsilon''(\omega) = 2\omega/\pi \int_0^\infty [\varepsilon'(x) - \varepsilon_\infty]/(x^2 - \omega^2)\, dx \tag{2.35b}$$

The above equations describe the *spectral dispersion* of the dielectric parameters ε' and ε" of a material and ω represents 2π × frequency.

2.11 Geometrical Representation of Dielectric Relaxation Process

The Debye relation, namely:

$$\varepsilon^*(\omega) = \varepsilon_\infty + (\varepsilon_s - \varepsilon_\infty)/(1 + j\omega\tau) \tag{2.36}$$

is derived from the concept of decay function as indicated before and the real and imaginary parts of the Debye relation is depicted in Figure 2.6.

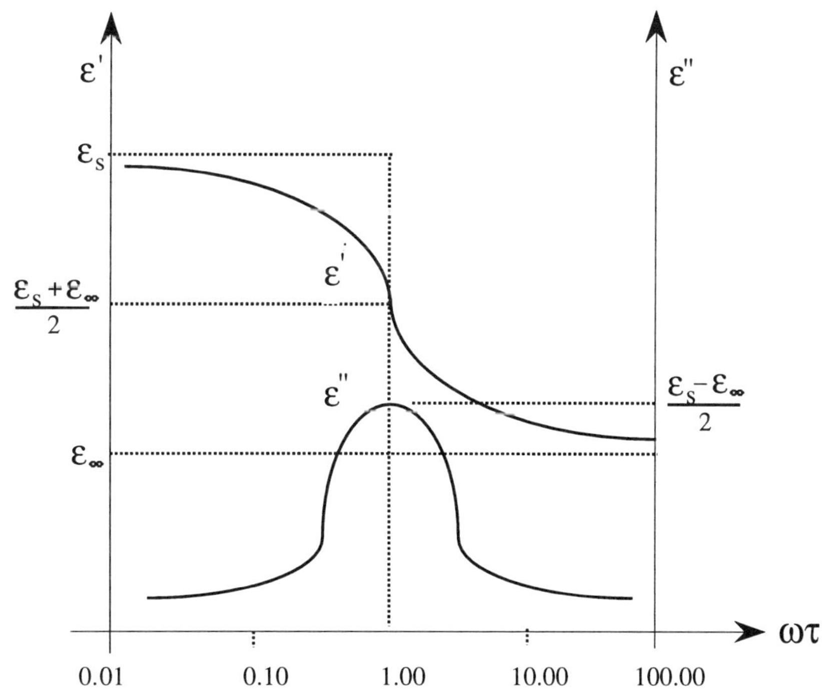

Figure 2.6 Dielectric relaxation: Debye relation.

In the Argand plane, the Debye relation can be represented as shown in Figure 2.7. This is known as the *Cole-Cole diagram*.

The simple Debye relation is derived on the assumption that (1) local field at any point in the medium is the same as the applied field; (2) d.c. conductivity of the material is negligible; and (3) all the dipoles have identical relaxation time τ.

The first assumption is rather valid only for dilute states such as in gases. In condensed matter, however, the local field deviates by an extent known as the *Lorentz field correction factor*. With the inclusion of this correction, the modified Debye relation has τ replaced by τ' given by:

$$\tau' = (\varepsilon_s - 2\varepsilon_0)/(\varepsilon_\infty + \varepsilon_0) \tag{2.37}$$

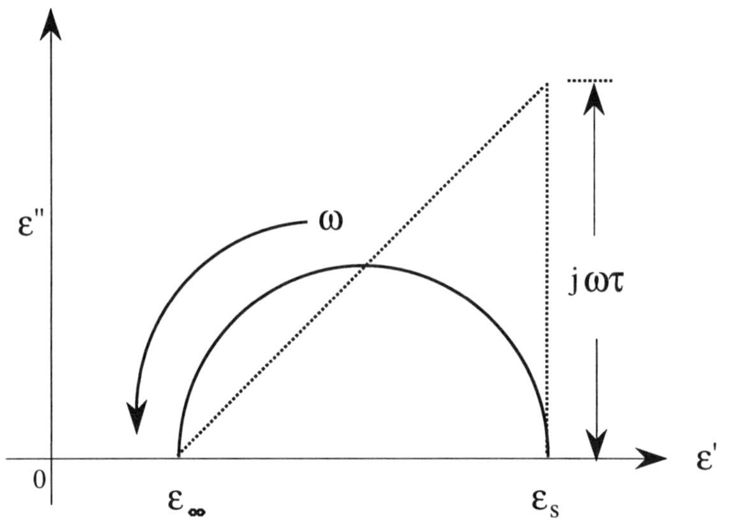

Figure 2.7 Cole-Cole diagram of a dielectric material.

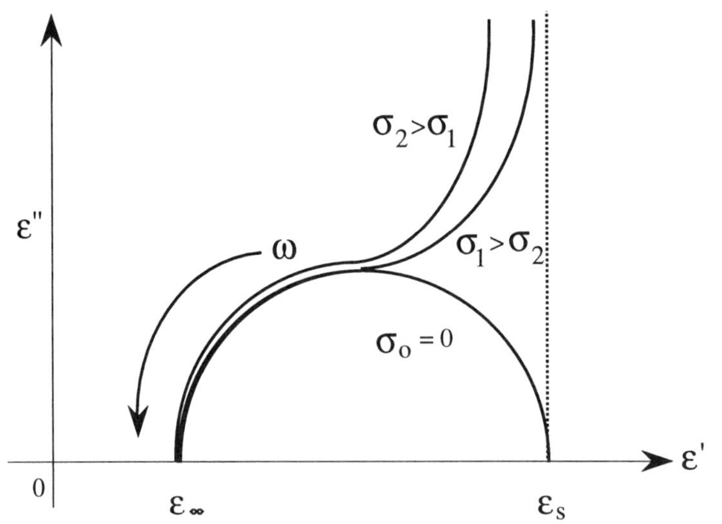

Figure 2.8 Effect of conducting on the Cole-Cole diagram.

The second assumption refers to the absence any d.c. conductivity. However, for a conducting polar dielectric with a conductivity σ, the modified Debye relation is:

$$\varepsilon''(\omega) = \varepsilon_\infty + (\varepsilon_s - \varepsilon_\infty)/(1+j\omega\tau) - j\sigma/\omega \qquad (2.38)$$

Illustrated in Figure 2.8 is the Cole-Cole diagram showing the effect of conductivity σ.

Alternatively, the *log(e") versus log(ω)* graph is presented in Figure 2.9.

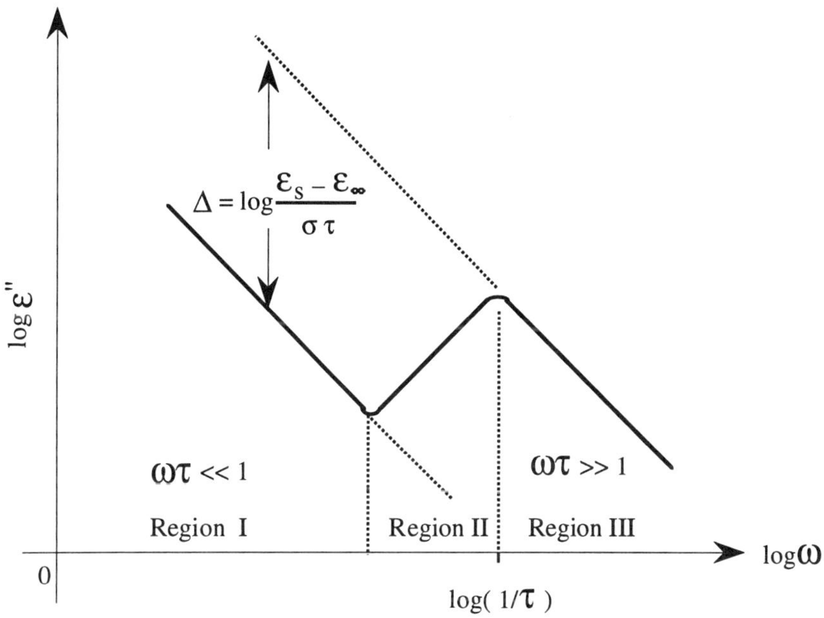

Figure 2.9 Lossy dielectric response *versus* frequency.

Pertinent to the third assumption, in certain materials there could be different species of dipoles so that instead of one relaxation time, a continuous distribution of τ is prevalent. Further even for single-type dipoles, if they have significant particulate eccentricity, with the application of an electric field only the longitudinal components of dipole moment are active in the relaxation process. Therefore, a large collection of such randomly dispersed dipoles would pose varying relaxation times. In such cases, the Debye relation is written as:

$$\varepsilon''(\omega) = \varepsilon_\infty + \sum_{i=1}^{N} \varepsilon_i/(1 + j\omega\tau_i) \qquad (2.39)$$

where N is the number of relaxational processes involved.

Further, imperfections in the material, contaminates, temperature etc. may also influence the Debye relation. The corresponding modified relation is given by:

$$\varepsilon''(\omega) = \varepsilon_\infty + (\varepsilon_s - \varepsilon_\infty)/(1 + j\omega\tau)^\alpha \qquad (2.40)$$

where α is a modifying factor with the limits $0 \le \alpha \le 1$. The presence of α skews the Cole-Cole diagram. This skewed representation is known as the *Davidson-Cole diagram* which is illustrated in Figure 2.10.

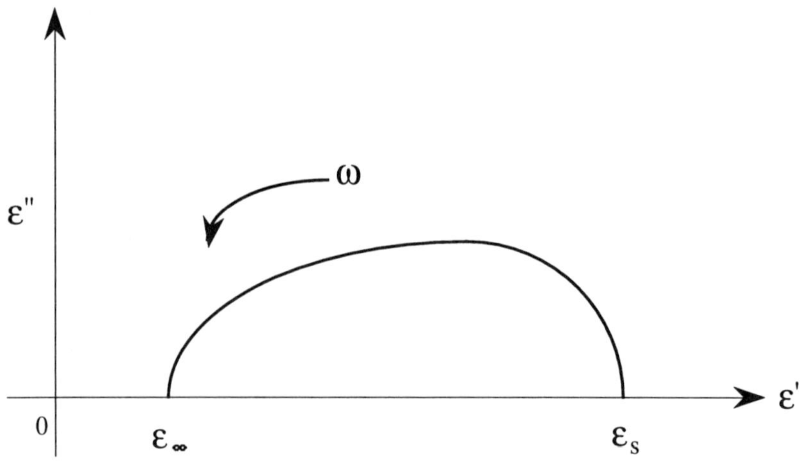

Figure 2.10 Skewed Cole-Cole plot: (Davidson-Cole diagram).

2.12 Double-Layer Relaxational Process

A *double layer* refers to a composite structure constituted by two lossy dielectrics as shown in Figure 2.11.

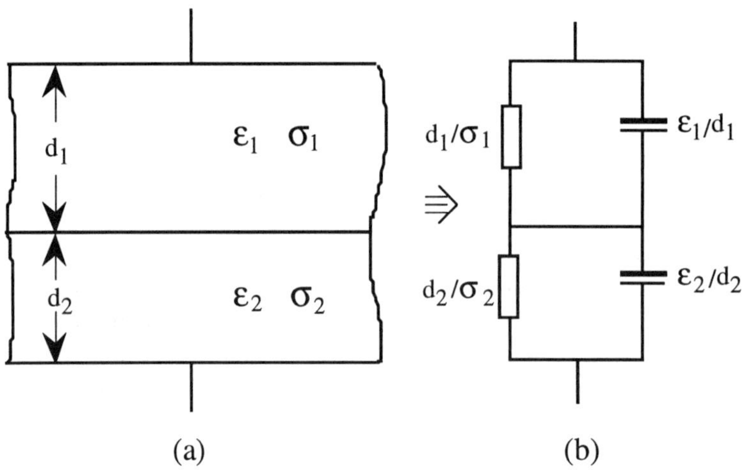

Figure 2.11 (a) A double-layer structure of two lossy dielectrics;
(b) Equivalent circuit representation of a double layer.

Due to distinct dielectric characteristics of the constituents, there is a specific type of relaxation process or the response to an alternating voltage associated with the material due to the field transition at the double-layer boundary. This is known as the *Maxwell-Wagner effect*.

2.13 Gaseous Dielectrics

Gaseous dielectrics, in general, include all gases or a mixture of gases and materials in the vapor state. Tables 2.A.1 and 2.A.2 summarize the basic characteristics of various dielectrics in the gaseous phase. The dielectric property of monoatomic gases (such as the rare gases helium or argon), which exhibit nonmolecular or atomic interactions, is governed by the electric dipole moment per unit volume specified by the polarization vector $\boldsymbol{P} = N\alpha_e\boldsymbol{E}$

where N is the number of atoms per meter^3 and $\boldsymbol{E}$ is the externally applied electric field intensity. The parameter α_e refers to the atomic constant, and from the definition of dielectric polarization $(\boldsymbol{P})$ the following relation can be specified:

$$\varepsilon_o(\varepsilon_r - 1) = N\alpha_e \tag{2.41}$$

As a simple model $(\varepsilon_r - 1)$ can be equated to $4\pi NR^3$ with the distance parameter R having the order of magnitude as the radius of an atom; and α_e remains relatively independent of temperature. Essentially the dielectric property of monoatomic gases is determined by the electronic polarizability of the atoms as dictated by their electronic structure.

The dielectric constant of polyatomic gases, on the other hand, is invariably temperature dependent inasmuch as the interactive forces between the atoms are not negligible. Further, the physical arrangement of the atoms in polyatomic gases would cause not only electronic and ionic polarizations but also a characteristic orientational polarization wherein the permanent dipoles polarize or tend to align along the direction of the external electric field (Figure 2.2). Correspondingly, the total polarization of polyatomic gases is the result and the effect of electronic polarization (quantified by the constant α_e), ionic polarization dictated by a parameter α_i, and orientational polarization decided by a factor α_p as controlled by the permanent dipole moment $(\boldsymbol{m}_p)$ and the Boltzmann thermal energy k_BT. Explicitly,

$$\boldsymbol{P}/\boldsymbol{E} = \varepsilon_o(\varepsilon_r - 1) = N(\alpha_e + \alpha_i + m_p^{\,2}/3k_BT) \tag{2.42}$$

The volt-ampere characteristic of a gaseous dielectric is shown in Figure 2.12. At low, applied voltage (or weak electric field), the electric forces acting on the charged particles are rather small, and the increase in current with voltage follows linear Ohm's law relationship (Region I). A further rise in the voltage (Region II) maintains a steady flow of current constituted by available free ions and electrons. However, in Region III, the excessive voltage provides enough kinetic energy to the charged particles, which upon collision with neutral particles would liberate additional electrons. In this ionized state, the current grows rapidly with voltage. The development of the collision-ionization process leads to a *breakdown* in the form of an *arc discharge*.

2.14 Dielectric Breakdown

Endurance of dielectrics under the action of an electric field is dictated by the various coexisting dielectric phenomena, namely, the dielectric polarization, electrical conduction and dielectric relaxation. It turns out that the dielectric behavior of a material cannot be sustained with indefinite increase of the applied external field. If the electric voltage (and hence the electric field) is increased progressively, there will be a "breakdown" at which the material will cease to be a dielectric and fail to exhibit electrical insulation properties.

Characteristically, the *breakdown phenomenon in a dielectric* is accompanied by a surge of current flowing through the material which increases sharply with the increase in the applied voltage. The point of inflection, in the I-V characteristic of a dielectric/insulating material refers to $dI/dV \rightarrow \infty$ or the occurrence of breakdown. The subsequent reduction in voltage can be attributed to a decrease in the resistance of the material facilitated by the causative mechanisms of the breakdown.

The theory and physics of dielectric breakdown can be explained by considering the various mechanisms of breakdown associated with dielectrics in solid, liquid, and gaseous phases.

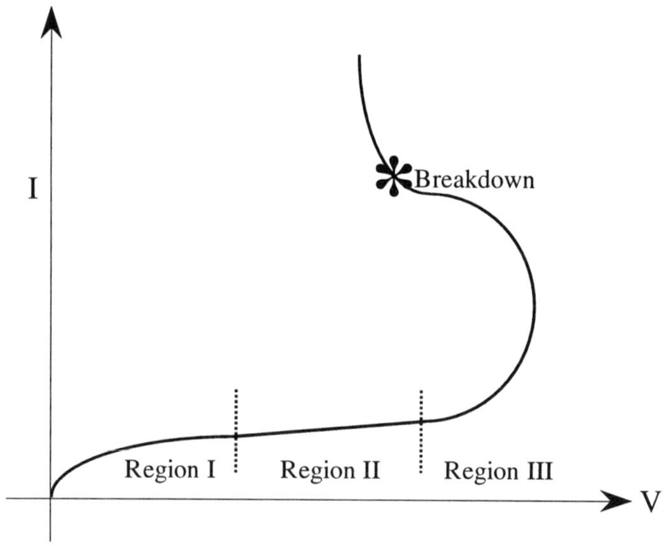

Figure 2.12 Voltage-current characteristic of a gaseous dielectric.

A. Breakdown in solid dielectrics: The mechanisms of breakdown in solid dielectrics are as follows:

- *Disruptive breakdown*: It refers to the effect of high-energy fields causing physical/molecular disintegration, mostly catastrophic and visible. It is commonly observed in very thin dielectrics subjected to excessive electric potential even at low temperatures. It is classified as a subset of *conductive breakdown* with higher energy release and follows no definite mathematical and/or physical model(s).
- *Thermal breakdown:* In heterogeneous dielectrics, the presence of nonuniform electric resistance across a dielectric could cause uneven leakage current to flow through it. In regions where the localized current is intense, the resulting heating would raise the temperature, lowering the electrical resistivity further. This cumulative effect spirals up to a condition resulting in the thermal breakdown by electrical conduction.
- *Conductive breakdown*: It is attributed to ionic charge carriers in the dielectric that move in an electric field and produce a leakage current. The electric field itself may produce additional ions by collision of electrons and molecules so that the current increases with increasing voltage stress. Eventually, the large flow of ions and electrons culminates in transforming the dielectric into a conductor which specifies the state of conductive breakdown. Conductive breakdown is amenable for theoretical analysis and mathematical modeling [1,5].
- *Electromechanical breakdown*: This refers to the mechanical failure (followed by electrical failure) in a dielectric due to the intense electrostatic pressure exerted on the dielectric. It often occurs in certain soft polymers (such as polyisobutylene and polyethylene) at low temperatures and seldom occurs in hard polymers like polystyrene or polymethylmethacrylate.
- *Electrochemical breakdown*: Contaminants such as ionic impurities in a dielectric could cause a leakage current which over a passage of time may induce thermal breakdown, causing a run-away buildup of heat. Dielectric aging under impure ambients often faces electrochemical breakdowns.

• *Breakdown due to bubble formation*: In moisture or liquid contaminated solid dielectrics, internal voids or cavities are susceptible to expansion with the associated liquid product being subjected to decomposition under electrical stresses and/or thermal aging. Thus, they are prone to bubble formation which could constitute a link of conducting chain across the electrodes causing a current surge and conduction breakdown.

• *Corona-induced breakdown*: Corona-induced high speed electrons generated in the surrounding air or within the void pockets of the dielectrics at high electrical stresses could initiate an erosion in the material with surface tracking and leakage across the electrodes manifesting as the breakdown.

B. Breakdown in liquid dielectrics: The breakdown in liquid dielectrics is due to two possibilities:

• *Bubble mechanism*: The electric stress on the liquid molecules could cause extreme agitation and formation of bubbles by heat or cavitation. Alignment of such bubbles along electric flux lines allow ion formation in chains. Such ionic chains carry conduction currents which lead to heavy ionization and subsequent arcing through the liquid registering a breakdown.

• *Conduction mechanism initiated by the contaminants in the liquid:*. The contaminants in the liquid could form a conducting bridge across the electrodes, thereby facilitating an excessive leakage current through the material leading to the conduction breakdown.

C. Breakdown in gaseous dielectrics: Breakdown in gases commences with the ionization of the gas molecules due to collisions with electrons. In the presence of the applied electric field, the free electrons in the gaseous system acquire sufficient kinetic energy to ionize the gas molecules *via* collisions. The ionization further accentuates the collisions due to electrons set free from the molecules. The secondary emission electrons from the electrodes could also participate in the collision ionization process. The breakdown is initiated with a spark discharge across the electrodes culminating into an arc discharge eventually resulting in a high current density and a short-circuit condition across the gaseous column (Figure 2.12).

The breakdown in a gas manifests as a spark discharge visible as a luminous spark across the electrodes which apply the electric field on the gaseous medium. The breakdown potential V_B obeys the following *Paschen's law*:

$$V_B = A_P ph \tag{2.43}$$

where p is the gas pressure, h is the length of gas column across the electrodes and A_P is the *Paschen parameter* dependent on p and h.

2.15 Dielectric Strength and Insulation Integrity

• *Intrinsic dielectric strength*: It is a parameter which refers to the inherent dielectric breakdown strength depicting the ultimate or maximum electric field which a dielectric material can sustain (without the breakdown taking place) prior to the ravages of aging, fatigue, erosion or any extraneous abuses invading the material.

• *Dielectric strength factor*: It is the ratio of the intrinsic dielectric strength of a material to that of air in the same spacing and electrode configuration used to subject the material to electrical stress.

Dielectric strength is an important factor that specifies the quality of a dielectric material as an electrical insulator. In practical insulators, the actual dielectric strength could be much lower than the intrinsic value of the constituting dielectric. This is due to the fact that the dielectric insulation material may acquire (during manufacture, storage or usage) additional ionic and moisture contents reducing the endurance of the dielectric to electrical stressings; also, factors like mechanical stresses, thermal influences or radiation exposures could reduce the dielectric strength significantly. Other influences such as electrical fatigue (due to cyclic electric stressing), corona erosions and electrostatic stresses may deteriorate the dielectric quality of insulating materials as being precursors of the breakdown.

2.16 Liquid Dielectrics

Extensive applications of liquid dielectrics are in electrical installations such as coolants in power transformers, reactors or circuit-breakers, and as dielectrics in high-voltage capacitors. They also serve as impregnates in winding insulations and as arc-discharge suppressants in disengaging metal contacts in relays, circuit-breakers, etc.

Invariably, petroleum (mineral) oils are used in abundance for electrical insulation applications. Organic (vegetable) and synthetic oils are also used, however, on a limited scale. Petroleum oil is a mixture of hydrocarbons of the naphthene, paraffin and aromatic series. Naphthene hydrocarbons which constitute about 75% of the petroleum oil are not susceptible to oxidation. The paraffin and aromatic hydrocarbons are also chemically stable. However, excessive presence of aromatic hydrocarbon would reduce the *flash point* impairing the electrical characteristics of the material. The disadvantages of petroleum oils are combustibility, low permittivity and low flash point temperature.

Synthetic dielectric liquids obtained by chlorinating the crystalline substance diphenyl (H_5C_6 - C_6H_5) are constituted by sovol molecular structure ($Cl_3H_2C_6$ - $C_6H_3Cl_2$). The sovol molecules are asymmetric and hence the synthesized (liquid) material is a polar dielectric with a relative permittivity in the range of 5. It is superior to a nonpolar petroleum oil dielectric. Also sovol is incombustible, but high toxicity and viscosity restrict its applications.

Other synthetic liquids like trichlorobenzene ($C_6H_3Cl_3$), chlorinated diphenyls (askarels), silicone liquids, and flurocarbon organic esters oils such as $(C_4F_9)_3N$, $C_8F_{16}O$ are also used in electrical insulation applications.

Table 2.3 provides data on the static and quasistatic properties of typical dielectric liquids used in electrical engineering applications and a comprehensive listing of the dielectric properties of various liquid dielectrics are furnished in Table 2.B.

Liquid dielectrics, in general, are highly hydroscopic and moisture and impurities deteriorate their electrical properties. Also, their dielectric characteristics are significantly temperature dependent. Typical dependency of the dielectric parameters *versus* temperature and moisture are illustrated in Figure 2.13.

Table 2.3 Static and Quasistatic Properties of Liquid Dielectrics Used in Electrical Engineering Applications

Dielectric Material	Volume Resistivity ohm/meter	Dielectric Constant (ε_r)	Loss-tangent ($\tan\delta$) at 50/60 Hz	Dielectric strength (volt × 10^6) per meter	Dissipation Factor at 50/60 Hz & 100°C
Mineral oils (Transformer oil)	10^{12}–10^{13}	2.2-2.4	0.003-0.005	18 at 20° C	0.10
Capacitor oil	12^{13}–10^{14}	2.2	0.002-0.003	20	—
Cable oil					
Low viscosity type	10^{12}–10^{13}	2.3	0.003	18	—
High viscosity typc	10^{13}	2.1	0.003	20	—
Polyethyl-silicone	10^{11}–10^{14}	2.6	0.0003	25	—
Askarel/ sovol	10^{14}–10^{15}	4–5	0.005-0.008	18 - 20	1.50
Silicon oil (dimethyl-phenylmethyl)	10^{15}	2.5	0.0004	35	0.04
Fluoro-carbon oil	10^{14}	2	0.0005	40	0.05
Organic esters					
Castor oil	3 x 10^{12}	3.8	0.02	—	6.00
Dibutyl sebacate	5 x 10^{14}	4.5	0.001	3.8	1.00
Butyl stearate	5 x 10^{14}	3.3	0.001	—	1.00
Silicate ester	10^{13}	2.7	—	2.7	—

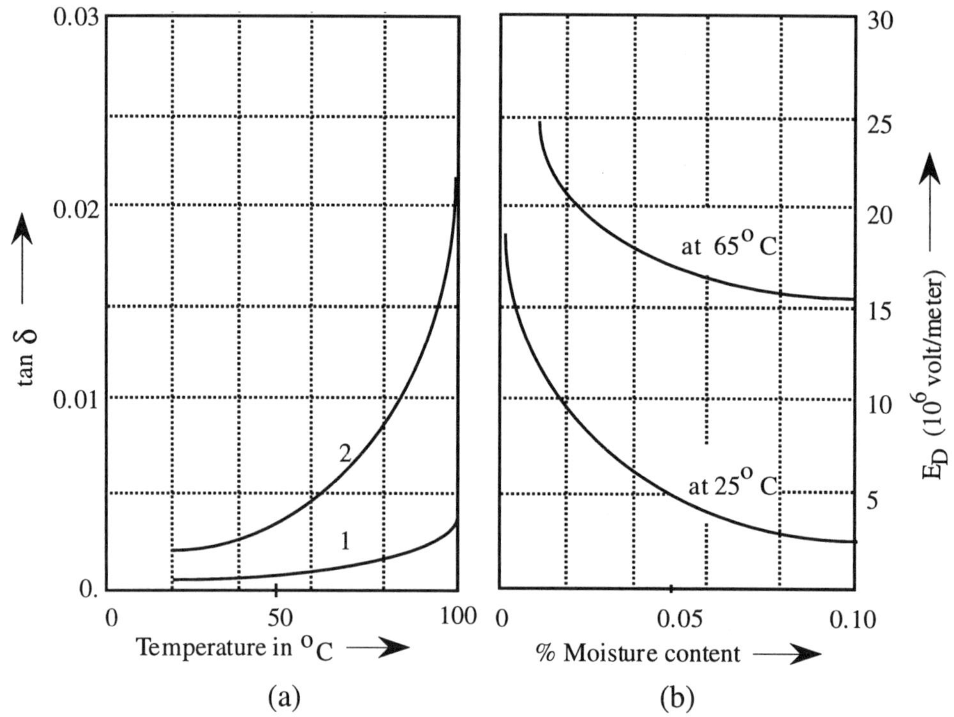

Figure 2.13 (a) Loss tangent *versus* temperature at 50/60 Hz: (1) Refined mineral oil; (2) used mineral oil;
(b) dielectric strength *versus* moisture content of typical mineral oils.

2.17 Dielectric Strength of Liquid Dielectrics

The flow of ions (*ionic conduction*) and colloidal suspension (*molionic conduction*) decide the dielectric loss in liquid dielectrics. Also, such conducting agents may form a chain across which an electric discharge could propagate causing a breakdown. Moisture and other contaminants are delinked at higher temperature conditions; and dielectric strength therefore increases at elevated temperatures (Figure 2.13b).

In the presence of dissolved gases, formation of bubbles dictate the breakdown conditions. Hence the dielectric breakdown is also pressure dependent as illustrated in Figure 2.14.

2.18 Miscellaneous Aspects of Dielectric Materials

Wetting properties of dielectrics: Several dielectrics absorb moisture which renders them hygroscopic. Permeation of moisture through the dielectrics "wets" them, impairing their electrical properties. Dielectrics come in contact with moisture at manufacturing, storage, processing and operational levels.

A host of solid dielectrics (such as organic polymers) absorbs moisture as per an empirically modified *Henry's law* given by:

$$C = ap^n \tag{2.44}$$

where C is the equilibrium humidity per unit volume of the material, a is a *solubility factor*, p is the pressure of water vapor to which the material is subjected to and n is an exponent greater than unity.

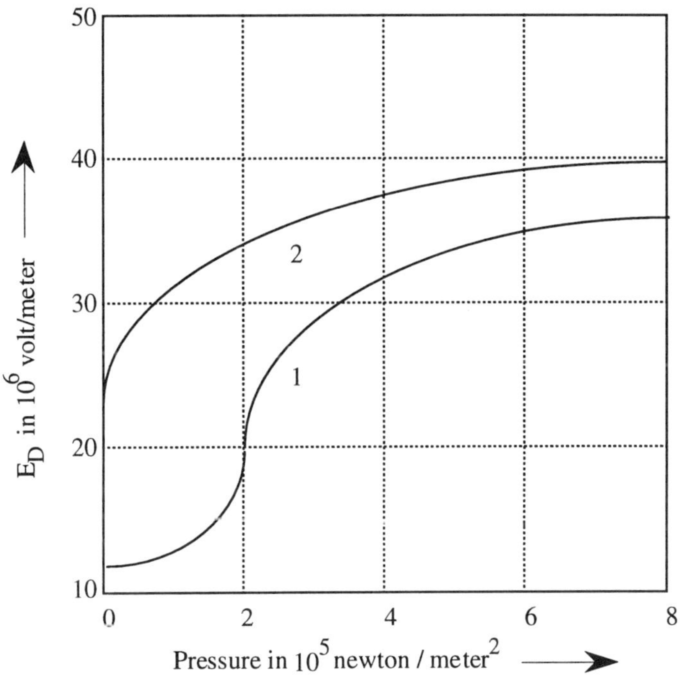

Figure 2.14 Dielectric strength *versus* pressure.
(1) Oil with gas content.
(2) Outgassed oil.

Moisture diffuses through a dielectric material consistent with a characteristic coefficient of the material known as the *diffusion constant*, K. Typically for polymeric materials K is in the order of 10^{-6} to 10^{-3} cm^2/hour and a has a range of values in the order 10^{-5} to 10^{-3} gram/cm^3 . mmHg. The product Ka is known as the *wetting parameter* of the material.

Wetting depends the macroscopic aspects of the dielectric. For example, a fibrous material could absorb more moisture than a compacted soild material. Therefore, in the design of composite dielectrics with two or more constituents, wetting property should be duly taken into account. The voids and interspatial regions accommodate moisture in such materials and the characteristics of wet dielectrics could vary considerably over temperature changes.

Hydrophobic (water repellent) additives can save a dielectric from becoming excessively hygroscopic. Such *hydrophobization* can be done with organosilicon compounds. Hydrophobization can be at bulk level or as a water-repelling surface protective process.

The wetting property of dielectrics is characterized by their electrical behavior under humid tropical climatic conditions, and specified by the *tropical resistance* parameter. It should be noted that under such climates, the dielectrics may also be affected by fungi. To combat against this dielectrics could be treated with fungicides such as 2, 4-dinitrophenol, paranitrophenol, pentachlorophenol, etc.

Thermal properties of dielectrics: Temperature could influence the dielectric properties significantly. In general, electrical properties of dielectrics deteriorate with increases in temperature. Constant use of dielectrics in elevated temperature ambients would lead to thermal aging. Dielectrics are also susceptible to damage with thermal shocks.

Usable dielectrics should possess adaquate heat resistant properties, noninflammable characteristics, and thermal shock withstandability. Dielectric materials are classified as different classes *vis-a-vis* thermal characteristics.

Dielectric materials under radiation environments: Like heat, high energy radiations (ionizing radiations) can also affect the dielectric properties of materials. Normally, X-rays, gamma rays, β-particles could cause radiation-induced aging of dielectrics. Aging selection of dielectric materials for applications in such radiation environments needs specific data on the material characteristics *versus* radiation dosage.

References

[1] A. R. Von Hippel: *Dielectric Materials and Applications.* (The M.I.T. Press, Cambridge, MA: 1966).

[2] B. Tareev: *Physics of Dielectric Materials.* (Mir Publishers, Moscow: 1975).

[3] R. Coelho: *Physics of Dielectrics for Engineers.* (Elsevier Scientific Publishing Co., Amsterdam: 1979).

[4] I. Bunget and M. Popescu: *Physics of Solid Dielectrics.* (Elsevier Scientific Publishing Co., Amsterdam: 1984).

[5] H. Frölich: *Theory of Dielectrics.* (Oxford University Press, London: 1958).

[6] J. B. Birks: *Modern Dielectric Materials.* (Heywood Publishing Co., London, 1960).

[7] J. C. Anderson: *Dielectrics.* (Chapman and Hall, London, 1963).

[8] J. B. Hasted: *Aqueous Dielectrics.* (Chapman and Hall, London, 1973).

Defining Terms

Absolute permittivity: Refers to the absolute extent of a dielectric medium in permitting the electric force field through it.

Capacitance: The charge holding capacity at two locales separated by a distance in a dielectric medium with a potential difference across them.

Complex permittivity: A complex parameter which represents both the conduciveness for eletric field permeation and the lossy behavior under time-varying electric field excitation of a dielectric material.

Coulomb's law: An experimental postulation quantifying the extent of the force of interaction between two charges separated by a distance in a dielectric medium.

Dielectrics: Materials which are electrical insulators and offer no free electrons at ordinary temperature for d.c. conduction process.

Dielectric constant: A measure of permittivity of a dielectric relative to free space in allowing electric force field across it.

Dielectric breakdown: Molecular rupture and establishment of a current path in a dielectric under the influence of excessive electric field across it.

Dielectric dispersion: Frequency response of a lossy dielectric material.

Dielectric strength: Maximum electric field across a dielectric without the breakdown being observed.

Dipole: Refers to an electrical dipole constituted by two equal and oppposite charges separated by a distance.

Dipole moment: The moment vector of a dipole of magnitude equal to the dipole charge dipole length and direction along the line joining the charges constituting the dipole.

Electrical conductors: Materials which exhibit a significant flow of electric current due to movement of electric charges.

Electrical insulators: Nonconductors of electric charge flow.

Electrical semiconductors: Materials whose electric conductivity property lies between those of conductors and insulators (dielectrics).

Loss-tangent: An entity which implicitly refers to the dielectric loss involved in the dielectric relaxation process.

Nonpolar dielectrics: Dielectrics with molecules which exhibit no electric dipole moment in the absence of an external electric field.

Polar dielectric: Dielectrics with molecules which exhibit a permanent dipole moment.

Polarization: In reference to dielectric materials, refers to the alignment of the molecules in a regular fashion along the direction of the applied electric field.

Polarization vector: An entity to quantify in magnitude and direction of the extent of polarization in a dielectric medium as decided by the dipole moment per unit volume.

Polarization process: Time-dependent electric dipole oscillation in a dielectric subjected to a time-varying EM field.

Relaxation time: Time constant of the dielectric relaxation process.

APPENDIX 2A

Table 2.A.1 Conventional Notations for Dielectric Parameters

Notations	Parameters
ε_s	Dielectric constant: Static/quasistatic (low-frequency) value (same as ε_r)
ε_∞	Optical limit of the dielectric constant $(\varepsilon_\infty)^{1/2} = n$, (refractive index of the medium)
$\varepsilon' - j\varepsilon''$	Complex permittivity, or $(k' - jk'')$
ε': $\varepsilon_o\varepsilon_r$	ε_o = Free-space permittivity equal to: $(1/36\pi) \times 10^{-9}$ farad/meter ε_r = Relative permittivity (dielectric constant)
ε''	$\sigma/\omega\varepsilon'$ (loss factor)
$\tan\delta$	Loss tangent
σ	Conductivity of the lossy dielectric (siemen/meter)
D	Dissipation factor (same as the loss factor)
PF	Power factor; $\tan\delta/(1 + \tan^2\delta)^{1/2}$
Q	Quality factor; $(1/\tan\delta)$
A_D	Dielectric attenuation, $(9.1 \times 10^{-8}$ f(Hz) $\tan\delta\sqrt{\varepsilon_r}$ dB/meter)
Z_D	Wave impedance of a dielectric medium; $(120\pi/[\varepsilon_r(1 - j\tan\delta)]^{1/2}$ ohm)
E_D	Absolute dielectric breakdown strength (volt/meter)
E_{Dr}	Breakdown strength relative to air/vacuum

Table 2.A.2 Properties of Gaseous Dielectrics [a]

Material Gaseous Dielectrics	ε' LF (KHz)	ε' HF (> MHz)	tan δ LF (KHz)	tan δ HF (MHz)	Relative Dielectric Strength $\frac{V/mil}{V/mil}$ E_{Dr}
Air					1.00
Helium	1.000065 at atmospheric pressure 20Å				0.15
Hydrogen	1.000253 at atmospheric pressure 20Å				0.65
Nitrogen	1.000548 at atmospheric pressure 20Å				0.90
Neon	1.000125 at atmospheric pressure 20Å				
Carbon dioxide	1.000922 at atmospheric pressure 20Å				0.80
Argon	1.000517 at atmospheric pressure 20Å				
Oxygen	1.000494 at atmospheric pressure 20Å				

Reference

[a] N. E. Hill, W. E. Vaughan, A. H. Price and M. Davies: *Dielectric Properties and Molecular Behavior.* (Van Nostrand and Reinhold Co., New York: 1969).

Table 2.A.3 Properties of Gaseous Dielectrics [a]

Material Gaseous Dielectrics	ε' LF (KHz)	 HF (> MHz)	$(10^{-7} \times p_{tr}^{*}) \tan \delta$ LF (KHz)	 HF (MHz)	Relative Dielectric Strength $\frac{V/mil}{V/mil}$ E_{Dr}
CH_2ClF					1.03
CF_4					1.10
$CHCl_2F$					1.33
CF_3Br					1.35
$CHClF_2$					1.40
$CClF_3$			7.08		1.43
C_2F_6					1.88
C_3F_8					2.00
C_2ClF_5					2.30
SF_6					2.35
C_4F_8					2.40
CCl_2F_2					2.42
C_4F_{10}					2.50
$C_2Cl_2F_4$					2.52
ClO_3F					2.73
CCl_3F					3.50
$CHCl_3$					4.24
CCl_4					6.33
CH_3F			7.08		
CH_3F_3			7.08		
CH_3Cl			7.08		
CH_3Br			7.08		
$(CH_3)_3N$			7.08		

Reference

[a] N. E. Hill, W. E. Vaughan, A. H. Price and M. Davies: *Dielectric Properties and Molecular Behavior.* (Van Nostrand and Reinhold Co., New York: 1969).

* Ptr: Pressure in torr.

Table 2.B Properties of Liquid Dielectrics

Material Liquid Dielectrics	ε_S	ε_∞	ε_r LF (KHz)	ε_r HF (> MHz)	tanδ LF (KHz)	tanδ HF (> MHz)	E_D (volt/mil)	Remarks
Hydrocarbons								
Mineral oils for transformers, cables, capacitors, switches	2.25	2.22	2.25	2.24	0.00300	0.0022	56–60 (KV)	Cable oil-5314™ Cable oil-p 1101270™ [b]
			2.20	2.70			56–60 (KV)	For transformers [c]
Naphthenic oils	2.16	2.12	2.16	2.16	< 0.00008	0.0020		Diala oil-15™ [petroleum hydrocarbons mainly naphthenes] [b]
Paraffinic oils	2.06	2.06	2.06	2.06	< 0.00010	0.0025		Bayol-d™ [b] [77.6% paraffins]
Synthetic hydrocarbons								
Mono-and diisopropyl biphenyl	2.60							Wemcol™ at 100⁰C [c]

(continued...)

Material Liquid Dielectrics	ε_s	ε_∞	ε_r LF (KHz)	ε_r HF (> MHz)	tanδ LF (KHz)	tanδ HF (> MHz)	E_D (volt/mil)	Remarks
Phenyl xylyl ethane [PXE]	2.60							PXE at 25^{0}C, and 2.5 at 100^{0}C [c]
Chlorinates hydrocarbons [PCB]								
Aroclor-1221™		2.65	(4.55–3.10)			0.0590		[b]
Pyranol-1467™	4.42	2.84	(4.42–4.40)	(4.40–3.19)	0.00360	0.0740		[b]
Pyranol-1476™	5.04	2.70	5.40	(3.85–2.81)	0.00006	0.0040		[b]
Aroclor-1254™		2.69	5.05	(3.70–2.70)	0.00420	0.0040		[b]
Aroclor-1262™	4.03	2.75	(4.03–2.76)	2.75	0.03920	0.0005		[b]
Fluorocarbons								
$(C_4F_9)_3N$		1.67	1.90 at 60 Hz, 100^{0}C				40(KV)	Dissipation factor at 60 Hz is < 0.05 [a]
$(C_3F_6O)_5C_2HF_5$		2.15	2.45 at 60 Hz, 100^{0}C				50(KV)	Dissipation factor at 60 Hz is < 0.006 [a]
$(C_8F_{10}O)$		1.63	1.86 at 60 Hz, 100^{0}C				30(KV)	Dissipation factor at 60 Hz is < 0.05 [a]

(continued...)

Material: Liquid dielectrics	ε_S	ε_∞	ε_r LF (KHz)	ε_r HF (> MHz)	tanδ LF (KHz)	tanδ HF (> MHz)	E_D (volt/mil)	Remarks
Silicone oils								
Polyorgano siloxanes	2.75	2.65	2.75	2.74	(0.0014–0.0006)	(0.0015–0.0092)	35(KV)	Ignition sealing compound #4 [b]
Dow Corning™ (i) DC500	2.20	2.19	2.20	2.20	0.0001	(0.0030–0.0060)		At 22^{0}C [b]
(ii)DC200	2.90	2.87	2.90	2.90	<0.0005	0.0103		At -17^{0}C [b]
Transformer retro filling fluids								
RTEmp, FR-15LO™ -Temp silicone oil, Japanese fluids			2.48 at 60 Hz,100^{0}C					Dissipation factor at 60 Hz for silicone oil is 0.040 [a]
Organic esters and ethers								
Castor oil			3.74 at 60 Hz, 100^{0}C					Dissipation factor at 60 Hz is 6 [a]
Dibutyl sebacate			4.40 at 60 Hz, 100^{0}C		0.0100		38(KV)	Dissipation factor at 60 Hz is 1 at 1MHz is 0.015 [a]
Butyl stearate		1.83	3.30 at 60 Hz, 100^{0}C					Dissipation factor at 60 Hz is 1 at 1MHz is 0.013 [a]

(continued...)

Material	ε_S	ε_∞	ε_r		$\tan\delta$		E_D	Remarks
Liquid Dielectrics			LF (KHz)	HF (> MHz)	LF (KHz)	HF (> MHz)	(volt/mil)	
Silicate ester			4.34 at 60 Hz,100^{0}C					[a]
			2.65 at 60 Hz, 100^{0}C				27(KV)	Dissipation factor at 1 MHz is 0.42 [a]
LIQUID GASES								
argon		1.52	1.51					[a]
helium		1.06	1.05					[a]
Water and water ethylene glycol mixture			78.00		300			[d]
Askarels		2.61	3.60 at 60 Hz, 100^{0}C				35(KV)	Dissipation factor at 60 Hz is 1.5 [a]
Miscellaneous liquids								
n-Hexane		1.89	1.89					
Cyclohexane		2.03	2.03					
Carbon tetrachloride		2.15	2.24					

(continued...)

Material	ε_S	ε_∞	ε_r LF (KHz)	ε_r HF (> MHz)	tanδ LF (KHz)	tanδ HF (> MHz)	E_D (volt/mil)	Remarks
Liquid Dielectrics								
Benzene		2.25	2.28					
Bromine		2.76	3.09					
Methane			1.66					
Oxygen		1.49	1.51					
Carbon dioxide		1.42	1.60					
Diethyl amine		1.96	2.42					
Chlorobenzene		2.33	5.71					
Acetic acid		1.88	6.15					
Hydrogen bromide		1.76	7.00					at 10⁰C
Iodo ethane		2.29	7.82					
Hydrogen sulfide		1.89	9.26					at 20⁰C
Water			80					80–0.4(T-20) T in ⁰C
Distilled water		34				0.3 at 25 GHz		
Sea water			81		9×10^4			

(continued...)

Material	ε_S	ε_∞	ε_r		tanδ		E_D	Remarks
Liquid Dielectrics			LF (KHz)	HF (> MHz)	LF (KHz)	HF (> MHz)	(volt/mil)	
Acetone		1.84	20.70					
Ethyl alcohol		1.85	24.30					
Propionitrile		1.87	24.30					
Nitrobenzene		2.42	34.82					
Water		1.77	78.54					
Sulfuric acid		2.04	101.00					
Hydrogen cyanide		1.16	115.00					

References

[a] N. E. Hill, W. E. Vaughan, A. H. Price and M. Davies: *Dielectric Properties and Molecular Behavior.* (Van Nostrand and Reinhold Co., New York: 1969).

[b] A. R. Von Hippel: *Dielectric Materials and Applications.* (M.I.T Press, Cambridge, MA: 1966).

[c] W. T. Shugg: *Handbook of Electrical and Electronic Insulating Materials.* (Van Nostrand Reinhold Co., New York:1986).

[d] P. F. Bruins: *Plastics for Electrical Insulation.* (Interscience Publications, New York: 1968).

Table 2.C Properties of Solid Dielectrics

Material Solid Dielectrics	ε_s	ε_∞	ε_r LF (KHz)	ε_r HF (> MHz)	tanδ LF (KHz)	tanδ HF (MHz)	E_D (volt/mil)	Remarks
Acetal resins	3.70		3.70	3.70			380	Homopolymer [c]
Acrylic resins Poly methyl meth - acrylate (PMMA)	3.70		3.70	2.20	0.060		500	[c]
Poly methyl acrylate (PMA)								
Methyl methacrylate Styrene copolymer								
Alkyd resins	1.22	1.20	1.22	(1.22–1.20)	(0.00198–0.00227)	0.0034		Alkyd, diisocyanate, foamed [b]
Alkyd resins mineral filled	4.17	3.78	4.13	(3.96–3.78)	(0.0102–0.0087)	(0.0115–0.0136)		Glastic grade MM™ [b]
	6.02	4.32	5.77	(5.19–4.63)	(0.034–0.024)	(0.031–0.0288)		Plaskon alkyd 411™ [b]
	5.02	4.34	4.89	(4.59–4.44)	(0.0212–0.014)	(0.0146–0.0141)		Plaskon alkyd 442™ [b]

(continued...)

Material Solid Dielectrics	ε_s	ε_∞	ε_r LF (KHz)	ε_r HF (> MHz)	tanδ LF (KHz)	tanδ HF (MHz)	E_D (volt/mil)	Remarks
Cellulosic polymers	3.82	3.10	(3.77–3.53)	(3.42–3.24)	(0.0095–0.023)	(0.023–0.04)		Acetates (1) LL-1™ [b]
	5.34	3.10	(5.28–4.90)	(4.57–3.20)	(0.0085–0.0135)	(0.054–0.048)		(2) Tenite I 008A S 4 ™ [b]
Cellulose calendered	10.8	3.74	(8.40–7.00)	(6.60–3.74)	(0.64–0.10)	(0.064–0.165)		Nitrate Pyrali [b]
	7.6	3.35	(6.80–6.10)		(0.128–0.04)	(0.065–0.10)		Methyl cellulose Methocel™ [b]
Cellulose paper	3.10	2.74		(5.70–4.30)	(0.005–0.0067)	(0.012–0.020)		Ethyl cellulose
			(3.06–2.99)	(2.92–2.8 0)				Lumarith#22361™ [b]
Ceramic, porcelain glass	6.84	5.5	(6.37–5.96)	(5.86–5.75)	(0.089–0.010)	(0.008–0.0047)		Steatite bodies (1)AlSiMag-35 (85T)™
Alumina ceramic	4.95	4.95	4.95	(4.95–4.91)	(0.004–0.0016)	(0.001–0.0009)		(2)AlSiMag393 (24T)™ [b]
		8.70		8.80			250	Type I [c]
		10.10		10.1			250	Type IV [c]
zirconia ceramic				5.00				[c]

(continued...)

Material Solid Dielectrics	ε_S	ε_∞	ε_r LF (KHz)	ε_r HF (> MHz)	tan δ LF (KHz)	tan δ HF (MHz)	E_D (volt/mil)	Remarks
Silica, soda, boro-silicate, alumino boro-silicate glass	4.05	4.05	4.05	4.05	(0.00136–0.00044)	(0.0005–0.0015)		Boro-silicate glass
					(0.0033–0.0016)	(0.0011–0.0022)		[b]
	3.88	3.76	3.86	3.85	(0.15–0.32)	(0.3180–0.0455)		Corning 7230™ [b] (aluminium boro-silicate) [b]
	90.00	5.46	(82.50–44.00)	(17.50–9.00)				Foam-glass (soda-lime) [b]
Elastomers and rubbers								
Butyl rubber (isobutylene copolymer)							600	GR-I (Butyl rubber) [b]
Chloroprene rubber, neoprene	2.39	2.35	(2.38–2.36)	2.35	(0.0034–0.0027)	(0.0010–0.0009)	400–700	[c]

(continued...)

Material Solid Dielectrics	ε_S	ε_∞	ε_r LF (KHz)	ε_r HF (> MHz)	tanδ LF (KHz)	tanδ HF (MHz)	E_D (volt/mil)	Remarks
							500--800	[c]
EPDM (ethylene propylene diene monomer)	6.00–8.00		(2.50–3.50)					
								[e]
Isoprene,cis-poly-isoprene elastomers	(2.50–3.50)							
			(2.30–3.00)				500	[e]
Fluoroelastomers							869	Wire aged 7 days at 300^0F tested 75^0F [e]
			15					Royalite149-11™ [b]
			7.16					(U.S. rubber)
NBR (nitrile butadine rubber)								Expanded Royalite M21982-1
		3.13						
								[b]
			(5.20–4.87)		(0.0320–0.0590)			
Silicone elastomers and varnishes	5.41			(4.41–3.62)		(0.108–0.020)		Silastic 120 [b]
			1.26	1.15	(0.0175–0.0110)			
	1.28	5.73				(0.0204–0.0184)		Silastic 125 [b]
		7.00	5.76	5.75	(0.0051–0.0010)			Silastic 160 [b]
	5.78					(0.0008–0.0254)		
		7.80	7.10	6.90	(0.0040–0.0013)			Silastic 250 [b]
	7.10					(0.0012–0.0430)		
		3.02	(9.27–9.00)	(8.8–7.6)	(0.0047–0.0150)			
	9.32					(0.025–0.140)		
			(3.18–3.16)	(3.1–3.0)	(0.0055–0.0106)			
	3.19					(0.0064–0.0190)		

Material Solid Dielectrics	ε_s	ε_∞	ε_r LF (KHz)	ε_r HF (> MHz)	tanδ LF (KHz)	tanδ HF (MHz)	E_D (volt/mil)	Remarks
EPOXY RESINS								
Urea formaldehyde epoxies	7.10	4.79	(6.70–6.20)	(6.00–4.65)	(0.038–0.022)	(0.0310–0.0782)		Plaskon urea, natural [b]
Epoxy Novolac™								
Epoxy resin mineral filled							2000	Coated glass cloth [e]
Epoxy resin glass fiber filled/reinforced	5.00		5.00	4.6			390	Glass filled epoxy molding resin [c]
Epoxy varnish, coating	3.90						2300	Solventless, rigid low viscocity 1 part at 100 Hz, 23°C [c]

(continued...)

Material Solid Dielectrics	ε_S	ε_∞	ε_r LF (KHz)	ε_r HF (> MHz)	tanδ LF (KHz)	tanδ HF (MHz)	E_D (volt/mil)	Remarks
Fluorocarbon homopolymers, and copolymers								
PTFE–polytetra-fluoroethylene	2.76		2.74	2.73	0.0009	0.0006	>1000	Dilecto™ [b]
	3.35	3.15	(3.24–3.17)	(3.16–3.1)	(0.0361–0.0108)	(0.0042–0.0038)		Dilecto™ [b]
					< 0.0005	< 0.0003		
	2.10	2.1	2.10	2.10	(0.0148–0.014)	(0.0096–0.0059)		Teflon™ [b]
PCTFE– poly chloro trifluoroethylene	2.82	2.34	(2.76–2.50)	(2.46–2.33)				Kel-f grade 300™ [b]
PVF– polyvinyl fluoride			8.50				3500	Type-20, (1mil)™ [e]
PVDF– polyvinylidene fluoride	8.40		8.40	6.60				[c]
FEP– fluorinated ethylene propylene			2.00	2.00			6500 1800	25⁰C, 60 Hz, (1 mil) 25⁰C, 60 Hz, (20 mil) [e]

(continued...)

Material Solid Dielectrics	ε_s	ε_∞	ε_r LF (KHz)	ε_r HF (> MHz)	tanδ LF (KHz)	tanδ HF (MHz)	E_D (volt/mil)	Remarks
Melamine formaldehyde and other formaldehydes	8.20	3.64	(7.15–5.90)	(5.4–3.52)	(0.135–0.056)	(0.06–0.037)		Bakelite BT-48-306™ [100%] [b]
	4.87	3.70	(4.74–4.50)	(4.36–3.55)	(0.030–0.021)			
			(3.70–7.20)	(5.30–5.00)		(0.028–0.041)		Bakelite BM-120™ [46% woodflour] [b]
Unfilled and filled with cellulose, glass fiber, mineral, asbestos, woodflour	3.70		(9.80–5.90)	(5.30–4.37)	(0.30–0.29)	(0.125–0.022)		Bakelite BM-250™ [66% asbestos fiber] [b]
			(4.95–4.66)	(4.51–3.21)	(0.250–0.088)			
	14.2	4.40				(0.046–0.029)		Laminated fiber glass BK-174™ [b]
			(7.80–9.20)	(7.20–8.40)				
	5.30	3.43	(7.00–7.50)	6.80	(0.076–0.027)	(0.036–0.041)		Micarta #254™ [b] [50% cellulose]
Melamines	(7.90–9.50)						(240–270)	[e]
C. ureas	(7.00–9.50)						(220–300)	[e]

(continued...)

Material Solid Dielectrics	ε_s	ε_∞	ε_r LF (KHz)	 HF (> MHz)	tanδ LF (KHz)	 HF (MHz)	E_D (volt/mil)	Remarks	
Laminates									
NEMA -- grades				6.0			700	Grade 'X'	[c]
Epoxy laminates				5.20			500	G-10	[c]
Glass cloth – epoxy or polymer laminates				5.20			500	FR-4	[c]
Paper – polypropylene laminates				4.60			600	FR-3	[c]
Paper phenolic laminate			4.10						[d]

(continued...)

Material Solid Dielectrics	ε_s	ε_∞	ε_r LF (KHz)	ε_r HF (> MHz)	tanδ LF (KHz)	tanδ HF (MHz)	E_D (volt/mil)	Remarks
Mica								
Muscovita mica	5.40	5.40	5.40	5.40	(0.0025–0.0003)	(0.0003–0.0002)	3000	Ruby mica [b]
Phlogopite mica, fluoro-phlogopite mica				5.60			3000	Phlogopite amber natural [c]
				6.50			3000	Fluoro-phlogopite synthetic [c]
Mica paper							1000	0.0026 inch glass + 0.002 inch mica

(continued...)

Material Solid Dielectrics	ε_S	ε_∞	ε_r LF (KHz)	ε_r HF (> MHz)	tanδ LF (KHz)	tanδ HF (MHz)	E_D (volt/mil)	Remarks
Papers								
Transformer paper, condenser/capacitor paper, cabel paper	2.40		2.40	1.30			730–310	Aramid paper [c]
Nomex paper	2.40		2.40				730	Aramid calendered [c]
Polyamid paper			2.00				450	Nomex type 410™, i) a.c. rapid rise thickness (2,5,10, 15,30) in mils [e]
			2.30				600	
			2.60				750	
			3.00				800	
			3.40				600	
			2.60				600	ii) a.c. 1-min hold thickness is 10 mils
	2.60						1200	iii) d.c. rapid rise thickness is 10 mils [e]
Aramid paper			1.30				310	Aramid uncalendered [c]
Tyvek paper	3.60		3.60				1000	Aramid and mica [c]

(continued...)

Material Solid Dielectrics	ε_S	ε_∞	ε_r LF (KHz)	 HF (> MHz)	tanδ LF (KHz)	 HF (MHz)	E_D (volt/mil)	Remarks	
Phenolics									
Phenolic resins	5.00–13.00		(4.40–9.00)	(4.00–6.00)			200–800		[e]
Polyamides	3.88	3.03	(3.75–3.45)	(3.33–3.16)	(0.0144–0.0254) (0.065–0.050)	(0.0257–0.0210) (0.038–0.022)		Nylon-66 (Dupont)	[b]
	4.50	2.85	(4.20–3.70)	(3.20–2.85)				Nylon-610 90% humidity	[b]
Nylons6,6/6,6/10,6/12	8.00		8.00	4.60			600	Nylon 6/6	[c]
Polyamide-imide	4.30			3.90			580		[c]
Polycarbonates									
Polycarbonate resins	3.20			3.00			380	Typical polycarbonate molding resins	[b]
polycarbonate, glass fiber reinforced	3.00		3.00	2.90			375–400	1/8 in. material	[e]
Lexan™	3.20		3.20	3.00			380		[c]

Material Solid Dielectrics	ε_S	ε_∞	ε_r LF (KHz)	ε_r HF (> MHz)	tanδ LF (KHz)	tanδ HF (MHz)	E_D (volt/mil)	Remarks
Polyesters	3.60	2.95	(3.57–3.42)	(3.35–3.31)	(0.0104–0.0150)	(0.020–0.0195)		Polydiallyl phthalate at 26.8^{0}C [b]
DAP – diallyl phthalate	(4.30–4.60)		(4.10–4.50)	(3.40–4.50)			380	[e]
PET – polyethylene terephthalate			3.20	3.00			7000–7500	[c]
PBT – polybutylene terephthalate	3.30		3.30	3.10			400	Short time [c]
Mylar™								
Polyester paper (epoxy)							1000	Short time, vpm
							13 [KV]	Wrapped electrode, [e]

(continued...)

Material Solid Dielectrics	ε_S	ε_∞	ε_r LF (KHz)	ε_r HF (> MHz)	tanδ LF (KHz)	tanδ HF (MHz)	E_D (volt/mil)	Remarks
Polyester mat on polyester film (epoxy)							1000	Short time, vpm
							9 [KV]	Wrapped electrode, [e]
Asbestos paper glass cloth (epoxy)							1100	Short time, vpm
							5 [KV]	Wrapped electrode, [e]
Glass cloth (epoxy)							900–1200	Short time, vpm
							3–11 [KV]	Wrapped electrode, [e]
Mica-glass cloth (silicone)							1000	Short time, vpm
							12 [KV]	Wrapped electrode, [e]
Acetate films			4.30	3.70			2800	0% RH, high plasticizer, p-903
			4.80	4.10			2300	50% RH, high plasticizer, p-903
			6.80	6.10			1300	100% RH, high plasticizer, p-903
			3.80	3.50			3600	0% RH, low plasticizer, p-904
			4.80	4.40			3000	50% RH, low plasticizer, p-904
			7.50	7.00			1800	100% RH, low plasticizer, p-904 [e]

(continued...)

Material	ε_s	ε_∞	ε_r		$\tan\delta$		E_D	Remarks
Solid Dielectrics			LF (KHz)	HF (> MHz)	LF (KHz)	HF (MHz)	(volt/mil)	
Poly ethylene	2.25	2.25	2.25	2.25	< 0.0005	< 0.0004		Polyethylene A-3305™ [100%] [b]
LDPE – low density polyethylene	2.20			2.20			5000	[c]
LLDPE – liner low density polyethylene	2.25		2.25	2.35			450--1000	[e]
XLPE – cross-linked polyethylene				2.30			550	[c]
Polyimides								
Kapton			3.50	3.40			7000	[c]
Polyether imide	3.2			3.10			480	[c]

(continued...)

Material Solid Dielectrics	ε_s	ε_∞	ε_r LF (KHz)	ε_r HF (> MHz)	tanδ LF (KHz)	tanδ HF (MHz)	E_D (volt/mil)	Remarks
Polypropylene								
Homopolymer	2.20			2.20			600	[c]
Copolymer of polyethyl				2.24			600	[c]
Polyphenylene								
Polyphenylene oxide	2.60			2.60			550	[c]
Polyphenylene sulfide	3.10			3.20			380	[c]
Polystyrene								
	2.55	2.54	2.55	2.55	<0.0003	<.0002		Styron C-176™ [b]
Edistir™, Bakelite™	1.03	1.03	1.03	1.03	<0.0002	<0.0001		Styrofoam 103.7™ [b]
	2.98	2.75	(2.95–2.90)	(2.86–2.75)	(0 0062–0.0074)	0.0065		Experimental plastic Q767.2™ [b]

(continued...)

Material Solid Dielectrics	ε_s	ε_∞	ε_r LF (KHz)	ε_r HF (> MHz)	tanδ LF (KHz)	tanδ HF (MHz)	E_D (volt/mil)	Remarks
Polysulfone								
UDEL	3.10			3.00			425	[c]
Polyaryl ether	3.14		2.85	3.10			430	1/8 in. thick [e]
Polyaryl sulfone	3.94	3.24	3.94	3.24			356	1/16 in. thick; short-time [e]
Polyurethane films and varnishes	5.71			3.42			645	Type-5,2-parts solventless flexible value [c]
Polyurethane poly-butadiene type	3.14			2.87			610	2 parts solventless polybutylene-based value [c]
Polyurethane polyether type							2000^* 1500^*	*in vpm *in vpm [e]
Polyurethane glass cloth								

(continued...)

Material / Solid Dielectrics	ε_s	ε_∞	ε_r LF (KHz)	ε_r HF (> MHz)	tanδ LF (KHz)	tanδ HF (MHz)	E_D (volt/mil)	Remarks
Polyvinyl compounds								
PVC – polyvinyl chloride	3.67		(3.65–3.52)	(3.42–3.39)	(0.0068–0.0145)	(0.013–0.012)	1800	[e]
	6.21		(5.52–3.96)	(3.53–3.00)	(0.073–0.107)	(0.072–0.050)		
Vinyl chloride acetate					(0.004–0.0088)	(0.0124–0.0111)		
Polyvinyl acetate	2.69	2.51	2.67	2.61	(0.0054–0.0190)	(0.0190–0.0165)		
Polyvinyl formal-urethane and -phenolic coating	3.16	2.76	(3.12–3.00)	(2.92–2.85)				
Styrene		2.40		2.40	0 0014–0.0037			

References
[a] N. E. Hill, W. E. Vaughan, A. H. Price and M. Davies: *Dielectric Properties and Molecular Behavior.* (Van Nostrand & Reinhold Co., New York: 1969).
[b] A. R. Von Hippel: *Dielectric Materials and Applications.* (M.I.T Press, Cambridge, MA: 1966).
[c] W. T. Shugg: *Handbook of Electrical and Electronic Insulating Materials.* (Van Nostrand & Reinhold Co., New York:1986).
[d] P. F. Bruins: *Plastics for Electrical Insulation.* (Interscience Publications, New York: 1968).
[e] S. L. Saums and W. W. Pendleton: *Materials for Electrical Insulating and Dielectrical Function.* (Hayden Book Co., Inc., NJ: 1973).

CHAPTER 3

Electrical Insulating Materials

3.1 Introduction

Electrical insulating materials ensure the integrity of desired paths of electromagetic power flow in electrical systems and equipment. They are materials at very high *resistivity* and can thus be used as isolators or separators between conductors having different potential (voltage) levels. In its use such as in isolators, an insulating material has the basic function of confining the current flow within the conductive circuit of a given device or part of equipment, thus protecting the latter from short circuits, current leakage, and similar undesirable malfunctions [1].

Insulating materials may be solid, liquid, or gaseous substances. They may be monolithic (discrete) materials or may be constituted by physically cohesive composites structured with multiple constituents. The electrical insulation property stems from the dielectric characteristics of the materials. Therefore, the term electrical insulating material, in general, encompasses the host of monolithic and composite dielectric materials having high electrical resistivity.

Based on the primary functions, electric insulation can be categorized as follows:

- Separation-type insulation
- Barrier insulation
- Creepage insulation

The *separation-type insulation* refers to the isolation of conductors at higher electrical potential from those at lower potential levels such as the ground. *Barrier insulation* is required to achieve a higher *breakdown strength* in specific applications. *Creepage insulation* is a special case of spacing insulation that avoids the creepage of electric flash-overs by virtue of having certain specified surface characteristics.

The secondary function of insulators can be grouped as follows :

- Mechanical supports for high-voltage parts.
- Protective enclosures for electrical systems or parts.
- Thermal dissipation adjuncts in heavy-current systems.
- Feed-through units serving as bushings for electrical leads at high potentials.

The general properties of interest for insulating materials are presented in Table 3.1.

3.2 Dielectric Characterization of Insulators

Microscopically, the forbidden gap energy in insulating materials is too wide for easy bridging of thermal excitation of bound electrons in the valence band to the conduction band. That is, in terms of band theory (Chapter 1), an insulating material is a medium of molecular units in which electrons are tightly bound to the atomic nuclei and are not free to move within the material. Termed also as a *dielectric*, such a material has molecules in which the atoms and their electrons and the nuclei are so arranged that one part of the molecule has a positive collection of electric charges while the other part is negatively charged. This separated set of opposite charges constitutes a *dielectric dipole*. Under the influence of external electric field force, these molecular dipoles turn or rotate to align (or polarize) themselves along the direction of the field force.

Table 3.1 General Engineering Characteristics of Insulating Materials

Mechanical Properties	Electrical Properties	Thermal Properties	Chemical Properties	Miscellaneous Properties
Tensile, compressive, shearing, and bending strengths	Electric breakdown strength in the bulk medium	Thermal conductivity	Resistance to chemical reagents	Specific gravity
Elastic moduli	Surface breakdown strength	Thermal expansion	Effects upon adjacent materials	Refractive index
Hardness	Liability to track	Primary creep	Electro-chemical stability	Transparency
Impact and tearing strengths	Volume and surface resistivities	Plastic flow	Stability against aging and oxidation	Color
Viscosity	Dielectric permittivity	Thermal decom--position	Solubility	Porosity
Extensibility	Dielectric loss tangent	Spark, arc, and flame resistances	Solvent crazing	Permeability to gases and vapors
Flexibility	Insulation resistance (bulk and surface resistances)	Temperature coefficients of other properties		Moisture adsorption

(continued)

Mechanical Properties	Electrical Properties	Thermal Properties	Chemical Properties	Miscellaneous Properties
Machinability	Frequency dependency of electrical properties	Melting point		Surface adsorption of moisture
Fatigue		Pour point		Resistance to fungus
Resistance to abrasion		Vapor pressure		Resistance to aging by light
Stress crazing		Low smoke generation		Degassing

Adapted from [2]: P. F. Bruins: Plastics for Electrical Insulation. (Interscience Publishers, New York: 1968). With permission of the author.

Should the field be alternating, the dipoles also reorient (or polarize) alternatively along the directional variation of the applied field at the same frequency. This *dielectric relaxation process* may be accompanied by a loss of energy, known as *dielectric loss.*

The orientation of dipoles along the applied electric field is referred to as *dielectric polarization*. The work done in establishing this polarization of dipoles is equal to $\varepsilon|\boldsymbol{E}|^2$ depicting that the dielectric permittivity (ε) of the material refers to the extent of *electrostatic energy* storage in the material; and the relaxation process if accompanied by any frictional forces, causes a loss in the dielectric energy storage.

Further, if the applied electric field force is very intense, it is possible that the molecular dipoles may rupture causing an *insulation breakdown*. This breakdown process is further augmented at elevated temperature ambients.

On the basis of the above outline on the dielectric characterization of insulators, the following parameters can be specified to quantify the dielectric properties of an insulator.

- Relative permittivity or dielectric constant, (ε_r)
- Complex (relative) permittivity, $(\varepsilon' - j\varepsilon'')$
- Dielectric loss tangent or dissipation factor $(tan\delta = \sigma/\omega\varepsilon = \varepsilon''/\varepsilon')$
- Dielectric Q factor $(= 1/tan\delta)$
- Dielectric attenuation, $(\alpha = \omega\varepsilon_r^{1/2}\ tan(\delta/2c)$ nepers/meter$)$
- Dielectric breakdown strength

In the above parameters, σ represents the a.c. conductivity of the material, $\omega = 2\pi \times$ *frequency*, and $c = 3 \times 10^8$ meters/sec represents the velocity of propagation of electromagnetic energy. Other quantities are pertinent to dielectric materials as elaborated in Chapter 2.

Apart from the microscopic aspects of the insulating materials being decided by their dielectric characterizations, there are macroscopic properties (both electrical and nonelectrical) of insulators of importance as normally conceived in engineering applications and indicated in Table 3.1.

3.3 Bulk Electrical Properties

Insulation resistance: This refers to the extent to which an insulator prevents the flow of electric charges through it. An ideal insulator has a bulk resistance of infinity with zero current flowing through it. In practice, insulation resistance can be divided into the following types :

- Volume resistance
- Surface resistance

Volume resistance (R) refers to the bulk resistance offered by the whole body of the insulating medium. In terms of bulk resistivity ρ_v (ohm-meter), an insulator of length ℓ meter and area of cross-section a meter^2, the bulk resistance for the flow of electric current across this cross-sectional area is given by:

$$R_v = \rho_v \ell / a \qquad \text{ohm} \tag{3.1}$$

Surface resistance (R_s) depicts the resistance offered by the surface of an insulating material to the sheet of surface current on it. It has the unit ohm per square.

- Both bulk and surface resistances are affected by temperature. A typical variation of R_v with temperature of an oil-impregnated paper is shown in Figure 3.1.
- Surface resistance (R_s) is influenced significantly by ambient humidity.
- Insulation resistance also depends on (to a small extent) the polarity of the applied voltage. This is due to the inhomogeneity of the material. Standard test methods to evaluate R_v or R_s duly take this into consideration.
- Insulation resistance decreases with the age of the material. The ambient conditions (such as thermal, moisture, chemicals, and mechanical stresses) plus the electrical overstressings (in terms of voltage and/or current) decide the life time and the aging rate.

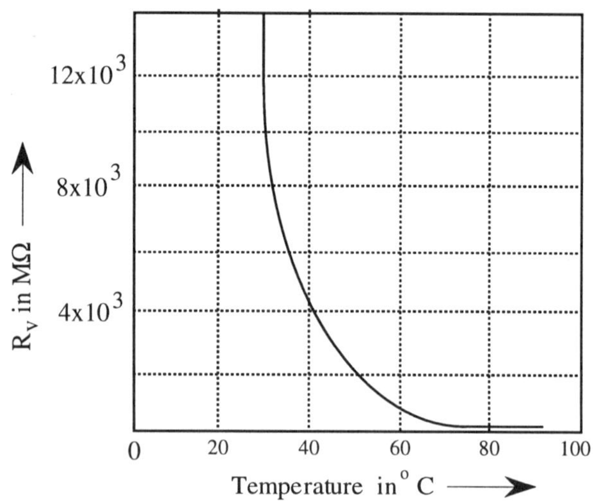

Figure 3.1 Bulk resistance *versus* temperature of an insulating material (oil-impregnated paper).

Dielectric strength: This is the *electrical breakdown strength* specified in terms of the maximum electric field strength that the material endures without experiencing molecular rupture. It is expressed in kilovolt/mm or volt/mil. Dielectric strengths of typical insulating materials are given in Table 3.2. These figures are only approximate and are significantly dependent on the physicochemical characteristics of the materials.

Factors which affect the dielectric strength of an insulator are:

- Temperature: The dielectric strength decreases with temperature, because the thermal energy accentuates the molecular rupture process.
- Humidity also reduces the dielectric strength.

Table 3.2 Dielectric Strength of Typical Insulating Materials

Materials	Dielectric Breakdown Strength (kilovolt/mm)
Porcelain (For low voltage applications)	2–4
Porcelain (For high voltage applications)	10–15
Natural rubber	20–25
Synthetic rubber	5–40
Laminated asbestos	~4
Mica	~80

Dielectric permittivity and dielectric loss (dissipation factor): Most of the insulating materials used in practice have *relative permittivity* (*dielectric constant*) in the range of 2 to 10 in their monolithic applications. Typical insulating materials and their dielectric constant and dielectric loss parameters are presented in Table 3.A.

Both the dielectric constant and the loss tangent of dielectric materials are affected by temperature and humidity. The dielectric constant is, however, influenced to a lesser extent only.

Electrical resistance: The indicators of *electrical resistance* properties of an insulation material are: (i) Insulation resistance which refers to the ratio of direct current applied to electrodes embedded in the test material to the total current between them; (ii) Volume resistivity which measures the electrical resistance between the opposite faces of a unit cube of material; and, (iii) surface resistivity which depicts the resistance between two opposite edges of a unit square of a material. Electrical resistance, in general, is decided by the inherent chemical composition and the homogeneity of the test material.

Arc resistance: *Arc resistance* is an indicator of the surface breakdown characteristics of an insulating material. Its values are determined by the time in seconds for breakdown along the surface of the test material. Surface contaminants and moisture are the determining factors in deciding the arc resistance value of a material.

Table 3.3 Classification of Insulating Materials on the Basis of Limiting Temperature Performance

Materials	Limiting Temperature ^{0}C
Cotton, silk, paper and similar materials without impregnation with oil, rubber, or polyvinyl chloride	~90
Same as above, but impregnated with polyimide resins	~105
Enameled wire insulations on bases of polyurethane and epoxy resins, and molding powder plastics	~120
Inorganic materials (mica, fiberglass, asbestos) impregnated with varnishes, or other compounds	~130
Polyester epoxies, varnishes, and other heat-resistant varnishes	~155
Composite materials with mica and fiberglass as base materials and asbestos impregnated with silicones or silicone rubber	~180
Mica, ceramics, glass, Teflon™	~255

3.4 Solid Insulating Materials

The existing gamut of solid insulating materials is extremely diverse in origin and poses significant vagaries in their properties. Essentially, these materials can be grouped as: (i) Natural organic substances such as paper, cotton, rubber, etc.; (ii) Inorganic natural materials like mica; and (iii) synthetic products like plastics.

For engineering applications, the choice of an electrical insulator depends upon the severity of electrical, thermal, and mechanical stresses it would face in its applications. Among these, thermal stresses are given primary considerations inasmuch as even small temperature changes would induce considerable damage to insulators *via* chemical degradation, cracking, melting, etc. causing eventual reduction in its life time. Solid insulators are classified on the basis of thermal considerations and grouped under each limiting temperature. A classification of insulating materials on the basis of limiting temperature is presented in Table 3.3.

The characteristics and applications of commonly used insulating materials are presented in Table 3.A.

3.5 Liquid Insulating Materials

These are mainly used with a common purpose as heat transfer media as well as electric insulators. They form adjunct insulating systems along with solid insulators. The general characteristics required for a good liquid insulator are:

- High breakdown strength
- High impulse strength
- High volume resistivity
- Compatible dielectric constant
- High specific heat and thermal conductivity
- High flash point
- High chemical stability
- Good gas absorbing properties
- Low viscosity
- Low density
- Low solubility
- Low solvent power
- Good arc quenching characteristics
- Nonflammable
- Nontoxicity

Typical liquid insulators and their properties are listed in Table 3.4.

Table 3.4 Typical Liquid Insulators and Their Applications

Liquid Insulator	Temperature Range (°C)	Applications
Mineral oil	–50 to 110	General, all-purpose
Askarels	–50 to 110	High voltage transformers, capacitors, switch gears
Silicone liquid	–95 to 210	High voltage transformers
Halogenated hydrocarbons (excluding askarels)	–50 to 200	General purpose electrical appliances, gas insulated cables
Synthetic hydrocarbons	–50 to 110	Cables, capacitors, and switch gears
Organic esters	–50 to 110	Electronic appliances
Vegetable oils	–50 to 110	(Obsolete)

The electrical insulation parameters of typical liquid insulating materials are furnished in Table 3.5.

Table 3.5 Electrical Insulation Parameters of Some Liquid Insulators

Electrical Insulation Parameters	Liquid Insulation Materials		
	Mineral oil	Synthetic Hydrocarbons	Askarels
Insulation breakdown strength in kilovolt per 2.5 mm	35-50	45-55	20-50
Impulse breakdown strength in kilovolt per 25 mm			
Positive impulse	75–100	–	–
Negative impulse	100–200		
Dielectric dissipation factor (loss tangent) at 100°C	0.003	< 0.0005	0.03
Bulk resistivity ($\times 10^9$ ohm-meter)	~1000	~300	~10
Dielectric constant	2.3	2.2	5.0

3.6 Gaseous Insulating Materials

Typically the following classes of gases are usable as insulating media:

- Pure elements: N, H, He, Ar
- Air
- Oxide gases: CO_2, SO_2
- Electronegative gases: CH_2Cl_2, SF_6
- Hydrocarbons: methane, ethane, propane and freon

The insulating characteristics and applications of gaseous insulators are listed in Table 3.6.

Electronegative gases have a higher dielectric constant than air. For example, the relative permittivities of CCl_4 and SF_6 are 6.4 and 2.4, respectively. These gases are nonflammable and nonexplosive. They are widely used in high voltage systems. SF_6 is colorless, nontoxic, and almost chemically inert. Also it has good thermal stability and arc-quenching properties.

Electronegative gases have a great affinity for free electrons and as a result, they show very high breakdown strength.

Hydrocarbon gases are hardly used for electrical insulations purposes inasmuch as they are highly inflammable.

Table 3.6 Gaseous Insulators and Their Relative Properties

Gaseous Insulators	Density	Thermal Conductivity	Thermal Capacitor	Breakdown Strength	Applications
	Parameters Relative to Those of Air				
Air	1	1	1	1	Used in low breakdown situations only (<3 kv/mm)
N	0.98	1.1	1.05	1	Non-oxidizing, noncorrosive applications (under pressure)
CO_2	1.5	0.6	0.9	0.9	Used in fixed capacitors; better dielectric constant than air
H	0.07	6.7	14.4	0.6	Used as a coolant

3.7 Composite Insulators

Composite insulators can be synthesized to offer outstanding electrical insulation properties. Typically, the following multiconstituent insulating materials are used in practice:

- Asbestos/polyethylene fiber
- Asbestos/paper/polystyrene resin
- Asbestos/varnish/cotton fiber
- Paper/polyester film
- Mica/polyester or epoxy bases/alkyd binder

The aforesaid materials, in general, can be broadly classified as:

- Resin-rich system
- Resin-poor system

3.8 Inorganic Insulation Materials

At high temperatures, normally organic insulators fail to perform as required. In such situations inorganic compounds offer better physical, thermal, and dielectric stability. Typical inorganic insulators are:

- Silica glass (silica + alkali + base)
- Nonalkaline glass
- Ceramics (silica + alumina + magnesia + boron oxide + titania or zirconia)
- TiO_2 based high permittivity ceramics
- Mica (silicates of alumina and potash)
- Micanites (composites made with mica plus binders)
- Asbestos (fiberous magnesium silicates)

3.9 Concluding Remarks

Electrical insulating materials, though generally dielectrics, have multiple roles to play in practical applications. They should offer desirable mechanical, thermal, and chemical characteristics under operating environments. Therefore, both monolithic and composite insulating materials constitute a select class of dielectrics either chosen from available generic dielectrics or synthesized appropriately to meet the operational requirements.

References

[1] W. T. Shugg: *Handbook of Electrical and Electronic Insulating Materials.* (Van Nostrand Reinhold Co., New York: 1986).

[2] P. F. Bruins (Ed.): *Plastics for Electrical Insulation.* (Interscience Publishers, New York: 1968).

[3] M. Clark: *Insulating Materials for Design and Engineering Practice.* (John Wiley and Sons, New York: 1962).

[4] T. Tanaka: Electrical insulation and its future. Proc. 21st Symp. Elec. Insulating Materials, 1968 (Scientific Publishing Division of MYU K. K, Japan), pp. 7-17.

Defining Terms

Arc Resistance: Characteristic parameter of an insulator in offering an inhibitory trend to the breakdown-induced arc discharges.

Bulk resistance: The volume resistance offered by a material of finite length and area of cross-section by virtue of its electrical resistivity characteristics.

Dielectrics: Covalently bonded materials with very large forbidden gap which disallow free electrons in the conduction band for electrical conduction.

Dielectric strength: Maximum electric field sustained in a dielectric without an electrical breakdown characterized by high current conduction.

Electrical insulating materials: Materials which offer high electrical resistivity between two locales at different electrical potential levels, thereby offering an electrical isolation between the locales.

Table 3.A Characteristics and Applications of Standard Insulating Materials

Material parameters

Volume resistivity in ohm-cm	ρ
Dielectric strength in volt/mil at room temperature (27°C)	E_M
Dielectric constant at room temperature and at 60 Hz	ε_r
Dissipation factor at room temperature (27°C)	$\tan\delta$

Types of plastics

TP = Thermoplastic (molding resin)
TS = Thermosetting plastics
ER = Extrusion resins
ER* = High temperature insulators
EP = Extrusion-type plastics
EL = Elastomers
EC = Embedding compounds
IC = Insulation coatings
I = Impregnants
NS = Solventless type
S = Solvent type

Insulating Material and Its Chemistry	Type	Characteristics and Applications	ρ	E_M	ε_r	$\tan\delta$
Acetals Homopolymer	TP	General purpose molding compound. Flammability and moisture absorption limit the electrical application of this compound	$\sim 10^{15}$	380	3.7	
Copolymer (Celcon™) (trioxane + dioxane ethers)	TP	General purpose molding compound. Flammability limits its use in electrical applications. Used for making coil forms, control knobs, etc.	$\sim 10^{14}$	380	3.7	0.001

(continued...)

Acrylics Polymerizing methyl methacrylate alone (PMMA); or combined with other unsaturated monomers	TP	High arc resistance. Decreasing dielectric constant with increasing frequency (suitable for high frequency applications). Flammability is relatively high.	$\sim 10^{18}$	500	3.7	~0.040
Amide polymers Nylon 6/6™ (adipic acid +hexamethylene diamine)	TP	Moisture absorption limits its use in electrical applications. Useful as jackets over primary wire insulation, coil forms, insulator blocks, and electrical connectors.	10^{13}	600	8.0	0.200
Nylon II™ $[NH(CH_2)_{10}CO]_x$	TP	Low moisture absorption. Limited electrical applications due to high cost and low thermal properties.	10^{14}	750	3.9	0.040
Imide polymers Polyamide-imide (PAI) (trimellitic anhydride + aromatic diamines)	TP	High price and processing difficulties limit electrical applications. High flame resistance and low smoke generation properties make the resin suitable for aerospace applications with weight reductions.	8×10^{16}	580	4.3	0.025
Polyether-imide (PEI) (imide molecule in amorphous polymer)	TP	High resistance to soldering, flame, and dimensional stability make these compounds suitable for circuit boards, terminal bases, connectors, fuses, and IC test devices. Also used in microwave applications.	6.7×10^{17}	480	3.2	0.004
Polyesters Polybutylene terephthalate (PBT) (poly condensation of 1,4 - butanediol and dimethyl terephthalate)	TP	Useful as coil forms, connectors business machine parts, TV components, terminal blocks. The compound has good room temperature resistance to water.	4×10^{16}	400	3.3	0.002

(continued...)

Insulating Materials and Its Chemistry	Type	Characteristics and Applications	ρ	E_M	ε_r	$\tan\delta$
Polycarbonate (PC) (polyester carbonic acid)	TP	Good physical dielectric and self-extinguishing properties. These are used in power tools, appliances, copiers and business machines. Flame retarding, low smoke emission types are available.	8.2×10^{16}	380	3.2	0.0009
Modified polyphenylene-oxide (PPO)	TP	Flame retardancy and good mechanical properties permit PPO resins attractive for business machines and household appliances, switch plates, connectors, and telecommunication equipment.	$\sim 10^{16}$	550	2.6	0.0004
Polyphenylene sulfide (PPS) (dichlorobenzene+ sodium sulfide)	TP	Used in wide ranges of electrical applications where parts are exposed to high temperatures: Oven wires, sockets, switches.	$\sim 10^{16}$	380	3.1	0.0003
Polypropylene (PP)	TP	Chemical resistancy permits their use in automative batteries. Motor vehicle electrical connections, fuse blocks, and other appliances use this resin widely.	$> 10^{16}$	600	2.2	<0.0005
Styrenics Polystyrene (PS)	TP	Flammability, low strength and susceptibility to attack by hydrocarbons limit their electrical applications, except for computer tapes, cassette reels, etc.	$< 10^{6}$	500	2.5	0.0001
Polystyrene butadine	TP	Higher strength permits its use for stronger units like computer cases, housings etc. Still in-flammable and low resistance to chemicals.	$< 10^{6}$	400	2.5	0.0020

(continued...)

Insulating Materials and Its Chemistry	Type	Characteristics and Applications	ρ	E_M	ε_r	tanδ
Styrenics Styrene-acrylonitrile (SAN)	TP	Applications include battery cases, instrument covers, transmitter caps, etc.	$< 10^6$	425	2.6	0.0070
Acrylonitrilebutadiene styrene (ABC)	TP	Strong mechanical properties permit heavy electrical applications as conductors and electrical apparatus housings.	2×10^{16}	425	2.6	0.0040
Sulfone Polymer Polysulfone	TP	High temperature withstandability and hydraulic stability allow this resin be used beyond the scope of other thermoplastics. Extensively used in circuit boards, IC carriers, coil bobbins, connectors, bustings, terminal blocks, etc.	10^{16}	425	3.1	0.0008
Polyether sulfone	TP	Same as above.	10^{17}	400	3.5	0.0010
Alkyds/Polyesters	TS	Used for brush holders, distributor caps, circuit-breakers, relays, switches, connectors, terminal boards, and housings. This plastic has excellent electrical and mechanical properties over a wide range of temperature.	10^{13}	375	5.6	0.1000
Allyls	TS	Extensively used for high reliability connectors, switches, terminal bonds, motor-starter blocks, etc.	10^{13}	400	4.2	0.0040

(continued...)

Insulating Material and Its Chemistry	Type	Characteristics and Applications	ρ	E_M	ε_r	tanδ
Aminos Urea formaldehyde and Melamine formaldehyde (cellulose filled or mineral filled)	TS	Heavy duty applications as in switch gear, wiring fences, appliance knobs, engine ignition parts, etc.	2×10^{11}	300	8.0	–
Epoxies (glass-filled epoxies)	TS	Extensively used in electrical applications as coil bobbins and connectors and in a variety of components. It has outstanding dimensional stability and no degassing until 260°C.	10^{13}	390	5.0	0.0100
Phenolics (Phenol plus formalde–hyde)	TS	Excellent for heavy duty uses and electrical insulations. Used in switch gears, electrical controls, wiring devices, and hermetically sealed electrical equipment.	10^{13}	380	6.0	–
Fluropolymers (High temperature insulators) polytetra-fluoroethylene (PTFE)	ER*	Excellent dielectric. Widest useful temperature range, low smoke, low–flame communication wire/cable, fire alarm cable.	$> 10^{18}$	500	2.1 at 1 MHz	0.0003 at 1MHz
Perfluoroalkoxy (PFA)	ER*	Comparable to PTFE in properties and applications	$> 10^{18}$	550	2.1	0.0003

(continued...)

Insulating Material and Its Chemistry	Type	Characteristics and Applications	ρ	E_M	ε_r	$\tan\delta$
Fluoropolymers Fluorinated ethylene-propylene copolymer (FEP)	ER*	Good dielectric. Narrow temperature range as compared to PTFE. Low smoke, low-flame communication cable, fire alarm cable, oil well logging cable, coaxial cable, computer back-panel wiring.	$>10^{18}$	550	2.1	0.0007
Ethylene tetrafluro-ethylene (ETFE)	ER*	Computer back-panel wiring, wiring nuclear power plants, hookup wire in aerospace and mass-transit circuits. Lower operating temperature than PTFE	$>10^{16}$	400	2.6	0.0050
Polyvinylidene fluoride (PVDF)	ER*	Dielectric and operational temperature ranges are not as good as PTFE. Back panel wiring of computers. hookup wiring in aerospace circuits.	2×10^{14}	260	6.1	0.1590
Ethylene chloro-trifluoroethylene copolymer (ECTFE)	ER*	Good dielectric. Low smoke, low-flame wiring, nuclear power plant wiring, aerospace circuit wiring, cathodic protection lead wire.	10^{15}	490	2.5	0.0090
Polychlorotri-fluoroethylene (PCTFE)	ER*	Comparable in performance and applications with ECTFE.	$>10^{18}$	500	2.5	0.0060
Ethylene Polymers Low density polyethylene (PE)	ER	Primary insulations jacket for low temperature applications. Controllable molecular weights (low to high). High density versions are tough and excellent oxidation resistant.	5×10^{13} – 2×10^{12}		4-6	~0.1000

(continued...)

Insulating Material and Its Chemistry	Type	Characteristics and Applications	ρ	E_M	ε_r	tanδ
Propylene Polymers (PP) Polypropylene ($CH_3CH - CH_2$)	EP	Primary insulation on wire and cable and for service under 5 kilovolts. Also used in injection molding battery cases. Useful in low temperature applications.	2×10^{14} to 2×10^{12}		4-6	0.1000
Polyvinylchloride (PVC) ($CH_2 = CHCl$)	EP	Low cost, low inflammability, and flame-retardant characteristics permit the wide-scale use of PVC in telephone wire inside insulation, building service wiring, low-tension automative wires, and apparatus cable. Useful in low temperature applications.	10^{16}	~500	~2.4	0.0003
Elastomers	EL	High cost limitations force extended heat curable silicone rubber only for expensive, high temperature insulation requirements. Applications include shipboard power/control cables, aircraft high-tension ignition/control cables, automotive ignition system insulation and multiconductor insulations.	10^{15}		3.0-3.6	
Organic elastomers (Organic compounds similar to natural rubber)	Butyl rubber	Useful as jackets on wires and cables and to a lesser extent primary insulators in low temperature applications.	10^{17}	600	2.1-2.4	0.0030
	Neoprene		10^{11}	400-700	6-8	0.0300
Epoxies (for encapsulation, potting, casting, and dipping processes) (aliphatic amines, aromatic amines, polyamides, urea formaldehydes, phenol formaldehydes)	EC	Epoxies are widely used for embedment of components for transformers, motors/generators, switch gears, coils, capacitors, resistors, and also for high voltage bushings. Excellent dielectric and good thermal, chemical/moisture resistant properties are common in most of the epoxies.	$\sim 4 \times 10^{14}$	500	3.9	0.0400

(continued...)

Insulating Material and Its Chemistry	Type	Characteristics and Applications	ρ	E_M	ε_r	tanδ
Polyurathenes (Diphenylmethane, diisocyanates, aliphatic cyanates, polyols, castor oils)	EC	Used for embedment of transformers, coils, switches, inductors, solid-state ignition systems, voltage regulators, printed circuit assemblies. They are outstandingly tough and resistant to abrasion. Fairly resistant to moisture and chemicals.	4×10^{11}	250	10	~0.08
Silicones Low viscous grade	EC	They have wide useful temperature range (-65°C to 265°C). Life expectancy is 100 times better than other encapsulants. Mechanical properties are however, lower. Initial cost is higher. Available in different viscous grades.	2×10^{14}– 5×10^{13}	400	4.5	0.02
Medium viscous grade			1×10^{14}– 3×10^{13}	400	4.5	0.02
High viscous grade			$\sim 10^{14}$	400	4.5	0.02
Varnishes resins Alkyd Polyesters	IC and I	These are early versions of insulating varnishes. made from vegetable oil (such as linseed oil) together with rosin, ester gum, etc. Improved versions include addition of petroleum asphalt, coal tars, and pitches. Some of these varnishes are still used in nonstandard applications.				
Alkyd Polyesters	IC and I	These are widely used coating and impregnants for applications to all types of electrical machinary, transformers, and electrical components.				
Baking type			$\sim 10^{6}$	4000 (dry) 2900 (wet)	4–8	0.04–0.08
Baking type NS				~2000		

(continued...)

Insulating Material and Its Chemistry	Type	Characteristics and Applications	ρ	E_M	ε_r	tanδ
Epoxies Unmodified epoxy resins, epoxy-novolac resins, epoxy ester resins	IC and I	These have outstanding bonding strengths with high resistance to humidity and chemicals. Due to high cost, uses are limited to critical applications such as hermetic motors and heavy-duty rotating equipment where use of polyester is not recommended.	10^6	~2000 ~4500	3.8 ~ 6.0 3.3	0.0040 0.0030
Parylenes Parylene N, (poly-para-xylene) parylene C parylene D	IC and I	Recommended for thin-coat applications as in circuit boards or modules for protection against contaminates, moisture, salt spray, and corrosive vapors. Other circuit components can also be coated with parylene compounds.	10^{17} parylene - N	7000	2.7	0.0002
Phenolics Modified phenolics, thermo-plastic phenolics, alcohol/ketone soluble phenolics, phenolic dispersion resins	IC and I	Useful as moisture and fungus resistant coatings. Blending with other varnishes improve the temperature withstandibility and chemical resistant properties.				

(continued...)

Insulating Material and Its Chemistry	Type	Characteristics and Applications	ρ	E_M	ε_r	tanδ
Polyurethanes (PURS)	IC and I	These have outstanding abrasion resistance when coated on surfaces. Used for coating circuit boards, and as enamel coatings on electrical equipment.		2000 (Dry) 1200 (Wet)		
Silicones	IC and I	Excellent dielectric properties over a wide temperature range. Long-term service at temperatures up to 250°C. Greater resistance to corona than other coating varnishes. Excellent water/moisture resistant. High cost prohibits its use except in critical applications as in rotating machines at high temperature operations, and equipment under hostile ambients.	10^{14}	2000	3.1	0.0016

Note: Data presented in this table were collected from various available sources. Numerical values indicated are typical and approximate.

CHAPTER 4

Composite Dielectric Materials

4.1 Introduction

Composite dielectrics represent, in general, a heterogeneous system of multiconstituent materials. Typically, a two-phase composite dielectric is constituted by a host material with an inclusion of another material. This host-inclusion system could be formed by a combination (or a mixture) of dielectric-dielectric, dielectric-conductor, and/or dielectric-semiconductor phases. The constituent phases may form structurally an embedment system consisting of multi-layer "layups" or random dispersion of the inclusions across the host medium; or there could be a structured matrix of specific type to yield certain desirable dielectric properties. Generically, a composite dielectric can be treated as as a mixture-medium, largely heterogeneous and could be anisotropic and nonlinear, as well.

4.2 Theory of Dielectric Mixtures

Mixture models describing the effective response of dielectric mixtures to electromagnetic stimulus have been addressed comprehensively in the literature [1]. Such mixtures are useful in several electromagnetic applications. *Radar absorbing materials* (RAMs), *surface coatings* non-reflective to electromagnetic waves, *electrostatic dissipative compounds*, *EMI shielding materials*, special purpose insulators, *bioelectromagnetic phantoms*, and *conductive adhesives* used in microelectronics are a few examples of composite dielectrics.

A dielectric mixture can be synthesized in a number of ways. Typically the following are the *host inclusion systems* considered in the dielectric mixture theory:

- Spherical/near-spherical inclusions dispersed randomly as "chunks" in a homogeneous host medium
- Elongated (ellipsoidal or spheroidal) inclusions dispersed randomly in a homogeneous receptacle
- Fibrous (needle-like) inclusions with random dispositions in the host medium
- Random dispersion of flaky (disk-like) inclusions in the host medium
- Multi-layer of laminates or orderly "stacked-up" individual laminae (of different phases)
- Orderly-oriented and laid stretches (*"tows"*) of fibers in a host medium
- Special geometrical inclusions (such as *honey-comb*, *mesh-like* structures) embedded in the host
- Multiphase composite with voids/porosity introduced deliberately

However, irrespective of the shape, size, physical state, volume fraction, or orientation of different phases, the effective dielectric response of the mixture and hence its constitutive dielectric parameter, namely, the *effective permittivity* (ε_{eff}), should always lie within two specified limits referred to as the *Wiener bounds* which will be discussed in detail in a later section.

In the following section, general considerations of *dielectric mixture theory* [1] pertinent to the foregoing types of dielectric mixtures are reviewed and state-of-the-art models describing the *effective permittivity* are addressed. Even in the case of the simplest type of mixtures with two constitutive phases, it is indicated that the dielectric characteristics are dependent not only on the dielectric polarizability of the constituents but also the stochastic attributes of the mixture medium.

4.3 Permittivity of Heterogeneous Mixtures

The earliest version of a dielectric model of a two-phase heterogeneous mixture is due to Clausius and Mossotti [2], who on similar considerations of Maxwell-Garnett theory [3,4] (applied to conducting spherical particulates dispersed in a dielectric host medium) derived the following expression for the effective permittivity ε_{eff} of the mixture assuming that the constituents of the mixture are electrostatically noninteracting:

$$(\varepsilon_{eff}/\varepsilon_2) = (1 + 3\theta X_d); \qquad X_d = (\varepsilon_1 - \varepsilon_2)/(\varepsilon_1 + 2\varepsilon_2) \tag{4.1}$$

where ε_1 and ε_2 are the relative permittivities of the inclusions and the host medium, respectively, and θ is the *volume fraction of the inclusions*. Equation 4.1 is popularly known as the *Clausius-Mossotti formula* [2] for the effective permittivity of a dielectric mixture.

In many cases, dielectric mixtures have more heterogeneous characteristics than the model potrayed by Clausius and Mossotti; and as such the dielectric formulations developed to describe many of such mixtures are empirical or semi-empirical in nature, decided by curve-fitting strategies to experimental data. However, rigorous formulations taking heterogeneity into account have also been developed by extending the binary phase model of Clausius and Mossotti. The following is the chronological account on the development of various dielectric mixture theories.

Following the Clausius-Mossotti approach, Rayleigh [5] obtained Equation (4.1) in another form for diluted dispersions (that is, for small volume fractions, $\theta << 1$). It is given by:

$$(\varepsilon_{eff} - \varepsilon_2)/(\varepsilon_{eff} + 2\varepsilon_2) = \theta(\varepsilon_1 - \varepsilon_2)/(\varepsilon_1 + 2\varepsilon_2) \tag{4.2}$$

Bruggeman [6] applied *Rayleigh's formula* to incremental changes in the volume loading of the composite medium sequentially and arrived at the so-called *one-third power law*, namely:

$$(\varepsilon_2/\varepsilon_{eff})^{1/3}(\varepsilon_1 - \varepsilon_{eff})/(\varepsilon_1 - \varepsilon_2) = (1 - \theta) \tag{4.3}$$

or, in the limiting case of $|\varepsilon_1| << |\varepsilon_2|$ and $|\varepsilon_{eff}|$, Equation 4.3 specifies that $\varepsilon_{eff} \cong \varepsilon_2(1 - \theta)^{3/2}$. An extension of Equation 4.3 written in the form $\varepsilon_{eff} = \varepsilon_2(1 - \theta)^d$ with $d \approx 1$ *to* 5 refers to *Archie's law* [7] widely used in studies on geophysical substances.

The electrostatic-based mixture formulations on electrical capacity of dispersed systems was extended by Fricke [8,9] to include a "shape-factor" in order to account for the particulate shape. Thus, *Fricke's formula* has an eccentricity term (x') based on the geometrical aspect of prolate and/or oblate spheroidal inclusions depicted in Figure 4.1. It is given by:

$$\varepsilon_{eff} = \{\varepsilon_1\varepsilon_2(1 + x'\theta) + \varepsilon_2^2 x'(1 - \theta)\}/\{\varepsilon_{eff}(1 - \theta) + \varepsilon_2(x' + \theta)\} \tag{4.4}$$

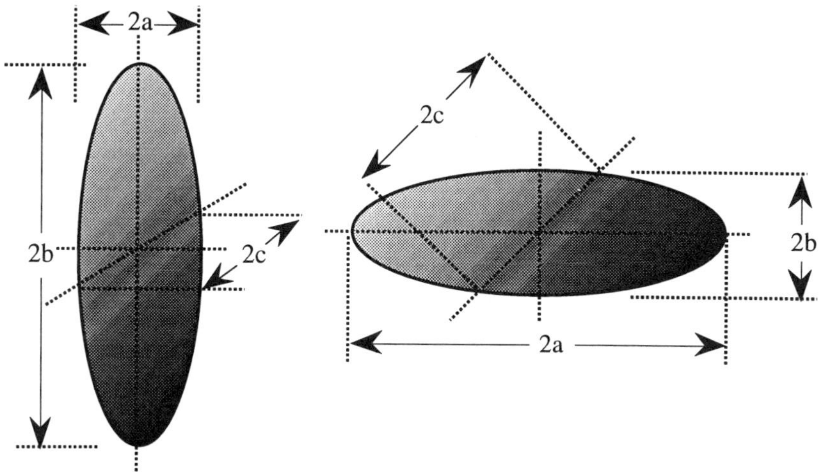

Figure 4.1 The spheroidal geometry.
A. Prolate spheroid; B. Oblate spheroid.

When the concentration of the included particles is high so that each particle is surrounded by a mixture rather than by the component, Böttcher [10] derived a mixture formula given by:

$$(\varepsilon_{eff} - \varepsilon_2)/\varepsilon_{eff} = 3\theta(\varepsilon_1 - \varepsilon_2)/(\varepsilon_1 + 2\varepsilon_{eff}) \tag{4.5}$$

Another interesting equation was developed by Looyenga [11] who assumed that on mixing two components such that $\varepsilon_1 = \varepsilon_{eff} \pm \Delta\varepsilon_{eff}$ and $\varepsilon_2 = \varepsilon_{eff} \mp \Delta\varepsilon_{eff}$ or *vice versa*, the effective permittivity ε_{eff} of the mixture can be written in the following form on the basis of *Rayleigh's equation* as:

$$\varepsilon_{eff} = \{\varepsilon_2^{1/3} + \theta(\varepsilon_1^{1/3} - \varepsilon_2^{1/3})\}^3 \tag{4.6}$$

The above expression appears to have also been independently derived by Landau and Lifschitz as reported in [12].

In all the aforesaid formulations, though the mixture was considered as a random entity, no stochastic attribute was, however, explicitly included. Analytical descriptions of dielectric characteristics of a mixture formed by the random volumetric dispersion of shaped inclusions in a continuous medium is, in general, complicated due to the statistical nature of the random spatial locations and orientation of the dispersed phase. That is, when such a mixture is placed in an electric field, the electrical conduction and polarization would depend on the random spatial dispersion (and relative orientation) of the shaped inclusions in the medium; and therefore, the dielectric or permittivity characteristics of the mixture would be essentially statistical in nature as determined by the random particle dispersion. Hence, a study on the electrical characteristics of the mixture should be concerned not only with the calculations on the electric field induction in the mixture but also should correspond to a probability problem. Such a stochastic attribution to dielectric mixtures was first developed by Lichtenecker [13] and Lichtenecker and Rother [14] on the basis of the following considerations.

Considering a two-component system in which the inclusions with a permittivity ε_1 are dispersed in a continuous medium of permittivity ε_2, the system can be regarded as a matrix in which the dispersing medium represents a receptacle for the mutually isolated (or out-of-contact) particulate inclusions as indicated by Zheludev [15]. The effective permittivity (ε_{eff}) of this matrix mixture would then depend on the permittivities of the mixture constituents, namely, ε_1 and ε_2, and on the volume fraction (θ) of the inclusions. The function which interrelates ε_{eff} and other quantities, namely, ε_1, ε_2, and θ, would be determined both by electric field induction in the mixture as well as by the statistical considerations arising from random volumetric dispersion (and relative orientation) of the inclusions. Thus, the value of the effective permittivity of a statistical mixture can be described by a certain function F_1 as follows:

$$\varepsilon_{eff} = F_1(\varepsilon_1, \varepsilon_2, \theta, q) \tag{4.7}$$

where q is a *form factor* which depends on the shape of the inclusions. In the above equation (Equation 4.7), though a two-component case is considered, the discussion can be extended to any number of mixture components without any loss of generality. (Such multiphase systems are discussed in Chapter 6.)

All the theoretical works on the topic under discussion aim at finding the explicit nature of the function F_1 in Equation 4.7 and in the determination of this function, certain conditions have to be observed relevant to statistical mixtures. They are: (i) If the values of the permittivity of all the components of the mixture change in one and the same ratio, the value of the effective permittivity of the mixture (ε_{eff}) should change identically (*Wiener's proportionality postulate* [16]). Hence, F_1 must be a homogeneous function of the first degree extracted from the set of independent variables ε_1 and ε_2. That is,

$$F_1(s\varepsilon_1, s\varepsilon_2, \theta, q) = sF_1(\varepsilon_1, \varepsilon_2, \theta, q) \tag{4.8}$$

where s is an arbitrary constant factor. The above postulation follows directly from the laws of electrostatics according to which the direction of the lines of electric field at the boundary of two dielectrics depends only on the ratio of the permittivities of these dielectrics and is independent of their absolute values. (ii) The permittivity of a heterogeneous system (mixture) is closely connected with the arrangement of the particles in the system in relation to the field direction as can be seen from the simple example of a two-component laminated dielectric system shown in Figure (4.2). When the directions of the field and of the laminations coincide (parallel combination, Figure (4.2a), it follows that:

$$\varepsilon_{eff} = \theta\varepsilon_1 + (1 - \theta)\varepsilon_2 \tag{4.9}$$

and when the field and lamination directions are perpendicular (series combination, Figure 4.2b), the following relation holds good:

$$\varepsilon_{eff} = 1/[\theta/\varepsilon_1 + (1 - \theta)/\varepsilon_2] \tag{4.10}$$

The true value of the effective permittivity (ε_{eff}) of a statistical mixture shown in Figure (4.2c) should in fact, lie between the extreme values determined by Equations 4.9 and 4.10. Hence, it is constrained by the following inequalities suggested by Wiener in 1912 [16]:

$$1/[\theta/\varepsilon_1 + (1-\theta)/\varepsilon_2] \le \varepsilon_{eff} \le \theta\varepsilon_1 + (1-\theta)\varepsilon_2 \qquad (4.11)$$

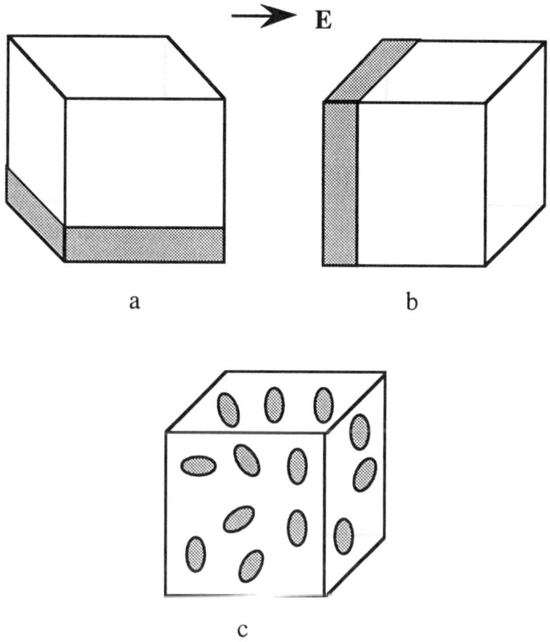

Figure 4.2 Two-component dielectric mixture systems.

a. Parallel arrangement of dielectrics (laminar dielectrics arranged parallel to external E-field).
b. Series arrangement of dielectrics (laminar dielectrics arranged serially in reference to external E-field).
c. Random relative dispositions of inclusions in the host medium.

Further, (iii) the function F_1 should be valid irrespective of the number of components in the mixture; and, if the values of ε_1 and ε_2 are the same, ε_{eff} should coincide with this single value. That is, if the mixture contains only one component (say, ε_1 and $\theta = 1$), the value of ε_{eff} should be the same as ε_1. Last, (iv) the components of a matrix system are geometrically dissimilar. That is, if the positions of the receptacle and inclusion are interchanged, the result is a system with a different permittivity. Analytically, this can be expressed as:

$$F_1\{\varepsilon_1,\theta;\ \varepsilon_2,(1-\theta)\} \ne F_1\{\varepsilon_1,(1-\theta);\ \varepsilon_2,\theta\} \qquad (4.12)$$

Explicit evaluation of the function F_1 is based on ascertaining the properties of the matrix system as determined from the characteristics of its components. All the investigations earlier to Lichtenecker [13] and Lichtenecker and Rother [14] on this aspect had resulted in a general functional relation of the following type:

$$F_2(\varepsilon_{eff}) = \theta F_2(\varepsilon_1) + (1-\theta)F_2(\varepsilon_2) \qquad (4.13)$$

so that $\varepsilon_{eff} = F_1(\varepsilon_1,\varepsilon_2,\theta)$. If the function F_2 in Equation 4.13 which determines the law of mixing is explicitly known, F_1 (and hence, the value of ε_{eff}) can be uniquely determined.

The analytical endeavor of evaluating the function F_2 (or the function F_1) for various types of dielectric mixtures has resulted in several mixture relations and a comprehensive review of them was published by Brown [17] and by van Beek [18]. The contents of these reviews have also been summarized and reported by Tinga et al. [19]. These existing dielectric mixture formulations can be grouped into three major categories with the following characteristics: (i) Formulations based on electric field induction in the mixture containing spherical inclusions which either mutually interact or do not interact; (ii) analyses based on the electric field induction in the mixture containing shaped inclusions such as *ellipsoids, oblate/prolate spheroids, needle-like,* or *disk-like particles,* etc. and depolarization effects due to particle shape are either considered or ignored and; last (iii) formulations based on the theory of mixture which is regarded as a probability problem.

The investigations on dielectric mixtures due to Lorenz [20], Rayleigh [5], Bruggeman [6], Böttcher [10], Meredith and Tobias [21], and Looyenga [11] fall under the first group cited above. Considering the second category, in which the particle shapes have been explicitly taken into account, the works of Wiener [16], Fricke [8,9], Sillars [22], Polder and Van Santen [23], Lewin [24], Hamon [25], Boned and Peyrelasse [26], and Sihvola and Kong [27] can be regarded as significant contributions.

However, studies dealing with the calculations (of the third type mentioned above) which are based on statistical considerations of the dielectric mixture are relatively few in number. In effect, reference can be made to only one work on the subject, that of Lichtenecker [13] who proposed the *logarithmic law of mixing* which can be summarized as follows.

Considering the theory of mixtures as a probability problem, Lichtenecker [13] and Lichtenecker and Rother [14] deduced the *logarithmic mixture law* from the general principles of statistics. For a mixture of two components, it is given by:

$$\varepsilon_{eff} = \theta\varepsilon_1{}^k + (1-\theta)\varepsilon_2{}^k \qquad (-1 \le k \le 1) \tag{4.14}$$

Here, where $k = 1$, Equation 4.14 gives the same expression for ε_{eff} as derived for a laminated dielectric (field parallel to the laminations, Figure 4.2a); and, for $k = -1$, Equation 4.14 gives the expression for a laminated dielectric (field perpendicular to the laminations, Figure 4.2b). For an unordered system k tends to zero and the formula of Equation 4.14 assumes the following form:

$$\varepsilon_{eff} = \varepsilon_1{}^{\theta}\varepsilon_2{}^{(1-\theta)} \tag{4.15a}$$

or

$$log(\varepsilon_{eff}) = \theta log(\varepsilon_1) + (1-\theta)log(\varepsilon_2) \tag{4.15b}$$

Experimental studies on heterogeneous dielectric systems support Lichtenecker's formula even for anisotropic media such as barium titanate-polystyrol mixture as indicated by Zheludev [27]. The author [28] applied Lichtenecker's formula successfully to describe the complex permittivity of a polycrystalline (organic) compound taken in powder form and to evaluate the dielectric constant of human blood [29]. Wallin [30] indicated that the logarithmic law (or its modified versions) fits closely the experimental data on oil shale. On the basis of these results, a major conclusion is that the logarithmic law of mixing holds good at all volume fractions in describing the dielectric behavior (static or dynamic) of statistical mixtures. For spherical inclusions and also for almost sphere-like particles with uneven and coarse surfaces (as in the case of polycrystalline powder samples), the geometrical shape of the particles does not play a significant role in determining the macroscopic dielectric behavior of the mixture. That is, any small *depolarization effects* which may arise due to the relative orientation of coarsely surfaced (almost sphere-like) particles are overwhelmed by the stochastic characterizations resulting from random dispersion of the inclusions in the volume of the mixture concerned. Hence, as depicted by Equation (4.15), the permittivity of such mixtures is solely a function of the permittivities and the relative volumes of the mixture constituents.

4.4 Dependence of Permittivity on Particulate Geometry

Considering a dielectric mixture containing shaped/aspherical inclusions randomly dispersed in the host, it is necessary to attribute a *shape* or a *form factor* to the particulates in question to account for the depolarization effects. The particles/inclusions are called shaped if two or more of the lateral dimensions of the particles are significantly different as in the case of ellipsoids, prolate/oblate spheroids, and disk-like or needle-like particles.

For a spheroidal geometry (Figure 4.1) with semiaxial lengths *a, b,* and *c* and taking $b=c$, the aspect ratio is equal to (a/b). When this aspect ratio is of significant value (either large or small compared to unity) the corresponding eccentricity (e) would play a dominant role in the polarization of the particles when the mixture is subjected to an external field; and the depolarization arising from the relative disposition of the particles due to the random nature of particle dispersion (and/or orientation) in the mixture would become another effective stochastic parameter to be considered. As stated earlier, the works of Wiener [16], Fricke [8,9], Sillars [22], Lewin [24], and Hamon [25] are the earliest contributions which explicitly take into account the particle shape.

The logarithmic law *per se* does not contain any term to account for the particulate shape. Hence, it predicts the mixture permittivity as independent of the particulate shape and is applicable only to spherical/near-spherical inclusions. This drawback (the "shapeless" aspect) of the logarithmic law was criticized as inconsistent and theoretically unsound by Reynold and Hough [31] and later by Dukhin [32]. However, the author [33] obviated this deficiency of the logarithmic law by combining it with the well-known Fricke's formula thereby giving a modified version of the logarithmic law which explicitly accounts for the particle shape as explained below.

Fricke in his two classical papers [8,9] developed an expression for the effective permittivity of a dielectric mixture with an explicit shape/form factor to account for the shape of the inclusions. His analytical description of the mixture was based on the electric field induction in the dispersed system. The effective permittivity (ε_{eff}) of the mixture was expressed in terms of the permittivities of the host ε_2 and inclusions ε_1, the volume fraction (θ) of the inclusions and a shape/form factor x'_o to account for the depolarization effects in the electrical induction flux as:

$$\varepsilon_{eff} = \{\varepsilon_1\varepsilon_2(1 + x'_o\theta) + \varepsilon_2^2 x'_o(1 - \theta)\}/\{\varepsilon_1(1 - \theta) + \varepsilon_2(x'_o + \theta)\} \tag{4.16}$$

where x'_o, the *shape/form factor* is dependent on the ratio of $\varepsilon_1/\varepsilon_2$. However, the results obtained on the basis of Fricke's formula deviated significantly from the measured data. Hence the author [33] included a statistical attribution to Fricke's formula on the basis of the logarithmic law to obtain a modified form factor x_o given by:

$$x_o = M\{\varepsilon_1^{\theta}\varepsilon_2^{(1-\theta)}[\varepsilon^{(1-\theta)} + \varepsilon_2^{\theta}] - \varepsilon_1\varepsilon_2\}/\{\varepsilon_2[\varepsilon_2^{(1-\theta)} + \varepsilon_1^{\theta} - \varepsilon_1^{\theta}\varepsilon_2^{(1-\theta)}]\} \tag{4.17}$$

where M is a function of the (a/b) ratio of the inclusions.

Considering the particulate inclusions, they could in general either be oblate spheroidal $(a > b)$ or prolate spheroidal $(a < b)$ as indicated in Figure (4.1). In the extreme cases $(a << b)$, they tend to be needles and for $(a >> b)$, they become disks. For an oblate spheroid, the eccentricity is given by $e = (1 - b/a)$ and correspondingly for prolate spheroidal inclusions the eccentricity, $e = (1 - a/b)$. The factor M in the above equation (Equation 4.17) can be expressed in terms of the eccentricity of the inclusions as $M = 2/(m-1)$ or $(m-1)/2$ depending on whether $\varepsilon_1 \geq \varepsilon_2$ or $\varepsilon_1 \leq \varepsilon_2$, respectively; and the parameter m is related to the eccentricity e as follows [22]:

$$m = e^2\{1 - (1 - e^2)^{1/2}[\arcsin(e)/e]\}^{-1} \quad (4.18)$$

The aforesaid modified version of the Fricke's formula as proposed in [33] has yielded results which correlate closely with the measured values pertaining to certain test mixtures.

Reynolds and Hough [31] succeeded in reducing all the existing mixture formulations except the logarithmic law to the generalized linear functional form of the type specified by Equation 4.13. In order to overcome this inconsistency pertaining to the logarithmic law, the author [34] also developed an improved version of the logarithmic law of mixing based on a weighted coefficient format that fitted into a generalized linear form. Further, it had been generally contended that the logarithmic law could not be extended to a mixture with lossy dielectric and metallic (conductor) inclusions. On the contrary, the author successfully applied the logarithmic law to mixtures with conducting inclusions as elaborated in [35] on the basis of electrical susceptibility considerations as will be discussed in Chapter 6.

Inasmuch as the logarithmic law of mixing is not amenable for representation by a sample, generalized linear function, Reynolds and Hough [31] doubted some error in the logarithmic formulation and later (in 1974) Dukhin and Shilor [32] attributed the observed inconsistency to an illogical assumption by Lichtenecker [13] who considered a disperse system as chaotic and ordered simultaneously.

Despite the prevalence of the aforesaid mathematical inconsistency, the logarithmic law of mixing has surprisingly gained recognition, supported by experimental data gathered on stochastic mixtures with near-spherical inclusions [28,29]. As such it was considered preferable to eliminate the persisting incompatibility of the logarithmic law with respect to the generalized linear form. This has been done by the modifications as suggested by the author in [34] and is described below.

Considering a stochastic mixture, the effective permittivity as given by the logarithmic law of mixing corresponds to a weighted geometrical mean of ε_{1S} and ε_{2S}, namely, $\varepsilon_{eff} = \varepsilon_{2S}^{\theta}\varepsilon_{1S}^{(1-\theta)}$.

The logarithmic relation can also be specified in a different form of weighted geometrical mean as presented below:

$$\varepsilon_{eff} = C(\varepsilon_1, \varepsilon_2, \theta)\,[\varepsilon_u]^{n(a/b)}\,[\varepsilon_l]^{1-n(a/b)} \quad (4.19)$$

where $\varepsilon_U = \theta\varepsilon_2 + (1-\theta)\varepsilon_1$ and $\varepsilon_L = [\theta/\varepsilon_2 + (1-\theta)/\varepsilon_1]^{-1}$ are *Wiener's upper and lower limits,* respectively (see Equations 4.9 and 4.10). In Equation 4.19, it is presumed that the nth fraction of the stochastic mixture system behaves as if polarized in the direction of the electric field induction and the remaining $(1-n)^{th}$ fraction is polarized orthogonally. Here, n is considered as a function of the axial ratio of the inclusions (namely, a/b) alone and C is a weighting factor depending on ε_1, ε_2 and θ.

The expression of Equation 4.19 should satisfy certain limiting conditions pertaining to n, θ, and ε_{eff}. The conditions are: (i) $0 \leq n \leq 1$; (ii) $0 \leq \theta \leq 1$; and (iii) for any finite values of ε_1 and ε_2, ε_{eff} must be bounded and lie within Wiener's limits. Hence it follows that:

$$\varepsilon_{eff} = \begin{cases} X(\theta)/2 & \varepsilon_2 > \varepsilon_1 \\ Y(\theta)/2 & \varepsilon_2 < \varepsilon_1 \end{cases} \qquad 0 \leq \theta \leq \phi_1$$

$$= \frac{1}{2}\left[\frac{A(\phi_1)}{2C(\phi_1)} + \frac{B(\phi_2)}{2C(\phi_2)}\right] C(\theta)\,Z(\theta), \quad \varepsilon_2 > \varepsilon_1 \qquad \phi_1 \le \theta \le \phi_2$$

$$= \frac{1}{2}\left[\frac{B(\phi_1)}{2C(\phi_1)} + \frac{A(\phi_2)}{2C(\phi_2)}\right] C(\theta)\,Z(\theta), \quad \varepsilon_2 < \varepsilon_1$$

$$= \begin{cases} Y(\theta)/2 & \varepsilon_2 > \varepsilon_1 \\ X(\theta)/2 & \varepsilon_2 < \varepsilon_1 \end{cases} \qquad \phi_2 \le \theta \le 1 \tag{4.20}$$

where $X(\theta) = Z(\theta) + 1/\varepsilon_L(\theta)$, $Y(\theta) = Z(\theta) + \varepsilon_U(\theta)$, $Z(\theta) = \varepsilon^n_U(\theta)\,/\varepsilon^{n-1}_L(\theta)$, $A(\theta) = 1 + 1/\varepsilon^n_U\,\varepsilon^n_U$, $B(\theta) = 1 + 1/\varepsilon^{n-1}_U\,\varepsilon^{n-1}_L$, and $C(\theta) = \sqrt{\varepsilon_L(\theta)/\varepsilon_U(\theta)}\;\varepsilon_2^{\theta}\,\varepsilon_1^{1-\theta}$.

Further, n is equal to $(5 - m)/4$ or $(m - 1)/4$ depending on $\varepsilon_2 > \varepsilon_1$ or $\varepsilon_2 < \varepsilon_1$, respectively. Here, m is a function of the a/b ratio which can be determined in terms of the eccentricity of the inclusions as indicated earlier (Equation 4.18).

In Equation. 4.20, ϕ_1 and $\phi_2 = (1 - \phi_1)$ denote the volume fractions at which the weighting coefficient C attains minimum and maximum values, respectively; and it can be shown that:

$$\phi_1 = (1/2) - (1/2)(1 - 4t)^{1/2} \tag{4.21}$$

where t is given by:

$$t = (\varepsilon_1 + \varepsilon_2)/[2(\varepsilon_2 - \varepsilon_1)\ln(\varepsilon_2/\varepsilon_1)] - (\varepsilon_1\varepsilon_2)/(\varepsilon_2 - \varepsilon_1)^2 \tag{4.22}$$

Since Equation 4.20 is in a linear form and is functionally related to the shape-dependent (depolarizing) parameter a/b, it is compatible with Reynolds-Hough's expression [31]. It has also been found valid for dynamic (time-varying) cases relevant to the complex permittivity of a mixture.

4.5 Orderly-Textured Mixtures

Orderly-textured dielectric mixtures refer to a specific class of dielectric composites consisting of a host dispersed with shaped inclusions that are aligned/oriented specific to the external (applied) field direction. Designing composite dielectrics for electromagnetic applications with such orderly-textured arrangements could yield certain special characteristics such as curtailing certain polarizations of the EM wave, etc. The ordered arrangements can be accomplished *via* interwoven fibers or stacks and piles of shaped inclusions such as ellipsoids, disks, needles, etc. Such shaped inclusions are used since they can be oriented or aligned conveniently specific to the external field. Spherical inclusions on the other hand will impart nondirectional attributes due to their geometrical symmetry, as their dispersion in the host medium is isotropically random.

When these ordered mixtures are subjected to an external electric field, the resulting polarizations refer to the alignment of the electric field flux either parallel or perpendicular to the ordered arrangement. Accordingly, the effective dielectric polarization of the medium differs significantly from that of a mixture with randomly scattered shaped inclusions; hence, orderly-textured mixtures have to be modeled differently.

Taylor [36] used the so-called *"average field hypothesis"* to evaluate the dielectric property (effective permittivity) of such ordered mixtures on an approximate, statistical ensemble average basis. However, this *field-averaging law* is applicable only to low volume concentrations of inclusions inasmuch as it is deduced by neglecting the interparticulate interactions. An alternative approach, known as the *effective medium approximation* [37,38] addresses a random mixture whose dielectric property is ascertained by discretizing the medium into independent cells. Again, relevant formulations apply to a small concentration of the inclusions only. The effective medium, in general, replaces the heterogeneous status of the medium by an effective region free of scattering effects [37].

4.5.1 *Logarithmic Law of Mixing and Orderly-Textured Mixtures*

Consider a simple orderly-textured dielectric mixture constituted by an orderly arrangement of shaped inclusions in a host medium. The ordered disposition of the inclusions would render the effective dielectric properties (effective permittivity) considerably different from that of a mixture consisting of randomly dispersed shaped inclusions. A weighted exponent strategy described in [35,39] models the orderly-textured test mixture using *Langevin's theory of dipole orientation**. This theory is judiciously applied by the author as described in [40] to extrapolate the disordered particulate state formulation so as to describe a test mixture having an ordered state of inclusions. That is, Langevin's function (which represents the monotonic growth of orientational polarizability with respect to the enhancement of ordered texture) is used as a weighting coefficient in the logarithmic law pertaining to a random system. The dependence of the effective permittivity of an orderly-textured mixture on the shape of the inclusions is thus predicted on the basis of the weighted exponent forms of the logarithmic law.

Therefore, the effective permittivity of a mixture with the prolate/oblate spheroidal inclusions being orderly-textured (parallel or perpendicular to an external field direction) can be specified as follows [40].

Let the spheroidal geometry be such that its axial dimension along the x-direction is 2a and its axial dimension along the y-direction is 2b. Then, if the electric field is parallel to the y-axis, the effective permittivity is given by:

$$\varepsilon_{eff} = \varepsilon_{log}(\varepsilon_U/\varepsilon_{log})^r \tag{4.23}$$

and, if the electric field is parallel to the x-axis, the corresponding effective permittivity is given by:

$$\varepsilon_{eff} = \varepsilon_{log}(\varepsilon_L/\varepsilon_{log})^n \tag{4.24}$$

with $r = (1 - N/L')$ and $n = (1 - L'/N)$; $\varepsilon_{log} = \varepsilon_1^{\theta}\varepsilon_2^{(1-\theta)}$ where ε_1 and ε_2 are the relative permittivities of the inclusions and the host medium, respectively; and θ refers to the volume fraction of the inclusions. ε_U and ε_L are Wiener's upper and lower limit values, respectively, of ε_{eff} and are given by $\varepsilon_U = \theta\varepsilon_1 + (1 - \theta)\varepsilon_2$ and $\varepsilon_L = 1/[\theta/\varepsilon_1 + (1 - \theta)/\varepsilon_2]$. Further, the parameter N refers to the extent of depolarization decided by the particulate asymmetry. It is equal to $N(\perp)$ or $N(\parallel)$ in the r-term depending on the particulates being oblate spheroidal $(a > b)$ or prolate spheroidal $(a < b)$, respectively. In Equation 4.24, involving the n-term, the corresponding values of N are interchanged. The quantities $N(\perp)$ and $N(\parallel)$ are explicitly given in terms of the particulate eccentricity e as follows:

$$N(\parallel)|_{Prolate\ Spheroid} = (1 - e^2)\{(1/2e)\ln[(1 + e)/(1 - e)] - 1\}/e^2$$

$$= 1 - 2N(\parallel)|_{Prolate\ Spheroid} \tag{4.25a}$$

---

* *Langevin's theory* of dipole orientation is presented in Appendix 4A.

$$N(||)|_{Oblate\ Spheroid} = \{1 - [(1 - e^2)^{1/2}/2e)\ sin^{-1}(e)\}/e^2$$

$$= 1 - 2\,N(||)|_{Oblate\ Spheroid} \qquad (4.25b)$$

When $b >> a$, the prolate spheroid represents needle-like (fibrous) inclusions and for $a >> b$ the oblate spheroid depicts disk-like (flaky) particles.

Further, the parameter L' in Equations 4.23 and 4.24 represents $dL(e)/de$; when $e = 0$ or $a = b$, the particles are spherical and the corresponding slope of $L(e)$ at $e = 0$, namely, $L'(0) = 1/3$, which is the well-known *order function* of the totally disordered state. As $e \rightarrow 1$, $L'(e) \rightarrow 0$, representing a totally ordered state corresponding to the particulates being fully aligned with respect to the electric field direction. Hence, Equations 4.23 and 4.24 decide the development of the electrical polarizability (and hence, the permittivity of the mixture) corresponding to the ordered state from the disordered statistics. It does not restrict the amount of particulates present in the mixture. Therefore, it is free from the constraint of dilute-phase approximation.

Calculated sample results pertinent to a set of data obtained from Equations 4.23 and 4.24 are depicted in Figure (4.3) wherein the relevant results derived thereof are compared with those due to Taylor [36]. The inferences are:

1. Referring to Figure (4.3), the results due to Taylor's method [36] give a value of $\varepsilon_{eff} = 14.2$ at $e = 0$, for $\varepsilon_1 = 78.3$, $\varepsilon_2 = 2$, and a volume fraction of $\theta = 0.4$. At $e = 0$, however, as indicated before, this result should correspond to randomly dispersed spherical/near spherical inclusions with an effective permittivity $\varepsilon_{eff} = 8.67$. Unlike Taylor's formulation, the method described above gives this value of 8.67 at $e = 0$. The reason for Taylor's result yielding an overestimated value is due to the fact that his results are based on Böttcher's [10] formula which has the inherent deficiency of short-range statistical variations being neglected.

2. Hence, Taylor's formulations are valid only for high volume concentrations of inclusions $(\theta > 0.4)$. The formulations of Equations 4.23 and 4.24 are, however, devoid of this deficiency.

3. For large values of e, the present as well as Taylor's formulations are bounded by Wiener's limits. That is, for absolute parallel or perpendicular orientation of the particles with respect to the electric field, both formulations would yield similar results.

4. Except for its values at $e = 0$ being different, the trends of the variation of ε_{eff} with respect to the aspect ratio as calculated by the present method as well as by Taylor's algorithms remain the same for $\varepsilon_1 > \varepsilon_2$ or $\varepsilon_1 < \varepsilon_2$.

5. For prolate (needle-like) inclusions, the value of the effective permittivity of the mixture tends towards Wiener's upper limit in the limiting case of all the inclusions being aligned parallel to the electric field direction. Likewise, when all the inclusions are antiparallel (perpendicular) to the applied electric field, the value of the effective permittivity of the mixture tends towards Wiener's lower limit.

6. For oblate (disk-like) inclusions the above trends are reversed.

7. When the needle-like inclusions are aligned with the needle-axis parallel to the electric field (or when the electric field is tangential to the surface of the disk-like inclusions), the effective permittivity saturates to Wiener's upper limit (or Wiener's lower limit, respectively) even at relatively low aspect ratios. On the contrary, with the

corresponding perpendicular orientations, the effective permittivity reaches Wiener's limits only at asymptotically large values of the aspect ratio. This could be anticipated inasmuch as the parallel orientation aids the permeation of the electric flux.

8. Use of Langevin's theory enables the construction of the ordered texture from the disordered dispersion regardless of particulate concentration and it also implicitly accounts for the interparticulate interaction within the macroscopic test mixture.

Taylor's approach is based essentially on "average field approximation", namely, the effective electrical properties are simply related to the average electric fields in the host and in the inclusions. Evaluation of these fields exactly is not possible because there is no method of dealing with the correlation between particles especially in the solid phase. However, Taylor proceeded to elucidate the average fields with the approximations of neglecting interparticulate correlations assuming that each inclusion is a single particulate entity in a homogeneous medium subjected to the electric force of the average field.

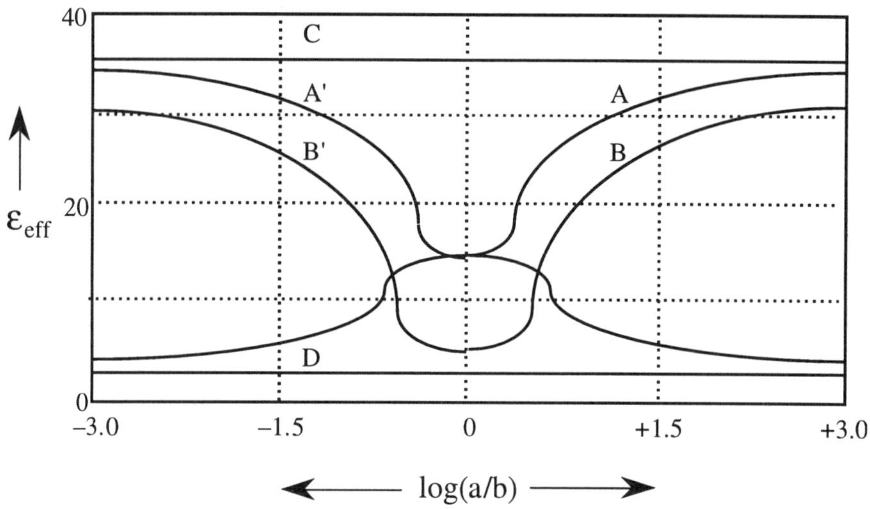

Figure 4.3 Effective permittivity (ε_{eff}) *versus* aspect ratio (a/b) of the inclusions of volume fraction, $\theta = 0.4$.

A,A': Taylor's formulation(s) –for the particulate orientation perpendicular (A) or parallel (A') to the applied electric field (**E**).
BB': Corresponding results due to Equations 4.23 and 4.24.
C: Wiener's upper limit.
D: Wiener's lower limit.
Data: $\varepsilon_1 = 78.3$; $\varepsilon_2 = 2.0$.

4.6 Interparticulate Interactions in Composite Dielectrics

Among the number of formulations developed to elucidate the effective dielectric behavior of composite dielectrics as detailed in the previous sections, determination of the effective permittivity of mixtures having shaped inclusions with explicit particulate interactions (when subjected to an external electric field) has been addressed only sparsely in the literature. Such mixtures are characterized by not only the particulate inclusions interacting with each other due to the applied electric field but also the shape of the particles that facilitate their dispersion being textured or ordered in the host medium [40]. When these ordered mixtures are subjected to an electric field, there is an induced polarization in the medium which renders the alignment of the electric flux either parallel or perpendicular to the ordered arrangement. Also, certain classes of electromagnetic *"soft materials"* constituted by a dispersion of dielectric particles in weakly conducting or dielectric fluids (known popularly as the *electrorheological fluids*; see Chapter 24) exhibit spontaneous alignment of the particles under the application of an external field. In such cases, the particles may assume dispositions of being nearly in physical contact with each other. This close proximity would result in significant mutual interactions between them which must be duly taken into account in describing the effective dielectric response of the composite material.

The analytical strategy to examine such interaction is to consider a *multipolar field expansion* in the vicinity of interacting particles and determine the coefficients of expansion for a subsequent use in formulating the effective dielectric response of the mixture. Lam [41] followed this technique using Rayleigh's [5] method for the multipolar field expansion and deduced the expansion coefficients *via* orthogonalization of the Legendre functions (in the potential expansion) in the case of spherical particles. Thus, in the presence of an external field inducing interparticulate interactions, the theoretical considerations in formulating the effective dielectric response of the mixture consisting of shaped inclusions can be specified by a multipolar electric field potential expansion method. This technique facilitates the calculation of electrostatic interaction forces between dielectric particles as a function of particle separation. It can be postulated that the change in electrostatic interaction force between the particles with respect to varying the relative spatial dispositions of the constituents can be considered as being proportional to a corresponding change in an order parameter u; and the effective permittivity of the mixture can be described by a functional relation which includes this order parameter to specify the implicit effects of particle interactions [42].

Pertinent to a mixture with spherical particles, the potential function expansion which is valid at all points outside the interacting spheres and includes an arbitrary number of multipolar moments can be expressed in terms of spherical harmonics as indicated by Morse and Feshbach [43]. Such a multipolar expansion includes induced contributions arising from the effects of particle proximity, and additional multipole contributions will accrue in the case of shaped particles. To comply with the particulate geometry, the spheroidal coordinate system shown in Figure 4.4 can be considered in describing the potential function.

Denoting the focal distance of a prolate spheroid as L, and distances from a field point (x_1, y_1, z_1) to the foci as r_1 and r_2, respectively, the spheroidal coordinates are defined as:

$$\xi = (r_1 + r_2)/2L, \quad \eta = (r_1 - r_2)/2L, \qquad \Phi = tan^{-1}\,(y/x) \tag{4.26}$$

where ξ is a constant that describes a prolate spheroid with an interfocal distance $2L$, major axis ξL, and minor axis $L(\xi^2 - 1)^{1/2}$; η is a constant which defines two sheets of a hyperboloid of revolution with foci at $z = \pm L$; and Φ is a constant plane through the z-axis at an angle ϕ to the x-z plane. A suitable expansion for the potential function V can be written to include multipole contributions from the neighboring particles as follows [42]:

$$V(\zeta,\eta,\phi) = \begin{cases} \displaystyle\sum_{n=0}^{\infty}\sum_{m=0}^{n} A_{nm}\, cosm\phi\, P_n^m(\eta)\, P_n^m(\xi_o)/\, P_n^m(\xi); & \zeta < \zeta_o \\ \displaystyle\sum_{n=0}^{\infty}\sum_{m=0}^{n} cosm\phi\, P_n^m(\eta)[B_{nm}\, Q_n^m(\xi_o)/Q_n^m(\xi) + C_{nm}\, P_n^m(\xi)]; & \zeta > \zeta_o \end{cases} \tag{4.27}$$

where the first term in the expression for the exterior potential represents the multipole components of the reference spheroid, and the second term provides multipole contributions from the adjacent spheroids. The contribution of associated Legendre functions has been selected to satisfy the boundary conditions of the potential. For instance, inside the spheroid

$(\xi - \xi_o)$, the functions are selected such that the potential is finite at a particle surface; and in the exterior of the spheroid $(\xi - \xi_o)$ such that the potential is matched at the particle boundary and asymptotically reduces to zero at infinity.

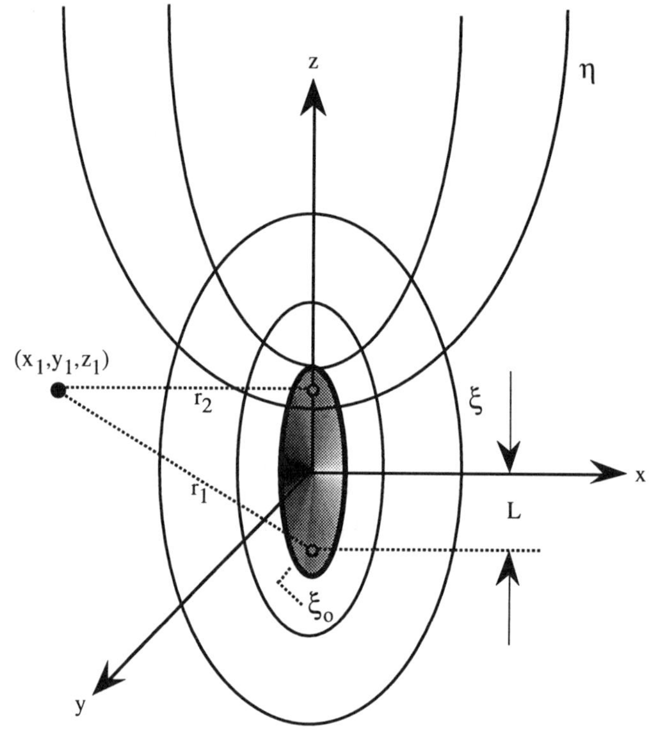

Figure 4.4 Spheroidal coordinate system. (The surface of the reference spheroid with the focal dimension L is defined by ξ_o.)

Evaluation of the coefficients in Equation 4.27 requires three independent equations relating three sets of coefficients. Two of them can be elucidated from the usual boundary conditions of continuous potential function and electric displacement at the particle boundary. Application of these boundary conditions eliminates one set of coefficients (A_{nm}), and allows an identity to be formed between the remaining two. The third interrelation between the coefficients required for the complete solution stems from the effective conductivity model of a cubic lattice of conducting spheres due to Rayleigh [5]. Rayleigh recognized that the terms of the potential expansion being not singular at the origin of the reference spheroid were due to terms originating at infinity (that is, the applied potential) and at the other spheroids. This allows a crucial identity which can be expressed as follows:

$$\sum_{n=0}^{\infty}\sum_{m=0}^{n} C_{nm} \cos m\phi\, P_n^m(\eta)\, P_n^m(\xi) =$$
$$\sum_{n=0}^{\infty}\sum_{m=0}^{n}\sum_{i=1}^{N_s} B_{nm} \cos m\phi_i\, P_n^m(\eta_i)\, Q_n^m(\xi_i)/Q_n^m(\xi_o) + E_o \xi \eta L \tag{4.28}$$

where E_o is the applied field and N_s is the number of spheroids interacting with the reference spheroid. The potential expansion in the prolate spheroidal system results in a Legendre function representation so that the orthogonalization can be pursued following the method due to Lam [41]. In order to facilitate the orthogonalization, Equation 4.29 can be transformed into spherical coordinates (r, θ) *via*: $r_1 = r + L\cos\theta$, $r_2 = r - L\cos\theta$ and the corresponding prolate spheroidal coordinates are given by $\xi = r/L$, $\eta = \cos\theta$. Hence, Equation 4.28 can now be expressed as:

$$\sum_{n=0}^{\infty} \sum_{m=0}^{n} C_{nm} \cos m\phi \, P_n^m(\cos\theta)\, P_n^m(r_L) =$$

$$\sum_{n=0}^{\infty} \sum_{m=0}^{n} \sum_{i=1}^{N_s} B_{nm} \cos m\phi_i \, P_n^m(\cos\theta_i)\, Q_n^m(r_{Li})/Q_n^m(r_o) + E_o r\cos\theta \tag{4.29}$$

where $r_L = r/L$ and r_o refers to the spheroid surface.

Application of the orthogonality properties eliminates one of the expansion coefficients and results in an infinite series solution for the remaining coefficient. By assuming that the position vector with respect to the reference sphere of the calculated field point is of smaller magnitude than the position vector of the ith interacting spheroid, and by neglecting the interactions from the particles not along the line of electric flux (which imposes an azimuthal isotropicity with $m = 0$ in the foregoing expressions), the expansion coefficient is expressed explicitly by:

$$[4\pi/(2n+1)]B_n H_n = \sum_{n'}\sum_{i} D_{n,n'}(R_{Li})B_{n'} + (4\pi/3)E_o\delta_{n1} \tag{4.30}$$

where

$$H_n = \{[1/P_n(r_o)] + Q_n^{-1}(r_o)/[Q_n(r_o)\, P_n^{-1}(r_o)]\}/[1 \quad \varepsilon_2/\varepsilon_1]$$

and

$$D_{n,n'}(R_{ni}) = [\kappa_n\,(n + n')!\,(-1)_n\, 4\pi\,(-R_{Li})^{-n-n'}][Y_{n+n',0}(\theta_i,\phi_i)]/[(2n + 1)\, n!n'!]$$

with

$$\kappa_n = [r_o^{\,n+1}/P_n(r_L)] \sum_{\ell=0}^{\infty} (r_o/r_{Li})^{2\ell}$$

Thus, Equation 4.30 is an infinite series depicting the final boundary condition required to evaluate the coefficients in the potential expansion. The algebraic expansion of Equation 4.30 can be cast in the form of a matrix equation which can be numerically approximated given a finite number of contributing terms N as indicated by the convergence of the series. Once the B_n's are determined, the electric field components are obtained *via* appropriate spatial differentiation of the potential functions resulting in explicit field components specified by:

$$E_r = -\sum_{n}^{N} B_n P_n(\cos\theta)[Q_n^1(r_L)/Q_n(r_o) + H_n P_n^1(r_L)]/[\sqrt{(R_L^2 - 1)}]$$

$$E_\theta = -1/r_L \sum_{n}^{N} B_n P_n^1(\cos\theta)\,[Q_n^1(r_L)/Q_n(r_o) + H_n P_n(r_L)]/[(\sqrt{(\cos^2\theta} - 1)] \tag{4.31}$$

at the points exterior to the spheroid.

4.6.1 *Electrostatic interaction forces*

The force experienced by a particle along the (particle) axis parallel to the electric field is specified as [42]:

$$F_z = [(\varepsilon_2/\varepsilon_1) - 1] \int_S E_r \bullet E_z \, dS \tag{4.32}$$

which dictates that the interparticle electrostatic force is proportional to the surface integral of the radial and vertical field product. Thus, for a given spatial geometry, the interparticulate force can be considered as being directly proportional to the sum of the product of contributing coefficients in the expansion of the electric field.

This interaction force arising from the external field will influence the dielectric behavior of the mixture. Therefore, to include such interaction effects implicitly in deciding the effective dielectric response of the mixture, an order parameter can be stipulated in terms of the interaction force quantified *via* Equation (4.32) as a function of the interparticle separation. Hence, an expression to determine the effective permittivity of the mixture involving the order parameter explicitly can be derived as indicated in the following section.

4.6.2 *Dielectric mixture model with finite interparticulate interactions*

When the mixture under consideration is subjected to a uniform electric field, the particles (either by design, or spontaneously) have a tendency to align to form a chain-like, orderly texture either parallel or perpendicular to the applied field (Figure 4.5). The effective permittivity (ε_{eff}) is then determined by the spatial hierarchy of particulate dispersion, the volume fractions of the constituents and the interactive effects among the particles along the lines of electric flux. The mixture formulation to determine the ε_{eff} of a N-component statistical mixture can be represented in a general form as:

$$\varepsilon_{eff} = \sum_{i=1}^{N} \theta_i F(\varepsilon_i) \tag{4.33}$$

where θ_i denotes the volume fraction of the ith constituent; and F is a function that implicitly includes the effects of particle orientation, shape, and interactions. The upper and lower bounds, respectively, over the span $0 \le \theta \le 1$ are specified by the corresponding functionals F_{up} *and* F_{lo}. Within these limits, the bounded value of ε_{eff} can be derived from the principles of statistical mixture theory, by deducing the homogeneous function F of Equation 4.33 explicitly.

Evaluation of such a functional relation is straightforward in the limiting cases of extreme spatial anisotropy (namely, parallel and series arrangements); however, it is rather difficult to obtain an explicit expression for ε_{eff} for random dispersion of the constituents characterized by the stochastic (spatial) attributes of the ε_i. In such cases, an algorithmic approach to describe the effective material response (in terms of the known limiting values) can be written as follows:

$$\varepsilon_{eff} = \left[\sum_{i=1}^{N} \theta_i F_{up}(\varepsilon_i)\right]^{u} \left[\sum_{i=1}^{N} \theta_i F_{lo}(\varepsilon_i)\right]^{1-u} \tag{4.34}$$

where the order parameter $u \in [0,1]$, weights the limiting values of the extreme anisotropic spatial arrangements to match the effective material response under a particular spatial configuration and thereby it also implicitly accounts for the particle interactions.

In the case of significant particle eccentricities, or when the mixture constituents tend to form laminar/columnar structures, the bounds F_{up} and F_{lo} as mentioned earlier are recognized commonly as the Wiener [16] limits. These limits bound all possible values of the effective parameter of an N-ary mixture and correspond to the two extreme anisotropic spatial dispositions of the material constituents, namely, the parallel $(u = 1)$ and series $(u = 0)$ connected cases, forming stacked parallel planes and columns of the mixture constituents, respectively.

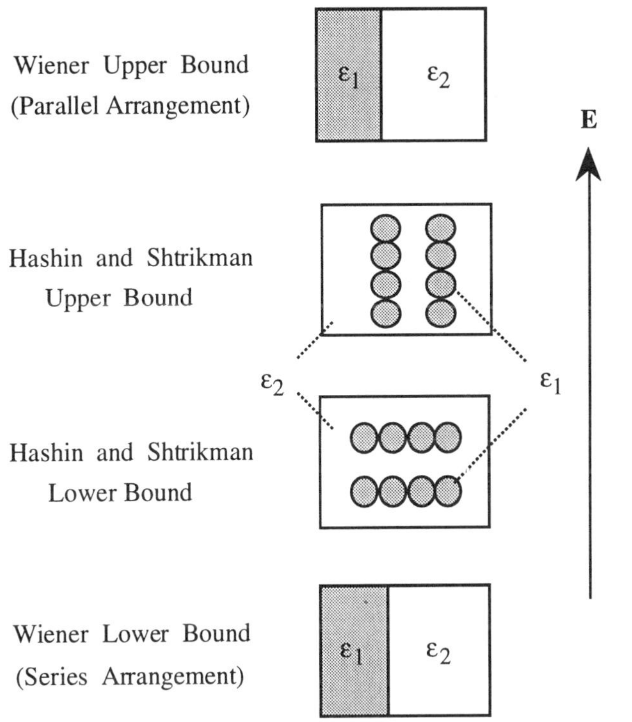

Figure 4.5 Extreme spatial anisotropic arrangements of the constituents of a two-phase dielectric mixture.

Alternatively, Hashin and Shtrikman [44] derived more restricted bounds for the limiting cases of spherical particle inclusions, These bounds are shown to be equivalent to those obtained by considering the material being composed of composite spheres (that is, a sphere with an inner radius r_a consisting of a material specified by ε_a, surrounded by a concentric shell of radius r_b composed of a material ε_b) dispersed in a host medium of material parameter ε. Corresponding to Equation 4.34, the effective permittivity pertinent to the test mixture under discussion as bounded by the Hashin and Shtrikman limits is given by:

$$\varepsilon_{eff} = \{\varepsilon_2 + \theta_1/[(\varepsilon_1 - \varepsilon_2)^{-1} + (\theta_2/3\varepsilon_2)]\}^u x\{\varepsilon_1 + \theta_1/[(\varepsilon_2 - \varepsilon_1)^{-1} + (\theta_2/3\varepsilon_1)]\}^{1-u} \quad (4.35)$$

In Figure (4.6), the types of ordered arrangements pertinent to the Wiener and the Hashin and Shtrikman bounds are illustrated.

4.7 Determination of the Order Function and Calculation of ε_{eff}

Pertinent to the test mixture under discussion, the order function u should reflect the overall effects as due to the change of associated energy and/or force for a given spatial hierarchy of interacting particles. In other words, the order function $(0 \leq u \leq 1)$ should correspond to a normalized electrostatic force *($0 \leq F/F_o \leq 1$)* of interacting particles with respect to particle separation. Here, the normalization of the interaction force is done with respect to a value F_o obtained when the particle separation tends to zero.

The specific algorithmic approach in modeling the effective permittivity of the test mixture can be outlined as follows:

- Determine the multipolar electric field potential around a reference particle (Equation 4.31).
- Calculate the electrostatic interaction force of the reference and adjacent particles as a function of varying particle spatial dispositions (Equation 4.32).
- Normalize the interaction force with respect to the limiting value when the particle separation tends to zero.
- Take values of the spatial order parameter u, for a given spatial configuration from the corresponding normalized interaction force.
- Apply the order parameter to the mixture formula (Equations 4.34 and 4.35).

The theoretical considerations presented here to evaluate the effective permittivity of the test mixture refers to both the nonspherical (spheroidal) particulate inclusions with eccentricity *($\varepsilon > 0$)* and spherical particles with $\varepsilon \rightarrow 0$. In either case, the relevant analysis includes the effects of interparticulate interactions.

4.8 Sample Results

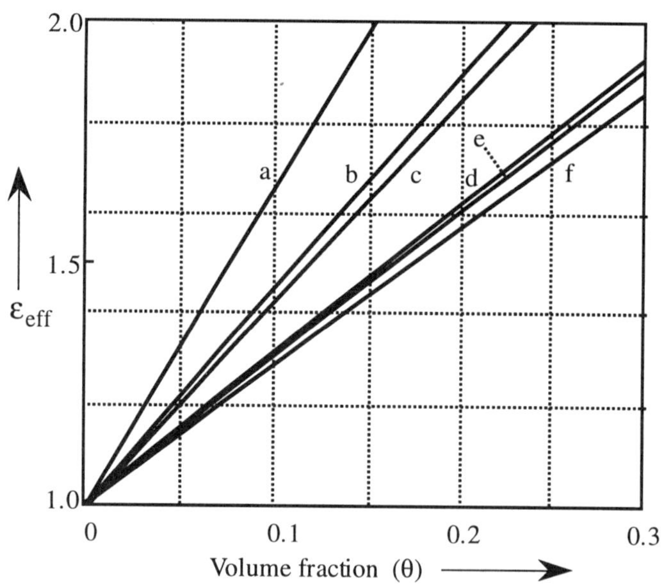

Figure 4.6 Effective permittivity (ε_{eff}) *versus* volume fraction (θ) of spherical particulate chains dispersed in a host medium. The center-to-center separation (r) normalized with respect to the particle radius (a): r/a = 2.1 and 3. Ratio of the dielectric constant of the host medium to that of the particles: $\varepsilon_1/\varepsilon_2 = 1/10$. Calculations refer to the formulations due to Miller and Jones [45] and by Equations 4.34 and 4.35. (a): Hashin and Shtrikman upper bound (u = 1). (b and e): Results *via* Equation 4.34 and 4.35 with u = 0.49 and u = 0.05, respectively. (c and d): Miller and Jones formulation with r/a = 2.1 and 3.0, respectively. (f) Hashin and Shtrikman lower bound (u = 0).

Spherical particles: With spherical particulate inclusions in the mixture, the sample computed data as per the present approach refer to two systems, namely, *r/a = 2.1* and *r/a=3*, where *r* is the center-to-center separation between spheres and a is the radius. The normalized force of interaction as a function of sphere separation refers to computation using Equation 4.32. The corresponding values of u obtained are 0.49 and 0.05 for *r/a*=2.1 and 3, respectively, assuming a ratio of the dielectric constants of the host medium to that of the particles as 1/10. Hence, the values of effective permittivity as a function of volume fraction in respect to these two systems as determined by Equation 4.35 are presented in Figure 4.6. Depicted in Figure 4.7 are also the corresponding results obtained by the method due to Miller and Jones [45]. The numerical values of the deviations in the sample results presented with respect to those of Miller and Jones are presented in Table 4.1. It can be observed that the maximum percentage of deviation for a sphere separation of *r/a*=3 (corresponding to dilute suspension) is only 0.87%. With a concentrated system represented by *r/a* = 2.1, the deviation is, however, larger and has a maximum value of 4.94% over all volume fractions. This larger deviation at higher concentrations can be attributed to the fact that the evaluation of *u* in Equation 4.35 takes into account only a single chain of particles in calculating the local electric field distributions and resulting interparticle forces. Miller and Jones, however, assumed regularly spaced multiple chains in calculating the local field contributions. However, their formulation is inapplicable to shaped particles and is confined to spherical particulate inclusions only.

Table 4.1 Percentage Error ($\Delta\varepsilon_{eff}$) between the Values of Effective Permittivity due to Miller and Jones and Those Obtained *via* Equations 4.34 and 4.35 as a Function of the Volume Fraction (θ)

θ	$\Delta\varepsilon_{eff}(\%)$	
	r/a = 3	r/a = 2.1
0.02	–0.19	0.74
0.04	–0.36	1.36
0.06	–0.52	1.83
0.08	–0.58	2.24
0.10	–0.71	2.66
0.12	–0.76	3.03
0.14	–0.87	3.30
0.16	-	3.54
0.18	-	3.80
0.20	-	4.03
0.22	-	4.19
0.24	-	4.33
0.26	-	4.45
0.28	-	4.56
0.30	-	4.69

The sample results presented (and those of Miller and Jones) above are bounded by the limits of Hashin and Shtrikman. It can also be observed that the results on dielectric permittivity at low volume fractions (with *r/a*=3) tend towards the lower bound. This implies that Hashin and Shtrikman's lower bound corresponds to a sparsely spaced chain of spherical particles in antiparallel direction to the applied electric field; and in this relative disposition of the chain orientation and applied field being perpendicular, the interactive forces can be regarded as minimal.

Nonspherical particles: Sample computations carried out refers to a system with nonspherical particles (prolate spheroids) of eccentricity e ranging from 0.01 to 0.1. The particle-to-host dielectric ratio is assumed to be 10, and the interaction force is computed using Equation 4.32 as a function of spheroid separation δ normalized by the focal distance L. Relevant results are presented in Figure 4.7. The interparticulate force is seen to decrease for particles of increasing eccentricity. It might be expected that this force would rather increase inasmuch as an increase in the field intensity predicated along the interparticle axis would result in greater interparticle forces. However, as indicated in Equation 4.32, the interparticle force involves integrating the field intensity over the entire particle surface and relevant calculations show that the significant eccentricities would actually decrease the field intensity over large angular distributions away from the interparticle axis. Additionally, as shown in the inset of Figure 4.8, increasing the particle eccentricity (for particles with a fixed dimension) decreases the particle volume, thereby decreasing the surface area available for the electrostatic interaction.

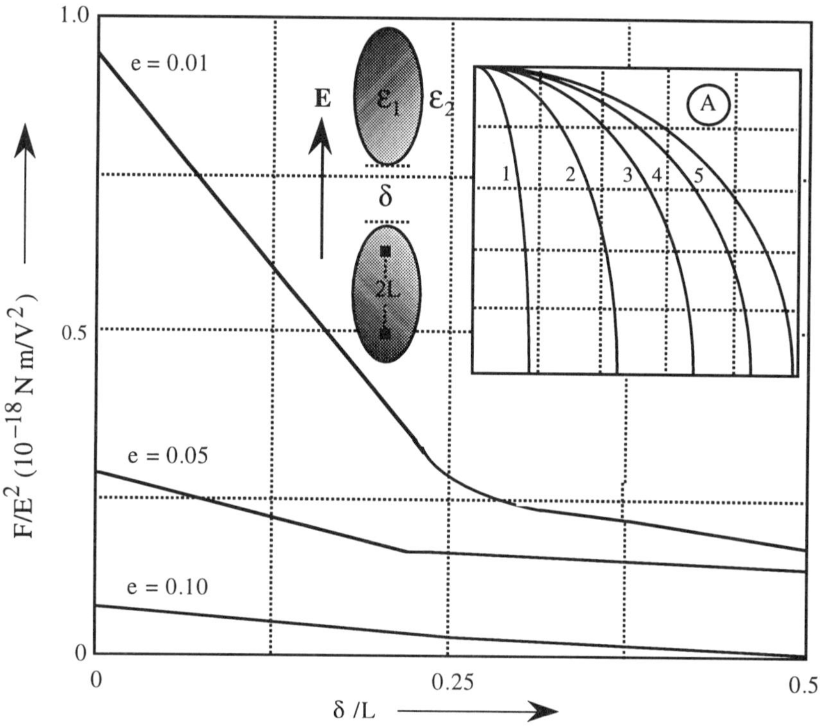

Figure 4.7 Interparticulate force *versus* interparticle separation. (The force is normalized with the values as particle separation $\rightarrow 0$; and the particle separation is normalized with respect to the distance between the foci.) Each spheroid has an equivalent vertex dimension of 1 and $\varepsilon_1/\varepsilon_2 = 10$.

A. Contours of prolate spheroids as a function of eccentricity: (1) e = 0.4; (2) e = 0.3; (3) e = 0.2; (4) e = 0.1; and (5) e = 0.01.

The order parameter u is obtained from the interaction force calculated *via* Equation 4.32 as a function of δ/L for a given particle eccentricity and normalized with respect to the value obtained when $\delta/L \rightarrow 0$. Figure 4.8 presents the resulting family of curves representing u *versus* δ/L with eccentricity as a parameter. As the eccentricity increases, the effect of increasing particle separation is diminished in reducing the relative interaction strength between the spheroids. Thus, as the particles attain significant shape along the lines of electric flux, the polarization anisotropy induced by the particle geometry tends to overwhelm the induced interactions between the particles.

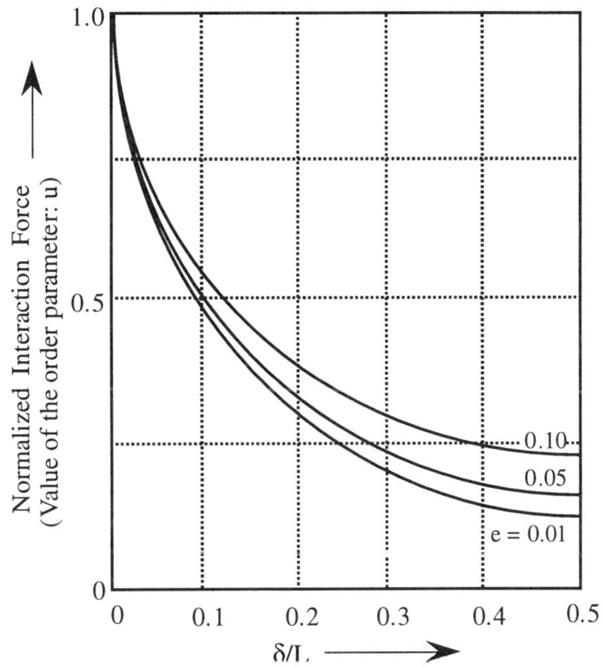

Figure 4.8 Electrostatic interaction force between prolate spheroids *versus* particle separation. (The interaction force is normalized with the value as the particle approach each other at double-layer separation and the particle separation is normalized with respect to the distance between the foci.)

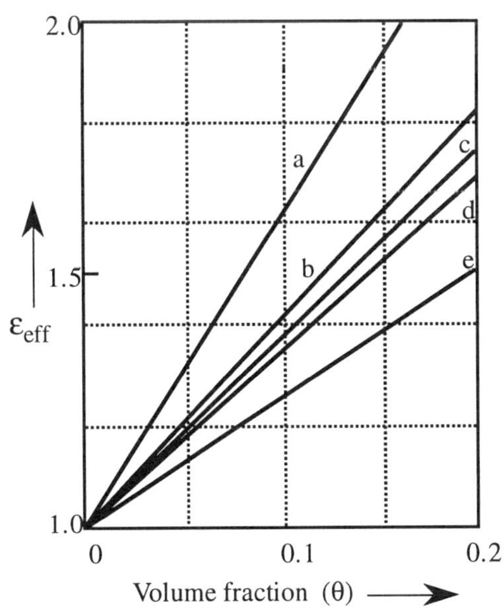

Figure 4.9 Effective permittivity (ε_{eff}) *versus* volume fraction of the spheroidal chains of particulate inclusions with a normalized particulate separation distance (δ/L) equal to 0.2.

Figure 4.9 illustrates the effective permittivity *versus* the volume fraction of spheroidal inclusions with eccentricity as the parameter in the case of a test mixture with spheroids having a dielectric constant of 10 and dispersed in a medium of dielectric constant equal to 1. It is evident that for a given δ/L ratio, as the eccentricity is increased, the effective permittivity *versus* the volume fraction of inclusions tends towards Hashin and Shtrikman's upper limit.

Experimental results: As an example, the dielectric mixture which corresponds closely to the test mixture discussed above can be constituted by a dispersion of erythrocytes (red blood corpuscles) in blood plasma. Such a mixture approximates an electrorheological fluid (see Chapter 24) constituted by shaped dielectric particles dispersed in a weakly conducting fluid. The erythrocytes are shaped particles with an aspect ratio of about 1/4. The measured data on the effective permittivity of the mixture constituting the total blood *versus* the volume fraction of erythrocytes as reported by Bianco et al. [46] are listed in Table 4.2. Also furnished in Table 4.2 are results computed by the present method which include the particle interactions, and by thc methods of Boned and Peyrelasse [26] and by the author elsewhere [29], in all of which the particle interactions are neglected. It is obvious from the tabulated data that the results obtained by the method discussed above are closer to the measured values than those of Boned and Peyrelasse [26] or the author [29]. In other words, inclusion of interaction effects in the permittivity formulation enhances the accuracy of the algorithm.

Table 4.2 Measured and Computed Values of Effective Permittivity for a System of Erythrocytes Dispersed in Blood Plasma

Volume Fraction of Erythrocytes	Measured Data [45]	Calculations as per Equations 4.34 & 4.35 (Interaction Included)		Method due to [29] (Interaction Neglected)		Method due to [26] (Interaction Neglected)	
(θ)	(ε_{eff})	(ε_{eff})	% Deviation	(ε_{eff})	% Deviation	(ε_{eff})	% Deviation
0.14	68.75	68.49	0.38	68.26	0.71	64.65	5.95
0.28	67.66	67.18	0.71	66.80	1.27	63.43	6.26
0.41	66.92	66.11	1.21	65.78	1.70	62.68	6.33
0.53	65.94	65.23	1.08	64.94	1.52	61.73	6.38
0.84	63.60	63.27	0.52	63.08	0.82	59.51	6.44

4.9 Conclusions

The composite dielectrics are of two specific types, namely, dielectric-dielectric and dielectric-conductor compositions. The essence of various modeling techniques pertinent to dielectric-dielectric composites is presented in this chapter along with an outline on the evolution of relevant formulations. A comprehensive listing of existing formulations in the literature is summarized in Chapter 5.

As regards the multiphase dielectrics, the theory and relevant formulations need a separate presentation which is provided in Chapter 6.

Inclusion of conductors in a dielectric host leads to a special class of dielectric-conductor composites. Relevant details and formulations are given in Chapter 7.

In addition to the various methods of classifying and modeling dielectric mixtures indicated in this chapter, the other apporaches envisaged in the existing literature are summarized below:

Multiscattering models: (i) In a model based on *weak fluctuation theory*, a mean permittivity value of the mixture has been defined in terms of the permittivities of the host and the inclusions and the volume fractions of the constituents. Then by using a spherical symmetric correlation function for the dielectric permittivity fluctuations, Tsang and Kong [47] have predicted the effective permittivity of the medium in terms of the normalized variance of the fluctuations (Δ) valid for any value of the volume fractions of the inclusions (θ_i), if $\Delta << 1$. Otherwise, the corresponding formulation is restricted to $\theta_i << 1$ (dilute phase approximation of the inclusions). (ii) Another approach advocated by Kong [48] evalutes the complex effective permittivity of the mixture by combining the Maxwell-Garnett formula and that derived by the Rayleigh scattering theory. The corresponding formulation reduces to the quasistatic Maxwell-Garnett-Rayleigh equation at low frequencies. (iii) Also, exclusive for spherical scatterers, the effective permittivity has been deduced *via* effective field approximation [37]. (iv) Using *strong fluctuation theory* for a random medium, the effective permittivity relevant to a continuous medium has been derived by Tsang et al. [49] assuming a symmetric correlation function. Corresponding results have been shown to degenerate to Böttcher's formula under low frequency limits [37].

Effective medium approach: This modeling strategy is based on replacing a heterogeneous medium by an effective medium in which the electromagnetic propagation constant is, on an average, assumed to be free from scattering effects. Corresponding quasistatic results on the effective permittivity have been deduced under low frequency limiting conditions. (Relevant formulations have also been extended to composites with conducting inclusions which will be described later in Chapter 6.)

References

[1] P. S. Neelakanta: Complex permittivity of chaotic dielectric mixtures: A review. J. Instr. Electron. Telecom. Engrs; vol. 37(4), 1994: 385-392.

[2] H. Fröhlich: *Theory of Dielectrics.* (Clarendon Press, Oxford: 1949).

[3] J. C. Maxwell-Garnett: Colours in metal glasses and metal films. Phil. Trans. A: Roy. Soc. London. vol. 203, 1904: 385-420.

[4] J. C. Maxwell-Garnett: Colours in metal glasses, in metallic films and in metallic solutions-II. Phil. Trans. A: Roy. Soc. London, vol. 205, 1906: 237-262.

[5] Lord Rayleigh: On the influence of obstacles arranged in rectangular order upon the properties of a medium. Phil. Mag., vol. 34, 1892: 481-502.

[6] D. A. G. Bruggeman: Berechnung verschiedener physikalischer konstantan von heterogenen Substanzen. Ann. Physik, vol. 24, 1935: 636-679.

[7] G. E. Archie: The electrical resistivity log as an aid in determining some reservoir characteristics. Trans. Am. Inst. Min. Met., Petrol. Engrs. vol. 146, 1942: 54-62.

[8] H. Fricke: A mathematical treatment of the electrical conductivity and capacity of disperse systems I. Phys. Review, vol. 24, 1924: 575-587.

[9] H. Fricke: A mathematical treatment of the electrical conductivity and capacity of disperse systems II. Phys. Review, vol. 26, 1926: 687-681.

[10] C. J. F. Böttcher: *Theory of Electric Polarization*. (Elsevier Science Publishing Co., Amsterdam: 1952).

[11] H. Looyenga: Dielectric constants of heterogeneous mixtures. Physica, vol. 31, 1965: 401-406.

[12] L. D. Landau and E. M. Lifschitz: *Course of Theoretical Physics*. (Pergamon Press: Oxford: 1960).

[13] K. Lichtenecker: Mischkörpertheori als Wahrscheinlichkeitsproblem. Phys. Zeitsch, vol. 30, 1929: 805-809.

[14] K. Lichtenecker and K. Rother: Die Herleitung des logarithmischen Mischungegesetzes aus allgemeinen Prinzipien der stationären Strömung. Phys. Zeitsch, vol. 32, 1938: 255-260.

[15] I. S. Zheludev: Piezoelectricity in textured media in *Solid State Physics, Advances in Research and Applications*, vol. 29, (H. Ehrenreich et al., Eds.), (Academic Press, New York: 1974).

[16] O. Wiener: Die Theori des Mischkörpers für das Feld der staionären Strömung. Abhdl. D. kgl. Akad. d. Wiss. Leipzig, vol. 32, 1912: 509-604.

[17] W. F. Brown: Dielectrics, in *Encyclopedia in Physics* V-XVII, (Springer-Verlag, Berlin: 1956).

[18] L. K. H. van Beek: Dielectric behaviour of heterogeneous systems, in *Progess in Dielectrics*, vol. 7, (J. B. Birks, Ed.), (CRC Press Cleveland: 1987).

[19] W. R. Tinga, W. A. G. Voss and D. F. Blossey: Generalized approach to multiphase dielectric mixture theory. J. Appl. Phys. vol. 44, 1973: 3897-3902.

[20] L. Lorenz: Über die Refractionsconstante. Ann. Phys. (Leipzig), vol. 11, 1880: 70-103.

[21] R. E. Meredith and C. W. Tobias: *Advances Electrical and Electronic Engineering*, vol. II, (Wiley Interscience: New York: 1962).

[22] R. W. Sillars: Properties of a dielectric containing semi-conducting particles of various shapes. J. Inst. Elect. Engrs. vol. 80, 1937: 378-392.

[23] D. Polder and J. H. Van Santen: The effective permeability of mixtures of solids. Physica, vol. 12, 1946: 257-271.

[24] L. Lewin: The electrical constant of a material loaded with spherical particles. J. IEEE part III, vol. 94, 1947: 65-68.

[25] B. V. Hamon: Maxwell-Wagner loss and absorption currents in dielectrics. Aust. J. Phys. vol. 6, 1953: 305-315.

[26] C. Boned and J. Peyrelasse: Some comments on the complex permittivity of ellipsoid dispersed in continuum media. J. Phys. D, vol. 16, 1983: 1777-1786.

[27] A. H. Sihvola and J. A. Kong: Effective permittivity of dielectric mixtures. IEEE Trans. Geoscience Remote Sensing, vol. 26(4), 1988: 420-429.

[28] P. S. Neelakantaswamy: Estimation of permittivity of a compact crystal by dielectric measurements on its powder: A stochastic mixture model for the powder dielectric. J. Phys. D, vol. 17, 1983: 1755-1799.

[29] P. S. Neelakantaswamy, K. Aspar and R. Rajaratnam: A dielectric model of the human blood. Biomed. Technik, vol. 28, 1983: 18-22.

[30] S. R. Wallin: Dielectric Properties of Heterogeneous Media. Ph.D. Thesis, University of Wyoming, 1985.

[31] J. A. Reynolds and J. M. Hough: Formulae for dielectric constant of mixtures. Proc. Phys. Soc. London, vol. LXX, 1957: 769-775.

[32] S. S. Dukhin and V. N. Shilov: *Dielectric Phenomena and the Double-Layer in Disperse Systems and Polyelectrolytes.* (John Wiley and Sons, New York: 1974).

[33] S. Kisdnasamy and P. S. Neelakantaswamy: Complex permittivity of a dielectric mixture, Modified Fricke's formula based on logarithmic law of mixing. Electron Letts. vol. 20, 1984: 291-293.

[34] P. S. Neelakantaswamy, R. I. Turkman and T. K. Sarkar: Complex permittivity of a dielectric mixture: Corrected version of Lichtenecker's logarithmic law of mixing. Electron Letts., vol. 21, 1985: 270-271.

[35] P. S. Neelakanta: Complex permittivity of a conductor-loaded dielectric. J. Phys.: Condens. Matter, vol. 2, 1990: 4935-4947.

[36] L. S. Taylor: Dielectric properties of mixtures. IEEE Trans. Antennas Propagat., vol. AP-13, 1965: 943-947.

[37] D. S. MacLachlan, A. Priou, I. Chenerie, E. Isaac and F. Henry: Modeling the permittivity of composite materials with a general effective medium equation. J. Electromag. Waves and Applns., vol. 6(9), 1992: 1099-1131.

[38] J. Jortner, I. Webman and M. H. Cohen: Theory of optical and microwave properties of microscopically inhomogeneous materials. Phys. Rev. B., vol. 15(12), 1977: 5712-5713.

[39] P. S. Neelakanta: Permittivity of dielectric-conductor mixture: Application of logarithmic law of mixing to electric susceptibility. Electron. Letts., vol. 25(12), 1989: 800-802.

[40] K. Subramaniam, P. S. Neelakanta and V. Ungvichian: Permittivity of orderly-textured mixture dielectrics. Electron. Letts., vol. 27(17), 1991: 1534-1535.

[41] J. Lam: Magnetic permeability of a single cube lattice of conducting magnetic spheres. J. Appl. Phys., vol. 60(12), 1986: 4230-4235.

[42] J. C. Park: Stochastical and Neuromimetic Aspects of Modeling Electromagnetic Composites. Ph.D. Dissertation, Department of Electrical Engineering, Florida Atlantic University, Boca Raton, FL. April 1994.

[43] P. M. Morse and H. Feshbach: *Methods of Theoretical Physics*. (McGraw-Hill Book Co., New York: 1953), 1285-1294.

[44] Z. Hashin and S. Shtrikman: A variational approach to the theory of the effective magnetic permeability of multiphase materials. J. Appl. Phys., vol. 33(10), 1962: 3125-3131.

[45] R. D. Miller and T. B. Jones: On the effective dielectric constant of columns or layers of dielectric spheres. J. Phys. D: Appl. Phys., vol. 21, 1988: 527-532.

[46] B. Bianco, G. P. Drago, M. Marchesi, C. Martini, G. S. Mela and S. Ridella: Measurements of complex dielectric constant of human sera and erythrocytes. IEEE Trans. Instrum. Meas., vol. IM-28, 1979: 290-295.

[47] L. Tsang and J. A. Kong: Scattering of EM waves from random media with strong permittivity fluctuations. Radio Sciences, vol. 16(3), 1981: 303-335.

[48] J. A. Kong: *Electromagnetic Waves Theory* (Wiley Interscience, New York: 1986).

[49] L. Tsang, J. A. Kong, and R T. Shin: *Theory of Microwave Remote Sensing*. (Wiley Interscience, New York: 1986).

[50] G. H. Wannier: *Statistical Physics* (Dover Publications, Inc., New York: 1966).

[51] R. Coelho: *Physics of Dielectrics for the Engineer.* (Elsevier Scientific Publishing Co., Amsterdam: 1979).

General Reading

[51] G. P. de Loor: Dielectric Properties of Heterogeneous Mixtures. Ph.D.Thesis, University of Leiden, 1956.

[52] B. Tareev: *Physics of Dielectric Materials.* (Mir Publishers, Moscow: 1979).

[53] A. von Hippel (Ed.): *Dielectric Materials and Applications*. (MIT Press, Cambridge, MA: 1954).

Defining Terms

Composite dielectrics: A combination of two or more lossy or lossless dielectrics or a combination of a dielectric host plus conducting/semiconducting inclusions.

Effective permittivity: Dielectric permittivtiy of a heterogeneous dielectric composite medium.

Form factor: This refers to the *shape factor* depicting a numerical value attributable to the regular and distinct shape of a particle.

Host-inclusion systems: Mixtures embodied by a host material acting as a receptacle for the dispersion of particulate inclusions.

Logarithmic law of mixing: As postulated by Lichtenecker and Rother, refers to the effective property of a stochastic mixture medium being decided by the individual properties of the constituents and their volume fractions through a logarithmic relation.

Order function: A factor which denotes the extent or state of orderliness of orientation/polarization/alignment of the particulates in a stochastic mixture medium.

Orderly-textured medium: A composite medium in which the inclusion(s) are physically arranged in an orderly fashion.

Particulate interactions: Force of interaction between the particles in a dispersed system.

APPENDIX 4A

Langevin's Theory of Dipole Orientations

In the presence of an applied electric field *(E)* the shaped particulate inclusions in a dielectric mixture are subjected to a force of alignment (similar to dielectric polarization). In this configuration, the statistical dispositions of the shaped inclusions can be specified by a mean value say, $< \alpha_E > = \alpha_p < \cos \beta >$ where β is the angle between the electric field *(E)* and the particle; and α_p refers to the positional attribute of the particle in the state of random dispersion. The statistical ensemble average of α_p can be specified explicitly with the following assumptions duly taken into consideration:

(i) Each particle considered is discretely positioned and the collection of them in the ensemble is regarded as isotropic.

(ii) The ergodicity hypothesis specifying the ensemble average taken at an instant being the same as the time average taken on any inclusion over a time interval, holds good.

The foregoing assumptions have the basis analogous to the statistics of molecular dipoles making an angle between β and $(\beta + d\beta)$ with the applied electric field. Hence the corresponding ensemble average $< \cos \beta >$ is given by the function $L(w) = \coth(w) - 1/w$, referred to as the *Langevin function* [50], where w is an implicit disorder function at an equilibrium condition which is being overcome by the applied electric field force in orienting the included particles in an aligned fashion. This factor w corresponds to the Boltzmann temperature which sets the disorder of the molecules being overcome by the polarizing action of any applied field. The statistical concepts of molecular dipole orientation due to Langevin have been effectively applied in elucidating the permittivity characteristics of condensed matter as described in [51]. The saturation of molecular/dielectric polarizability with the increase in the disorder function is depicted by the nonlinear saturation function $L(w)$ *versus* w. Further, on the basis of Maxwell-Boltzmann statistics applied to dipole polarization, the author [35] established the upper and lower bounds of the order function concerning a dielectric mixture with conducting inclusions in terms of Langevin's function.

CHAPTER 5

Complex Permittivity of Dielectric Composite Materials

5.1 Introduction

As discussed in Chapter 4, dielectric mixtures constitute a class of composite materials. Specific to a two-phase dielectric mixture constituted by lossless and/or lossy dielectric host-inclusion system, the various models and analytical considerations projected in Chapter 4 have led to the emergence of a collection of formulations to predict the effective (complex) permittivity of such materials.

Table 5.1 lists the various formulations available in the literature concerning heterogeneous composite dielectric materials. The expressions presented refer to two-phase host-inclusion systems. For multiphase systems, relevant formulas are presented in the next chapter (Chapter 6) and for the class of materials which are distinctly identifiable and constituted by a dielectric host (lossless or lossy) forming a receptacle for the dispersion of discrete conducting (or semiconducting) particles, the pertinent formulas are presented in Chapter 7. (Formulas specific to ferroelectric and ferrielectric composites are not, however, furnished here. They are presented in Chapter 12 separately.)

5.2 Descriptions of the Symbols and Notations

$\varepsilon_1 = \varepsilon_1' - j\,\varepsilon_1''$	(Complex) dielectric permittivity of the host medium*
$\varepsilon_2 = \varepsilon_2' - j\,\varepsilon_2''$	(Complex) dielectric permittivity of the inclusions*
ϑ_2, θ	Volume fraction of the dispersed component*
$(1 - \vartheta_2), (1 - \theta)$	Volume fraction of the host receptacle*
A_a	Depolarizing factor
$\overline{\varepsilon_1}$	Mean permittivity of mixture around a particulate inclusion
ε_{eff}	Effective (complex) permittivity of the mixture
σ_1	Conductivity of the host medium*
σ_2	Conductivity of the inclusions*
ω	$2\pi \times$ frequency

* Two-phase composite dielectric mixture.

ε_∞	High-frequency value of permittivity as controlled by Maxwell-Wagner effect (also known as optical limit of the permittivity equal to N^2, N being the refractive index of the medium)
ε_s	Static permittivity (low-frequency/quasistatic limit)
a/b	Semiaxial aspect ratio of ellipsoidal particles
m	Sillars' shape parameter
e	Eccentricity of ellipsoidal/spheroidal particles
x	Form factor (shape factor)
n	Fractional number of particles polarized along E-field
ε_i	(Complex) dielectric permittivity of ith component in the composite dielectric mixture
σ_i	Conductivity of ith component in the composite dielectric mixture
$\varepsilon_r = \varepsilon_r' - j\,\varepsilon_r''$	Relative (complex) permittivity
$\hat{\sigma} = \left.\begin{matrix} \sigma + j\omega\varepsilon \\ \sigma' + j\sigma'' \end{matrix}\right\}$	Complex conductivity
Im	Imaginary part of ...
Re	Real part of ...

It should be noted that the formulations presented in Table 5.1 as mentioned earlier are largely refer to two-phase mixtures. However, some of them could be extended to multicomponent mixtures as well. In such cases, the component is indicated with i^{th} subscript as appropriate.

The expressions of Table 5.1 are taken from various sources in the literature and relevant reference(s) of the sources are duly indicated.

Table 5.1 Effective Permittivity Parameter of Composite Dielectrics: Summary of Formulations

Type of Composite Dielectric	Particulate Shape of Inclusions	Effective Permittivity (ε_{eff})
Mixture of nonpolar dielectrics	Spherical particles	$\frac{\varepsilon_{eff}-1}{\varepsilon_{eff}+2} = \sum_{i=1}^{m}\theta_i\left(\frac{\varepsilon_i-1}{\varepsilon_i+2}\right)$ • m: Number of inclusions • $\sum_{i=1}^{m}\theta_i = 1$ • Clausius-Mossotti-Lorentz-Lorenz theory [1-6] • Interparticulate interactions neglected
	Spherical particles	$\left(\varepsilon_{eff}\right)^{1/3} = \sum_{i=1}^{m}\theta_i\left(\varepsilon_i\right)^{1/3}, \quad \sum_{i=1}^{m}\theta_i = 1$ • Landau and Lifshitz [7] • Interparticulate interactions neglected
	Spherical particles	$\left(\varepsilon_{eff}\right)^{1/2} = \sum_{i=1}^{m}\theta_i\left(\varepsilon_i\right)^{1/2}, \quad \sum_{i=1}^{m}\theta_i = 1$ • Beer's formula as stated in [8] • Interparticulate interactions neglected
	Spherical particles	$\frac{\varepsilon_{eff}-1}{\varepsilon_{eff}+x} = \sum_{i=1}^{m}\theta_i\left(\frac{\varepsilon_i-1}{\varepsilon_i+x}\right)$ • Wiener's formula [9] • x = Mixture characterizing parameter ($\neq 2$) • Interparticulate interactions neglected
Two-phase dielectric matrix mixture with sparcely spaced spherical inclusions	Spherical particles	$\varepsilon_{eff} = \left[\frac{2\varepsilon_1+\varepsilon_2+2\vartheta(\varepsilon_2-\varepsilon_1)}{2\varepsilon_1+\varepsilon_2-\vartheta(\varepsilon_2-\varepsilon_1)}\right](\varepsilon_1)$ • Maxwell's formula [10]; also based on solution to Wagner's theory • Diameters of the spheres << spacing between the spheres • ϑ = Volume fraction of the inclusions • Matrix-mixture refers to the host medium being a continuous phase and the inclusions are discrete, isolated particles

(continued...)

Type of Composite Dielectric	Particulate Shape of Inclusions	Effective Permittivity (ε_{eff})
	Spherical particles	$\left(\varepsilon_1 - \dfrac{3\vartheta\varepsilon_1}{\vartheta + \dfrac{2\varepsilon_1+\varepsilon_2}{\varepsilon_2-\varepsilon_1} - 0.523\dfrac{\varepsilon_1-\varepsilon_2}{\frac{4}{3}\varepsilon_1+\varepsilon_2}\vartheta^{10/3} + \ldots}\right)$ • Extension of Maxwell's formula • ϑ = Volume fraction of the inclusions • Diameter of the spheres need not be relatively small in comparison with the sphere spacings
Two-phase dielectric mixture with constituents having large differences in the permittivities	Spherical particles	$\varepsilon_{eff} = B + \sqrt{B^2 + \dfrac{\varepsilon_1\varepsilon_2}{2}}$ $B = \dfrac{(3\vartheta_1 - 1)\varepsilon_1 + (3\vartheta_2 - 1)\varepsilon_2}{4}$ • $(\vartheta_1 + \vartheta_2) = 1$ • Odelevski's formula [11]
	Spherical particles	$\varepsilon_{eff} = \varepsilon_1 \dfrac{2\varepsilon_1 + \varepsilon_2 + 2\vartheta_2(\varepsilon_2 - \varepsilon_1)}{2\varepsilon_1 + \varepsilon_2 - \vartheta_2(\varepsilon_2 - \varepsilon_1)}, \quad \vartheta_2 < 0.20$ • Complete solution of Maxwell-Wagner's theory • Rayleigh [12]
	Spherical particles	$\varepsilon_{eff} = \dfrac{(1 + 2\vartheta_2)}{(1 - \vartheta_2)}, \quad \varepsilon_1 = 1$ (Free space) • Poisson, Lorenz [3]; see also [13]
	Spherical particles	$\varepsilon_{eff} = \varepsilon_1\left[1 + 3\vartheta_2\dfrac{(\varepsilon_2 - \varepsilon_1)}{(2\varepsilon_1 + \varepsilon_2)}\right], \quad \vartheta_2 < 0.05$ • Wagner [14]; Rayleigh[12]

(continued...)

Type of Composite Dielectric	Particulate Shape of Inclusions	Effective Permittivity (ε_{eff})
Dielectric host plus randomly dispersed dielectric inclusions	Spherical particles	$\dfrac{(\varepsilon_2 - \varepsilon_{eff})}{(\varepsilon_2 - \varepsilon_1)}\left(\dfrac{\varepsilon_1}{\varepsilon_{eff}}\right)^{1/3} = (1 - \vartheta_2)$ • Bruggeman [15], Hanai [16]
	Spherical particles	$\varepsilon_{eff} = \varepsilon_1 + 3\vartheta_2\,\varepsilon_{eff}\dfrac{(\varepsilon_2 - \varepsilon_1)}{(2\varepsilon_{eff} + \varepsilon_2)}$ • Böttcher [17]
	Spherical particles	$\dfrac{(\varepsilon_{eff} - \varepsilon_1)}{(4\varepsilon_{eff} - \varepsilon_1)} = \vartheta_2\dfrac{(\varepsilon_2 - \varepsilon_1)}{(2\varepsilon_{eff} + \varepsilon_2)}, \quad \vartheta_2 < 0.20$
	Spherical particles	$\dfrac{(\varepsilon_{eff}^{1/3} - \varepsilon_1^{1/3})}{(\varepsilon_2^{1/3} - \varepsilon_1^{1/3})} = \vartheta_2$ • Looyenga [18]
	Spheres and near-spherical particles	$\varepsilon_1^{\vartheta_1}\,\varepsilon_2^{\vartheta_2}$ (Logarithmic law), $\vartheta_1 + \vartheta_2 = 1$ • Lichtenecker [19], Lichtenecker and Rother [20], • Neelakantaswamy et al. [21-23] • Kisdnasamy and Neelakantaswamy [24]

(continued...)

Type of Composite Dielectric	Particulate Shape of Inclusions	Effective Permittivity (ε_{eff})
Dielectric host plus randomly dispersed conducting inclusions	Conducting spheres	$\varepsilon_{eff} = \varepsilon_1 \frac{(1 + 2\vartheta_2)}{(1 - \vartheta_2)}$, $\varepsilon_2 >> \varepsilon_1$, $\overline{\varepsilon_1} = \varepsilon_1$, $\vartheta_2 < 0.25$ • Corkum [25]
	Conducting spheres	$\varepsilon_{eff} = \varepsilon_1/(1 - 3\vartheta_2)$, $\varepsilon_2 >> \varepsilon_1$, $\overline{\varepsilon_1} = \varepsilon_{eff}$
Dielectric host plus anisotropically or isotropically dispersed shaped dielectric inclusions	Spheroids ǁ field	$\varepsilon_{eff} = \varepsilon_1 \frac{\varepsilon_1 + [A_a(1 - \vartheta_2) + \vartheta_2](\varepsilon_2 - \varepsilon_1)}{\varepsilon_1 + A_a(1 - \vartheta_2)(\varepsilon_2 - \varepsilon_1)}$, $\vartheta_2 < 0.20$ • Complete solution of Sillars' theory [26]
	Spheroids ǁ field	$\varepsilon_{eff} = \varepsilon_1 \left[1 + \vartheta_2 \frac{(\varepsilon_2 - \varepsilon_1)}{\varepsilon_1 + A_a(\varepsilon_2 - \varepsilon_1)} \right]$, $\vartheta_2 < 0.10$ • Sillars [27]
	Ellipsoids random orientation	$\varepsilon_{eff} = \varepsilon_1 \{1 + \frac{1}{3} \vartheta_2 \sum_i \overline{\varepsilon_1} (\varepsilon_2 - \varepsilon_1)/[\overline{\varepsilon_1} + A_a(\varepsilon_2 - \overline{\varepsilon_1}]\}$ • Polder, and Van Santen [28], de Loor [29]

(continued...)

Type of Composite Dielectric	Particulate Shape of Inclusions	Effective Permittivity (ε_{eff})
	Ellipsoids random orientation	$\varepsilon_{eff} = \varepsilon_1 \{1 + \frac{1}{3} \vartheta_2 \sum_i (\varepsilon_2 - \varepsilon_1)/[\varepsilon_1 + A_i (\varepsilon_2 - \varepsilon_1)]\}$, $\vartheta_2 < 0.10$ • Frike [45,46]
	Prolate ellipsoids ǁ field (rods, cylinders)	$\varepsilon_{eff} = \varepsilon_1 + \frac{1}{3} \vartheta_2 \frac{(\varepsilon_2 - \varepsilon_1)(5\varepsilon_1 + \varepsilon_2)}{(\varepsilon_1 + \varepsilon_2)}$, $\overline{\varepsilon_1} = \varepsilon_1$ • van Beek [48]
	Prolate ellipsoids ǁ field (rods, cylinders)	$\varepsilon_{eff} = \varepsilon_1 + \frac{1}{3} \vartheta_2 \frac{(\varepsilon_2 - \varepsilon_1)(5\varepsilon_{eff} + \varepsilon_2)}{(\varepsilon_{eff} + \varepsilon_2)}$, $\overline{\varepsilon_1} = \varepsilon_1$ • van Beek [48]
	Prolate ellipsoids ǁ field (rods, cylinders)	$\left[\frac{\varepsilon_2 - \varepsilon_{eff}}{\varepsilon_2 - \varepsilon_1}\right]\left[\frac{5\varepsilon_1 + \varepsilon_2}{5\varepsilon_{eff} + \varepsilon_2}\right]^{5/2} = (1 - \vartheta_2)$ • Nielsen [30]; van Beek [48]
Dielectric host plus anisotropic or isotropic dispersion of shaped dielectric inclusions	Cylinders ⊥ field (regular array)	$\varepsilon_{eff} = \varepsilon_1 \frac{\varepsilon_1 + \varepsilon_2 + \vartheta_2 (\varepsilon_2 - \varepsilon_1)}{\varepsilon_1 + \varepsilon_2 - \vartheta_2 (\varepsilon_2 - \varepsilon_1)}$, $\vartheta_2 < 0.20$ • Rayleigh [12]; van Beek [48]
	Oblate spheroids, lamellae, disks	$\varepsilon_{eff} = \varepsilon_1 + \frac{1}{3} \vartheta_2 \frac{(\varepsilon_1 + 2\varepsilon_2)(\varepsilon_2 - \varepsilon_1)}{\varepsilon_2}$, $\vartheta_2 < 0.10$, $\overline{\varepsilon_1} = \varepsilon_1$ • van Beek [48]

(continued...)

Type of Composite Dielectric	Particulate Shape of Inclusions	Effective permittivity (ε_{eff})
	Oblate spheroids, lamellae, disks	$\varepsilon_{eff} = \varepsilon_2 \dfrac{3\varepsilon_1 + 2\vartheta_2(\varepsilon_2 - \varepsilon_1)}{3\varepsilon_2 - \vartheta_2(\varepsilon_2 - \varepsilon_1)}, \qquad \overline{\varepsilon_1} = \varepsilon$ • Bruggeman [15]
	Ellipsoidal particles with semiaxial aspect ratio, a/b	$\varepsilon_{eff} = \dfrac{\varepsilon_{1s}\varepsilon_{2s}(1 + x\vartheta_2) + \varepsilon_{1s}^2 x(1-\vartheta_2)}{\varepsilon_{2s}(1-\vartheta_2) + \varepsilon_{1s}(x + \vartheta_2)}$ • $x = \dfrac{M\varepsilon_{2s}^{\vartheta_2}\varepsilon_{1s}^{(1-\vartheta_2)}(A) - \varepsilon_{1s}\varepsilon_{2s}}{\varepsilon_{1s}(B) - \varepsilon_{2s}^{\vartheta_2}\varepsilon_{1s}^{(1-\vartheta_2)}}$ • $A = [\varepsilon_{2s}(1-\vartheta_2) + \varepsilon_{1s}\vartheta_2]$ • $B = [\varepsilon_{2s}\vartheta_2 + \varepsilon_{1s}(1-\vartheta_2)]$ • Kisdnasamy and Neelakantaswamy [24], Frike [45,46] • $M = \left(\dfrac{2}{m-1}\right), \quad \varepsilon_{2s} > \varepsilon_{1s}$ $= \left(\dfrac{m-1}{2}\right), \quad \varepsilon_{2s} < \varepsilon_{1s}$ • Sillars' formula [26,27] • $m = e^2\left[1 - (1-e^2)^{1/2}\dfrac{\sin^{-1}(e)}{e}\right]^{-1}$ • $e = \left(1 - \dfrac{b}{a}\right), \quad a > b$ $= \left(\dfrac{a}{b} - 1\right), \quad b > a$

(continued...)

Type of Composite Dielectric	Particulate Shape of Inclusions	Effective Permittivity (ε_{eff})
	Ellipsoidal particles with semiaxial aspect ratio, a/b	$\varepsilon_{eff} = \begin{cases} X(\vartheta_2)/2 & \varepsilon_2 > \varepsilon_1 \\ Y(\vartheta_2)/2 & \varepsilon_2 < \varepsilon_1 \end{cases} \quad 0 \le \vartheta_2 \le \phi_1$

$$\varepsilon_{eff} = \begin{cases} \frac{1}{2}\left[\frac{A(\phi_1)}{2C(\phi_1)} + \frac{B(\phi_2)}{2C(\phi_2)}\right] C(\vartheta_2)\, Z(\vartheta_2) & \varepsilon_2 > \varepsilon_1 \\ \frac{1}{2}\left[\frac{B(\phi_1)}{2C(\phi_1)} + \frac{A(\phi_2)}{2C(\phi_2)}\right] C(\vartheta_2)\, Z(\vartheta_2) & \varepsilon_2 < \varepsilon_1 \end{cases} \quad \phi_1 \le \vartheta_2 \le \phi_2$$

$$\varepsilon_{eff} = \begin{cases} Y(\vartheta_2)/2 & \varepsilon_2 > \varepsilon_1 \\ X(\vartheta_2)/2 & \varepsilon_2 < \varepsilon_1 \end{cases} \quad \phi_2 \le \vartheta_2 \le 1$$

- $X(\vartheta_2) = Z(\vartheta_2) + 1/\varepsilon_L(\vartheta_2)$
- $Y(\vartheta_2) = Z(\vartheta_2) + \varepsilon_U(\vartheta_2)$
- $Z(\vartheta_2) = \varepsilon_U^{\,n}(\vartheta_2)/\varepsilon_L^{\,n-1}(\vartheta_2)$
- $A(\vartheta_2) = 1 + 1/(\varepsilon_U^{\,n}\,\varepsilon_L^{\,n})$
- $B(\vartheta_2) = 1 + 1/(\varepsilon_U^{\,n-1}\,\varepsilon_L^{\,n-1})$ and
- $C(\vartheta_2) = \sqrt{\varepsilon_L(\vartheta_2)/\varepsilon_U(\vartheta_2)}\,(\varepsilon_2^{\vartheta_2}\varepsilon_1^{1-\vartheta_2})$
- n: Fraction of particles polarized along E-field $0 < n < 1$
- $\varepsilon_U, \varepsilon_L$: Weiner's upper and lower limits
- $n = \left(\frac{5-m}{4}\right), \quad \varepsilon_2 > \varepsilon_1$

  $= \left(\frac{m-1}{4}\right), \quad \varepsilon_2 < \varepsilon_1$
- $\phi_1 = \frac{1}{2} - \frac{1}{2}\sqrt{1-4t}$
- $t = \dfrac{(\varepsilon_1 + \varepsilon_2)}{2(\varepsilon_2 - \varepsilon_1)\ln\left(\dfrac{\varepsilon_2}{\varepsilon_1}\right)} - \dfrac{\varepsilon_1\varepsilon_2}{(\varepsilon_2 - \varepsilon_1)^2}$
- $\varepsilon_U = \vartheta_2\varepsilon_2 + (1 - \vartheta_2)\varepsilon_1$

(continued...)

Type of Composite Dielectric	Particulate Shape of Inclusions	Effective Permittivity (ε_{eff})
		$\varepsilon_L = \left[\frac{\vartheta_2}{\varepsilon_2} + \frac{1-\vartheta_2}{\varepsilon_1} \right]^{-1}$ • Neelakantaswamy et al. [23]
Dielectric host plus randomly dispersed conducting shaped inclusions	Conducting ellipsoids	$\varepsilon_{eff} = \frac{\varepsilon_1}{1 - \frac{1}{3}\vartheta_2 \sum_i (1/A_i)}$, $\varepsilon_2 >> \varepsilon_1$, $\overline{\varepsilon_1} = \varepsilon$ • Altschuller [31]; van Beek [48]
	Conducting ellipsoids	$\varepsilon_{eff} = \varepsilon_1 \left[1 + \frac{1}{3}\vartheta_2 \sum_i (1/A_i) \right]$, $\varepsilon_2 >> \varepsilon_1$, $\overline{\varepsilon_1} = \varepsilon_1$ • de Loor [29]; van Beek [48]
	Conducting ellipsoids (needles)	$\varepsilon_{eff} = \varepsilon_1 + \frac{1}{3}\vartheta_2 \varepsilon_2$, $\varepsilon_2 >> \varepsilon_1$, $\varepsilon_2 >> \overline{\varepsilon_1}$ • Polder and Van Santen [28]; van Beek [48]
	Conducting disks	$\varepsilon_{eff} = \varepsilon_1 + \frac{2}{3}\vartheta_2 \varepsilon_2$, $\varepsilon_2 >> \varepsilon_1$, $\varepsilon_2 >> \overline{\varepsilon_1}$ • Polder and Van Santen [28]; van Beek [48]

(continued...)

Type of Composite Dielectric	Particulate Shape of Inclusions	Effective Permittivity (ε_{eff})
Lossy spherical particles randomly dispersed in a lossy homogeneous host medium	Spherical particles	$\varepsilon_{eff} = \frac{\sigma''_{eff}}{\omega}$, $\sigma''_{eff} = Im(\hat{\sigma}_{eff})$ $\sigma_{eff} = \left[\frac{1 + 2\vartheta\chi_o}{1 - \vartheta\chi_o}\right]\sigma$ • Maxwell [10] and Maxwell-Garnett [5,6] • $\chi_o = \left(1 - \frac{\hat{\sigma}_1}{\hat{\sigma}_2}\right)\left(1 + 2\frac{\hat{\sigma}_1}{\hat{\sigma}_2}\right)^{-1}$, $\hat{\sigma}_1 = \sigma_1 + j\omega\varepsilon_1$, $\hat{\sigma}_2 = \sigma_2 + j\omega\varepsilon_2$, $\vartheta < 0.10$
	Spherical particles	$\varepsilon_{eff} = \frac{\sigma''_{eff}}{\omega}$, $\sigma''_{eff} = Im(\hat{\sigma}_{eff})$ $\left(\frac{\hat{\sigma}_{eff} - \hat{\sigma}_1}{3\hat{\sigma}_1}\right) \cong \vartheta\left(\frac{\hat{\sigma}_2 - \hat{\sigma}_1}{\hat{\sigma}_2 + 2\hat{\sigma}_1}\right)$, $\hat{\sigma}_1 = \sigma_1 + j\omega\varepsilon_1$, $\hat{\sigma}_2 = \sigma_2 + j\omega\varepsilon_2$, $\vartheta < 0.10$ • Rayleigh [12]
	Spherical particles	$\varepsilon_{eff} = \frac{\sigma''_{eff}}{\omega}$, $\sigma''_{eff} = Im(\hat{\sigma}_{eff})$ $\left(\frac{\hat{\sigma}_{eff}}{\sigma_1}\right)^{1/3}\left(\frac{\hat{\sigma}_1 - \hat{\sigma}_2}{\hat{\sigma}_{eff} - \hat{\sigma}_2}\right)(1 - \vartheta) = 1$ *and* $\hat{\sigma}_{eff} \approx \hat{\sigma}_1 (1 - \vartheta)^{3/2}$, $\lvert\hat{\sigma}_2\rvert << \lvert\hat{\sigma}_1\rvert$ $\varepsilon_{eff} \approx \varepsilon_1/(1 - \vartheta)^3$, $\lvert\hat{\sigma}_2\rvert << \lvert\hat{\sigma}_{eff}\rvert$ • Bruggeman [15], Looyenga [18]

(continued...)

Type of Composite Dielectric	Particulate Shape of Inclusions	Effective Permittivity (ε_{eff})
	Spherical particles	$\varepsilon_{eff} = \frac{\sigma''_{eff}}{\omega}, \ \sigma''_{eff} = Im(\sigma_{eff})$ $\vartheta = \frac{(\hat{\sigma}_{eff} - \hat{\sigma}_1)(2\hat{\sigma}_{eff} + \hat{\sigma}_2)}{3\hat{\sigma}_{eff}(\hat{\sigma}_2 - \hat{\sigma}_1)}$ • Looyenga [18], Bo ¨ttcher [17]
Lossy coarse spherical particles randomly dispersed in a lossy homogeneous host-medium	Spherical particles	$\varepsilon_{eff} = \frac{\sigma''_{eff}}{\omega}, \ \sigma''_{eff} = Im(\hat{\sigma}_{eff})$ $\hat{\sigma}_{eff} \approx \hat{\sigma}_1 (1 - \vartheta)^{m_A}, \ m_A = 1 \text{ to } 5$ • Archie [32]
Lossy spheroidal particles in a lossy dielectric host-medium	Spheroidal particles	$\varepsilon_{eff(S)} = \varepsilon_{2(S)}/(1 - \vartheta)^B$, (S: Static value) • $B = \frac{1}{3}\left[\frac{3A_c + 1}{A_c(1 - A_c)}\right]$ • Frame and Tedford [33] • $A_c = 1/3$ for spherical particles with aspect ratio, $a/b = 1$ • $A_c = \frac{1}{(1 - a_r^2)} - \frac{a_r \cos^{-1}(a_r)}{(1 - a_r^2)^{3/2}}$, for oblate spheroids with $a_r = a/b < 1$ $= \frac{a_r}{(a_r^2 - 1)^{3/2}} \ln(A) - \frac{1}{(a_r^2 - 1)}$, for prolate spheroids with $a_r > 1$ $A = [a_r + (a_r^2 - 1)]$ • Frame and Tedford [33]

(continued...)

Type of Composite Dielectric	Particulate Shape of Inclusions	Effective Permittivity (ε_{eff})
	Ellipsoidal particles	• $B = \frac{1}{3} \sum_{i=abc} A_i$, for Ellipsoidal inclusions with Semiaxial lengths a, b, and c, $(a > b > c)$ = Depolarization along the ith Axis $= \frac{abc}{2} \int_0^{\infty} \frac{ds}{(s+i)^2 [(s+a)^2 (s+b)^2 (s+c)^2]^{1/2}}$ • $A_a + A_b + A_c = 1$ • Frame and Tedford [33]
Conducting particulate dispersion in a dielectric host medium	Chain-like arrangement of discrete spherical or near-spherical particles	$\varepsilon_{eff} = \frac{\sigma_{eff}}{\omega}$ $\sigma_{eff} = C^2 \sigma_2 \vartheta(\vartheta)^{\vartheta - 2/3}$ • C^2 : Area occupied by chain ends For a given ϑ, $2C^3 - 2C^2 + \vartheta = 0$ • Scarisbrick [34]
	Chain-like arrangement of discrete shaped ellipsoidal particles	$\varepsilon_{eff} = \frac{\sigma_{eff}}{\omega}$ $\sigma_{eff} = C^2 \sigma_2 \vartheta(\vartheta)^{\rho\vartheta - 2/3}$ • ρ = Order function or shape factor $= \frac{(bc/a^2)^{2/3}}{1 + (b/a)^2 + (c/a)^2}$ • Kusy [35]

(continued...)

Type of Composite Dielectric	Particulate Shape of Inclusions	Effective Permittivity (ε_{eff})
Lossy particulate dispersion in a lossy host medium	Spherical particles	$\left(\frac{\sigma_2}{\sigma_{eff}}\right)^{1/3}\left(\frac{\sigma_1-\sigma_{eff}}{\sigma_1-\sigma_2}\right) = (1-\vartheta)$ $\varepsilon_{eff} = \varepsilon_1\frac{\sigma_{eff}}{\sigma_1} + \left[\varepsilon_2 - \varepsilon_1\frac{\sigma_2}{\sigma_1}\right]\frac{\partial\sigma_{eff}}{\partial\sigma_2}$ If $\sigma_1 = \sigma_2$, $\sigma_{eff} = \sigma_1 = \sigma_2$, and $\partial\sigma_{eff}/\partial\sigma_2 = (1-\vartheta)$ • Peyrelasse et al. [36]
	Spherical particles	$\varepsilon_{eff}\left[\frac{3}{(\sigma_{eff}-\sigma_1)} - \frac{1}{\sigma_{eff}}\right] = 3\left[\frac{(\varepsilon_2-\varepsilon_1)}{(\sigma_2-\sigma_1)} + \frac{\varepsilon_1}{(\sigma_{eff}-\sigma_1)}\right] - \frac{\varepsilon_2}{\sigma_2}$ • Hanai [16]
	Ellipsoidal particles	$\varepsilon_{eff} = \varepsilon_1 + (\varepsilon_2-\varepsilon_1)(1-\vartheta)f_{20} - (1-\vartheta)(L_o)$ $L_o = \lim_{\omega\to 0}\left[(\varepsilon_2'' - \varepsilon_1'')f_2'\right]$ • f_2 : Function of [complex permittivity $\varepsilon_1^*(\omega)$ for $\omega \to 0$, $\varepsilon_2^*(\omega)$, volume fraction ϑ; shape of the dispersed particles and spatial distribution of the particles] $= f_2' - jf_2''$ $\vartheta f_1 + (1-\vartheta)f_2 = 1$ $f_1 = \sum_{k=1}^{3}\cos^2\alpha_k/\{1 + A_k[(\varepsilon_1^*/\varepsilon_2^*) - 1]\}$ • α_k: Angles between the ellipsoids' axes and applied external field • A_k: Depolarization factors • Reynolds and Hough [37]; Peyrellasse et al. [36] • ϑ: Volume fraction of the inclusions * Asterik indicates complex value

(continued...)

Type of Composite Dielectric	Particulate Shape of Inclusions	Effective Permittivity (ε_{eff})
	Ellipsoidal particles	$\varepsilon_{eff}^* = \varepsilon_2^* - \sum_n \frac{F_n \varepsilon_2^*(\varepsilon_2^* - \varepsilon_1^*)}{\varepsilon_2^*(1 - q_n) + \varepsilon_1 q_n}$ At $\omega \to 0$, $(\sigma_{eff})_{dc} = \sigma_2 - \sum_n \frac{F_n \sigma_2(\sigma_2 - \sigma_1)}{\sigma_2(1 - q_n) + \sigma_1 q_n}$ $(\varepsilon_{eff})_{static} = \varepsilon_2 - \sum_n \frac{F_n \{\varepsilon_2 [C] - \varepsilon_1 \sigma_2^2\}}{[D]}$ $C = [\sigma_2^2 - q_n(\sigma_1 - \sigma_2)^2]$ $D = [\sigma_2(1 - q_n) + \sigma_1 q_n]^2$ $\left(1 - \frac{\varepsilon_{eff}^*}{\varepsilon_2^*}\right) = \sum_n \frac{F_n}{(q - q_n)}, \quad q = \frac{\varepsilon_2^*}{(\varepsilon_2^* - \varepsilon_1^*)}$ • q_n: Poles $(0 < q_n < 1)$ • F_n: Residues $(0 < F_n < 1)$ • $\sum_n^n F_n = \vartheta$ (Component 1 volume fraction) • Formulation due to Bergman [38], stated by Peyrellasse et al. [36]
	Arbitrary	$\frac{(\varepsilon_{eff} - \varepsilon_1)}{(\varepsilon_{eff} + K_1)} = \vartheta_2 \frac{(\varepsilon_2 - \varepsilon_1)}{(\varepsilon_2 + K_1)}$ • K_1 = Empirical parameter • Wiener [9]; van Beek [48]

(continued...)

Type of Composite Dielectric	Particulate Shape of Inclusions	Effective Permittivity (ε_{eff})
Dielectric host plus randomly dispersed dielectric inclusions of arbitrary shape	Arbitrary	$\varepsilon_{eff} = K_2 [(1 - \vartheta_2) \ln \varepsilon_1 + \vartheta_2 \ln \varepsilon_2]$ • K_2 = Constant, $\cong \frac{1}{3}$ • Lichtenecker [19], Lichtenecker and Rother [20], Reynolds [39]
	Arbitrary	$\varepsilon_{eff} = \varepsilon_1 + \vartheta_2 \frac{(\varepsilon_2 - \varepsilon_1)(1 - K_3)}{1 - \vartheta_2 K}$ • k_3 = Constant • Pierce [40]; van Beek [48]
	Spheres	$(1 - \vartheta_2)\frac{\varepsilon_1 - \varepsilon_{eff}}{\varepsilon_1 + 2\varepsilon_{eff}} + \vartheta_2 \frac{\varepsilon_2 - \varepsilon_{eff}}{\varepsilon_2 + 2\varepsilon_{eff}} = 0$ • Derived for conductivity of metallic mixtures • Landauer [41]; van Beek [48]
	Arbitrary	$\varepsilon_{eff} = \frac{2\varepsilon_I \varepsilon_{II}}{\varepsilon_I + \varepsilon_{II}}$ • $\varepsilon_I = \frac{2\varepsilon_1 + \varepsilon_2 + 2\vartheta_2(\varepsilon_2 - \varepsilon_1)}{2\varepsilon_1 + \varepsilon_2 - \vartheta_2(\varepsilon_2 - \varepsilon_1)}$ • $\varepsilon_{II} = \frac{2\varepsilon_2 + \varepsilon_1 + 2(1 - \vartheta_2)(\varepsilon_1 - \varepsilon_2)}{2\varepsilon_2 + \varepsilon_1 - (1 - \vartheta_2)(\varepsilon_1 - \varepsilon_2)}$ • $\vartheta_2 \approx \frac{1}{2}$ • Kamiyoshi [42], de Loor [29], van Beek [48]

(continued...)

Type of Composite Dielectric	Particulate Shape of Inclusions	Effective Permittivity (ε_{eff})
	Arbitrary	$log\,(1-\vartheta_2) = \frac{3\varepsilon_2}{(2.19\varepsilon_2 + 0.81)} log\left[\frac{\varepsilon_2 - \varepsilon_{eff}}{\varepsilon_2 - \varepsilon_1}\right] - \frac{(2.19\varepsilon_2 - 0.81)}{1.19\,(2.19\varepsilon_2 + 0.81)} log\left[\frac{(2.19\varepsilon_{eff} + 0.81)}{(2.19\varepsilon_1 + 0.81)}\right]$ • Derivation follows Bruggeman's formulation with $\varepsilon_2 \to 1$ • Kubo and Nakamura [43], van Beek [48]
Dielectric host plus dielectric dispersions	Spherical	$\Phi(\varepsilon_{eff}) = \sum_{i=1}^{m} \theta_i \Phi(\varepsilon_i)$ • $\sum_{i=1}^{m} \theta_i = 1$ • Φ: Generalized function • $\Phi = log(\varepsilon) \Rightarrow$ Lichtenecker's formula • $\Phi = (\varepsilon)^{1/3} \Rightarrow$ Landau-Lifshitz formula • $\Phi = (\varepsilon)^{1/2} \Rightarrow$ Beer's formula • $\Phi = \left(\frac{\varepsilon - 1}{\varepsilon + 2}\right) \Rightarrow$ Lorentz-Lorenz theory • $\Phi = \left(\frac{\varepsilon - 1}{\varepsilon + x}\right) \Rightarrow$ Wiener's formula • $\theta_i = M_i \frac{D_m}{D_i}$ $M_i = \theta_i \frac{D_i}{D_m}$: Mass fraction of the ith constituent

(continued...)

Type of Composite Dielectric	Particulate Shape of Inclusions	Effective Permittivity (ε_{eff})
		D_i: Density of the ith constituent D_m: Density of the mixture $= \sum_{i=1}^{m} \theta_i D_i$ • $\sum_{i=1}^{m} M_i = 1$
Random dispersion of lossy/conducting particles in a homogeneous dielectric host material	Lossy/conducting spherical particles	$\frac{\varepsilon_{eff} - \varepsilon_1}{3\varepsilon_{eff}} = \vartheta \frac{\varepsilon_2 - \varepsilon_1}{\varepsilon_2 + 2\varepsilon_{eff}}$, ϑ: Volume fraction of the inclusions $\varepsilon_{eff} = \left(\varepsilon'_{eff} - j\varepsilon''_{eff}\right)$ $\varepsilon_1 = \left(\varepsilon'_1 - j\varepsilon''_2\right)$ $\varepsilon_2 = \left(\varepsilon'_2 - j\varepsilon''_2\right)$ $X = \frac{\varepsilon'_2}{\varepsilon_1}$; $Y = \frac{\varepsilon''_2}{\varepsilon_1}$ $R = \frac{\varepsilon'_{eff}}{\varepsilon_1}$; $S = \frac{\varepsilon''_{eff}}{\varepsilon_1}$ For $X = 1$, • $32R^4 - 32R^3 + 2[(1 - 3\vartheta)^2 Y^2 - 3]R^2 - [(1 - 3\vartheta)^2 Y^2 - 7]R - (1 + 3\vartheta)^2 Y^2 - 1 = 0$ • $S = \frac{[1 - (1 - 3\vartheta)R]Y}{(4R - 1)}$ • Approximation for: $\vartheta \leq \frac{1}{3}$, $Y >> 1$ • $R = \frac{1}{[1 - 3\vartheta + 2/Y^{1/2}]}$ • Böttcher [17]

(continued...)

Type of Composite Dielectric	Particulate Shape of Inclusions	Effective Permittivity (ε_{eff})
	Lossy/conducting spherical particles	$\frac{\varepsilon_{eff}}{\varepsilon_1}(1-\vartheta)^3 = \left(\frac{\varepsilon_2 - \varepsilon_{eff}}{\varepsilon_2 - \varepsilon_1}\right)^3$ • For $X = 1, \quad R = 1 + r, \quad C = 1 - \delta$ • $64\,r^9 + 48\delta^3 Y^4 r^6 - 12\,\delta^6 Y^6 r^5 - 36\,\delta^6 Y^6 r^4 - 3\,\delta^6 Y^6(9 + 5Y^2)r^3 - 12\,\delta^9 Y^{10} r^2 - \delta^9 Y^{10}(9 + \delta^3 Y^2)r - \delta^9 Y^{12}(\delta^3 - 1) = 0$ • Approximation for $y >> 1, 0 \le \vartheta \le 0.9$ • $R = 1/(\ 1 - \vartheta)^3$ • $S = 3(1 - \delta^3)/\delta^3 Y$ • Bruggeman [15]; Rothwell [47]
	Lossy/conducting spherical particles	$\frac{\varepsilon_{eff}}{\varepsilon_1} = \frac{\varepsilon'_{eff}}{\varepsilon_1} = 1 + \frac{3\vartheta}{(1 - \vartheta - 1.65\vartheta^{10/3})}$ $\sigma_2 = \infty$ $S = 0$ • Rayleigh octopole theory [12]; Rothwell [47]
	Lossy/conducting spherical particles	$\frac{(\varepsilon_{eff} - \varepsilon_1)}{(\varepsilon_{eff} + 2\varepsilon_1)} = \vartheta\frac{(\varepsilon_2 - \varepsilon_1)}{(\varepsilon_2 + 2\varepsilon_1)}$ $R = \frac{(1 + 2\vartheta)Y - [(1 - \vartheta)X + \vartheta + 2]S}{(1 - \vartheta)\,Y}$ $S = \frac{9\vartheta Y}{[(1 - \vartheta)X + (2 + \vartheta)]^2} + (1 - \vartheta)^2 Y^2)$ • Lorenz [3]; Kharadly and Jackson dipole theory [13]; Rothwell [47]

(continued...)

Type of Composite Dielectric	Particulate Shape of Inclusions	Effective Permittivity (ε_{eff})
	Lossy/conducting spherical particles	$\frac{(\varepsilon_{eff}-\varepsilon_1)}{(3\varepsilon_1)} = \vartheta\frac{(\varepsilon_2-\varepsilon_1)}{(\varepsilon_2+2\varepsilon_1)}$ $R = (1+3C) - \frac{(X+2)S}{Y}$ $S = 9\vartheta Y/[(X+2)^2+Y^2]$ • Wagner theory [44]; Rothwell [47]
Lossy dielectric plus lossy dielectric	Coarse compacted particles as in water-bearing rocks	$\frac{\varepsilon_{eff}-1}{\varepsilon_{eff}+2} = \frac{\varepsilon_1-1}{\varepsilon_2+2}\theta_1 + \frac{\varepsilon_2-1}{\varepsilon_2+1}\theta_2 + \ldots$ $(\theta_1+\theta_2) = 1$ • Based on Lorentz-Lorenz formula
Two component porous dielectric mixture	Series connected layers of dielectric with air voids example: cellulose, paper	$\varepsilon_{eff} = \frac{1}{1-\frac{D_p}{D_c}\left(1-\frac{1}{\varepsilon_c}\right)}$: Dry nonimpregnated mixture with air voids $= \frac{\varepsilon_{ic}}{1-\frac{D_p}{D_c}\left(1-\frac{\varepsilon_{ic}}{\varepsilon_c}\right)}$: Mixture impregnated with a liquid through air voids $= \frac{\varepsilon_c\varepsilon_{ic}}{\varepsilon_c\theta_p(1-k)+\varepsilon_c\varepsilon_{ic}\theta_p k+\varepsilon_{ic}(1-\theta_p)}$: Mixture impregnated with a solid material • ε_c : Permittivity of the core material (e.g. cellulose) • ε_{eff} : Permittivity of the mixture (e.g. paper) • $\varepsilon_{i,c}$: Permittivity of the impregnating material • D_p : Volume mass of the mixture (e.g. paper) • D_c : Density of the core material (e.g. cellulose) • k : Volume-shrinkage factor • θ_p : Relative volume of air voids = $(1-D_p/D_c)$ • Renne's formula as given in [8, Equation 2.80]

(continued...)

5.3 Multiphase Dielectric Composite Materials

As indicated earlier composite dielectric materials can also be constituted of multiple constituents instead of a simple, two-phase medium. Most generally, a multiphase dielectric composite refers to multiple inclusions forming a system of confocal ellipsoidal shells dispersed randomly in a host dielectric.

Studies devoted to find the (effective) complex permittivity of a multiphase dielectric mixture are comprised of two types. In the first case, the interaction between the neighboring inclusions are ignored. It refers to the low volume contents of the included materials. In the second case, the first-order inclusion interaction effects are implicitly contained in the relevant formulations. The expressions available in the literature to deduce the effective dielectric parameters of a multiphase dielectric mixture are presented in Table 5.1. These formulations are presented to supplement and complement the host of expressions listed in Table 5.1. Again, the listed formulations are restricted to the dielectric host plus multiphase dielectric inclusions only. For the inclusions which are discrete conductors, relevant results are given in the next chapter (Chapter 6).

5.4 Notations

ϑ_i:	Volume fraction of the inclusions
ε_i:	Dielectric constant of inclusions
ε_h:	Dielectric constant of the host medium or the receptacle
ε^*:	Effective dielectric constant of the medium immediately surrounding an included particle
ε:	Effective dielectric constant of the multiphase mixture
$\varepsilon_1, \varepsilon_2, \varepsilon_3 \ldots$:	Dielectric constants of confocal ellipsoidal shells as illustrated in Figure 6.1
J:	Index to denote the jth phase
a, b, c:	Semiaxial lengths of an ellipsoid (a > b > c)
S'_{e^*}:	Shape function
$n(u)$:	Depolarization coefficient
$U = U_1, U_2, U_3$:	Specific parametric quantity defining the ellipsoidal boundaries as governed by:

$$\frac{x^2}{\left(u + a^2\right)} + \frac{y^2}{\left(u + b^2\right)} + \frac{z^2}{\left(u + c^2\right)} = 1, \quad a > b > c$$

Table 5.2 Summary of Salient Macroscopic Multiphase Dielectric Mixture Formulations

Inclusion Type	Mixture Formulations	Remarks and References
Ellipsoids two phase	$(\varepsilon - \varepsilon_h) = \frac{\vartheta_i}{1 - \vartheta_i}(\varepsilon_i - \varepsilon)\, S'_{\varepsilon *}$ $S'_{\varepsilon *} = \sum_{j=1}^{3} \{1 + n_j[(\varepsilon_2/\varepsilon^*) - 1]\}^{-1}$ $\varepsilon^* = \varepsilon_1$ *and* $n_j = n_j(u) = \frac{1}{2} V(u_j)\, F(u_j)$ $V(u_j) = \frac{4}{3}\pi\left\{(u_j + a_j^2)(u_j + b_j^2)(u_j + c_j^2)\right\})^{1/2}$ $F(u_j) = \int_{u_j}^{\infty} \frac{ds}{(s + a_j^2)V(s)}$	All interaction effects assumed negligible [5, 6, 64]
Arbitrary two phase	$\frac{(\varepsilon - \varepsilon_h)}{(\varepsilon + U)} = \frac{\vartheta(\varepsilon_i - \varepsilon_h)}{(\varepsilon_i + U)}$ $U = (\varepsilon_i S'_{\varepsilon *} - \varepsilon_h)/(1 - S'_{\varepsilon *})$; $\varepsilon^* = \varepsilon_h$	Semi-empirical [9]
Arbitrary multiphase	$\frac{(\varepsilon - 1)}{(\varepsilon + U)} = \varepsilon_i\, \vartheta_i \left(\frac{\varepsilon_i - 1}{\varepsilon_i + U}\right)$ $\varepsilon_i\, \vartheta_i = 1$ U is defined as above	[65]
Arbitrary two-phase	$\frac{\varepsilon}{\varepsilon'} = 1 + \left[\frac{\vartheta_i(1 - \vartheta_i)}{3}\right]\left[\frac{(\varepsilon_i - \varepsilon_h)}{\varepsilon'}\right]^2 + \ldots$ where $\varepsilon' = \varepsilon_i\, \vartheta_i + (1 - \vartheta_i)\, \varepsilon_h$	[61]

(continued...)

Inclusion Type	Mixture Formulations	Remarks and References
Spheres two-phase	$\frac{\varepsilon - \varepsilon_h}{\varepsilon + 2\varepsilon_h} = \vartheta_i \left(\frac{\varepsilon_i - \varepsilon_h}{\varepsilon_i + 2\varepsilon_h} \right)$	[12]
Spheres two phase	$\left(\frac{3\varepsilon_i}{C_1\varepsilon_i + C_2} \right) ln \left(\frac{\varepsilon_i - \varepsilon}{\varepsilon_i - \varepsilon_h} \right) - \left(\frac{C_1\varepsilon_i - 2C_2}{C_1\varepsilon_i + C_2} \right) ln \left(\frac{C_1\varepsilon + C_2}{C_1\varepsilon_h + C_2} \right) = ln\left(1 - \vartheta_i\right)$ $C_2 = 4\pi/9\sqrt{3}$ and $C_1 = (3 - C_2)$	
Arbitrary two phase	$\frac{\varepsilon - \varepsilon_h}{\varepsilon_i - \varepsilon_h} = \varepsilon' = \frac{(1 - k)\vartheta_i}{1 - k\vartheta_i}$ k is an empirical factor	Empirical [40]
Spheres two phase	$(\varepsilon - \varepsilon_h) = \frac{3\vartheta_i \varepsilon_h}{[(\varepsilon_i + 2\varepsilon_h)/(\varepsilon_i - \varepsilon_h)] - \vartheta_i}$	Interactions allowed [66]
Needles two phase	$(1 - \vartheta_i) = \left(\frac{\varepsilon_i - \varepsilon}{\varepsilon_i - \varepsilon_h} \right) \left(\frac{\varepsilon_i + 5\varepsilon_h}{\varepsilon_i + 5\varepsilon} \right)^{2/5}$	Interactions partially accounted for in the derivation
Disks two phase	$(1 - \vartheta_i) = \left(\frac{\varepsilon_i - \varepsilon}{\varepsilon_i - \varepsilon_h} \right) \left(\frac{2\varepsilon_i + \varepsilon_h}{2\varepsilon_i + \varepsilon} \right)$	[15, 29, 30]
Spheres two phase	$(1 - \vartheta_i) = \left(\frac{\varepsilon_i - \varepsilon}{\varepsilon_i - \varepsilon_h} \right) \left(\frac{\varepsilon_h}{\varepsilon} \right)^{1/3}$	

(continued...)

Inclusion Type	Mixture Formulations	Remarks and References
Arbitrary multiphase	$\ln \varepsilon = \vartheta_h \ln(\varepsilon_h) + \sum_{i=1}^{N} \vartheta_i \ln(\varepsilon_i); \sum (\vartheta_i + \vartheta_h) = 1$	Empirical [19, 20] Extension of two-phase logarithmic law
Spheres two phase	$\frac{\varepsilon - \varepsilon_h}{3\varepsilon} = \vartheta_i \left(\frac{\varepsilon_i - \varepsilon_h}{\varepsilon_i + 2\varepsilon} \right)$	Short-range statistical variations neglected [17,67]
Ellipsoids multiphase	$\left(\varepsilon - \varepsilon_h \right) = \sum_{i=1}^{N} \vartheta_i \left(\varepsilon_i - \varepsilon_h \right) S'_{\varepsilon*}$	Short-range statistical variations neglected [28]
Ellipsoids two phase	$\left(\varepsilon - \varepsilon_h \right) = \vartheta_i \left(\varepsilon_i - \varepsilon_h \right) S'_{\varepsilon*}$ ε_h or $\varepsilon_i < \varepsilon^* < \varepsilon$	ε^* contains Interaction Effects [29]
Ellipsoids multiphase anisotropic media	$(\bar{\varepsilon} - \bar{\varepsilon}_h) = \sum_{i=1}^{N} \vartheta_i (\bar{\varepsilon}_i - \bar{\varepsilon}_h) \bar{T}_i$ For example, $\bar{T}_i = \sum_{j=1}^{3} [1 + n_j (\varepsilon_i/\varepsilon - 1)]^{-1}$ $\bar{T}_i$ is a dyadic relating the internal to the average electric field	Interactions can be accented for in $\bar{T}_i$; solution is feasible if statistical variations are neglected [68]

(continued...)

Inclusion Type	**Mixture Formulations**	**Remarks and References**
Confocal ellipsoidal shells multiphase anisotropic mixtures	$\bar{\varepsilon}_j - \bar{\varepsilon}_h = \sum_{i=1}^{N} \vartheta_i (\varepsilon_i - \varepsilon_h)\ \bar{T}_i$ (see Figure). For thin shells, $\bar{T}_i \leq 1$, $T_2 = \varepsilon_1 [\varepsilon_2 + \frac{1}{2}(\varepsilon_3 - \varepsilon_2)(n'_j + n_j\vartheta'/\vartheta)] \times \{[\varepsilon_2 + (\varepsilon_2 - \varepsilon_3)(n_j\vartheta'/\vartheta - n_j)] \times [\varepsilon_1 + n_j(\varepsilon_2 - \varepsilon_1)] - n_j(\vartheta'/\vartheta)\,\varepsilon_2(\varepsilon_2 - \varepsilon_3)\}^{-1}$ and $T_3 - \varepsilon_2\,\varepsilon_1 \{[\varepsilon_2 + (\varepsilon_2 - \varepsilon_3)(n_j\vartheta'/\vartheta - n'_j)] \times [\varepsilon_1 + n_j(\varepsilon_2 - \varepsilon_1)] - n_j(\vartheta'/\vartheta)\,\varepsilon_2(\varepsilon_2 - \varepsilon_3)\}^{-1}$ $\vartheta' = \frac{4}{3}\pi a'\,b'\,c'$; inner ellipsoid axes: a', b', c' $\vartheta = \frac{4}{3}\pi a\,b\,c$; outer ellipsoid axes: a, b, c	Interaction effects can be approximately calculated through $\bar{T}_i$; $\bar{T}_i$ is solved for shells [20]

References

[1] O. F. Mossotti: Discussione analitica sull influenza che lazione di un mezzo dielettrico ... . *Mem. di. Matem. e. Fisica di Modena:* II, vol 24, 1846: 49-74.

[2] R. Clausius: *Die Mechanische Behandlung der Elektricität*, Braunschweig, vol. II, (Vieweg, 1897), pp. 62-97.

[3] L. Lorenz: Über die Refractionsconstante. *Ann. Phys.* (Leipzig), vol. 11, 1880: 70-103.

[4] H. A. Lorentz: *Theory of Electrons*, (Dover Publications, New York: 1952).

[5] J. C. Maxwell-Garnett: Colours in metal glasses and metal films. Phil. Trans. A: Roy. Soc. London, vol. 203, 1904: 385-420.

[6] J. C. Maxwell-Garnett: Colours in metal glasses and metallic films and in metallic solutions-II. Phil. Trans. A: Roy. Soc. London, vol. 205, 1906: 237-262.

[7] L. D. Landau and E. M. Lifshitz: *Electrodynamics of Continuous Media.* (Pergamon Press, Oxford: 1960).

[8] B. Tareev: *Physics of Dielectric Materials.* (Mir Publishers, Moscow: 1973), Chapter 2, Equation 2.87.

[9] O. Wiener: Die Theori des Mischkörpers für das Feld der stationären strömung. *Abdhl. d. kgl. Akad. d. Wies. Leipzig,* vol. 32, 1912: 509-604.

[10] J. C. Maxwell: *A Treatise on Electricity and Magnetism.* (Dover Publishing Co., NY: 1954).

[11] V. I. Odelevski: Raschet obobshchenoi provodimosti geterogenie sistem. (Effective conductivity calculations of heterogeneous systems). Zh. Tekh. Fiziki (Russian), vol. 21, 1951: 675-685.

[12] Lord Rayleigh: On the influence of obstacles arranged in rectangular order upon the properties of a medium. Phil. Mag., vol. 34, 1892: 481-502.

[13] M. M. Z. Kharadly and W. Jackson: Properties of artificial dielectrics comprising arrays of conducting elements. Proc. Inst. Elec. Eng., vol. 100, 1953: 199-212.

[14] K. W. Wagner: *Die Isolierstoffe der Electrotechnik.* (H. Schering, Ed.), (Springer-Verlag, Berlin: 1924).

[15] D. A. G. Bruggeman: Berechnung verschiedenier physikalischen konstanten von heterogenen Substanzen. Ann. Phy., vol. 24, 1935: 636-679.

[16] T. Hanai: Theory of the dielectric dispersion due to the interfacial polarization and its applications to emulsions. Kolloid Z., vol. 171, 1960: 23-31.

[17] C. J. F. Böttcher: *Theory of Electric Polarization.* (Elsevier Science Publishing Co., Amsterdam: 1952).

[18] H. Looyenga: Dielectric constants of heterogeneous mixtures. Physica, vol. 31, 1965: 401-406.

[19] K. Lichtenecker: Mischkörpertheori als Wahrscheinlichkeitsproblem. Phys. Zeitsch., vol. 30, 1929: 805-809.

[20] K. Lichtenecker and K. Rother: Die Herleitung des logarithmischen Mischungsgesetzes aus allgemeinen Prinzipien der stationären Strömung. Phys. Zeitsch., vol. 32, 1938: 255-260.

[21] P. S. Neelakantaswamy, B. V. R. Chowdary and R. Rajaratnam: Estimation of permittivity of a compact crystal by dielectric measurements on its powder: A

stochastic mixture model for the powder dielectric. J. Phys. D, vol. 17, 1983: 1755-1799.

[22] P. S. Neelakantaswamy, K. Asper and R. Rajaratnam: A dielectric model of the human blood. Biomed. Technik, vol. 28, 1983: 18-22.

[23] P. S. Neelakantaswamy, R. Turkman, and T. K. Sarkar: Complex permittivity of a dielectric mixture: Corrected version of Lichtenecker's logarithmic law of mixing. Electron. Letts., vol. 21, 1985: 270-271.

[24] S. Kisdnasamy and P. S. Neelakantaswamy: Complex permittivity of a dielectric mixture: Modified Frick's formula based on logarithmic law of mixing. Electron. Letts., vol. 20, 1984: 291-293.

[25] R. W. Corkum: Isotropic artificial dielectric. Proc. Inst. Radio Engrs., vol. 40, 1952: 574-587.

[26] R. W. Sillars: Properties of a dielectric containing semiconducting particles of various shapes. J. Instn. Elec. Engrs. (UK), vol. 80, 1937: 378-392.

[27] R. Coelho: *Physics of Dielectrics for the Engineer.* (Elsevier Publishing Co., Amsterdam: 1979).

[28] D. Polder and J. H. Van Santen: The effective permeability of mixtures of solids. Physica, vol. 12, 1946: 257-271.

[29] G. P. de Loor: Dielectric Properties of Heterogeneous Mixtures. Ph.D. Thesis, University of Leiden, 1956.

[30] W. Nielsen: Ann. Phys. (Leipzig), vol. 10, 1952: 336- .

[31] A. P. Alschuller: The shapes of particles from dielectric studies of suspensions. J. Phys. Chem., vol. 58, 1954: 544-547.

[32] G. E. Archie: The electrical resistivity log as an aid in determining some reseviors characteristics. Trans. Am. Inst. Min. Met., Petrol. Engrs., vol. 146, 54-62, 1942.

[33] R. I. Frame and D. J. Tedford: Long-term eletrical conduction in films of alkyd resin and graphite mixtures. IEEE Trans. Elec. Insulation, vol. EI-21, 1986: 23-29.

[34] R. M. Scarisbrick: Electrically conducting mixtures. J. Phys. D., vol. 6, 1973: 2098-2110.

[35] A. Kusy: Chains of conducting particles that determine the resistivity of thick resistive films. Thin Solid Films, vol. 43, 1977: 243-250.

[36] J. Peyrelasse, C. Boned, G. Canadas and R. Roger: Theoretical study of the static permittivity of conductive component binary mixtures. Phys. Rev. A, vol. 30(2), 1984: 994-998.

[37] J. A. Reynold and J. M. Hough: Formulae for dielectric constant of mixtures. Proc. Phys. Soc., vol. LXX, 1957: 769-775.

[38] D. J. Bergman: The dielectric constant of simple cubic array of identical spheres. J. Phys. C. (Solid State Phys.), vol. 12, 1979: 4947-4960.

[39] J. A. Reynolds: The Dielectric Constant of Mixtures, Ph.D Thesis, University of London, 1956.

[40] C. A. R. Pearce: The permittivity of two phase mixtures. Brit. J. Appl. Phys., vol. 6, 1955: 358-361.

[41] R. Landauer: The electrical resistance of binary metallic mixtures. J. Appl. Phys. vol. 23, 1952: 779-784.

[42] K. Kamiyoshi: A new deduction formula for determining the dielectric constant of powder dielectric: Part II, Sci. Rep. Res. Inst., Series A (Physics, chemistry and metallurgy), Tohoku Univ., vol. A.2, 1950: 180-192.

[43] M. Kubo and S. Nakamura: The dielectric constant of dispersion of spherical particles. Bull. Chem. Soc. Japan, vol. 26(6), 1953: 318-322.

[44] K. W. Wagner: Über dielektrische Nachwirkungs vorgange. Arch. Elektrochem. vol. 2, 1914: 371- ; ibid., vol. 3, 1914: 100- .

[45] H. Fricke: A mathematical treatment of the electrical conductivity and capacity of disperse system I. Phys. Rev., vol. 24, 1924: 575-587.

[46] H. Fricke: A mathematical treatment of the electrical conductivity and capacity of disperse system II. Phys. Rev., vol. 26, 1926: 678-681.

[47] W. S. Rothwell: Complex permittivity of conductor dielectric mixtures. IEEE Trans. Microwave Theory Tech., vol. MTT-19, 1971: 413-414.

[48] L. K. H. van Beek: Dielectric behavior of heterogeneous systems, in *Progress in Dielectrics,* (J. B. Birks, Ed.), (CRC Press, Cleveland: 1987).

Additional References

[49] L. S. Taylor: Dielectric properties of mixtures. IEEE Trans. Antenna Propagat., vol. AP-13(6), 1965: 943-947.

[50] K. Lal and R. Parshad: Permittivity of conductor-dielectric heterogeneous mixtures. J. Phys. D: (Appl. Phys.), vol. 6, 1973: 1788-1792.

[51] K. Subramaniam, P. S. Neelakanta and V. Ungvichian: Permittivity of orderly textured mixture dielectrics. Electron. Letts., vol. 27(17), 1991: 1534-1535.

[52] P. S. Neelakanta: Complex permittivity of a conductor-loaded dielectric. J. Phys. Condens. Matter, vol. 2, 1990: 4935-4947.

[53] B. U. Felderhof and R. B. Jones: Effective dielectric constant of dilute suspensions of spheres. Phy. Rev. B, vol. 39(9), 1989: 5669-5677.

[54] A. H. Sihvola and J. A. Kong: Effective permittivity of dielectric mixtures. IEEE Trans. Geo Science Remote Sensing, vol. 26(4), 1988: 420-429.

[55] W. R. Tinga, W. A. G. Voss and D. F. Blossey: Generalized approach to multiphase dielectric mixture theory. J. Appl. Phys., vol. 44(9), 1973: 3897-3902.

[56] J. H. Calderwood and B. K. P. Scaife: On the estimation of the relative permittivity of a mixture. Proc. 1979 IEE Conf. on Dielectric Materials, Measurements and Applications.

[57] S. A. Paipetis, G. M. Tsangaris and J. M. Tsangaris: Dielectric properties of metal-filled epoxies. Polymer Commn., vol. 24, 1983: 373-375.

[58] L. Poladian: Asymptotic behavior of the effective dielectric constant of composite materials. Proc. Roy. Soc. London. A., vol. 426, 1989: 343-360.

[59] C. Boned and J. Peyrelasse: Some comments on the complex permittivity of ellipsoids dispersed in continuum media. J. Phy. D: Appl. Phys., vol. 16, 1983: 1777-1784.

[60] J. B. Hasted: *Aqueous Dielectrics* (Chapman and Hall, London: 1970), Chapter 5, pp. 117-135.

[61] W. F. Brown: Solid mixture permittivity. J. Chem. Phys., vol. 23(8), 1955: 1514-1517.

[62] C. Grosse: On the calculation of the static permittivity of heterogeneous conductive systems. J. Phys. D: Appl. Phys., vol. 18, 1985: 1883-1889.

[63] D. S. McLachlan, A. Priou, I. Chenerie, E. Issac and F. Henry: Modeling the permittivity of complex materials with a general effective medium equation. J. Electromag. Waves Applns., vol. 6(9), 1992: 1099-1131.

[64] J. C. Maxwell: *A Treatise on Electricity and Magnetism.* vol. I & II (Dover Publications, Inc., New York: 1954).

[65] O. Wiener: Zur Theorie der Refraktionkonstanten. Ber. Sächs. Ges. Wiss, (Math. Phys. KL.), vol. 62, 1910: 256- 277.

[66] L. Lewin: The electrical constants of a material loaded with spherical particles. Proc. IEEE, vol. 94, 1947: 65-68.

[67] C. J. F. Böttcher: The dielectric constant of crystalline powders. Rec. trav. chim. Pays-Bas (Leiden), vol. 64(1), 1945: 47-51.

[68] J. D. Jackson: *Classical Electrodynamics.* (John Wiley and Sons, New York: 1962) pp. 98-131.

[69] W. R. Tinga: Multiphase Dielectric Theory Applied to Cellulose Mixtures. Ph. D. Thesis, Department of Electrical Engineering, University of Alberta, Canada, 1969.

CHAPTER 6

Composite Dielectric Materials with a Discrete Conducting Phase

6.1 Introduction

A class of composite dielectric materials refers to a two-phase, host-inclusion system in which the inclusion is a discrete-phase of conducting medium (such as metals, semiconductors, or solid electrolytes) dispersed randomly or textured as an orderly embedment in the host medium which forms a dielectric receptacle. Such composites are essentially dielectric-conductor mixtures and have unique (effective) dielectic properties due to the fact that the constituent phases have extremely opposite characteristics as regards their electrical susceptance and the associated conduction phenomena. In view of the high electrical conductivity of the inclusions and predominant dielectric susceptance (lossy or lossless) of the host medium, prediction of effective dielectric permittivity and/or the conductivity of the composite medium is not simple or trivial.

Since the time of Maxwell-Garnett [1,2] attempts have been made, however, to evaluate the effective electrical characteristics of such conductor-insulator mixture systems. The primary factors which decide the effective parameters of such mixtures are: (i) Conductivity or complex permittivity of the inclusions; (ii) complex permittivity of the host medium; (iii) shape of the particulate inclusions; (iv) frequency; and (v) spatial arrangment (random or textured) of the inclusions in the receptacle.

In modeling conductor-insulator mixtures, two basic approaches have been pursued in general: One involves the treatment of the dielectric behavior of the mixtures entirely independent of the conductive effects of the inclusions and the other uses the expressions for complex dielectric properties of dielectric-dielectric mixtures with the conductive phase being treated as an extremely high-loss dielectric. In the second case, the mixture theories concerning dielectric-dielectric mixtures have been generalized to high-loss materials to account for the conductor inclusions; and the complex permittivity of the mixture is calculated with the surmise that the permittivity of the conductor inclusions approaches infinity. Thus, a mathematical way of decoupling dielectric considerations from the conductive phenomenon was formalized, and, in most cases, it appears that only the static (d.c.) or quasistatic (low frequency) behavior of such dielectric-conductor mixtures has been investigated to deduce the (lossless) static permittivity ε_S and/or the d.c conductivity σ_{dc} of the mixture. Pertinent to these existing models details on the evolution of background concepts are presented in the next section.

6.2 Evolution of Dielectric-Conductor Mixture Formulations

6.2.1 *Maxwell and Maxwell-Garnett formula*

Maxwell-Garnett [1,2] used the quasistatic (potential) approach to elucidate the effective (complex) permittivity $\hat{\sigma}_{eff}$ of the medium containing a volume loading θ of identical spherical particles of complex conductivity $\hat{\sigma}_2$ dispersed randomly in a homogeneous dielectric host medium of complex conductivity $\hat{\sigma}_1$. (Here, the complex conductivity $\hat{\sigma}$ is defined as $(\sigma + j\omega\varepsilon)$, ε being the permittivity and $\omega = 2\pi \times$ frequency.) The effective conductivity is then given by:

$$\hat{\sigma}_{eff} = \hat{\sigma}_2 (1 + 2\,\theta \mathcal{M}_o)/(1 - \theta \mathcal{M}_o) \tag{6.1}$$

where $\mathcal{M}_o$ is referred to as the normalized dipole moment. It has been deduced as [3]:

$$\mathcal{M}_o = (1 - \hat{\sigma}_1/\hat{\sigma}_2)(1 + 2\,\hat{\sigma}_1/\hat{\sigma}_2)^{-1} \tag{6.2}$$

6.2.2 *Rayleigh's formula*

In Maxwell's mixture formulation, the interaction between the particles is neglected. This is true only when θ is very small (on the order of 0.1). Under such low volume fraction of spherical conductor loading, Rayleigh [4] derived the following alternative relation:

$$(\hat{\sigma}_{eff} - \hat{\sigma}_1)/3\hat{\sigma}_1 \approx \theta(\hat{\sigma}_2 - \hat{\sigma}_1)/(\hat{\sigma}_2 + 2\hat{\sigma}_1) \tag{6.3}$$

6.2.3 *Fricke's formula*

Fricke extended the Clausius-Mossotti theory [5,6] for the dielectric constant and Lorentz-Lorenz theory [7,8] for the index of refraction to determine the effective conductivity of a dilute suspension of conducting particles in a homogeneous medium as given above. He also elucidated on similar considerations, the effective electric conductivity of a mixture constituted by shaped conducting inclusions (of spheroidal shape) suspended in a homogeneous medium. The relevant formulations are:

$$[(\sigma_{eff} - \sigma_1)/(\sigma_2 - \sigma_1)]\,[(\sigma_2/\sigma_1) - 1] = \beta\theta/(1 - \theta) \tag{6.4}$$

where

$$\beta = \frac{1}{3}\left[\frac{2}{1 + [(\sigma_2/\sigma_1) - 1]\,(M/2)} + \frac{1}{1 + [(\sigma_2/\sigma_1) - 1]\,(1 - M)}\right]\left(\frac{\sigma_2}{\sigma_1} - 1\right)$$

$M = [\phi - (\sin^3\phi)/2]/\sin^3\phi$ with $\cos\phi = a/b$, and $a < b$ (oblate spheroid); or
$M = [(1/\sin^2\phi') - \cos^2\phi'/2\sin^3\phi']\log_e[(1 + \sin\phi')/(1 - \sin^2\phi')]$ with $\cos\phi' = b/a$, and $a > b$ (prolate spheroid).

Here, *a/b* or *b/a* refers to the *aspect ratio* of the spheroidal particle.

Defining a shape parameter, x, as:

$$x = -\left\{\left(\frac{\sigma_2}{\sigma_1} - 1\right) - \left(\frac{\sigma_2}{\sigma_1}\right)\beta\right\} \Big/ \left\{\left(\frac{\sigma_2}{\sigma_1} - 1\right) - \beta\right\} \tag{6.5}$$

Equation 6.4 can be written as:

$$[(\sigma_{eff}/\sigma_1) - 1)]/[(\sigma_{eff}/\sigma_1) + x] = \theta\,[(\sigma_2/\sigma_1) - 1]/[(\sigma_2/\sigma_1) + x]$$

For a spherical case, $x = 2$; and, applied to the case of spherical (conducting) particulate suspensions in a dielectric host medium, the effective conductivity of the mixture reduces to:

$$[(\sigma_{eff}/\sigma_1) - 1)/[(\sigma_{eff}/\sigma_1) + 2] = \theta\,[(\sigma_2/\sigma_1) - 1]/[(\sigma_2/\sigma_1) + 2] \tag{6.6}$$

This is the well-known form of *Fricke's formula* [9,10], for a dielectric with spherical conducting inclusions.

6.2.4 *Bruggeman's formula*

Bruggeman assumed that for a given volume loading of spherical particles, the effective conductivity has a unique value $\hat{\sigma}_{eff}$; and every unit volume of this particulate loaded medium of a small volume δvol, with additional spherical particles of conductivity $\hat{\sigma}_2$ would augment the effective conductivity to $(\hat{\sigma}_{eff} + \delta\hat{\sigma}_{eff})$. Using this concept of proportional increments of effective conductivity *versus* volume loading of the inclusions, Bruggeman modified Rayleigh's formula and arrived at the following expression:

$$\sigma_{eff} = (\hat{\sigma}_{eff}/\sigma_1)^{1/3} [(\hat{\sigma}_1 - \hat{\sigma}_2)/(\hat{\sigma}_{eff} - \hat{\sigma}_2)] (1 - \theta) = 1 \tag{6.7}$$

With $|\hat{\sigma}_2| << (|\hat{\sigma}_1| \text{ and } |\hat{\sigma}_{eff}|)$, Equation 6.7 reduces to:

$$\hat{\sigma}_{eff} \approx \hat{\sigma}_1 (1 - \theta)^{3/2} \tag{6.8}$$

This equation corresponds to the case where the particles or grains act as insulators. Further, it suggests that if $\theta = 1$, $\sigma_{eff} \to 0$ and if $\theta = 0$, $\hat{\sigma}_{eff} \to \hat{\sigma}_1$, that of the host medium.

6.2.5 *Archie's law*

Equation 6.8 written in more generalized form as $\hat{\sigma}_{eff} \approx \hat{\sigma}_1 (1 - \theta)^{m_A}$ is found to be a good empirical fit for brine-saturated sedimentary rocks (with m_A in the range 1 to 5). This is known as *Archie's law* [12].

6.2.6 *Looyenga's formula/Böttcher's formula*

Using the concept of two concentric spheres of different conductivities, one enclosed within the other to represent a two-phase mixture, Looyenga [13] deduced the following mixture formula:

$$\theta = [(\hat{\sigma}_{eff} - \hat{\sigma}_1)(2\hat{\sigma}_{eff} + \hat{\sigma}_2)]/[3\,\hat{\sigma}_{eff}(\hat{\sigma}_2 - \hat{\sigma}_1)] \tag{6.9}$$

Independently, Böttcher [14] arrived at the same formulation from the considerations of internal field(s) associated with spherical particles.

6.2.7 *Lal and Parshad formula [15]*

Lal and Parshad [15] extended the dielectric-dielectric mixture theory to a dielectric-conductor composite, and derived an expression for the static (relative) permittivity of the mixture (ε_s) in terms of the volume ratio θ of the conductor inclusions:

$$\varepsilon_s = \varepsilon_{2s}/(1 - \theta)^B \tag{6.10}$$

In this relation, ε_{2s} is the static permittivity of the host dielectric and B is a shape parameter of the conducting particles related to the depolarization factors A_i *via* the relation given by:

$$B = (1/3) \sum_i A_i \tag{6.11}$$

Assuming the included particles as ellipsoids, the subscript i refers to the ith axis of an ellipsoid with semiaxial lengths a, b, and c (see Figure 4.1). For a spheroidal particle $b = c$ and defining $x = a/b$, $x = 1$ specifies spherical particles. Prolate spheroidal particles with $x > 1$ will become needle-like fibers when $x >> 1$. Likewise, oblate spheroids with $x < 1$ will represent flaky, disk-like lamellae when $x << 1$. A_i can be evaluated by an integral relation due to Wallin [16].

It may be noted that Equation 6.10 cannot be extended to frequency-dependent, dynamic conditions; or is evaluation of B straightforward. Results presented in [15] are therefore based on an empirical value of B obtained *via* curve fitting to a set of test data.

6.2.8 *Scarisbrick and Kusy model*

Concerning the effective conductivity of a dielectric-conductor mixture, Scarisbrick [17] developed a random-chain model, to which Kusy [18] added a shape-dependent order function (U) established *via* probabilistic considerations. Again, the relevant expression refers only to d.c. conductivity of the mixture and is given by :

$$\sigma_{dc} = K^2 \sigma \theta (\theta)^{U\theta^{-2/3}} \tag{6.12}$$

where K is a constant decided by the conductive path cross-section. Further, σ and θ are the conductivity and volume fraction of the conducting inclusions.

6.2.9 *Frame and Tedford model*

Frame and Tedford [19] used Equations 6.10 and 6.12 to evaluate the static permittivity and d.c. conductivity of a composite made of alkyd resin loaded with graphite lamellae. In the relevant studies, the exponent B of Equation 6.10 was obtained by best-fitting the experimental data on static permittivity to the algorithm of Equation 6.10.

6.3 Complex Susceptibility Model: Neelakanta's Formula

Let ε and σ denote the relative (effective) permittivity and conductivity of the mixture. Subscripts 1 and 2 are used to specify the corresponding variables of the conducting inclusions and the dielectric matrix, respectively. The volume fraction of the inclusions is denoted by θ and U is an order function decided by the geometrical aspect ratio $(x = a/b)$ of the inclusions.

The electrical characteristics of a mixture formed by a random volumetric dispersion of shaped inclusions in a continuous host medium can be specified by the following functional relations:

$$F(\varepsilon) = \theta F(\varepsilon_1) + (1 - \theta) F(\varepsilon_2) \tag{6.13a}$$

and

$$G(\sigma) = \theta G(\sigma_1) + (1 - \theta) G(\sigma_2) \tag{6.13b}$$

Here, the functions F and G determine the law of mixing; and if they are known explicitly, the values of ε and σ can be determined uniquely. The law of mixing pertaining to a statistical mixture is constrained by: (1) Wiener's proportionality postulate* [20]; (2) Wiener's upper and lower bounds** on ε and σ; (3) the limiting values of $0 \le \theta \le 1$ and $0 \le U \le 1$; and (4) geometrical dissimilarity of the components in the mixture matrix.

---

* *Wiener's proportionality postulation*: If the values of ε or σ of the constituents change in one and the same ratio, the values of ε or σ of the mixture should change identically.

** Wiener's upper and lower bounds: In an m-component mixture:

$$[1/\sum_{i=1}^{m}\theta_i p_i] \le p_{mixture} \le \sum_{i=1}^{m}\theta_i p_i$$

$$(p \Rightarrow \varepsilon \text{ or } \sigma)$$

The analytical endeavor of evaluating the functions F and/or G for various types of pure dielectric-dielectric mixtures resulted in several formulations; a comprehensive review of them has been published by Brown [21] and van Beek [22]. The contents of these reviews have also been reported by Tinga and Voss [23]. Relevant details are summarized in Chapter 5.

These formulations, however, ignore the statistical aspects of the mixture except the so-called logarithmic law of mixing due to Lichtenecker [24] and Lichtenecker and Rother [25]. This logarithmic law, however, does not take the particulate geometry into account and it lacks a linear form [26]. These deficiencies have, however, been offset by the author and others as reported in [27] and explained in Chapter 5.

The following analysis presented here refers to the author's contribution reported in [28], in which the logarithmic law is extended to a generalized electric susceptibility parameter χ pertaining to a dielectric plus conductor mixture subjected to complex field considerations. Hence, a dynamic model is presented to calculate the effective complex permittivity of dielectric-conductor mixtures.

The complex susceptibility of a conductor-loaded mixture can be specified as a logarithmic model in the following form:

$$\log \chi = \theta \log \chi_1 + (1 - \theta) \log \chi_2 \tag{6.14}$$

In terms of explicit parameters of the mixture constituents, namely, $(\varepsilon_1, \sigma_1)$ and $(\varepsilon_2, \sigma_2)$, Equation 6.14 can be written as

$$\chi = (\sigma_1/\omega\varepsilon_o)^\theta \exp(i\pi\theta/2) \{[(\varepsilon_2 - 1)^2 + (\varepsilon_2 \tan\delta_2)^2]^{1/2} \exp(-i\varphi)\}^{(1-\theta)} \tag{6.15}$$

where ε_o is the free-space permittivity, $\omega = 2\pi \times$ frequency, $\tan\delta_2$ $(= \sigma_2/\omega\varepsilon_o\varepsilon_2)$ is the loss tangent of the host medium and $\varphi = \tan^{-1}[\varepsilon_2 \tan \delta_2/(\varepsilon_2 - 1)]$.

Using the relation $(\varepsilon - 1) \equiv$ real part of χ, it follows that:

$$\varepsilon = (\sigma_1/\omega\varepsilon_o)^\theta [(\varepsilon_2 - 1)^2 + (\varepsilon_2 \tan\delta_2)^2]^{(1-\theta)/2} \cos[\pi\theta/2 + \varphi(1 - \theta)] + 1 \tag{6.16}$$

The conductivity (σ) of the mixture can be extracted from the imaginary part of χ. Thus,

$$\sigma/\omega\varepsilon_o = (\sigma_1/\omega\varepsilon_o)^\theta [(\varepsilon_2 - 1)^2 + (\varepsilon_2 \tan\delta_2)^2]^{(1-\theta)/2} \sin[\pi\theta/2 + \varphi(1 - \theta)] + 1 \tag{6.17}$$

Equations 6.16 and 6.17 should, however, be "weighted" to meet the limiting conditions, namely, $(\varepsilon, \sigma) = (\varepsilon_2, \sigma_2)$ at $\theta = 0$ and $(\varepsilon_1, \sigma_1)$ at $\theta = 1$.

Further, by including the geometrical dependence *via* an order function U in the logarithmic formulation of Equation 6.14, the following modified expressions for ε and σ are obtained on the basis of the arguments given by the author in [28].

$$\varepsilon_{mod} = C_1 [(\varepsilon - 1)^{U_\varepsilon} + 1] \tag{6.18}$$

and

$$\sigma_{mod}/\omega\varepsilon_o = C_2 [(\sigma_1/\omega\varepsilon_o)^\theta (\varepsilon_2 - 1)^{1-\theta} \sin(\pi\theta/2)]^{U_\sigma} + (\sigma_2/\omega\varepsilon_o) \tag{6.19}$$

under the valid assumptions that $\sigma_2 << \sigma_1$ and $\varepsilon_2 tan\delta_2 << (\varepsilon_2 - 1)$. Here, the coefficients C_1 and C_2 are parameters decided by the limiting conditions, namely, $\varepsilon = \varepsilon_2$ at $\theta = 0$ and $\sigma = \sigma_1$ at $\theta = 1$. Hence, they are specified explicitly as:

$$C_1 = \varepsilon_2 [(\varepsilon_2 - 1)^{U_\varepsilon} + 1]^{-1} \tag{6.20a}$$

and

$$C_2 = [(\sigma_1 - \sigma_2)/\omega\varepsilon_o] (\omega\varepsilon_o/\sigma_1)^{U_\sigma} \tag{6.20b}$$

The order functions U_ε and U_σ implicitly determine the dependence of ε and σ, respectively, on the geometrical aspect ratio x $(=a/b)$ of the particulate inclusion. Defining the particle eccentricity e $(=1 - b/a)$ when $b < a$ or $(a/b - 1)$ when $a < b$, the value of $e = 0$ corresponds to spherical particles; U_ε and U_σ should therefore be expressed in terms of e. That is, for a given eccentricity, the U_ε^{th} fraction of the stochastic system can be regarded as being polarized along the electric field and the $(1 - U_\varepsilon)^{th}$ fraction along the orthogonal direction. Likewise, U_σ should represent the fraction corresponding to current percolations.

On the basis of similarity to Maxwell-Boltzmann statistics applied to dipole orientation, the upper and lower bounds of the order function can be specified as follows:

$$U_U = [1 - L(e)/e]/2 \qquad \cong (1/3) \qquad \text{when } e \to 0 \tag{6.21a}$$

$$U_L = [(L(e)/e]/2 \qquad \cong (1/6) \qquad \text{when } e \to 0 \tag{6.21b}$$

where $L(e)$ is the Langevin function equal to $[coth(e) - 1/e]$. The functions U_ε and U_σ can be equated to U_L or U_U depending on the following states of the test mixture: For large values of $(\sigma_1/\omega\varepsilon_o\varepsilon_2)$, the composite can be considered as conductivity dominant; and for low values of $(\sigma_1/\omega\varepsilon_o\varepsilon_2)$, the mixture becomes permittivity dominant. Accordingly, the permittivity of the mixture as a function of frequency can be sketched as shown in Figure [6.1], indicating three zones, namely, the low-frequency, the high-frequency, and the intermediate (quasistatic) regions. It is, however, to be noted that the region-to-region transition is not abrupt. For calculation purposes, two corner frequencies, namely, ω_L and ω_H, can be approximately assigned marking the transitions as shown in Figure (6.1) since the regions are distinguishable in terms of $d\varepsilon'/d\omega$ slope. In summary, the complex permittivity spectra of a conductor-loaded mixture can be specified by the following:

(1) Complex permittivity of the mixture (ε):

$$\varepsilon = (\varepsilon' - i\varepsilon'') \tag{6.22a}$$

$$\varepsilon' = \{\varepsilon_2/[1 + (\varepsilon_2 - 1)^{U_\varepsilon}]\} \{[(\sigma_1/\omega\varepsilon_o)^\theta (\varepsilon_2 - 1)^{1-\theta} cos(\pi\theta/2)]^{U_\varepsilon} + 1\} \tag{6.22b}$$

$$\varepsilon'' = \sigma/\omega\varepsilon_o \tag{6.22c}$$

$$\sigma = (\sigma_1 - \sigma_2)\{[\omega\varepsilon_o (\varepsilon_2 - 1)/\sigma_1]^{1-\theta} sin(\pi\theta/2)\}^{U_\sigma} + \sigma_2 \tag{6.22d}$$

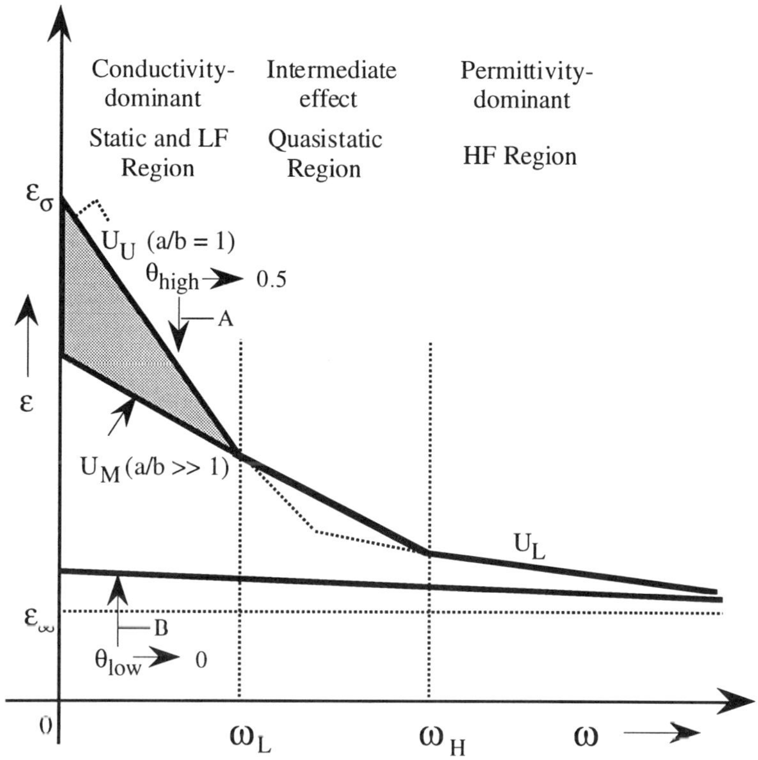

Figure 6.1 Permittivity *versus* frequency of a conductor-loaded dielectric.
A: For large volume fraction θ of inclusions.
B: For low volume fraction θ of inclusions.
The shaded region refers to the bounded range of values that ε may assume as described by the aspect ratio of the conducting inclusions ($1 \le a/b \ge \infty$).

(2) Order functions U_ε and U_σ at low frequencies ($\omega < \omega_L$)

$$U_\varepsilon = U_\sigma = \begin{cases} U_U = (1/2)[1 - L(e)/e] & e \cong 0 \\ U_M = (1/M) + (2 - 3M)/3 & e >> 1 \end{cases} \qquad (6.23)$$

where $M = [(2 \pm e)/(3 \pm e) - L'(e)]^{-1}$ and $L'(e) = dL/de$; the positive sign here refers to $a/b < 1$ and the negative sign is for $a/b > 1$.

(3) Order functions U_ε and U_σ at high frequncies ($\omega > \omega_H$):

$$U_\varepsilon = U_\sigma = U_L = (1/2)\,[(L(e)/e] \quad \text{for all values of } e \qquad (6.24)$$

(4) Calculation of approximate values of ω_L and ω_H (Figure 6.1)

$$\omega_H \text{ is the solution of:} \qquad \varepsilon_{mod}(\text{with } U_L) \cong \varepsilon_{mod}(\text{with } U_U) \qquad (6.25a)$$

and

ω_L is the solution of: $\varepsilon_{mod}(with\ U_L) \cong \varepsilon_{mod}(with\ U_M)$ (6.25b)

(5) Complex permittivity in the quasi-static range $(\omega_L < \omega < \omega_H)$:

$$\varepsilon' = \varepsilon_L' - (\varepsilon_L' - \varepsilon_H')\ [ln(\omega/\omega_L)]/[ln(\omega_H/\omega_L)] \quad (6.26a)$$

$$\varepsilon'' = \sigma/\omega\varepsilon_o\varepsilon' \quad (6.26b)$$

$$\sigma = \sigma_L - (\sigma_L - \sigma_H)\ [ln(\omega/\omega_L)]/[ln(\omega_H/\omega_L)] \quad (6.26c)$$

where $(\varepsilon_L', \sigma_L)$ and $(\varepsilon_H', \sigma_H)$ refer to values of (ε', σ) at ω_L and ω_H, respectively.

6.4 Direct-Current Conductivity

Equation 6.19 can be rewritten to represent the static conductivity (σ_{dc}) of the mixture. The d.c. condition refers to the limiting case of $\omega \to 0$, or a factor τ_o (which is extremely large) should replace $\omega/2\pi$ in Equation 6.19. The factor to can be evaluated under the condition $x \to 1$, $\theta \to 1/2$, $U_\sigma \to 1$ and $\sigma_{dc} \cong (\sigma_1\sigma_2)^{1/2}$, representing the weighted-average value. It is found that

$$\sigma_{dc} \cong (\sigma_1 - \sigma_2)\ \{[2\pi\varepsilon_o(\varepsilon_2 - 1)/\sigma_1\tau_o]^{1-\theta}\ sin(\pi\theta/2)\}^{U\sigma} + \sigma_2 \quad (6.27a)$$

with

$$\tau_o = \frac{\pi\varepsilon_o(\varepsilon_2 - 1)}{\sigma_2} \frac{\sigma_1^{1/2} + \sigma_2^{1/2}}{\sigma_1^{1/2} - \sigma_2^{1/2}} \quad (6.27b)$$

and

$$U_\sigma = U_U = 1/2[1 - L(e)/e] \quad (6.27c)$$

6.5 Results Pertinent to Complex Susceptibility Model

The formulations presented in Section 6.3 and 6.4 have been verified by comparing the computed results obtained for a set of dielectric-conductor mixtures with the corresponding measured data available in the literature. Relevant results are presented in Table 6.1 and 6.2. Pertinent conclusions have been comprehensively discussed in [28].

6.6 Percolation Model(s)

The critical behavior of the dielectric permittivity of metal-insulator composites near the percolation threshold of conduction has been studied by Grannan et al. [30] using samples of a KCl matrix dispersed with small silver particles. Relevant studies indicate empirically that the dielectric constant obeys a scaling relation with a critical exponent factor, s. The following is the expression for the effective dielectric comstant of the metal-insulator mixture:

$$\varepsilon_{eff} = C[(\theta_c - \theta)/\theta_c]^{-s} \quad (6.28)$$

where C is a constant prefactor, θ is the volume fraction of the metal in the composite, θ_c is the critical volume fraction at which conduction begins and s is a critical exponent. The above scaling relation (Equation 6.28) is characterized by the critical exponent which resembles that observed in thermodynamic phase transitions.

Doyle and Jacobs [31] developed an effective cluster model to described the dielectric enhancement in metal-insulator composities. The basis for their model is as follows: Disordered suspensions contain a wide range of particle clusters of various sizes and shapes, composed of varying numbers of spheres in different spatial arrangements. Relevant to this type of metal-insulator mixtures, the effective permittivity (ε_{eff}) was deduced by modifying the Clausius-Mossotti equation with the inversion of a polarization parameter, β. The relevant expression is given by:

$$\varepsilon_{eff} = \varepsilon_n[1 + 3\beta/(1 - \beta) \quad (6.29)$$

where ε_n is the permittivity of the host medium and β is explicitly given by:

$$\beta = \theta\{1 + (\theta/\theta_c)\,[(1/\theta_c) - 1]\} \quad (6.30)$$

Again, the value of θ_c is empirically deduced from experimental data. The above mode is shown to fit a disordered suspension or mesosuspension of isolated conducting spheres and localized spherical, closely packed metallized clusters with a wide range of radii suspended in a background host dielectric.

While the above models refer to effective permittivity of metal-insulator mixtures, electrical resistivity (or conductivity) of such mixtures has been modeled by McLachlan et al. [32,33] *via* a general effective media approach combining percolation theory principles and effective media theories. The resulting formulations are elaborated in [32,33].

Chen and Johnson [34] have developed a model to describe the a.c. electrical properties of random metal-insulator composites wherein the metallic inclusions are filamentary or modular shapes. Again their model is based on power-law considerations pertinent to percolation principles.

6.7 Sillars' Model

Sillars [35] developed a model to describe the properties of a dielectric containing semiconducting particles of various shapes. His study reports Wagner's model and indicates the significant influence of conducting shaped particles (such as spheroidal particles) on the effective conductivity of dielectric-conductor mixtures.

6.8 Multilayered Conducting Dielectrics

A pertinent model to describe the effective permittivity characteristics of multilayered conducting dielectrics has been developed by Ongara [36] by exactly solving for Debye-like relaxation in such composites.

6.9 Granular Films of Conductor-Insulator Mixtures

Cohen et al. [37] portrayed the electromagnetic characteristics of granular silver and gold films by deducing their electrical properties *via* generalized Maxwell-Garnett theory.

The microstructural effects on the dielectric properties of granular composite films has been considered by Sheng [38] who deduced the effective dielectric function of such materials using Maxwell-Garnett theory.

6.10 Conclusions

From the various models discussed, it can be observed that the studies concerning dielectric-conductor mixtures are not totally comprehensive. Most of the models are empirical and are approximate. Further, the frequency dependency characteristics of such mixtures are far more incomplete. Closed-form expressions available offer results over only a limited range of frequencies and/or volume fractions. Studies on the shape dependency of the effective parameters are also significantly limited. Considering the fact that conductor plus insulator composites have wide applications in electromagnetic technology, the research in this area (though a century old) is rather incomplete, and offers a niche for futuristic in-depth studies.

References

[1] J. C. Maxwell-Garnett: Colours in metal glasses and metal films. Phil. Trans. A: Roy. Soc. London, vol. 203, 1904: 385-420.

[2] J. C. Maxwell-Garnett: Colours in metal glasses and metallic films and in metallic solutions I. Phil. Trans. A: Roy. Soc. London, vol. 205, 1906: 237-262.

[3] J. R. Wait: *Electromagnetic Wave Theory.* (Harper and Row Publishers, New York: 1985), pp. 69-71.

[4] Lord Rayleigh: On the influence of obstacles arranged in rectangular order up to the properties of a medium. Phil. Mag.; vol. 34, 1892: 481-502.

[5] R. Clausius: *Die Mechanische Behandlung der Elektricität*, Braunschneig, vol. II, (Vieweg, 1879),: pp. 62-97.

[6] O. F. Mossotti: Discussione analitica sull influenza che lazione di un mezzo dielectrico. Mem. di. Matem. e. Fisica di Modena: II, vol. 24, 1846: 49-74.

[7] L. Lorenz: Über die Refractionsconstante. *Ann. Phys.* (Leipzig), vol. 11, 1880: 70-103.

[8] H. A. Lorentz: *Theory of Electrons*. (Dover Publications, New York: 1952).

[9] H. Fricke: A mathematical treatment of the electrical conductivity and capacity of disperse systems I. Phys. Rev, vol. 24, 1924: 575-587.

[10] H. Fricke: A mathematical treatment of the electrical conductivity and capacity of disperse systems II. Phys. Rev, vol. 26, 1926: 678-681.

[11] D. A. G. Bruggeman: Berechnung verschiedener physikalischer konstanten von heterogenen subatanzen. Ann. Phys. vol. 24, 1935: 636-664.

[12] G. E. Archie: The electrical resistivity log as an aid in determining some reservoir characteristics. Trans. Am. Inst. Min. Met., Petrol. Engrs., vol. 5, 1942.

[13] H. Looyenga: Dielectric constant of heterogeneous mixtures. Physica, vol. 31, 1965: 401-406.

[14] C. J. F. Böttcher: *Theory of Electric Polarization.* (Elsevier Science Publishing Co., Amsterdam: 1952).

[15] K. Lal and R. Parshad: Permittivity of conductor-dielectric heterogeneous mixtures. J. Phys. D: (Appl. Phys.), vol. 6, 1973: 1788-1792.

[16] S. R. Wallin: Dielectric Properties of Heterogeneous Media. Ph.D. Thesis, University of Wyoming, 1985.

[17] R. M. Scarisbrick: Electrically conducting mixtures. J. Phys. D., vol. 6, 1973: 2098-2110.

[18] A. Kusy: Chains of conducting particles that determine the resistivity of thick resistive films. Thin Solid Films, vol. 43, 1977: 243-250.

[19] R. I. Frame and D. J. Tedford: Long-term electrical conduction in films of alkyd resin and graphite mixtures. IEEE Trans. Elec. Insulation, vol. EI-21, 1986: 23-29.

[20] O. Wiener: Die Theori des Mischkörpers für das Feld der stationären Strömung. *Abdhl. dy. kgl. Akad. d. Wiss. Leipzig,* vol. 32, 1912: 509-604.

[21] W. F. Brown: Dielectrics, in *Encyclopaedia in Physics*, V-XVII (Springer-Verlag, Berlin: 1956).

[22] L. K. H. van Beek: Dielectric behavior of heterogeneous systems, in *Progress in Dielectrics* (J. B. Birks, Ed.), (CRC Press, Cleveland: 1987).

[23] W. R. Tinga, W. A. G. Voss and D. F. Blossey: Generalized approach to multiphase dielectric mixture theory. J. Appl. Phys., vol. 44(9), 1973: 3897-3902.

[24] K. Lichtenecker: Mischkörpertheori als Wahrscheinlichkeitsproblem. Phys. Zeitsch., vol. 30, 1929: 805-809.

[25] K. Lichtenecker and K. Rother: Die Herleitung des logarithmischen Mischungsgesetzes aus allgemeinen Prinzipien der stationären Strömung. Phys. Zeitsch., vol. 32; 1938: 255-260.

[26] J. A. Reynold and J. M. Hough: Formulae for dielectric constant of mixtures. Proc. Phys. Soc., vol. LXX, 1957: 769-775.

[27] P. S. Neelakantaswamy, R. Turkman and T. K. Sarkar: Complex permittivity of a dielectric mixture. Corrected version of Lichtenecker's logarithmic law of mixing. Electron. Letts., vol. 21, 1985: 270-271.

[28] P. S. Neelakanta: Complex permittivity of a conductor-loaded dielectric. J. Phys. Condens. Matter, vol. 2, 1990: 4935-4947.

[29] S. A. Paipetis, G. M. Tsangaris and J. M. Tsangaris: Dielectric properties of metal-filled epoxies. Polymer Commn., vol. 24, 1983: 373-375.

[30] D. M. Grannan, J. C. Garland and D. B. Tanner: Critical behavior of the dielectric constant of a random composite near the percolation threshold. Phys. Rev. Letts., vol. 46(5), 1981: 375-378.

[31] W. T. Doyle and I. S. Jacobs: Effective cluster models of dielectric enhancement in metal-insulator composities. Phy. Rev. B., vol. 42(15), 1990: 9319-9327.

[32] D. S. McLachlan, A. Priou, I. Chenerie, E. Issac and F. Henry: Modeling the permittivity of composite materials with a general effective medium equation. J. Electromag. Waves Applns., vol. 6(9), 1992: 1099-1131.

[33] D. S. McLachlan, M. Blaszkiewicz and R. E. Newnham: Electrical resistivity of composites. J. Ann. Ceram. Soc., vol. 73(8), 1990: 2187-2203.

[34] I.-G. Chen and W. B. Johnson: Alternating-current electrical properties of random metal-insulater composites. J. Matl. Sci., vol. 26, 1991: 1565-1576.

[35] R. W. Sillars: Properties of a dielectric containing semiconducting particles of various shapes. J. Instn. Elec. Engrs. (UK), vol. 80, 1937: 378-392.

[36] R. Ongara: Exact solutions for Debye-like relaxations in multilayered conducting particles. IEEE Proc., vol. 133 pt. A. (5), 1986: 275-279.

[37] R. W. Cohen, G. D. Cody, M. D. Coutts and B. Abeles: Optical properties of granular silver and gold film. Phy. Rev. B, vol. 8(8), 1973: 3689-3701.

[38] P. Sheng: Microstructure and dielectric properties of granular composite film. Opt. Laser Tech., vol. 13(5), 1981: 253-260.

General Reading

[39] D. J. Bergman: Hierarchies of Stieltjes functions and their application to the calculation of bounds for the dielectric constant of two component composite medium. SIAM. J. Appl. Math., vol. 53(4), 1993: 915-930.

Defining Terms

Aspect ratio: The ratio of the largest dimension to the smaller dimension of a two- and/or three dimensional body.

Dielectric-conductor mixture: A two-phase mixture system constituted by a dielectric receptacle hosting a dispersion or a textured arrangement of discrete conducting inclusions.

Form factor (shape factor): A numerical value denoting the shaped extent of a nonspherical particle.

Percolation: A phenomenon in which the flux percolates or proliferates across a medium along random paths.

Table 6.1 Measured and Computed Data: Permittivity of Dielectric-Conductor Mixture

Mixture				Calculated Data on ε				Measured Data on ε	
				Method of [28]		Other Method(s)			
Host Dielectric Medium (ε_2)	Conducting Inclusions (σ)	Frequency (Hz)	Volume Fraction of Inclusions, θ	ε	Remarks	Semiempirical Best-fit data (ε)	Remarks	ε	Remarks
Alkyd resin (3.86)	Graphite lamellae	10^3	0.007	4.02	a/b = 1/13	4.01	b/a = 13	4.01 ± 6%	
	(200 S m^{-1})		0.200	4.34	$U_\varepsilon = U_M$	4.37	Shape factor	4.37 ± 6%	[19]
	Aspect ratio		0.060	5.63		5.66	B = 6.2	5.66 ± 6%	
	a/b<<1		0.130	9.39		9.15	[3]	9.15 ± 6%	
			0.170	12.87		12.25		12.25 ± 6%	
			0.250	25.09		22.97		22.97 ± 6%	
Epoxy (3.81)	Aluminum needles	1.6×10^6	0.050	4.56	B = 4.42	4.59	Logarithmic model [16]	4.71	
	(3.77×10^7 S m^{-1})		0.100	5.57	a/b = 1.5	5.55	$\varepsilon_1 = 165$	5.60	[29]
	a/b>1		0.150	6.94	$U_\varepsilon = (U_L U_U)^{1/2}$	6.70	Empirical value unjustifiably presumed in [16]	7.10	
			0.200	8.78		8.09	$\varepsilon_2 = 3.81$	8.18	
			0.250	11.25		9.77		9.16	
			0.300	14.50		11.80		12.05	

(continued..)

Table 6.1 Measured and Computed Data: Permittivity of Dielectric-Conductor Mixture

Mixture				Calculated Data on ε				Measured Data on ε	
				Method of [28]		Other Method(s)			
Host Dielectric Medium (ε_2)	Conducting Inclusions (σ)	Frequency (Hz)	Volume Fraction of Inclusions, θ	ε	Remarks	Semiempirical Best-fit data (ε)	Remarks	ε	Remarks
Aetna oil (2.21)	Mercury drops (10^6 S m^{-1}) a/b ≅ 1	10^3	0.015	2.29	a/b = 1	2.31	a/b = 1	2.31	Data due to Guillein as reported in [15]
			0.049	2.50	$U_\varepsilon = U_L$	2.57	Shape factor	2.57	
			0.171	3.80		3.88	B = 3	3.94	
			0.250	5.04		5.24	[14]	5.34	
			0.329	6.98		7.30	(Bruggeman's	7.15	
			0.401	9.45		10.27	formula)	9.67	
Mineral oil (2.1)	Iron spheroidal particles (10^7 S m^{-1}) a/b > 1	10^3	0.050	2.46	a/b = 1.3		a/b > 1	2.60	Data due to Nasuhoglu as reported in [15]
			0.100	2.93	$U_\varepsilon = U_L$		Shape factor	3.30	
			0.150	3.57			B ≅ 3.96	4.00	
			0.200	4.47			[14]	4.90	
			0.250	5.56			(Bruggeman's formula)	6.25	

Table 6.2 Measured and Computed Data: Dynamic Response of Dielectric-Conductor Mixture

Mixture			Calculated and Measured Data on Mixture Permittivity (ε)												
			θ = 0.007[a]		θ = 0.020		θ = 0.060		θ = 0.130		θ = 0.170		θ = 0.250		
Host Dielectric Medium (ε_2)	Conducting Inclusions (σ)	Frequency (Hz)	I[b]	II[c]	I	II	I	II	I	II	I	II	I	II	U_ε
Alkyd resin (3.86)	Graphite lamellae (200 S m^{-1})	10^1	4.60	4.03	4.80	4.40	6.60	5.87	10.00	10.46	13.20	14.98	28.60	31.86	U_M
	Aspect ratio a/b << 1	10^2	4.50	4.02	4.70	4.35	6.40	5.63	9.70	9.44	13.00	13.00	27.70	25.55	
	In the calculations a/b = 1/13	10^3	4.40	4.02	4.60	4.34	6.20	5.63	9.40	9.39	12.80	12.87	26.80	25.09	U_L
	In data of [19]	10^4	4.30	4.00	4.50	4.30	6.00	5.60	9.10	9.30	12.50	12.80	25.40	25.00	
	a/b = 1/13	10^5	4.10	4.00	4.40	4.30	5.90	5.60	8.70	9.30	12.00	12.80	25.00	25.00	

a Volume fraction of the inclusion: θ.
b Experimental results due to Frame and Tedford [19]: I.
c Calculated data as per the author's method [28]: II.

CHAPTER 7

Conductor-Loaded Polymeric and Ceramic Materials

7.1 Introduction

Conductor-filled polymers and ceramics are specific subsets of the conductor-loaded dielectrics discussed in Chapter 6. The primary reason for adding conducting particles in polymers, plastics, or ceramics is to enhance the electrical conductance of the medium. In general, polymers and ceramics when loaded with conductors become very good conductors of electricity and are useful in a wide range of electromagnetic applications. However, inclusion of a metallic constituent in a polymer matrix may affect the low density and high strength or impact resistance properties of the plastics or the ceramics. However, the particulate inclusions would increase the thermal conductivity of the matrix. In the following sections, the theoretical considerations in synthesizing conductor-loaded composites and specific applications of these composites are detailed.

7.2 Theoretical Considerations

Essentially a conductor-loaded polymer or a ceramic is a lossy dielectric. The electrical insulation property of the polymer or ceramic is affected (when the conducting particles are included) to an extent as controlled by the degree of filling and proximity of the conductive particles in the matrix. The particulate dispersion, in general, could be classified as: (i) Non-contact dispersion; (ii) particles at close proximity; and (iii) particulates in physical contact with each other forming a *chain* or a *web*. Depending on the three situations, the dielectric and conducting properties of the composite may vary significantly. The composites with isolated (non-contact) inclusions are predominantly dielectrics; and with the particles in contact bridging each other, the composite becomes predominantly a conductor. In the composites where the particles are in close proximity, the electrical conduction is controlled by *electron hopping* facilitated by the applied electric field force. Studies indicate that gaps as large as 10 nm could be jumped. The electron hopping is also termed as *tunneling* when the electrons can jump from valence band of ions/molecules to the conduction band without energy exchange (simple hopping, on the other hand, refers to situations where energy exchange is involved). Mott [1] developed a hopping model and deduced the probability p that an electron may hop from one energy state to another through an insulator given by:

$$p = A\ \tau\ exp[\Delta E/k_B T] + exp[\Delta E_p/k_B T] \qquad (7.1)$$

where A is a constant, ΔE is the *activation energy* required to hop the energy gap, $k_B T$ is the Boltzmann thermal energy, ΔE_p is the local polarization energy across the barrier, and τ is the tunneling factor. The tunneling factor exponentially decreases with the spatial distance through which the electron must tunnel; or, for tunneling to take place, close proximity of the particles is rather implicit. It should be noted that the hopping mechanism is also controlled by the temperature, T.

With the conducting particles physically in contact with each other, a random network or web of conducting chain is formed in the matrix; and the material exhibits *band-type conduction*. Band conduction and electric conduction behavior could be distinguished by studying the electric conduction behavior of the matrix under d.c. and a.c. excitations. Hopping conduction invariably exhibits higher a.c. conductivity which increases with frequency [2].

Composites with a large volume fraction of conductor inclusions show linear, Ohm's law type current-voltage relation, though some nonlinearity (presumably attributed to the presence of hopping mechanism) has also been observed in some composites. This nonlinearity has been declared as the effect of *localized polarization* and *space-charge induced effects*. Overall, electrical conductivity of conductor-loaded polymers and ceramics show variabilities depending on filler loading, particulate shape, processing conditions, polymer-metal wettability, and temperature.

There is a *critical volume fraction* of the conductor loading at which the resistivity of the composite falls sharply and the composite becomes predominantly a conductor. With the addition of conducting inclusions, a network of conducting links (either by direct contact or *via* hopping mechanism is formed permitting a current flow through the *percolation process* [3,4]. The formation of random network of conducting links and the associated current percolation is a stochastic process tangible for analysis through *Monto-Carlo simulations*.

The critical point at which the conduction sets in, the probability of continuous network of link formation (p_c) is given by [5]:

$$p_c = C_p/z \tag{7.2}$$

where C_p is the critical number of contacts per particle and z is the maximum number of possible contacts, or the *coordination number.*

The average number of contacts per filler particle is a function of the volume-fraction of the filler (θ) in the matrix. Denoting the maximum (possible) volume-packing fraction of the filler in the matrix as θ_m, the critical volume fraction, θ_c has been deduced in a functional form in terms of θ_m, C_p and z; and relevant theoretical results have also been compared with the experimental data. For example, considering a random dispersion of spherical conducting particles in a polymeric matrix, Gurland [6] determined experimentally the critical volume fraction (θ_c) as 0.38. With $\theta_m = 0.637$ for spherical particles, the coordination number (z) is 6. However, for random dispersion of the particles, the critical number of contacts per particle (C_p) is not definable inasmuch as the type of network of contacts established is rather random. Gurland determined the critical contacts for a random dispersion of well-defined types of lattice structures (such as *face-centered cubic*, *body-centered cubic*, *simple cubic*, and *diamond lattice*) of the conductor inclusions by measuring the corresponding critical values of θ_c. Relevant results indicate a near constancy of the value C_p as close to 3/2. Corresponding to this value of C_p, a semiempirical formulation to calculate the critical volume fraction of conducting sphere randomly dispersed in insulating medium has been established. It is given by:

$$\theta_c = \left\{1 + (z/Cp)[(1 - \theta_m)/\theta_m]\right\} \tag{7.3}$$

With $z = 6$ (for spheres), $C_p = 3/2$, and $\theta_m = 0.637$, θ_c is obtained as 0.35, which is close to the experimental result (0.38) due to Gurland [6].

On the basis of the consideration that proportionate volumetric change of conducting inclusions would cause a proportionate change in the conductivity of the composite, a stochastic model was developed by the author [7] to determine the effective conductivity of the composite. It is given by:

$$\sigma_{eff} = \sigma_m z \exp\left[-\exp\left(-\theta/\theta_m\right)\right] - ze_x \tag{7.4}$$

where $e_x = (1/e)^{1/e} = 0.692200627$. Further, in terms of the coordination number z, the maximum volume fraction of the filler has been deduced as:

$$\theta_m = -1/ln\,\{ln[\frac{1}{(e_x + 1/z)}]\} \tag{7.5}$$

The above results are also valid when the nonconducting phase refers to air voids. In Table 7.1 a comparison is made between the results on θ_m as obtained by the above formulation and that due to bond percolation theory.

Table 7.1: Maximum Volume Fraction *versus* Coordination Number

Type of Lattice	z	θ_m Results due to [7]	Bond Percolation Theory [5]
Face-Centered Cubic	12	0.7301	0.740
Body-Centered Cubic	8	0.625	0.680
Simple Cubic	6	0.531	0.520
Diamond	4	0.354	0.340

Theories concerning the elucidation of electrical conductivity of conductor-loaded polymers and ceramics when the particulate inclusions are flakes, fibers, and other irregular shapes are rather sparse. For almost spherical but irregular particulate dispersion, the critical volume fraction (θ_c) can be taken as that of spherical inclusions.

For shaped inclusions such as ellipsoids, spheroids, flakes, and coarsely spherical particles, the mixture formulations developed by the author as reported in [8] facilitate the evaluation of the effective conductivity of the mixture both at d.c. and a.c. conditions (relevant theory and results are presented in Chapter 6). These formulations can be readily extended to polymer matrix or ceramic matrix with shaped conducting inclusions, as well.

Yamaki et al. [9] developed a model for the effective conductivity of a three-dimensional polymeric matrix dispersed with conducting fibers. This model presumes that:

- The fiber inclusions are all alike with the same aspect (length/radius) ratio.
- The dispersion of the fibers across the matrix is random and uniform.
- Fiber centers coincide with lattice points.
- Conductivity of the polymer is zero (that is, the polymer is a perfect dielectric).
- Electrical conduction is facilitated by the bridging network of fibers in any spatial direction (isotropic randomness).
- The conductive paths multiply across the matrix and terminate at the two faces of the composite cube.

Their theoretical model is based on *percolation theory* [3,4]. However, no comparison of results on the predicted conductivity has been made. On the contrary, the assumptions pertinent to the theoretical approach due to the author [8] are less restricted and the results obtained thereof have been compared with existing experimental and other theoretical data. Further, the relevant formulation(s) permit the host medium to be a lossy dielectric.

Milewski [10] has deduced the maximum volume fraction of fiber inclusions *versus* the aspect ratio of the fiber in a completely random distribution. The relationship between the aspect ratio and maximum volume fraction has been shown to be exponential.

7.3 Application Potentials of Conductor-Loaded Ceramics and Plastics

Conductor-filled polymers/plastics and ceramics have significant application potential in the electromagnetic area as listed below [11]:

- Electromagnetic interference (EMI) shielding.
- Electrostatic dissipative/conductive materials.
- Electromagnetic phantom materials.
- Radar absorbing materials (RAM) for the radar stealth technology.
- Electrically conducting plastic-like materials such as tapes, surface heating conductors, cable-sleeve terminations, etc.

The state of the relevant composite materials and their use in the aforesaid applications are detailed below.

7.4 Constituent Materials

Invariably, the conductor-loaded polymers or ceramics are *diphasic*. That is, they are constituted of simple, discrete conducting inclusions dispersed in a polymeric/ceramic host. Apart from conventional metals like copper, aluminum, etc., carbon is also widely used (in graphite form) for the same purpose due to its high electrical conductivity and low cost, low density considerations. Other than aluminum, other metals have higher density than carbon. Semiconductors, solid-state electrolytes, and salts are other possible conducting inclusions compatible for synthesizing the composites under discussion.

Classically, to study the electrical conduction phenomenon in conductor-loaded composites, the combinations such as aluminum needles plus epoxy, Bakelite™ plus spherical silver particles, alkyd resin plus graphite lamellae have been used to constitute such composite materials. In addition, for the same purpose, use of aetna oil plus mercury drops, mineral oil plus spheroidal iron particles, and paraffin wax plus metallic powder have also been considered.

To synthesize conductor-loaded composites, three classes of conductors are used in practice. They are:

- Metallic inclusions (aluminum, copper, iron, stainless steel, and silver)
- Nonmetallic inclusions (carbon black, graphite, ferrous oxide, salts of copper and aluminum with and without binders, solid electrolytes, and semiconductors)
- Metal-coated dielectrics (metal-coated glass, nickel-coated fiberglass)

The general characteristics and the availabilities of these conducting inclusions are detailed in Table 7.2.

The host material used in the fabrication of conductor-loaded composites are selected on the basis of:

- Minimum degradation of the host medium due to catalyzation and/or oxidation of the conducting inclusions
- Wetting of metallic inclusions under heavy current operations

- Wetting property between the host and the inclusions
- Morphology of the polymer or ceramic material (amorphous, semicrystalline)
- Compatibility towards the processing technique(s) involved

As a consequence of the above considerations, the choice of polymers or ceramics for conductive composite applications is rather limited. A few examples, however, adopted in practice are:

- ABS or polycarbonate: Popularly adopted in molding business machines. With copper inclusions, however, this host material may suffer degradation due to catalyzation of copper.

- ABS and polyolefins: Poor wetting property with most metals. Addition of polypropylene with aluminum fiber/flake improves the wetting behavior.

- Graphite: This is a popular matrix material in constituting *cermet* composites. Metal-included graphites are useful as electrical contact materials such as brushers for electrical machines. It is conducive for baking the molded (powdered) metals at high temperatures (1000-1400°C). Copper-graphite cermet contacts are noted for a minimum welding tendency in heavy current circuit breakers.

- Cadmium oxide: Oxide-based cermets are extensively used in low voltage applications in tropical climates. Silver, nickel or tungsten are the metals which constitute the conducting inclusion components.

- Polymer concrete

- Polycarbonate

7.5 Characteristics of Conductor-Loaded Polymers and Ceramics

The characteristics of a conductor-loaded polymer or ceramic are decided by their requirements as end-= products. In general, the following properties are of engineering importance:

A: Electrical properties

- Bulk or volume resistivity (effective d.c. conductivity)
- Surface resistivity
- Effective permittivity (at static and at optical limits)
- Effective permeability (in case of the conductive filler is a magnetic material)
- Effective loss tangent of the material at a given frequency of operation
- Effective dielectric response of the material under time-varying field specified by the effective complex permittivity, $(\varepsilon'_{eff} - j\varepsilon''_{eff})$ (dielectric relaxation)
- Effective permittivity of the material with the volume fraction of the conducting fillers less than the critical volume fraction at which the electrical conduction occurs
- Effective dielectric breakdown strength

- Degradation of electrical parameters with shelf life, aging, and under repeated electrical overstressing
- Electrostatic (charge) bleed-off characteristics
- Variation of electrical characteristics with ambient conditions such as temperature, humidity, corrosive agents, etc.
- Effect(s) of the shape of the end product (in bulk form, thick film or thin film) of the composite in dictating its effective electrical characteristics
- Effect(s) of the resiliency of the end product (being soft, flexible, hard, tough, etc.) as decided by the type of polymeric or ceramic chosen in characterizing the effective electrical properties of the composite
- Effects of the heavy currents (particulate welding), and arc-induced erosion

B: Nonelectrical properties

- Cost-effectiveness
- Density
- Thermal endurance
- Corrosion resistant
- Resistance to nuclear radiations, ionizing radiations, and electromagnetic pulsing (EMP)
- Mechanical strength, hardness, durability, etc.
- Brittleness and flexibility
- High strength-to-weight ratio
- Aesthetic appearance in terms of color and surface finishes of the end products
- Porosity and defect sites
- Compatibility for use with other conventional monolithic or composite materials with adhesive and/or fastening feasibilities

C: Characteristics pertinent to processing and manufacturing

These characteristics are essentially decided by the constituent phases of the composite. The type of the conducting inclusion (in terms of shape, size and electrical conductivity) and the properties of the host medium (such as its complex permittivity, wettability, etc.) decide jointly not only the effective electrical parameters of the composite, but also the nonelectrical characteristics mentioned before as well as the processing/manufacturing method(s) involved.

Spherical or irregular shaped fillers permit conventional processing of being mixed into the polymer by batch mixing (using Banbury mixer) and extrusion compounding. For theoretical prediction of the effective conductivity of these composites, it can be well assumed that the particulates are randomly dispersed. The classical effort of Gurland [6] on Bakelite™, (thermosetting phenol-formaldehyde resin) plus silver composite, and other studies on oxide of iron particles plus dimethyl formamide indicate that the corresponding random mixtures exhibit varying extents of critical volume fractions (ranging from 0.2 to 0.4). The reason is that the critical volume fraction is not controlled by the concentration of the fillers, but also by the particle-size distribution [12]. Some confirming evidence to this observation refers to the experiments on aluminum or iron particles plus styrene-acrylomitride copolymer composites.

Broad particle size distributions increase the maximum loading of the filler particles and lower the probability of contacting a neighboring particle warranting increased volume fraction to start the threshold of conduction.

Another factor that influences the critical volume fraction is the electrical overstressing of the composite. Studies indicate that mixtures which are nonconducting (with volume fraction of fillers less than the critical value) become highly conducting after subjecting them to a high (critical) voltage. This is referred to as the *switching* phenomenon.

The processing and realizing of a composite with specified electrical characteristics are even more complex with shaped particles such as flakes and fibers.

With flakes, the processing could affect the effective conductivity of the composite significantly. Compared to compression molding, injection molding may align the flakes in the flow direction, thereby creating a directional anisotropy of conductivity. Synthesizing with conducting flakes as inclusions, in general, would facilitate more stable conducting composites.

7.6 Applications, Fabricational Aspects, and Characteristics

Pertinent to various generic properties and characteristics described in the previous sections, a number of application-specific conductor-loaded composites have been developed in the past and such materials have posed extensive potentials for the present and future developments in material technology *vis-a-vis* electromagnetic applications. Both polymer-based and ceramic-based matrices are used in practice. Likewise, the conducting particulate inclusions of different shapes and different materials (metallic and nonmetallic) are considered as potential candidates in developing viable composites. In the following section, specific applications are identified and compatible composites are elaborated in terms of their characteristics, processing, merits, and demerits.

7.6.1 Metallo-Plastics [13]: These are substitutes for metals in certain electrical engineering applications. They are constituted by the addition of a metal (in particulate, fiberous, or flaky form) in a polymeric (plastic) host material. The *metallo-plastics* have therefore considerably larger electrical and thermal conductivities in comparision with a pure plastic material. They are lightweight and present aesthetic appearance of plastics. Due to their high electric conductivity, they provide electromagnetic shielding (Chapter 21) and prevent electrostatic buildup on plastics (Chapter 20). They are superior to plastics with metallic surface coatings in terms of cost factor and damage resistance. They are also better than carbon-filled plastics considering the plastic strength, impact resistance, and color (options are not just restricted to black only).

The metal filling in plastics refers to the use of chunky fragments, near-spherical particulates, or fibrous whiskers. Of these, the use of fibers provides more effectiveness in controlling the electrical resistivity and provides better thermal conductivity as well.

The controlling of electrical conductivity of metallo-plastic depends on the volume fraction of the inclusions such that the average number of contacts with neighboring particles (coordination number) reaches a minimum value as observed by Gurland [6]. More theoretical insight, on the electrical behavior of conductor plus insulator compositions have been presented in Chapter 6. The critical volume fraction at which the conductivity of the composite begins to increase depends on the particulate shape, namely, chunky, fibrous, spheroidal, or near-spherical. Invariably, a low volume fraction is required with fibers to achieve a specified conductivity because the geometry of the fibers provides greater points of contacts with neighbors even at lower concentrations of the fibers included.

In general, controlling the conductivity of the metallo-plastic is rather a stringent design consideration. This due to the fact that resistivity of the metallo-plastics alters drastically even with small changes in the volume fraction.

Applications of metallo-plastics include:

- EMI shields
- Electrostatic control media

- Plastic housings for electrical equipment (such as computers) to provide good thermal dissipation, electrical grounding feasibilities, electrostatic discharge and shielding against electromagnetic influence to meet compliance requirements.
- Lightweight electrical designs as may be required in aerospace applications.
- Lightning protection devices on the leading edges wings and tail section of aircrafts.
- Electrical applications *in lieu* of metals susceptible for corrosions.

Electrical performance of state-of-the art metallo-plastics relative to metals and plastics is summarized in Figure 7.1.

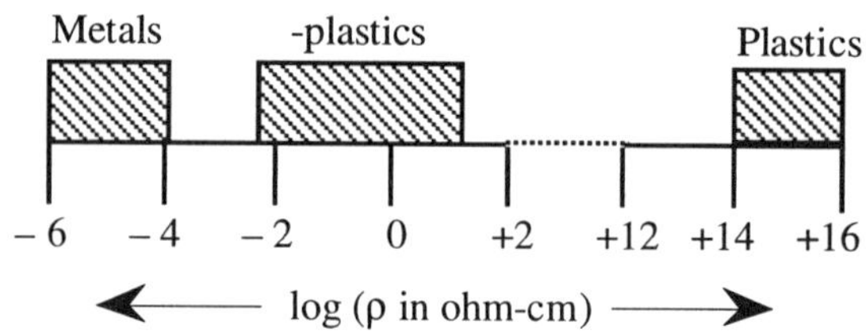

Figure 7.1 Electrical conductivity ranges of metals, plastics and metallo-plastics.

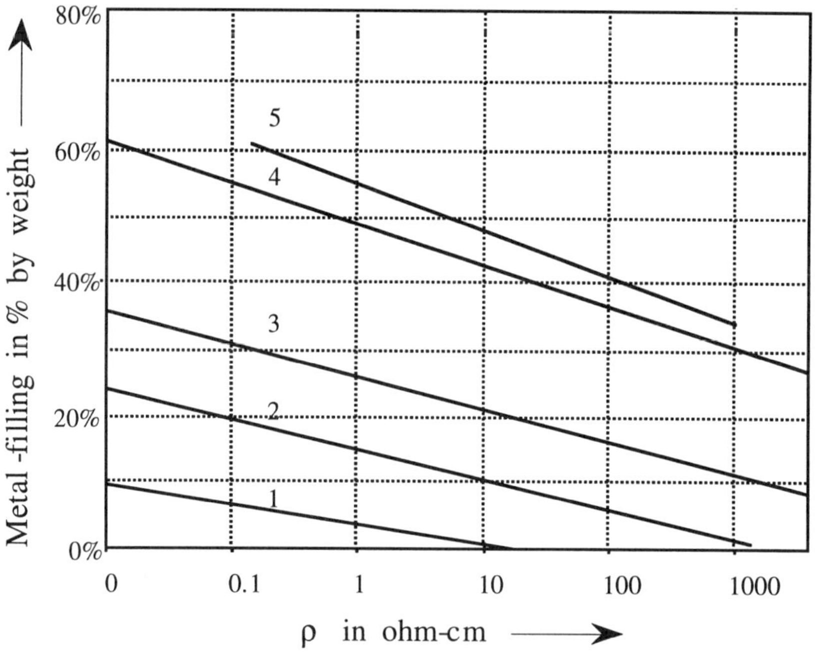

Figure 7.2 Resistivity *versus* weight fraction of metal filling in a plastic material. 1. Cross-link formulated composites; 2. 3/4 in. length metal fiber dispersion; 3. 0.05 in. metal fiber dispersion; 4. 3/4 in. clumped metal (chunky) dispersion; 5. 2000 Å carbon particulate dispersion.

In processing of metallo-plastics, low-cost, high conductivity materials can be realized with metal fibers (of 20-30 mil lengths) in conducive injection molding and/or extrusion processes. A wide range of plastic host materials and the conventional plastic-making technologies permit effectively designed end products. Uniformity of dispersion, fiber length, and concentration with special types of interlinking the fibers (such as *Cross-link Process*™ of MBAssociates) are fabricational considerations in developing metallo-plastics.

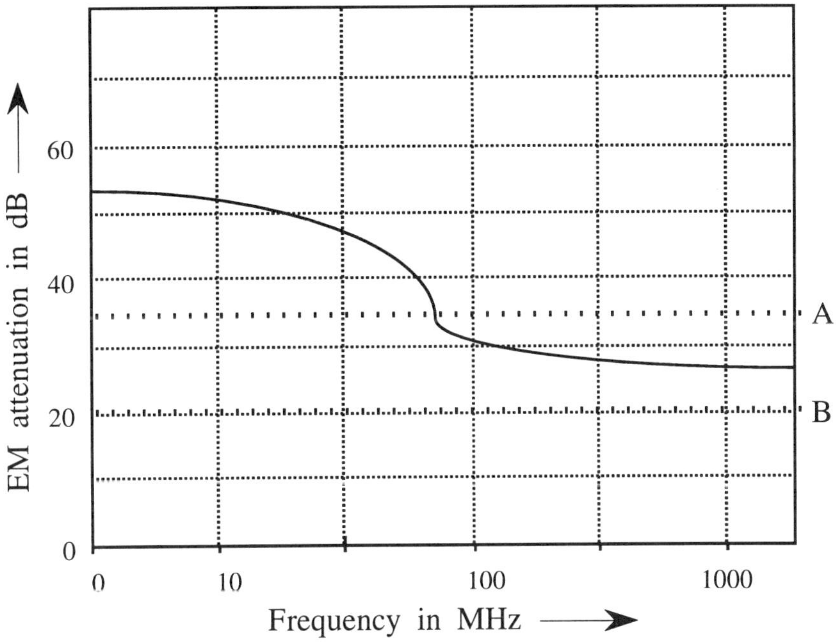

Figure 7.3 Plane-wave electromagnetic transmission loss (attenuation) *versus* frequency (10% metallic-filling in fibrous form); A: Stringent EMC requirements; B: Normal EMC requirements.

B. Conducting Polymers: These are polymeric materials doped with substances to increase their electrical conductivity. The doping is done by chemical, electrochemical, or ion implantation methods. Such materials are useful in electromagnetic interference (EMI) shielding, microwave absorbers, display units, and junction devices. They could be fabricated in sheet and/or film forms.

The family of conducting polymers includes:

- Polyacetylene
- Poly-p-phenylene-benzobis-thiazole (PBT)

Potential applications of conducting polymers cover high frequency uses enclaving EMI shielding and microwave absorbtion. Materials like polyacetylene and PBT could provide shielding effectiveness less than –40 dB for frequencies above 4 GHz. They are then useful as lightweight shielding of cabinets housing high-data-rate electronic equipment such as supercomputers. Conducting polymers can be used for microwave absorption using configurations such as the *Salisbury screen*. Owing to the higher conductivity/weight ratio over carbon-impregnated nonmetallic composites, conducting polymers are attractive as radar absorbing materials (RAM) in stealth technology. At room temperatures, it has been observed for a material like PBT that the d.c. and microwave conductivities are within a few percent; this suggests that the frequency dependency of the microwave conductivity of PBT is perhaps insignificant, which implies desirable broadband applications. More details on conducting polymers are furnished in the next chapter (Chapter 8).

7.7 Conclusions

Metal-loaded plastics and ceramics constitute a vital class of electromagnetic composites with current applications in many areas and have a promising trend in futuristic technology.

References

[1] N. F. Mott: Electrons in disordered structures. Adv. Phys. (Phil. Mag. Suppl.), vol. 16, 1967: 49-144.

[2] R. E. C. Read and C. D. Stow: An experimental investigation of charge transport through rubber. J. Phys. D: Appl. Phys., vol. 2 (Sr. 2), 1969: 567-576.

[3] G. Giraud, J. M. Laugier, J. Clerc and J. Roussenq: A.C. conductivity of a random medium: A percolation approach. IEEE Trans. Elec. Insulation, vol. EI-1(3), 1984: 205-209.

[4] B. Bridge and H. Tee: Large-scale simulation of the critical volume fraction for the percolation threshold in metal-fiber-loaded polymer composites. Int. J. Electronics, vol. 69(6), 1990: 785-792.

[5] D. M. Bigg: Electrical properties of metal-filled polymer composites, in S. K. Bhattacharya (Ed.): *Metal-Filled Polymers.* (Marcel-Dekker, Inc., New York: 1986).

[6] J. Gurland: An estimate of contact and continuity of dispersions in opaque samples. Trans. Metall. Soc. AIME, vol. 236, 1966: 642- .

[7] P. S. Neelakantaswamy and S. Kisdanasamy: Electrical conductivity of sintered materials: A random phase model independent of the shape factor. Proc. Int. AMSE Conf. "Modeling and Simulation" (Athens, Greece, June 27-29, 1984), vol. 2.1, pp. 69-92.

[8] P. S. Neelakanta: Composite permittivity of a conductor-loaded dielectric. J. Phys. (Condensed Matter), vol. 2, 1990: 4935-4947.

[9] J. Yamaki, O. Maeda and T. Katayama: Electrical conductivity of conductive filler-polymer composites. Rev. Electr. Commun. Lab. (Tokyo), vol. 26, 1978: 616-628.

[10] J. V. Milewski: The combined packing of rods and spheres in reinforcing plastics. Ind. Eng. Chem. Prod. Res. Dev., vol. 17, 1978: 363-366.

[11] R. P. Kusy: Applications, in S. K. Bhattacharya (Ed.): *Metal-Filled Polymers* (Marcel-Dekker, Inc., New York: 1986).

[12] A. Mallaris and D. T. Turner: Influence of particle size distribution on the electrical resistivity of compacted mixtures of polymeric and metallic powders. J. Appl. Phys., vol. 42, 1971: 614-618.

[13] D. E. Davenport: Metallo-plastics-high conductivity materials, in R. B. Seymour (Ed.): *Conductive Polymers*. (Plenum Press, New York: 1981).

Defining Terms

Conducting polymers: Polymeric materials doped with certain impurities to exhibit enhanced electrical conductivity (also known as *organometallics*).

Critical volume fraction: Minimum volume fraction of the conducting fillers required at which the conductor-filled composite exhibits abrupt transition in electrical resistivity.

Metallo-plastics: Specific class of composite constituted by a host dielectric with a dispersion of flaky or fiberous metal or metal-coated inclusions.

Percolation theory: Study concerning the spreading of entities like fluid, electric current, etc. through a randomly blocked medium.

Table 7.2 Conductive Fillers for Synthesizing Conductive Composites [5]

Filler Type	Material	Shape	Particulate Size	Resistivity ohm-cm	Density g/cm^3	Remarks	Source
Metal	Copper	Dendritic	~ 44 μm	1.70×10^{-6}	8.82	Susceptible for catalyzation and degradation of host polymer	U. S. Bronze Powder, Inc.
		Dendritic	~ 100 μm				
		Flake	~ 58 μm				Belmont Smelting & Refining Co.
		Electrolytic	< 50 μm				Assam Carbon Products
		Spherical	< 100 μm			Surface oxidization	
	Nickel	Spiky, equiaxial	4–7 μm	6.84×10^{-6}	8.90	Low rate oxidization	Inter, Nickel Co.
		Irregular	52–74 μm				Glidden
		Spherical	~ 4 μm				
		Irregular	10–30 μm				
		Irregular	~ 20 μm				
		Flakes	0.001mm × 2 μm × 2 μm				

(continued...)

Filler Type	Filler Material	Shape	Particulate Size	Resistivity ohm-cm	Density g/cm^3	Remarks	Source
	Aluminum	Irregular	~ 5 µm	2.80×10^{-6}	2.70	Surface oxidization	Alcan Metal Powders
		Smooth anisodiametric	< 15 µm			Low cost	
		Irregular	~ 100 µm			Low density	
		Spherical	< 100 µm				
		Ribbon	0.001 mm × 0.01 mm × 3–300 µm				M. B. A. Tracor
		Fibers	0.1 mm×1.25 µm				Transmet Corp.
		Flakes	0.025 mm × 1.0 mm ×1.25 mm				Transmet Corp.
	Iron	Irregular	20-100µm	10×10^{-6}	7.86	Surface oxidization	
	Stainless steel	Fibers	4–50 µm Dia.	74×10^{-6}	7.80		N. V. Bekaert, Brunswick
		Fibers	4–79 µm Dia.				
	Silver	Spherical	~ 100 µm	1.59×10^{-6}	10.5		
		Spherical	~ 100 µm				
		Electrolyzed polyhedral	75–100 µm				

(continued...)

Type	Filler Material	Shape	Particulate Size	Resistivity ohm-cm	Density g/cm^3	Remarks	Source
Non-metallic	Carbon black	Special	20-100 nm				Cabot Corp.
	graphite						
	Salts						
	Solid electrolytes						
Metal-Coated	Aluminum or nickel coated on glass fibers	Roving with less than 100% coated area					IBM Lundy Electronics
	Aluminum coated E-glass fiber	Continuous roving or chopped fiber	2.5 mm coating on 12.5 mm diameter fiber				M. B. A. Tracor
	Silver coated glass particles	Spherical	Dia. < 44 mm 53 to 105 coating thickness 0.02 to 0.1 mm silver content 4 to 12%				Potters Industries

CHAPTER 8

Conducting Polymeric Materials

8.1 Introduction

A highly *conjugated organic* polymer (having unsaturated bonds spaced along the polymer chain intermittently) such as polyacetylene has been known to exhibit high electronic conductivity when oxidized by suitable reagents. Likewise, a number of conjugated hydrocarbon and aromatic heterocyclic polymers (such as poly-p-phenylene, poly-p-phenylene-vinylene, poly-p-phenylene sulfide, polypyrrole and polythiophene as well as radical salts like tetracyanoquinodimethane (TCQN) complexes have also been identified as *conducting polymers* or *organometallics*. Approximate conductivities of various organic materials are presented in Table 8.1. Also, shown are the conductivities of typical inorganic materials for comparison.

Table 8.1 Conductivities of Organic and Inorganic Materials

Inorganic Materials	Polymers	Conductivity siemen/cm
Cu		$>10^5$
	Graphite	10^5
	Pyropolymers	10^0
Bi	Polyacetylene* $[(CH)_x]$	1
Si*	Poly-p-phenylene* $[(C_6H_4)_x]$	1
ZnO*	Polyphenylene sulfide* $[(C_6H_4 - S)_x]$	1
	Polypyrrole*	
	Polythiophene	1
Si		1
	Polyphthalocyanine	10^{-5}
	Polyacetylene	1
ZnO		10^{-10}
	Polydiacetylene	10^{-10}
	Polythiophene	
	Polypyrrole	10^{-15}
	Poly-(-phenylene	10^{-15}
	Polyphenylene sulfide	
	PVC	
SiO_2	Polyimide (Kapton™)	
Diamond	PTFE (Teflon™)	10^{-20}

* Heavily doped materials.

In terms of technological applications of these materials, they have been deployed as rectifiers, sensors, solar energy conversion elements, fuel cell components, switching devices, photoresist elements, chemoselective electrodes, electrophotographic devices, and durable synthetic materials replacing metals. Speculative applications on possible high temperature superconductive applications of these materials also prevail.

The conducting polymers are prepared in general *via* molecular doping of nominally insulating saturated backbone polymer matrices. The nature of the electronic states in such materials has been modeled to describe these states, addressing specifically two issues, namely: What is the influence of the molecular architecture of various polymer chains on their high energy π-electron valence states and, what are the consequences of differing backbone conformations on these states? The complexity of answers to these questions has set the understanding the fundamental aspects of structural and electronic characteristics of these materials which govern the charge transport only at a primitive level, despite of the intense activity in this field in recent times.

8.2 Requisites of Conducting Polymers

In realizing high conductivity in a polymer there are two necessary considerations:

- Conversion of an organized collection of molecules into a conductive molecular array by positioning them in close proximity with sufficient intermolecular orbital overlaps to provide a continuity of electronic paths for carrier delocalization (or migration) and in crystallographically similar environments.
- The arrayed molecules must exist in formal fractional oxidation states dubbed as *mixed valence* or *incomplete* charge transfer or *partial oxidation* states.

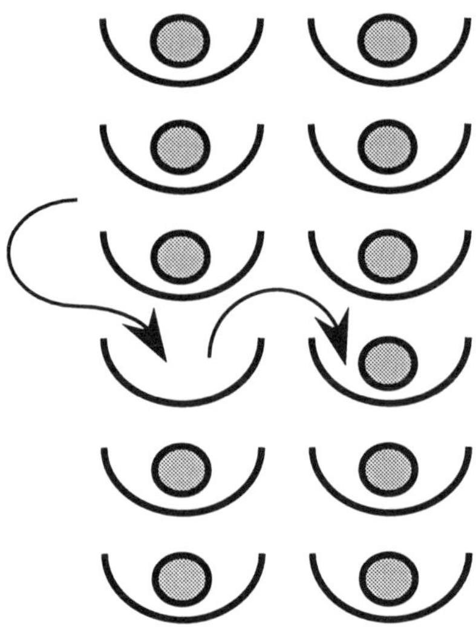

Figure 8.1 Charge mobility due to partial oxidation in a simple molecular stack.

The above requirements warrant the molecular entities be connected in series with fractionally occupied electronic valence shells. A simplified valence bond of this situation is depicted in Figure 8.1 illustrating the influence of partial oxidation which facilitates charge mobility *via* the creation of numerous electronic vacancies.

Pertinent to the one-dimensional molecular stack shown in Figure 8.1 (Hubbard model), the conductive molecular property dictates large *on-site coulombic repulsions** and relatively small *transfer integrals* or narrow *bandwidths***.

8.3 Electronic State in Polymers

To understand the concept of electronic conductivity in polymers with the requisites at the molecular level as stated above, the electronic states in polymers should be formalized.

A polymer solid, in general, consists of an assembly of very long molecules such that within the molecular chain there are strong covalent bonds but that between chains only weak bonding (usually of van der Waals type) occurs. The chains are made up of a very large number (10^5 or more) of small identical units repeatedly bonded together. Each unit can be regarded as a separate molecule with electronic states consisting of the molecular orbits of the molecule. These molecular orbits degenerate on each unit, overlap in space, and lift their degeneracy by forming a striation of extended electronic states known as *energy bonds*. These bonding and nonbonding units lead to valence and conduction bonds, respectively, with an energy gap of separation between these bands.

Apart from the band structure of polymers, structural disorders (or defects) in the polymers also influence the electronic conduction phenomena in these solids. Both extrinsic defects such as chain-end groups, leftover impurities from the fabricational process, deliberate commercial additives as well as intrinsically ever-present oxygen (in molecular or bound in oxidized products) can be crucial in dictating the electronic transport in polymer materials.

The polymer is considered as an infinite periodic array of sites with which are associated atomic *wave functions* implicitly representing the potential energy of the sites. These wave functions overlap from site-to-site in a simple tight-binding band structure (as modeled by Bloch). Such overlaps lead to nearest-neighbor interactions with interaction energies (called the *transfer integrals* or the *resonance*) as decided by the bond length between the interacting units.

Associating momenta to correspond to the eigenenergies of the band structure, movement of electrons in a band can be described by an effective mass (correlated to its momentum) and a mobility (correlated to its velocity) as dictated by the associated energy function; and the conductivity of the material qualitatively describes the ease with which the electron movement is facilitated in the interacting environment. Depending on the strength of coupling between the charge carrier (say, electron) and the polymer backbone, the electron-lattice collision/scattering time decides the mobility and hence the conductivity of the material.

A high mobility is anticipated for *π-electrons* (valence band electrons) moving along a polymer chain due to interamolecular coupling of orbitals. Considering an ideal infinite chain of polyacetylene $(CH)_x$, π-electrons from an half-filled band lead to metallic or conductive behavior. This simple model arising from an one-dimensional molecular stacking has, however, been subjected to scrutiny leading to newer concepts and speculations about charge transport mechanism supplemented by rigorous experimental studies on conductive polymers.

---

* *On-site coulombic repulsions*: The electrostatic repulsion is pertinent to the sites of infinite array of atoms in a tight-binding band structure of polymers.

** *Bandwidth/transfer integral/resonance:* Considering the tight-binding between the sites representing the infinite periodic array of atoms, the interaction between nearest-neighbors refers to the *transfer integral*.

8.4 Conducting Polymeric Materials: Characteristics

8.4.1. Doped polyacetylene

These materials are highly conducting. By control of dopant type and concentration, properties ranging from those of insultors and semiconductors to metals can be obtained with variations in conductivity of 12 orders of magnitude. Dopants include Li, I_2, $H_2IrCl_6 \cdot 6H_2O$, AsF_5 and others. Dopants are added to the polymer from solution or the vapor phase, or electrochemically. In vapor-phase doping, a polyacetylene film is placed in a glass vessel to which a bulb containing iodine is facilitated. At a maintained temperature of the bulb, the film is exposed to I_2 vapor (at a constant vapor pressure). Doping level is controlled *via* vapor pressure and exposure time.

The conductivity of iodine-doped polyacetylene is determined by the morphology and microstructure of the starting material. The geometrical configuration of the polymer backbones seems to play only a moderate role in determining the conductivity. Materials with an essentially 100% *cis* configuration have conductivities ranging from a low of 150 to a high of 700 siemen/cm. The introduction of a moderate amount of the *trans*-isomer lowers the conductivity by a factor of three or more to 50 to 80 siemen/cm. Further increase in *trans*-isomer, however, does not have significant effect. Considering the morphology of the material, it is generally observed that the conductivity increases as the diameter of the fibrils increases (200Å to 300Å). Globular morphology yields lowest conductivity.

Typical variation of conductivity of polyacetylene as a function of iodine concentration is shown in Figure 8.2.

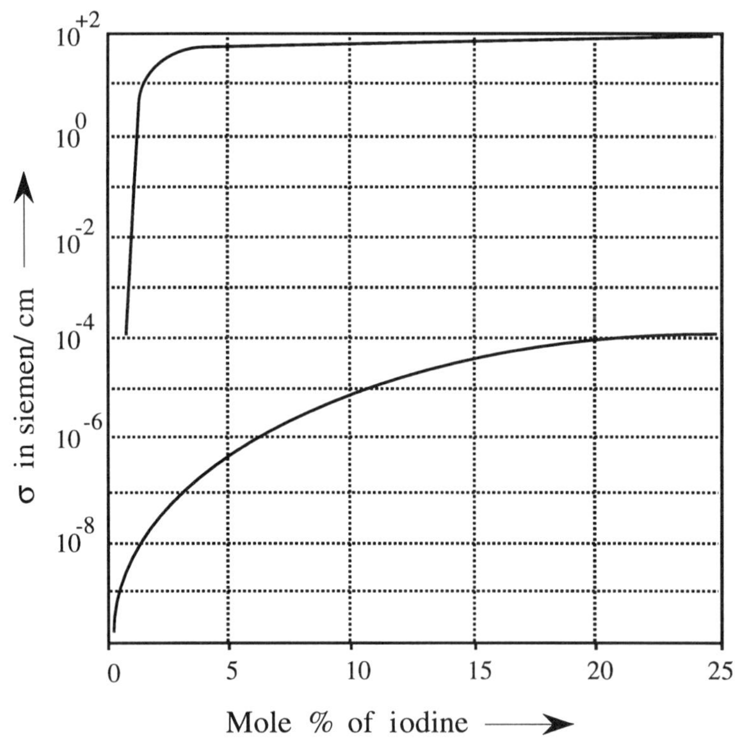

Figure 8.2 Conductivity of polyacetylene as a function of doping concentration.
A: *Cis* content 74%, fibrillar morphology.
B: *Cis* content 0%, fibrillar morphology.

Polyacetylene is the best example of a covalent organic material chemically doped either with p or n material to give a series of semiconductors and ultimately *organic metals.*

8.4.2. Doped aromatic polymers

These are PPP-poly-p-phenylene, PPV-poly-p-phenylene-vinylene, PPS-poly-p-phenylene sulfide, PYR-polypyrrole, and polythiophene compounds which have been made highly conducting *via* doping. They are built with either six-membered phenylene rings or the five-membered hetro rings which are linked together directly or indirectly by a C_2H_2 group and a S atom in the case of PPV and PPS, respectively.

Undoped polymers of this kind have conductivities in the range 10^{-10} to 10^{-16} siemen/cm. Upon doping with acceptors or donor-type dopants, the conductivity of these polymers increases to 1 to 10^3 siemen/cm. This metallic state conductivity is temperature dependent. It decreases by a factor of two from room temperature to 20 ^oK except for PPS for which σ decreases by a factor $10^3 - 10^4$ between 300 to 400 ^oK. Conductivity *(σ)* is found to be proportional to $(\text{temperature})^{-1/4}$.

8.4.3. Ionically conducting polymers

These are organic counterparts of *inorganic solid electrolytes* discussed in Chapter 16, though ionic polymers are themselves sometimes referred to as solid electrolytes. More specifically ionic-conducting polymers are also designated as *polymer complex.*

These compounds are of two types: The first type refers to the *homopolymers* which do not have high ionic conductivity (or electronic conductivity) unless doped or blended with other compounds. The second version refers to these blended or complexed polymers with high conductivities approaching those of liquid electrolytes.

Typical ionic (relative) conductivities of polymer complexes are shown in Figure 8.3.

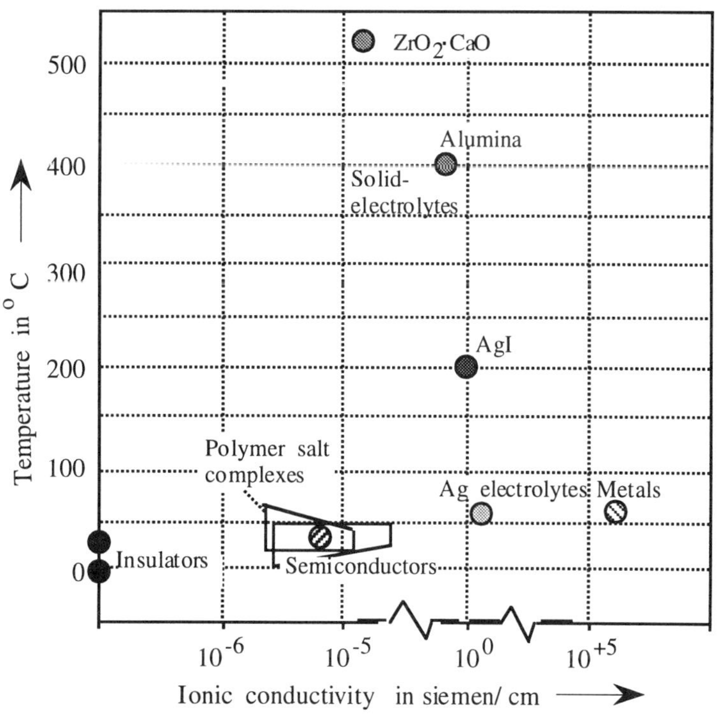

Figure 8.3 Relative ionic conductivities of polymer complexes at working temperatures.

Considering the ionic transport mechanism, it is controlled by the operating temperature relative to the glass transition temperature of the material. Accordingly, there are two groups of ionic conducting polymers: The first group includes complexes of polyethers and alkali metal salts which show appreciable increase in ionic conduction only at temperatures above their glass transition. Also included in this family are the polyphosphazene backbone polymers to which ion-solvating groups such as polymer MEEP have been grafted. In this first group of polymers, the ionic conduction process involves a cooperative interaction between the mobile ionic species and the polymer matrix.

The second group of materials have appreciable ionic conductivity at temperatures (typically room temperature) below their glass transition temperatures. Nafion and blends of poly(vinyl alcohol) and H_3PO_4 (denoted as PVA/H_3PO_4) are examples of this group. In these systems, the conduction mechanism is attributed to the percolation phenomenon pertinent to a conductive phase embedded in a nonconductive phase. Until a critical volume fraction of highly conducting species is reached, the compound does not conduct. Above this critical volume fraction, a continuous percolation path is established for the ionic transport and hence a high conductivity is perceived.

8.5 Theory of Ionic Conductivity of Polymers

In a simple model, ionic conductivity is governed by a jump diffusion mechanism with a *diffusion constant D* which is dictated by the *activation energy* ΔH, and the temperature T, *via* the *Arrhenius relation* given by:

$$D = D_o \exp(-\Delta H/RT) \qquad (8.1)$$

where D_o is the reference value and R is the gas constant. Explicitly, the corresponding relation for the ionic conductivity is given by:

$$\sigma(T) = \sigma_o \exp(-\Delta H/RT) \qquad (8.2)$$

where $\sigma_o = (\lambda^2 n^2 \omega F^2/NR) \exp(S/R)$ and the various parameters are:

λ is the jump length associated with diffusion
n is the number of mobile species
ω is the frequency of jump attempts
F is the Faraday constant
N is the number of sites participating in the jump diffusion and
$\exp(S/R)$ is the *entropy* term due to the internal energy (*enthalby*) which facilitates the ionic transport (*via* jump diffusion) by surmounting the energy barrier with a finite probability.

The conductive and dielectric behavior of polymer complex render them to exhibit an admittance parameter (Y) which varies with temperature or frequency. Representing $Y = G + jB$, the plots of (G, jB) as function of temperature or frequency in a complex plane are semicircles representing typical *Cole-Cole diagrams*.

Studies indicate the following as general conductivity characteristics of typical ionic complexes. (Table 8.2)

Table 8.2 General Characteristics of Ionic Conducting Polymers

Ionic Complexes	Characteristics
PVA/H_3PO_4	Follows Arrhenius temperature behavior with low temperature activation energy of 3.6 KCal/mol and high temperature activation energy of 22 KCal/mol. The reason for a break in Arrhenius plot at –40 °C is not known
Nafion	Displays Arrhenius behavior
Amorphous ionic-conducting polymers Examples: • PEO $LiClO_4$ (12:1) • Cross-linked PEO $LiCF_3SO_3$ (8:1) • PPO $LiCF_3SO_3$ (9:1) • PEA $LiCF_3SO_3$ (4:1) • PES $LiBF_4$ (6:1) • Polyphosphazcnc $LiCF_3SO_3$ (4:1) • Poly (N-methylaziridine) $LiClO_4$ (8:1)	Ionic conduction occurs through amorphous phase and Arrhenius plot (σ *versus* 1/T) does not follow a linear relation

The most common ionic conducting polymers, their glass-transition temperatures, and their conductivity at specific temperatures are presented in Table 8.3.

Table 8.3 Typical Polymer Complexes and Their Parameters

Polymer Complex	Homopolymer T_g (°C)	Conductivity (siemen/cm) at Temperature T (°C)
PPO-$ZnCl_2$	-64	
$(PPO)_8$-LiI	-64	10^4 at 85
$(PPO)_9$-$LiCF_3SO_3$	-64	10^{-4} at 95
$(PPO)_{12}$-$NaCF_3SO_3$	-64	10^{-4} at 95
(PPO)triol-sodium tetraphenylboride	-64	10^{-4} at 132
PEO-KSCN	-60	

(continued...)

Polymer Complex	Homopolymer T_g (°C)	Conductivity (siemen/cm) at Temperature T (°C)
PEO-LiSCN	–60	
$(PEO)_4$-NaI	–60	
$(PEO)_{10}$-NaI	–60	10^{-4} at 50
$(PEO)_8$-LiI	–60	10^{-4} at 55
$(PEO)_{4.5}$-$LiClO_4$	–60	10^{-4} at 110
$(PEO)_8$-$LiClO_4$	–60	10^{-4} at 50
$(PEO)_{4.5}$-$LiCF_3SO_3$	–60	10^{-4} at 120
$(PEO)_8$-$LiCF_3SO_3$	–60	10^{-4} at 100
$(PEO)_{14}$-$LiCF_3SO_3$	–60	10^{-4} at 85
$(PEO)_{4.5}$-KCF_3SO_3	–60	10^{-4} at 70
Polyphosphazene	–83	9.7^{-3} at 70
(MEEP)-$(AgSO_3CF_3)_{0.25}$		
(MEEP)-$[Sr(SO_3CF_3)_2]_{0.25}$		
(MEEP)-$[NaSO_3CF_3)_{0.25}$		
(MEEP)-$[LiSO_3CF_3)_{0.25}$		
Poly(pentamethylene sulfide)-$(AgSO_3CF_3)_{0.25}$		5×10^{-8} at 45
PEI-NaI		
PEI-$LiClO_4$	–35	10^{-4} at 109
PEI-$LiBF_4$	–35	10^{-4} at 74
Poly(vinyl pyrrolidone)-PEG_2-$LiSO_3CF_3$	180*	5×10^{-5}
Poly(ethylene succinate)$_3$-$LiBF_4$-1		3.4×10^{-6} at 65
PVA/H_3PO_3	70	10^{-5} at 24
Nafion	>150	10^{-3} at 24

From J.M. Margolis (Ed.): Conductive Polymers and Plastics (Chapman & Hall: 1989). Reprinted with permission of Chapman & Hall, NY.

* PEG plasticizes the PVP and reduces the T_g to -55°C.

8.6 **Mobility in Polymer Complexes**

Ionic conduction in polymers is controlled by the *mobility* (μ_i) and *transfer number* (τ_i) of the mobile species. The transfer number refers to the fraction of the total current carried by the mobile ions (both cations and anionic species). Therefore,

$$\tau_{i+} = \mu_{i+}/(\mu_{i+} + \mu_{i-}) \quad (8.3a)$$

$$\tau_{i-} = \mu_{i-}/(\mu_{i+} + \mu_{i-}) \quad (8.3b)$$

where μ_{i+} and μ_{i-} are anionic and cationic mobilities, respectively. The corresponding ionic conductivity is expressed as:

$$\sigma = \mu_{i+}\, en_{+} + \mu_{i-}\, en_{-} \tag{8.4}$$

where n_+ and n_- are number of anionic and cationic carriers, respectively, and e is the electronic charge.

8.7 Percolation in Ionic-Conducting Polymers

As mentioned earlier, ionic-conducting polymers such as Nafion swollen with an electrolyte solution and PVA/H_3PO_3 blends fall into the broad category of polymers which exhibit appreciable conductivities at temperatures below their glass-transition temperatures. In contrast, polyether-alkali metal complexes, polysuccinate-metal complexes, and poly-phosphazine-metal complexes exhibit high ionic conductivity at temperatures well above the glass transition of the polymer complex. In addition, unlike Nafion and blends of PVA/H_3PO_4, the polyether complexes have nonlinear Arrhenius behavior.

Considering Nafion or blends of PVA/H_3PO_4, they represent a multiphase system composed of ionic conducting clusters dispersed in a continuous matrix of insulating polymer. Therefore, only at a critical volume fraction of the conducting inclusions, does the ionic conducting commence. This refers to a *percolation phenomenon*. The percolation of the conduction is due to the transport of carriers *via* hopping between sites that are either forbidden or allowed with specific probabilities. That is, the sites may accept or reject the migrating ions with certain probabilities. Below the glass-transition temperature, the density and distribution of hopping sites and the availability of a potential hopping site that accepts a carrier may not appreciably change with time (*static percolation model*). If they do change with time as a result of the structural evolution of the host polymer, the corresponding percolation phenomenon is designated as the *dynamic-bond percolation.*

8.8 Complex Permittivity of Doped-Conducting Polymers

For high frequency applications, it is of interest to know the *dielectric relaxation* effects in conductive polymers. As a bulk material characteristic, the relaxation behavior manifests as the complex dielectric property specified in terms of the relative complex permittivity, $\varepsilon_r^* = \varepsilon_r' - j\varepsilon_r''$. For certain conductive polymers, the values of ε_r^* measured at microwave frequencies (X-band) are listed in Table 8.4.

8.9 Demerits of Doped-Conducting Polymers

Doping renders most polymers air and moisture sensitive. This may lead to oxidation which would degrade the electrical conductivity properties of the polymers upon aging. Further, polymers being virtually infusible and insoluble, processing of doped polymer films poses difficulties.

Many organometallic polymers which have good conducting properties have poor physical characteristics. For example, conducting organometallic polymers like poly-viny(lbis-fulvalenediiron) with tetracyanoquinodimethane being "brick dust"-like, is not useful in practice.

Selected organometallic polymers and their bulk resistivities are listed in Table 8.5.

Table 8.4 Complex Permittivity of Conductive Polymers

Material	Doping	Frequency (GHz)	ε_r^*
Poly-phenylene-benzobis-thiozole (PBT)	Iodine, 10^{16}ion/cm^3 (*via* ion implantation)	9.89	3 – j 8380
PBT	Iodine, 10^{16} ion/cm^3	9.89	3 – j 1158
Polyacetylene *cis*-$(CHI_{0.045})_x$	Electrochemical 4.5% I_2 by weight	8.90	5 – j 6070
Polyacetylene *cis*-$(CHI_{0.045})_x$	Electrochemical 4.5% I_2 by weight	8.90	5 – j 9090
cis-$(CHI_{0.045})_x$	Electrochemical 4.5% I_2 by weight	8.90	5 – j (4×10^5)

Adapted from [2].

Table 8.5 Resistivities of Selected Organometallic Polymers

Reactants		Bulk Resistivities (ohm-cm)
Cp_2TiCl_2	1,1' -Bis(A-hydroxyethyl) ferrocene	3.6×10^5
$UO_2(H_2O)_2(NO_3)_2$	Terephthalic acid	3.6×10^5
Cp_2ZrCl_2	1,1'- Bis(carboxyl) cobalticinium hexafluorophosphate	6.2×10^5
Cp_2TiCl_2	1,1'-Dihydroxymethyl ferrocene	7.4×10^5
$(Ph)_2SnCl_2$	Polyethyleneimine	9.0×10^5
$(Ph)_3SbCl_2$	1,1'- Bis(carboxyl) cobalticinium hexafluorophosphate	1.1×10^6
Cp_2TiCl_2	Terephthalic acid	3.4×10^6
Cp_2ZrCl_2	1,1' - Ferrocene dicarboxylic acid	4.5×10^6
Cp_2HfCl_2	1,1' - Ferrocene dicarboxylic acid	2.9×10^7
Cp_2TiCl_2	Azaleic acid	7.0×10^7
Cp_2TiCl_2	p-Phenylenediamine	1.0×10^8
Cp_2ZrCl_2	Terephthalic acid	1.7×10^8
$(Octyl)_2SnCl_2$	Polyethyleneimine	1.9×10^8

(continued...)

Reactants		Bulk Resistivities (ohm-cm)
$(C_4H_9)_3SnCl$	Poly(-o-acrylamideoxime) -80% inclusion	3.2×10^6
Cp_2TiCl_2	1,1' -Dibenzylferrocene dioxime	5.0×10^6
Ph_3SbCl_2	p-Benzoquinone dioxime	3.2×10^7
Cp_2HfCl_2	Poly(vinyl alcohol)- 37% inclusion	5.1×10^8
Ph_3SnCl_2	Polyethyleneimine- 73% inclusion	1.0×10^9
$(Bu)_2SnCl_2$	1,6- Hexanediol	1.25×10^{10}
Ph_3SbCl_2	Poly(acrylic acid)-42% inclusion	6.3×10^{11}
Ph_3BiCl_2	1,1'- Bis(carboxyl) cobalticinium hexafluorophosphate	1.25×10^{12}

8.10 Typical Dopants Used in Making Conducting Polymers

Both p-type and n-type dopants are used in realizing conducting polymers. Pertinent to polyacetylene polymer typical dopants used and the extent of conductivities realized are summarized in Table 8.6.

Table 8.6 Dopants Used in Fabricating Conducting Polyacetylene

Dopant		Conductivity siemen/cm
Acceptor p-type	Donor n-type	
Br		10^{-3}
Cl		10^{-4}
I		10^2
AsF_6		10^3
SbF_5		5×10^2
ClO_4		10^3
PF_5		10^2–10^3
	Li	2×10^2
	Na	10^2
	K	50

8.11 Other Groups of Polymeric-Conducting Materials

Conducting polymeric materials are broadly classified as polymers with modified conductivities *via* physical inclusion of metals and salts and those with enhanced conductivities envisaged chemically by doping treatment. Adding carbon, metal (powder, fibers, or flakes), and salts lead to the first category of conducting plastics described in Chapter 7.

The chemically modified polymers to yield higher electrical conductivities are discussed here. Apart from those described before, the following are other materials forming new horizons in conducting polymer technology.

- Macrocyclic tetrazannulene complexes with metal ions such as Ni^{2+} and Pd^{2+}
- Cationic and anionic charged microgel polymers, Group IV B metallocene polyoximes and metallic salts of ionomers
- Polymers quenched in sulfur nitride vapor
- Ditetramethyltetrathioseleno fuvalene hexafluorophosphate $(TMTSF)_2\ PF_6$ superconductor (zero resistivity at high pressures)
- β-carotene photoconductive semiconductor

8.12 Microgel-Conducting Polymers

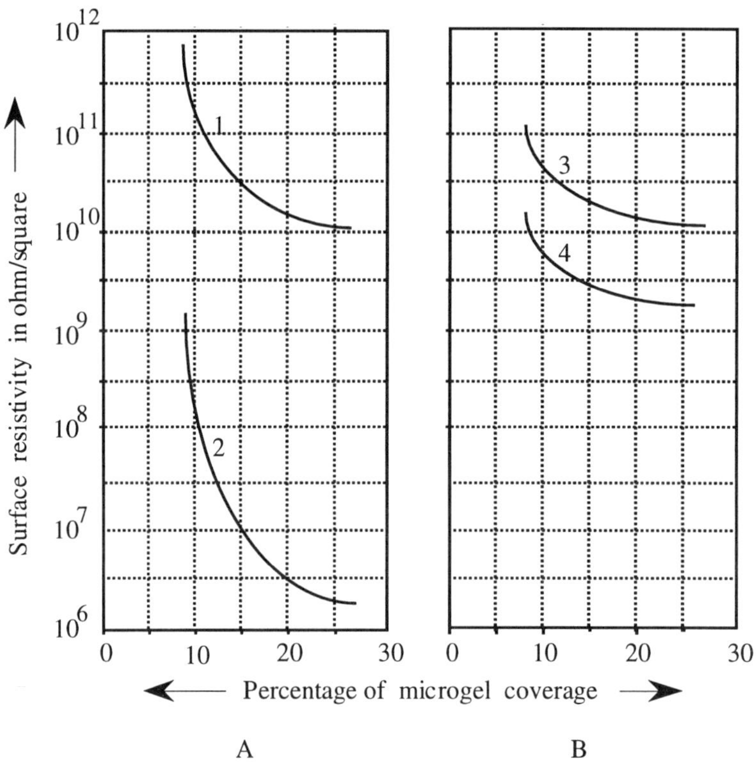

Figure 8.4 Percentage coverage of microgel *versus* surface resistance.
A: Cationic microgel;
B: Anionic microgel.
1. At 20% relative humidity; 2. at 50% relative humidity; 3. at 42% relative humidity;
4. at 71% relative humidity.

Conductive thin films are applicable for controlling the accumulation of static electricity on insulating substrates. Historically, water-soluble polyionomers mixed with *hydrophilic binder polymers* have been used for this purpose. The functions of the binder are to provide flexibility, improve adhesion, impart a degree of permanence, and act as a moisture sump. Conductive films, however, have poor abrasion resistance and tackiness at high temperatures and relative humidities.

Conducting *microgels* were developed to overcome the dependency of hydrophilic binders. These are cross-linked spherical polyelectrolyte particles dispersed in such solvents in which non-crosslinked polymers of similar composition would be soluble. Microgel polymers are 20–99 mole percentage ionic and their colloidal particulate sizes range from 0.01 to 1.0 micrometer. The gel could be both highly anionic and cationic. Microgels structurally take the following chemical form:

$$---(\text{-A})_X---(\text{-B})_Y---(\text{-CH}_2-\text{CH})_Z--- \tag{8.5}$$

where moiety A is a copolymerizable monomer and B is a di- or multifunctional monomer capable of cross-linking the polymer. Mole percentage of X is generally less than 20, Y is generally 5–10, and Z is 60–95 (other ratios are also feasible). These polymers are prepared by emulsion copolymerization of monomers A and B with another water-insoluble monnomer which can be transformed by a post-polymerization reaction to an ion-containing species. Vinylbenzyl chloride is copolymerized as above to yield cationic microgels. Microgels can be isolated by precipitation in nonpolar organic solvents, and yet redispersed in water methanol or in other polar solvents yielding wide formulations.

Microgels combined with latex polymers binders or with organic-solvent soluble polymeric binders have proved to be good conducting polymers. The surface resistivity of microgel-coated films are dependent on the area covered by the microgel and relative humidity. The choice of binder for the microgel system depends on its compatibility with conductive polymers and with the solvent system. Hydrophilic polymers useful as binders are invariably at the positive end of triboelectric series. The significance of this property is that these materials on contact and separation with dissimilar surface are more likely to be static propensive than other materials that occupy a more neutral position in the triboelectric series (see Chapter 20). Characteristics of two typical microgels are depicted in Figure 8.4.

8.13 Applications of Conducting Polymers

The ability to tailor the electrical properties of conducting polymers coupled with improved stability and processability have opened new avenues in the applications of these materials in several areas which include: Electrodes for batteries, fuel cells and capacitors, electrochromics, chemical and biochemical sensors, EMI and power cable shielding, ion exchange and release devices, and neutron detection.

8.13.1 *Uses of electronic-conductive polymers*

Battery electrodes and fuel cell electrodes: Products using conducting polymers as electrode materials include rechargeable storage batteries with characteristics comparable to Ni-Cd cells. Design optimization of such batteries refers to increasing parameters such as cell voltage, energy density, and power output by controlling the charge storage and discharge characteristics of conducting polymers used. The redox doping reactions involved in the charge-discharge behavior of conducting polymeric electrode materials are inherently different from those used in conventional batteries because the electrode is not dissolved and redeposited. On the contrary, the conducting polymeric electrode cell (such as polypyrrole-lithium battery) operates by the oxidation and reduction of the conjugated polymer backbone. During the charging cycle, a positive potential is applied to the polypyrrole electrode relative to the lithium electrode. As the polypyrrole oxidizes, anions in the electrolyte enter the porous polymer to balance the charge released. Concurrently, lithium ions in electrolyte are electrodeposited at the lithium surface.

When the battery terminals are connected to the load, the external current flow removes electrons from the lithium, causing lithium ions to reenter the electrolyte and to pass through the load into the oxidized polymer. The positive sites on the polymer are reduced releasing the charge-balancing anions back to the electrolyte. This cycle is repeatable as in any secondary battery cell.

In *fuel cells* conductive polymers are used as *catalytic materials*. For example, a polyacetylene $(C_2H_4)_x$ film upon oxidation in the presence of HBF_4 leads to conductive $(CH)_x$ with BF_4^- dopant anions as per the following reaction:

$$4(CH)_x + 4xyHBF_4 + xyO_2 \quad \rightarrow \quad 4[CH^{y+}\,(BF_4)_y^-]_x + 2xyH_2O \qquad (8.6)$$

Other reagents like perchloric acid and benzoquinone have also been used to oxidize $(CH)_x$ to the highly conducting state in aqueous solutions. The electrical connection between $[CH^{y+}\,(BF_4)_{y-}]_x$ electrode and a Pb electrode in a 7.4 M aqueous HBF_4 solution sets up an electrochemical reaction in which the polymer is reduced to the neutral state emitting BF_{4-} into the electrolyte, and the Pb is oxidized to Pb^{2+}. This reaction leads to the realization of a fuel cell using O_2 as the fuel. During the discharge of $[(CH)^{y+}(BF_4)_{y+}]_x\ |Pb(BF_4)_2\,|Pb^o$ cell, O_2 is introduced and the electrochemically reduced polymer gets simultaneously (chemically) oxidized. This leads to a continuous supply of electricity. In this process, the conducting polymer behaves catalytically undergoing no chemical change and the Pb electrode and oxidizing agent are converted into products.

Biosensors using polymer complexes: When a conducting polymer is exposed to oxidants in solution, its resistance changes. This property is used to in realizing *biosensors*. For example, triiodide is used to oxidize polyacetylene and the corresponding charge in resistance is used to array glucose concentration in solution. This technique begins with the glucose oxidase-mediated oxidation of glucose in solution leading to the generation of hydrogen peroxide which in the presence of catalytic lactoperoxide oxidizes iodide in buffer solution to triiodide. The triiodide is a compatible dopant for polyacetylene and it oxidizes the polymer, causing a resistance change that is proportional to the concentration of glucose in solution. The glucose content is arrayed by the percentage change in conductivity over a specified time. Sensitivity of these sensors is augmented *via* surface-confined enzyme systems and other uses of conducting polymers in biosensor applications include solid-phase enzyme immunoassays.

Conductive polymers for remote-reading devices: These are ambient-responsive elements usable as remotely readable indicating or display devices. The ambient conditions for which the elements are responsive are: Temperature *versus* time, temperature extremities, relative humidity, radiation dosage, mechanical stress/abuse, and chemical contaminations. Conjugated polymers with appropriate redox agents (dopants) are sensitive to these ambient conditions and exhibit conductivity changes arising from relevant chemical reactions or instability due to the presence of moisture and heat.

A common application of polymer complexes for this purpose refers to radio frequency (RF) radiation sensing and antitheft devices. With suitable coupling to an RF circuit, the change in the electrical properties of the conducting polymer when exposed to RF energy is detected through resonance offset measurements.

Conductive polymers as electromagnetic shields: Doped polymers such as pristine increase conductivity several orders of magnitude over a wide temperature range. This conductivity characteristic combines with high weight and high mechanical strength of such polymers to make them highly compatible as high frequency electromagnetic shielding materials.

The plane wave shielding characteristics of conducting polymers depend on doping, thickness of the shield and on the polarization and frequency of the electromagnetic radiation. General observation on the high frequency shielding performance of conducting polymers are:

- Conductive polymeric sheets of thickness much less than a skin depth offer a shielding effectiveness (SE) in excess of 40 dB at radio and microwave frequencies.
- Conductive latex polymers can be sprayed on computer cabinets or similar equipment to prevent EMI effects.
- Typical polymers such as PBT with iodine ions implanted (with a doping level of 10^{17} ions/cm^3) and polyacetylene *cis*-$(CHI_{0.8})_x$ doped electrochemically with 80% I_2 by weight exhibit frequency-dependent SE as depicted in Figures 8.5.
- SE increases monotonically with thickness/wavelength *(d/λ)* ratio over certain value of *d/λ*, and steeply increases thereafter.
- Conductive polymers can also be used in multilayer shield fabrication. In such applications, unlike carbon impregnated multilayer composite shields, conductive polymer-based multilayer shields show less resonances due to high absorptions which damp out the multiple reflections. Hence, conductive polymer multiple shields are useful for wideband applications.
- Compared to carbon-impregnated composite shields, conductive polymeric shields are lightweight.

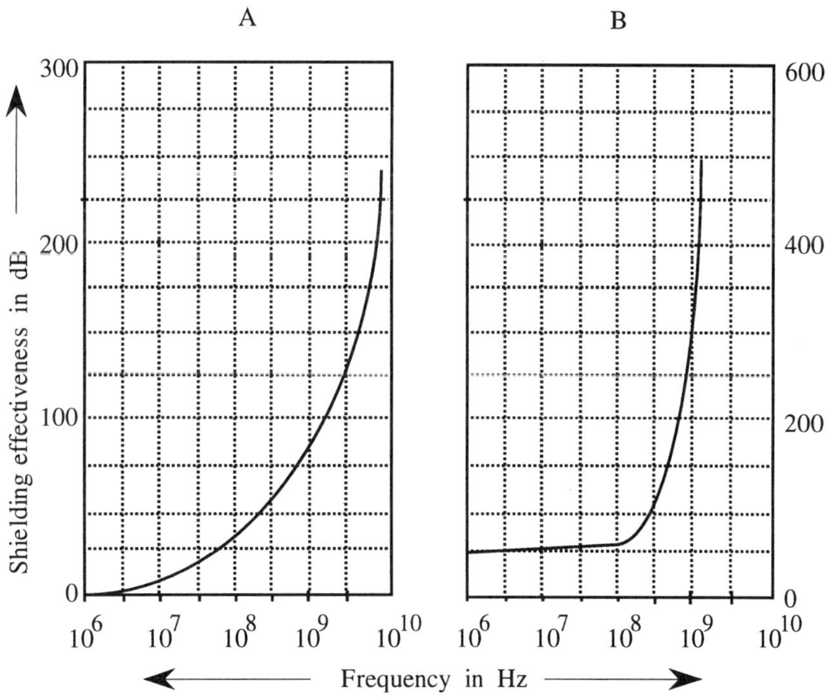

Figures 8.5 Shielding effectiveness *versus* frequency offered by typical conducting polymer EM shields.
A: PBT with ion implanted iodine. (Doping level: 10^{17} ion/cm^3.)
B: Polyacetylene: *Cis* –$(CHI_{0.8})_x$ doped electrochemically with 80% iodine by weight.

Conductive polymer based light-emitting diodes: Conjugated polymers have been incorporated as active materials into several kinds of electronic device such as diodes, transistors, and light-emitting diodes. Poly(phenylene vinylene) or PPV is a very promising material for such applications.

8.13.2 *Applications of ionically conducting polymers*

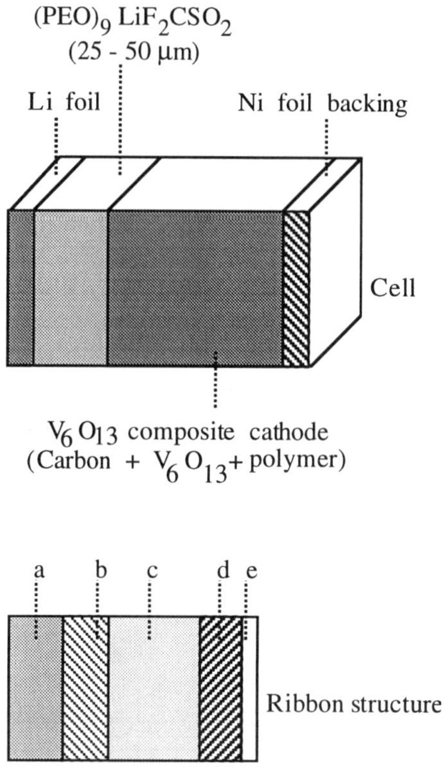

Figure 8.6 Polymer electrolyte-based battery.

These materials are also usable in batteries, fuel cells, chemical sensors, EMI shielding, photoelectric cells/devices, electrochromic displays and ion pumps. Ionically conducting polymers pose viable engineering applications due to the following reasons:

- They are highly conducting.
- They have favorable processing parameters.
- They are flexible and have conformability.
- They accommodate layer volume charges during the ion-electron exchange process.

Consistent with the aforesaid properties, ionically conducting polymers are being deployed in a variety of technologies as described below.

Polymer electrolyte-based batteries: These are all solid-state batteries in which the ionically conducting polymers are used as the electrolytic material. The reasons for using these polymeric materials as electrolytes are:

- They yield high energy density up to 200 watt-hour/kg in high rate/demand secondary systems.

- They can be fabricated with ease.
- They have a wide range of operating temperature.
- Their self-discharge rate is low.
- They are flexible (low elastic modulus), and possess good interfacial properties maintaining a good contact with the electrodes during charging and discharging sequences. Normally, solid-state batteries pose dimensional changes of the electrodes during charging (discharging) events. The polymer electrodes conveniently accommodate such changes and yield reversible deformations to stresses at the electrode-electrolyte interface. If such changes are not flexibly accommodated, the effective contact between electrode and electrolyte will be lost, causing malfunctioning of the cell.
- Being solid, they offer a rugged construction of the batteries.

Their main disadvantage, however, stems from the reduction in conductivity that occurs below 80^oC, permitting only modest load currents at room temperature. Batteries compatible for room temperature operations employing polyphosphazenes or PVA/H_3PO_4 blend have shown encouraging trends. In comparison with hard crystalline inorganic solid electrolytes, the polymeric electrodes provide reliable contact at the electrolyte interface over repeated charging cycles despite of the electrode deformation due to stresses, as mentioned earlier.

Polymer electrolytes can be cast into thin films, permitting battery size miniaturization. Since these thin films occupy a low volume or weight fraction of the battery, the volume of other active electrodes can be enhanced, allowing an increase in the amount of energy stored per unit volume or weight.

Typical structure of a polymer electrolyte battery is shown in Figure 8.6. The electrolyte used is a complex of poly(ethylene oxide) and LiF_3CS_3 and the cathode is a composite phase composed of the electrolyte together with V_6O_{13} and acetylene black. The anode is made of metallic lithium foils. Typical cathode ampacity hour falls in the range 2–2.5 milliampere-hour/cm^2.

The trend in realizing ionically conducting polymer based batteries refers to optimizing the following characteristics: High energy density, safety, ruggedness, ease of fabrication, variable geometry with series or parallel connection of cells, low self-discharge, lack or passivation, and requirements concerning ambient or subambient temperatures.

Polymer electrolyte-based photoelectrochemical cells: Ionic-conducting polymer, such as PEO complex together with KI and I_2 generate the iodide-triiodide redox couple which can be incorporated into a photoelectrochemical cell yielding a photovoltage upon excitation by photons *($h\nu$)*. A typical geometry is illustrated in Figure 8.7.

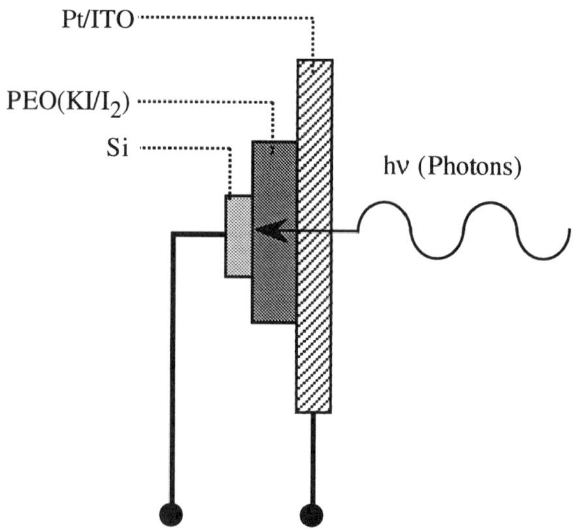

Figure 8.7 Photoelectrochemical cell.

Hydrogen sensor: Blends of poly(vinyl alcohol) and H_3PO_4 are efficient proton conductors over a wide range of temperature (–40 to 50°C), and as a result can be used for hydrogen sensing. A typical hydrogen sensor is depicted in Figure 8.8.

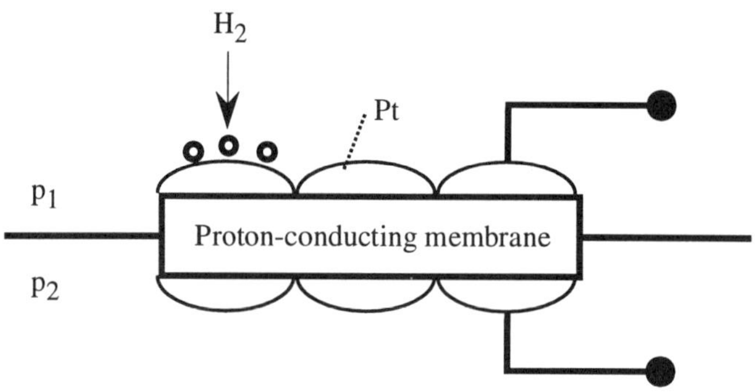

Figure 8.8 Hydrogen cell.

Essentially, the sensor is a potentiometric cell and induces an EMF depending on the difference in fuel energy at each electrode. The polymer screens two regions with differing hydrogen pressures (or concentrations) P_1 and P_2. Platinum or palladium electrodes on the polymeric film permit hydrogen disassociation and association as dictated by the partial pressure of the hydrogen.

The EMF is generated across the electrodes proportional to the difference in partial pressures P_1 and P_2 and is mediated by the electrical charge transport across the polymer electrolyte. The potential (V) induced is given by:

$$V = (RT\tau_1/2F)\ ln(P_2/P_1) \tag{8.7}$$

where R is the gas constant, T is the temperature, F is the charge transported across the electrolyte, and τ_I is the *ionic transfer number* given by $\sigma_i/(\sigma_i + \sigma_e)$ where σ is the conductivity and i and e represent the ionic and electronic species, respectively.

Hydrogen sensors based on conducting polymers are accurate and have high resolution and fast response time. With platinum electrodes, these sensors are highly selective of hydrogen in the presence of other common gases such as methane, ethane, butane, CO_2, N_2, etc.

Electrochromic display using conducting polymers: Electrochromic displays are devices that undergo a reversible color change as a result of an applied electric field or current. An electrochemical material exhibits significant, reversible optical absorption at visible wavelength. Ionic conducting polymers used in electrochromic displays separate the electrochronic film from an electrode, which is a source of ions, and mediate the injection of electrons and ions (from the source) to the electrochromic film. More details on the type of ionic-conducting polymers used for this purpose are discussed in Chapter 18.

8.14 Concluding Remarks

Conducting polymers or organometallics constitute micro- and bulk level applications in electrongenetic material technology. Synthesizing such new materials can meet various material crises posed in upcoming high-tech needs. For further reading, relevant references are indicated below.

References

[1] J. M. Margolis (Ed.): *Conductive Polymers and Plastics.* (Chapman and Hall, New York: 1989).

[2] K. Naishadham: Shielding effectiveness of conductive polymers. IEEE Trans. Elec. Mag. Compat., vol. 34(1), 1992: 47-50.

[3] C. U. Pittman, Jr. and C. E. Carraher, Jr.: Applications of organometallic polymers, in R. B. Seymour and F. Mark (Eds.): *Applications of Polymers.* (Plenum Publishing Corp., New York: 1988).

[4] D. Baeriswyl, G. Harbeke, H. Kiess and W. Meyer: Conducting polymers: Polyacetylene, in J. Mort and G. Pfister (Eds.): *Electronic Properties of Polymers.* (John Wiley and Sons, New York: 1982).

[5] R. B. Seymour (Ed.): *Conductive Polymers.* (Plenum Press, New York: 1981).

[6] K. F. Schoch, Jr.: Update on electrically conductive polymers and applications. IEEE Elec. Insulation Magazine, vol. 10(3), 1994: 29-32.

Defining Terms

Conductive polymers: Organometallic polymers which exhibit high electronic and/or ionic conductivity.

Doping: Addition of impurities to control the conductivity of the materials.

Mobility: Velocity of the electric charge carrier per unit electric field applied.

CHAPTER 9

Conductive Materials

9.1 Introduction

An electromagnetic material of utmost engineering importance is an *electrical conductor* which is characterized by the property of finite electrical conductivity (σ siemen/meter) depicting its ability to transport electric charges under the influence of an applied electric field force. The magnitude of σ spans a wide range of values (about 20 orders of magnitude) specifying availability of an innnumerable variety of conductive materials. A generic classification of conductors based on the magnitudes of σ is as follows:

- Semiconductors ($\sigma \sim 10^{-2}$ siemen/meter)
- Conductors (metals and alloys) ($\sigma \sim 10^1$–10^8 siemen/meter)
- Superconductors ($\sigma \rightarrow \infty$)

The charge transport across such conductive materials in general is mediated by the free electrons, the theory of which is presented in Chapter 1. (However, in certain materials, holes and/or ions could as well be the carriers of electricity.) Among the three classes of conductors indicated above, the present chapter is concerned largely with metals and their alloys and other conductors such as carbon commonly used in electrical technology. Semiconductors are discussed in Chapter 10 and superconductors are briefed in Chapter 11.

9.2 Metals and Alloys

As elucidated in Chapter 1, the electrical conductivity *(σ)* of a material refers to the constant of proportionality between the electric current density *(**J**)* (or the rate of flow of electric charges across a unit area of cross-section) in the material and the electric field force *(**E**)* to which the material subjected to. That is:

$$J = \sigma E \qquad \text{ampere/meter}^2 \qquad (9.1)$$

where $\boldsymbol{J}$ is in ampere/meter^2, $\boldsymbol{E}$ is in newton/columb or volt/meter and σ is specified by siemen/meter. Explicitly, σ is given by:

$$\sigma = ne^2 \tau/m \qquad \text{siemen/meter} \qquad (9.2)$$

where n is the number of electrons per unit volume, e is the electronic charge, m is the electronic mass and τ refers to the *collision time* depicting the average interval between successive collisions of the electrons in the material.

The conductivity *(σ)* can also be written in terms of *electronic mobility (μ_e)* defined as the magnitude of the average drift velocity of the electrons per unit electric field applied. That is,

$$\sigma = ne\mu_e \qquad \text{siemen/meter} \qquad (9.3)$$

Inasmuch as finite conductivity of a material depicts implicitly the collision process involved, it is possible to write the corresponding power dissipation per unit volume (or *power density, W*) as:

$$W = \sigma|\boldsymbol{E}^2| \qquad \text{watts/meter}^3 \qquad (9.4)$$

This also refers to the energy per unit time which the electrons transfer to the lattice per unit volume in the collision process and is converted into heat (*Joule's law*).

9.3 Resistivity and Ohm's Law

The reciprocal of σ is known as the *electrical resistivity* (ρ) with the unit ohm-meter. Since the collision time (τ) could depend on the thermodynamic state of the material (expressed in terms of the temperature T), σ (or ρ) is a temperature-dependent parameter.

At a given temperature, the current flow (I) across an area of cross-section (A) of the material of length ℓ due to the application of a potential V can be deduced as:

$$I = V/R \qquad \text{ampere} \qquad (9.5)$$

where $R = \rho\ell/A$ ohm, known as the *resistance* of the test piece of material considered. Equation 9.5 is well known as *Ohm's law* (Chapter 1), and Equation 9.1 corresponds to Ohm's law at the microscopic level and hence known as the *microscopic Ohm's law.*

9.4 Statistical Aspects of Electronic Conduction

The collision time or the average time between successive collisions introduced earlier is decided by the statistical aspects of the collision involved in the so-called *relaxation process.* The collisions are caused by thermal and/or structural imperfections in the crystal lattice. Corresponding to an initial state of disturbance in the electron gas, a characteristic time which governs the establishment of equilibrium (through collision) can be defined as the *relaxation time.* For isotropic materials, it is the same as the collision time and depicts the mean time of collision. With the average velocity of the electrons denoted by v_e, the *mean free path* (λ_e) of an electron can be determined by:

$$\lambda_e = |v_e|\tau \qquad \text{meter} \qquad (9.6)$$

In the case of metals, the conductivity σ related to a single velocity (*via* Equation (9.3) in terms of the corresponding mobility) is decided by *Fermi statistics.* That is, as the electrons are known to occupy discrete energy levels at the absolute zero temperature, an energy level W_F can be specified below which all the levels are occupied (and above which all the levels are empty). This level W_F is known as the *Fermi level* of electrons.

In terms of W_F, the statistical probability of a state corresponding to an energy level W being occupied by an electron at a temperature T is given by:

$$f(W) = \{\exp[(W - W_F)/k_BT] + 1\}^{-1} \qquad (9.7)$$

9.5 Physical Factors *versus* Electrical Conductivity of Metals

In metals as well as in dilute alloys, the total resistivity (ρ) is decided by the sum of two components, namely, the *thermal component* ρ_T due to lattice vibrations and *residual resistivity* ρ_r resulting from impurities and structural imperfections (which are independent of T). That is:

$$\rho = \rho_T + \rho_r \qquad \text{ohm-meter} \qquad (9.8)$$

Equation 9.8 is known as *Matthiessen's rule.* This rule, however, is less accurate at extreme temperatures and/or at high impurity/imperfection levels.

Consistent with Equation 9.3, the following factors can be regarded as those which influence the conductivity of metals:

1. Temperature: Inasmuch as the conductivity depends on mean collision time (Equation 9.3), the electrical resistivity of most metals increases with increase in temperature. (That is,

metals invariably pose a *positive temperature coefficient.*) The thermal component of resistivity ρ_T of conductors is approximately linear above a certain temperature known as the *Debye temperature* T_D (examples: T_D (for Cu) = 315 K and (for Aℓ) = 398 K). With reference to the resistivity (ρ_{RT}) at room temperature T_R, the resistivity (ρ) at any temperature T can be specified by:

$$\rho \cong \rho_{RT}[1 + \alpha_R(T - T_R) + \ldots] \tag{9.9}$$

where α_R is known as the *temperature coefficient of resistivity*. For pure metals, $\alpha_R \cong 0.004$ per °C.

Electrical resistivities of typical metals and the corresponding thermal coefficients of resistivities are presented in Table 9.1.

Table 9.1 Electrical Resistivities and Temperature Coefficients of Some Metals Near the Room Temperature

Element	ρ (ohm-cm) $\times 10^6$	α_R (ohm/ohm-deg) $\times 10^3$
Aluminum	2.6548	4.290
Beryllium	4.000	25.00
Cadmium	6.8300	4.200
Chromium	12.900 (0°C)	3.000
Cobalt	6.2400	5.300
Copper	1.6730	4.300
Gold	2.3500	3.500
Iridium	5.3000	3.930
Iron	9.7100	6.510
Lead	20.648	3.680
Magnesium	4.4500	3.700
Mercury	98.400 (50°C)	0.970
Molybdenum	5.2000 (0°C)	5.300
Nickel	6.8400	6.920
Palladium	10.800	3.770

(continued...)

Element	ρ (ohm-cm) $\times 10^6$	α_R (ohm/ohm-deg) $\times 10^3$
Platinum	9.8500	3.927
Rhodium	4.5100	4.300
Silver	1.5900	4.100
Tantalum	13.500	3.830
Thorium	15.700 (25°C)	3.800
Tin	11.000 (0°C)	3.640
Titanium	42.000	3.500
Tungsten	5.3000 (27°C)	4.500
Zinc	5.9160	4.190

As the temperature T approaches 0 K, most of the conductors "freeze" to exhibit zero conductivity and typically become superconductors (details of which are presented in Chapter 11).

2. Alloying: When two or more metals are mixed to form an alloy phase, the regular structure of pure metals diminishes. As a result, the electrical conductivity of a solid-solution alloy drops off rapidly with increase in alloy content. Resistivity of an alloy (ρ_{alloy}) can be expressed as:

$$\rho_{alloy} = \rho_{pure\ metal} + S\,\rho_f \qquad \text{ohm-meter} \qquad (9.10)$$

where S is the atomic percentage of added impurities and ρ_f is an incremental factor denoting the increase in resistivity per atomic percentage of impurity addition. For example, for Ni, $\rho_f = 1.3 \times 10^{-6}$ ohm-cm.

3. Annealing: Cold work (annealing/quenching) may set up local strains altering the conductivity of metals.

4. Age hardening: Resistivity of alloys increases with age hardening process.

9.6 Conductivity of Pure Metals

Common metals in a pure state have resistivity in the range of 1.5 to 150 ohm-cm at room temperature. As indicated earlier, the charge transport in pure metals is effected by the drift of free electrons. Though this is true for most of the metals, in some metals like beryllium and zinc, the movement of charge is considered as due *to holes*. The temperature dependency of conductivity (σ) in pure metals as dictated by the mobility of electrons can be specified by:

$$\sigma \cong A/T \text{ (in the high temperature regime)} \qquad \text{siemen/meter} \qquad (9.11a)$$

$\sigma \cong B/T^5$ (in the low temperature regime) siemen/meter (9.11b)

where A and B are constants for a metal.

The theoretical aspects of evaluating σ are based on quantum mechanical considerations in which free-electron assumption is rather implicit. However, at large atomic masses, the conducting electrons can be regarded no longer "free" since they interact with the lattice. Therefore, quite often the measured and theoretical values of σ could differ significantly. Examples of low and high atomic mass elements along with their predicted and measured values of σ are as presented in Table 9.2.

Table 9.2 Measured and Predicted Values of σ

Metal	Example	σ in siemen/meter	
		Predicted Values	Measured Values
With low atomic mass	Na	22	23
	K	19	15
With high atomic mass	Rb	20	8
	Cu	100	64
	Ag	90	67
	Au	107	68

9.7 Conductivity of Alloys

In the mixture of two metals (or metal alloys), again their electric current carrying capacity is decided by the carrier concentration which is independent of temperature and the specific resistivity of the alloys is determined by the temperature dependence of carrier mobility. Alloys with the highest specific electric resistance (ρ_0) have the lowest temperature coefficient of electrical resistance.

The effects of various alloying elements on the conductivity of alloys are different and depend on the type of the phases they constitute.

With the formation of *solid solutions*, the specific electric resistance at 20°C varies depending on the composition as per a nonlinear law shown in Figure 9.1 (for a Cu-Ni alloys.)

An alloy acquires the highest electrical resistance in most cases at the concentration of elements equal to 50 atomic weight percentage. It can be presumed that the impurity scattering due to distortions of the crystal lattice and disturbances of the periodicity of the energy levels attain the highest values in such alloys. Further in alloys in which at least one of the components is a transition metal, the temperature coefficient of electric resistance can acquire negative values as well; that is, their electric resistance would decrease on heating. In cases where a material should have a high electric resistance, alloys with appropriate structure of solid solutions should be considered.

The temperature dependency of resistivity in alloys can be ideally modeled as follows. When some sites of an ideal metal lattice (for instance, copper lattice) with a periodic potential are randomly replaced by atoms of some other element, say, gold, the lattice potential will cease to be strictly periodic. It will be distorted by the disordered impurity atoms. Naturally, such distortion will lead to carrier scattering leading to increased electrical resistance.

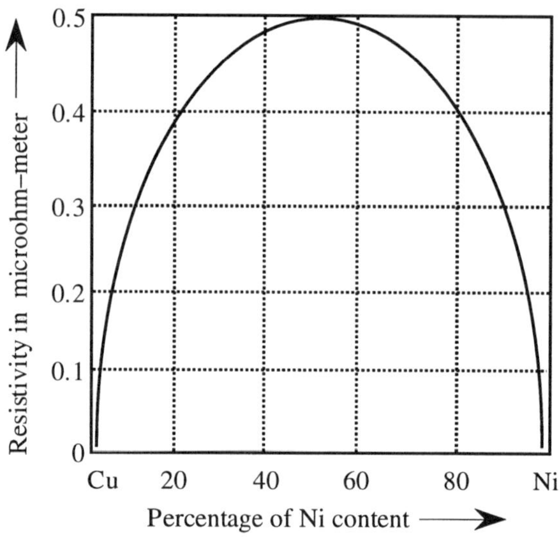

Figure 9.1 Resistivity *versus* percentage of nickel in a Cu-Ni alloy.

It has been demonstrated that in the simplest case of binary alloys (dilute solution of metal B in metal A) of solid-solution type, the carrier mobility due to scattering on lattice imperfections is described by the following approximate relation:

$$\mu_{al} \propto x(1-x) \tag{9.12}$$

where μ_{al} is the carrier mobility in the alloy and x and $(1-x)$ are fractional parts of the metals constituting the alloy.

The expression for the specific resistance of a binary alloy can then be written as:

$$\rho_{al} = \beta[x(x-1)] \qquad \text{ohm-meter} \tag{9.13}$$

where β is a proportionality factor.

The function $x(x-1)$ is maximum when the concentrations of both components are equal. For example, at room temperature $\rho_{Cu} = 1.7 \times 10^{-8}$ ohm-meter and $\rho_{Au} = 1.56 \times 10^{-8}$ ohm-meter, whereas $\rho_{50\%Cu\,+\,50\%Au} = 15 \times 10^{-8}$ ohm-meter. This is quite expected because the disorder in the lattice has a much more deterimental effect on the lattice periodicity than the thermal vibrations. If, however, the alloyed metals are taken in appropriate proportions from ordered alloys or metallic compounds with an ordered structure, the lattice periodicity is recovered and the resistance due to impurity scattering almost totally vanishes. For the copper-gold alloys the appropriate concentrations are those which correspond to the stoichiometric composition of Cu_3Au and CuAu. This confirms the validity of the quantum theory of electrical conductivity which maintains that the cause of the electrical resistance of solids is not the collision of free electrons with the lattice atoms but their scattering by the lattice defects which distort the periodic lattice potential. An ideal, regular, imperfection-free lattice with a strictly periodic potential would be incapable of scattering free-charge carriers and will therefore have a zero resistance. This conclusion is supported by numerous experiments carried out with pure metals in the low temperature range. As the degree of purity of a metal is increased, its specific resistance near absolute zero diminishes continuously tending to zero. (This observation does not, however, correspond to superconductivity but refers to the characteristic of all metals.)

For small impurity contents one may set $(1 - x) \cong 1$. Then $\rho_{al} \propto x$. This specific resistance is independent of the temperature and does not vanish at absolute zero. It is termed as *residual resistivity*, ρ_{res}.

At temperatures other than zero, a resistivity ρ_T due to electron scattering by the lattice vibrations can be added to the residual resistivity (pure component) so that the total resistivity becomes:

$$\rho = \rho_{res} + \rho_T \tag{9.14}$$

This relation as stated earlier expresses *Matthiessen's rule* and gives the additivity property of specific resistance.

Figure 9.2 gives the resistivity of copper alloys as a function of temperature demonstrating Matthiessen's rule.

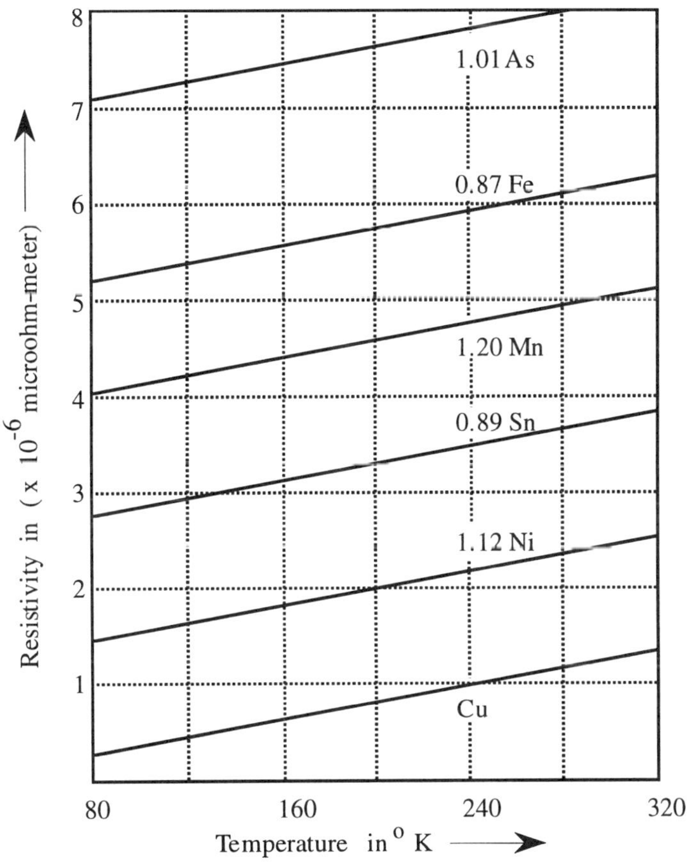

Figure 9.2 Resistivity *versus* temperature of typical copper alloys.

The *temperature coefficient of resistivity* of a conductor is expressed as the relative variation of specific resistance when the temperature is raised by 1 K. For pure metals, $\rho = \rho_T$ and therefore:

$$\alpha = 1/\rho_T \cdot (d\rho_T/dT) \tag{9.15}$$

Experiments show the α is approximately given by:

$$\alpha \cong (1/273)/^{o}K \cong 0.004/K \tag{9.16}$$

For alloys $\rho = (\rho_{res} + \rho_T)$. Hence, it follows that:

$$\alpha_{al} = 1/\rho\,(d\rho/dT) = [1/(\rho_{res} + \rho_T)] \cdot (d\rho_T/dT) \tag{9.17}$$

where ρ_{res} is independent of temperature.

The expression of Equation 9.17 can be transformed as follows:

$$\alpha_{al} = [1/(1 + \rho_{res}/\rho_T)] \cdot 1/\rho_T\,(d\rho_T/dT)$$

$$= \alpha_p/(1 + \rho_{res}/\rho_T) \tag{9.18}$$

where α_p is the temperature coefficient of resistivity of pure metals. α_{al} is less than α_p of pure metals. Table 9.3 gives the temperature coefficients of some metals and alloys for comparision.

Table 9.3 Temperature Coefficients of Resistivity of Some Metals and Alloys

	Cu	Sn	Ni	Bronze (88% Cu +8% Sn +4% Pb)	Nichrome (80% Ni +20% Cr)	Constantan (54% Cu +46% Ni)
α_p or α_{al} ($\times 10^3$/K)	4.1	4.2	6.2	0.5	0.13	–0.004

In most alloys, the dependence of resistance on temperature is much more involved than that which follows the simple additive rule given by Equation 9.14, and the temperature coefficient of resistance of some alloys may be much less than one could expect from Equation 9.18. Further, it does not remain constant over a wide temperature range and may in some cases even become negative, for instance, as in the case of constantan and some other alloys.

A higher specific resistance, together with a low temperature coefficient of resistivity, has made alloys valuable materials for the production of various wires and film resistors and variable resistors (rheostats) widely used in electrical engineering practice.

To summarize, conductivity materials mostly refer to metals which are crystalline substances. Metal conductors (silver, copper, aluminum, etc.), in general, have high conductivity, that is, low resistivity on the order of (0.0150 to 0.0283) $\times 10^{-6}$ ohm-meter. These pure metals are invariably used for electric wires and cables. Apart from high-conductivity metals often in use are conducting alloys of high resistivity, about (0.4 to 0.2) $\times 10^{-6}$ ohm-meter. These materials are used in standard resistors, heating coils, etc. All metallic conductors as indicated earlier have the property of electric conduction which is mainly by electrons. That is, the conduction current in metals is an ordered directional movement of free electrons. With increasing temperature, as explained before, the ohmic resistance of metallic conductors increases.

Typical metals used in electrical engineering practice and their characteristics are presented in Table 9.4.

Table 9.4 Properties of Metals/Alloys and Their Electromagnetic and/or Electronic Packaging Applications

Metal	Properties and Applications
	Atomic number: Z Atomic weight: M_a Density: d Melting point: M.P. Coefficient of thermal expansion: α_T Bulk electrical resistivity at 12°C: ρ_V Thermal conductivity: C_T Temperature coefficient of resistivity: $(1/\rho_V)(d\rho_V/dT)$: α_R
Aluminum (Al) (and its alloys)	Z: 13 M_a: 26.98 d: 2.71 gm/cm^3 M.P.: 659.7°C α_T: 0.24×10^{-4}/°K ρ_V: 2.76×10^{-6} ohm-cm C_T: 2.4 watt/cm/°C α_R: 40×10^{-4}per °C Aluminum is widely used as shielding cans, component mounts/chassis, power line conductors, heat sinks, lightweight mechanical fixtures etc. It is a versatile metal with combinations of lightweight and tensile strength up to 125,000 lb/in.^{2}. Its excellent resistance to corrosion in many environments is due to the protective, highly adherent oxide film which develops in air, oxygen, or oxidizing media. High-strength aluminum alloys generally are not as corrosion resistant in comparison with the high-purity or moderate-strength aluminum alloys. Aluminum is available in a variety of alloys and tempers. This metal can be fabricated economically by all common processes and can be cast by simple and known foundry techniques. Next to copper as a good conductor is aluminum. Aluminum occurs in abundance on the earth's surface. It is available in various forms such as oxides, sulfates, silicates, phosphates, etc. On a commercial scale, it is mainly produced from *bauxite* ($Al_2O_3.2H_2O$). For the extraction of aluminum from bauxite, the ore is purified and then dissolved in fused cryolite which is a double fluoride of aluminum and sodium ($AlF_3.3NaF$). The resulting solution is transferred to an electric furnace and aluminum is separated out by electrolysis.

Aluminum is a white metal with a bluish tinge and is very light in weight, which is its chief asset. It is 3.5 times lighter than copper. Pure aluminum is softer than copper and thus, it can be rolled into thin sheets (foils). Due to low mechnical strength, it cannot, however, be drawn into very thin wires. The resistivity of aluminum is about 1.65 times greater than copper. In view of the factors, namely, resistivity, mechnical strength, and density, for the same resistance and length of wire, an aluminum conductor should have a cross-sectional area 1.6 times the copper conductor; and the weight of aluminum is 0.48 times the weight of copper.

Aluminum is replacing copper these days in certain applications due to a number of economic and engineering reasons. It is used as coil wires for motors. Also, it is used in making bus bars, both soft and hard. Alloys of aluminum with magnesium, silicon, iron, etc. have greater mechanical strength and therefore they find applications as overhead transmission lines. Such alloys have comparatively high conductivity. Steel-cored aluminum cables, made by laying up aluminum wires around a core of steel wire, are also extensively used. The steel reinforced aluminum conductor (A.C.S.R) has great potentials for use in long-span transmission lines.

Aluminum wires as stated before have greater resistance to corrosion. In open air, aluminum appliances become coated with a thin film of aluminum oxide which protects the metal from further oxidation. Oxide film insulation on aluminum is profitably used in electrolytic capacitors designed with aluminum cans.

The natural oxide film (Al_2O_3) on aluminum is very thin but protects the metal from the air/oxygen. At the same time, this film has high electrical resistance. Because of this, the points of connection in aluminum wires that are poorly stripped and dressed may become seats of very high contact resistance. Moisture at the points where aluminum wires are connected to wires of other metals may give rise to galvanic (voltaic) couples that will cause aluminum to deteriorate. Because of this, such connections are often protected from moisture (for example, with a coat of varnish). The higher the chemical purity of aluminum, the better is its resistance to corrosion.

Commercially, pure aluminum is available in grades of various purity: High-conductivity aluminum, and aluminum which contains not more than 0.005% impurities (iron, silicon, zinc, titanium, and copper). It is used for electrodes in electrolytic capacitors, and for aluminum foils. Wire conductors are made from aluminum that has not more than 0.3 and 0.5% impurities. Wires of aluminum have semihard and hard tempers with diameters ranging from 0.08 to 10 mm; and aluminum bus bars are also made with a rectangular section.

Aluminum wires and conducting parts can be joined together by hot and cold welding (bonding), and by brazing or soldering using special brazing solders and fluxes. They may be cold-bonded together in special welding machines in which properly dressed surfaces of aluminum parts are mated together by applying a pressure of about $10{,}000 \times 10^5$ newton/meter^2 to cause interdiffusion of crystals across the mating parts and assure a firm and impervious joint. Sheet aluminum is used widely for electromagnetic shielding or screening purposes.

Beryllium (Be)

$Z = 4$
$M_a = 9.013$
$d = 1.85$ gm/cm^3
M.P. = 1278±5°C
$\alpha_T = 0.1 \times 10^{-4}$/K
$\rho_v = 0.36 \times 10^{-6}$ ohm-cm
$C_T = 2.0$ watt/cm/°C

Beryllium provides high stiffness-to-density ratios, high strength-to-density ratios, and excellent dimensional stability. Modulus of elasticity of 44×10^6 lb/in.^{2} and density of 0.066 lb/in.^{3} are common to all forms of beryllium. High thermal conductivity and low thermal expansion coefficient contribute to dimensional control under temperature variations.

Beryllium-Copper

Beryllium-copper combines strength, wear resistance, electrical and thermal conductivity, and ease of fabrication properties. Age-hardened beryllium-copper alloys can provide tensile strengths up to 215,000 lb/in.^{3} with endurance strengths (at 10^8 cycles) of approximately 40,000 lb/in.^{2}. A Rockwell hardness up to C-45 broadens the alloy's usefulness in applications requiring resistance to wear. Electrical conductivities reach 50% IACS, with a thermal conductivity range 775-1600 Btu/(h.ft^2.°F). Beryllium-copper alloys can first be fabricated or machined in the unhardened state. Desired properties can then be imparted to the finished product by a simple low-temperature thermal treatment.

Beryllium-Nickel

Beryllium-nickel is the highest strength nickel alloy at temeperatures from room to 900°F. After precipitation hardening, tensile values over 270,000 lb/in.^{2} with 230,000 yields can be obtained in wrought alloys. Like wrought forms of beryllium-copper, beryllium-nickel is normally supplied to a fabricator in a relatively soft condition. The alloy can be formed using conventional methods. The parts are then allowed to age by a simple cycle in an ordinary furnace to achieve outstanding mechanical properties. The alloy exhibits the general corrosion resistance of nickel.

Cobalt (Co) (and its alloys)

$Z = 27$
$M_a = 58.94$
$d = 8.9$ gm/cm^3
M.P. = 1495°C;
$\alpha_T = 0.13$ to 0.16×10^{-4}/K
$\rho_v = 5.80 \times 10^{-6}$ ohm-cm
$C_T = 1.2$ watt/cm/°C
$\alpha_R = 66 \times 10^{-4}$/°C

The retention of hardness and strength at high temperature is a distinctive feature not only of cobalt-based alloys but also of other cobalt-containing alloys. Materials are available for use under stress up to 2000°F and, at no load, up to 2400°F. Cobalt is the element with the highest Curie temperature (2050°F). The metal is used by itself or in alloys (cobalt-nickel, cobalt-phosphor, cobalt-nickel-phosphor) for memory and other magnetic devices.

It is an important alloying element in a series of permanent magnets as well as soft magnetic materials. Cobalt and its alloys are remarkable for their low coefficient of friction and nongalling characteristics. They can be used as high-temperature bearings without lubrication.

Copper (Cu) (and its alloys)

$Z = 29$
$M_a = 63.54$
$d = 8.94\ gm/cm^3$
$M.P. = 1083°C$
$\alpha_T = 0.167 \times 10^{-4}/K$
$\rho_V = 1.68 \times 10^{-6}$ ohm-cm
$C_T = 4.2$ watt/cm/°C
$\alpha_R = 39 \times 10^{-4}/°C$

Copper has the highest electrical conductivity of any metal except pure silver. Copper alloys are easily fabricated and set the standard for nonferrous alloys in most fabrication operations. The brasses (alloys of copper) are ideally suited to cold-forming operations such as deep-drawing, bending, spinning, and stamping. The solderability of copper is vital in realizing a good electrical continuity. A fair corrosion resistance in natural environments accounts for the wide use of copper and copper alloys. These metals tarnish superficially in moist air, but the pleasing colors of the surface films developed after further exposure often are a plus factor. Copper and its alloys are resistant to corrosion due to water, both fresh as well as brackish.

Copper is most popular as an electrical conductor and is extensively used in electromagnetic shielding (Faraday cage) applications. Also it is useful as heat sinks. Copper and brass are used with silver or gold-plating in the fabrication of microwave plumbing and magnetron anodes.

Copper is available in nature and its principal ores are cuprite (Cu_2O), copper glance (Cu_2S), copper pyrites ($CuFeS_2$), malachite ($CuCO_3 \cdot Cu(OH)_2$), and azurite ($2CuCO_3 \cdot Cu(OH)_2$). In the modern processing of copper, the ores, usually pyrites, are crushed and then calcined in a reverberatory furnace. The calcined ores are mixed with silica and a small quantity of coke. The mixture is then smelted in a blast furnace. The melted metal is oxidized in a Bessemer converter. It gives blister copper. Impurities in blister copper are removed by melting it in a reverberatory furnace in the presence of air. After removing the slag, copper with a purity of 99.7% is obtained. Close to one hundred percent pure copper could be obtained by the process of electrolysis.

Copper and copper-based alloys are unique in their physical and mechanical properties. They have the following properties: Moderate to high strength and hardness; excellent corrosion resistance; ease of workability both in primary and secondary fabrication operations; high electrical and thermal conductivity; non-magnetic properties; superior properties at subnormal temperatures; ease of finishing by polishing; plating; good to excellent machinability; excellent resistance to fatigue, abrasion, and wear; relative ease of joining by soldering, brazing, and welding; and availability in wide variety of forms and tempers.

Chief elements alloyed with copper are zinc, tin, lead, nickel, silicon, and aluminum, and to lesser extent beryllium, phosphorus, cobalt, boron, arsenic, zirconium, antimony, cadmium, iron, manganese, chromium, and mercury. Cast copper has a tensile strength of 150-170 MN/m^2, but rolling, drawing, or other hot and cold working can increase the tensile strength to 280 MN/m^2 in the case of annealed materials and to a maximum of 450 MN/m^2 hard-drawn wires (1 N = 0.101972 kg-force).

Copper is a strong metal, it is hard, durable and highly ductile. It has strong resistance against corrosion, oxidation and pitting. Hard, cold-drawn copper is obtained by drawing a cold bar of copper. When annealed (that is, first heated to about 300-350°C and then cooled), hard copper becomes soft and has 2-3% more conductivity than hard copper. Hard copper is springy in nature, in contrast to annealed copper, and can be easily shaped.

Cold-drawn copper is used as conductors for overhead transmission lines, bus bars, etc. where high mechanical strength is needed. Annealed copper (due to its flexibility) is used as insulated conductors in cables, windings, and coils. Rectangular copper bars are used in electrical machines as strips and in bus bars. Hard copper which can withstand wear on the parts of the brushes is used for commutator segments in electrical machines.

Alloys of copper with cadmium, chromium, silver, and tellurium in a small percentage have increased mechanical strength at the expense of reduced conductivity and are used as rotor bars, transmission line conductors, and traction collector wires. Copper containing 0.7 to 1.0 percent cadmium has greater strength under both static and alternating stresses and offers better resistance to wear making it suitable for contact and telephone wires. Copper with 1 percent chromium retains hardness and mechanical properties at reasonable high temperature and is used as welding electrodes, light current carrying springs upto 400°C. Silver bearing copper is a special type of high conductivity copper which is used for rotor conductors in large turbogenerators and commutators.

Copper in its pure state is a metal yellowish red in color, tough, ductile, and malleable. It melts at 1083°C and has a temperature coefficient of linear expansion of 17×10^6 per ^{o}C. It has, in general, excellent mechanical properties and ductility. Copper can be drawn into a wire of up to 0.03-0.01 mm in diameter and into thin strips. A thin film of oxide (CuO) which copper forms in the air prevents further penetration of air oxygen to copper, thereby improving corrosion resistant properties.

Conductive copper is marketed in different grades of various purity. The impurities copper may contain are bismuth, antimony, iron, lead, tin, zinc, nickel, phosphorus, sulfur, arsenic, and oxygen. The total amount of impurities in the highest grade conductive copper does not exceed 0.01%; that is, such graded copper is 99.9% pure. Copper grades that are used for the manufacture of conductive parts (magnet and hookup wires, and cables, etc.) contain not more than 0.05-0.1% impurities. Copper wire is normally drawn in round and rectangular sections. Round wire comes in diameters from 0.02 to 10 mm. Rectangular section wire (bus bar copper) may have the smaller side from 0.8 to 4 mm and the larger side, from 2 to 30 mm wide. Copper wire may be made from soft or annealed (at optimal temperature) grades and from hard-drawn grades of copper as indicated earlier.

Smaller diameter wire has a higher tensile strength and higher resistivity. Oxygen-free copper known for highest purity, is used in making very thin wires (0.01 mm diameter) that are intended for use at elevated temperatures (above 300 ^oC).

Brass

Basically brass is an alloy of copper and zinc which has a greater mechanical strength and wear resistance than copper but has considerably lower electrical conductivity. Brass with 30% zinc has only 25 percent electrical conductivity compared to copper. Brass has good weldability and solderability. It is easily moldable and machinable and has good resistance to corrosion.

It is widely used in the manufacture of electrical apparatus and instrument parts as a current carrying and structural material. Silicon, aluminum, and manganese brasses have electrical resistance in comparison with other brasses. They are used in resistance welding. Aluminum brass is used in capacitors.

In leaded brass with 0.5 to 0.2 percent lead, the machinability improves and friction properties become better. They are used extensively in industries. In tin brasses the tin content increases the corrosion resistance of Cu-Zn alloys. Typical parameters of a brass are:

$d = 8.5$ gm/cc

M.P. = 1027^oC

$\rho_V = 8 \times 10^{-6}$ ohm-cm

$\alpha_T = 0.18 \times 10^{-4}/K$

$\alpha_R = 15 \times 10^{-4}/^oC$

Bronze(s)

Copper alloys containing tin, cadmium, beryllium, and certain other metals are called bronzes which have fairly high conductivity. Cadmium-copper (0.9% Cd) has about 85-95% conductivity compared to copper. Bronze containing 0.8% Cd and 0.6% Sn has about 50-80% conductivity compared to copper. All bronzes have high mechanical strength. Cadmium-copper is used for making contact wires and commutator segments. Beryllium bronze is used for making current carrying springs, brush holders, sliding contacts, and knife-switch blades.

Phosphor bronze with 1.25 to 10 percent tin and a small quantity of phosphorus has excellent cold-winding characteristics, hardness and endurance properties, low coefficient of friction, and excellent corrosion resistance. It is used in making springs, diaphragms, and bearing plates.

Aluminum bronzes with less than 8 percent aluminum have good hot and cold rolling properties, high strength, good wear resistance as well as excellent corrosion resistance. *Cupronickel bronzes* with 2.5-30 percent Ni have high corrosion resistance, are moderately hard, but quite tough and ductile. They are used in tubular form capacitors. *Silicon bronzes* with 1.5 to 3 percent silica and *aluminium-silicon bronze* constitutes an extremely versatile series of alloys. They have high strength, exceptional corrosion resistance, and excellent cold and hot workability.

Bronzes are noted for low shrinkage which is 0.6-0.8% compared to 1.5-2.5% of steel or cast iron. Typical compositions of some bronzes are: (tin 10%, copper 90%); (tin 6-7%, phosphorus 0.15%, copper 93.85-92.85%); (aluminum 6-8%, copper 94-92%); and (beryllium 2-2.2%, nickel 0.2-0.5%, copper 97.8-97.3%).

Bronzes have good processing properties; that is, they can be cut, extruded, soldered, or brazed. They can be worked into strips and wires which are usable as spring contacts, conducting springs, and other current-conducting as well as structural parts.

The parts and products made from bronzes are subjected to heat treatment to harden; that is, they are quenched and tempered at optimum temperatures.

Bronzes are inferior to copper in electrical conductivity but superior in strength, elasticity, wear, and corrosion resistance. A bronze and copper comparison follows:

Copper 99.95% pure:
Kind of temper: Soft
Relative conductivity: 100%
Tensile strength: $(20\text{-}27) \times 10^7$ N/m^2
Elongation: 35%

Copper 99.95% pure:
Kind of temper: Hard
Relative conductivity: 98%
Tensile strength: $(36\text{-}44) \times 10^7$ N/m^2
Elongation: 2%

Beryllium bronze (2% beryllium; 0.5% nickel; 97.5% copper):
Kind of temper: Soft
Relative conductivity: 36%
Tensile strength: $(70\text{-}79) \times 10^7$ N/m^2
Elongation: 20%

Beryllium bronze (2% beryllium; 0.5% nickel; 97.5% copper):
Kind of temper: Hard
Relative conductivity: 26%
Tensile strength: $(160\text{-}175) \times 10^7$ N/m^2
Elongation: 9%

Phosphor bronze (6-7% tin; 0.15% phosphorus; 93.85-92.85% copper):
Kind of temper: Soft
Relative conductivity: 15%
Tensile strength: $(40\text{-}45) \times 10^7$ N/m^2
Elongation: 60%

Phosphor bronze (6-7% tin; 0.15% phosphorus; 93.85-92.85% copper):
Kind of temper: Hard
Relative conductivity: 10%
Tensile strength: $(95\text{-}105) \times 10^7$ N/m^2
Elongation: 3%

Lead (Pb)
(and its alloys)

$Z = 82$
$M_a = 207.21$
$d = 11.34$ gm/cm^3
M.P. = 327.43°C
$\alpha_T = 0.29 \times 10^{-4}$/K at 12°C
$\rho_v = 21 \times 10^{-6}$ ohm-cm

$C_T = 0.35$ watt/cm/°C

$\alpha_R = 43 \times 10^{-4}$/°C

Lead is one of the most easily formed metals. At room temperature it approaches the plastic state (melting point about 620°C) and is easily rolled, extruded, cast, or often shaped manually. Lead is resistant to most active chemicals. It is quite stable as a metal or in compounds because of its chemical family and high atomic weight. Lead, with a density of 0.41 lb/in.^{3}, is the lowest cost high-density material. The element has the rare, balanced combination of neutrons and protons that make it an excellent shield against ionizing radiation such as gamma rays and X-rays.

Magnesium (Mg) (and its alloys)

$Z = 12$

$M_a = 24.32$

$d = 1.74$ gm/cm^3

M.P. = 651°C

$\alpha_T = 0.25 \times 10^{-4}$/K at 12°C

$\rho_V = 4.3 \times 10^{-4}$ ohm-cm

$C_T = 1.6$ watt/cm/°C

$\alpha_R = 43 \times 10^{-4}$/°C

Magnesium, with a specific gravity of 1.74, has long been recognized as the lightest structural metal. Magnesium is well suited to modern die-casting processes. The metal flows easily and readily fills complex dies and thin sections. The most common magnesium alloys in regular use are in the magnesium-aluminum-zinc system for all cast and wrought forms. Magnesium-zinc-zirconium alloys, both cast and wrought, provide improved properties. Magnesium-thorium-zirconium, magnesium-thorium-zinc, and magnesium-thorium-manganese are the alloy systems used for retention of usable properties at elevated temperatures. Magnesium-lithium alloys are the lightest commercial magnesium alloys. A magnesium-lithium alloy, with specific gravity of only 1.35, can be obtained in various mill forms. Magnesium-lithium alloys have improved cold workability. They are fusion weldable by the inert gas shielded arc and also can be electrical resistance spot welded. They are receptive to the same chemical treatment as other magnesium alloys. Although magnesium is sometimes used unpainted, protective and decorative finishes are applied in most cases. Application of electrolytic coatings is sensitive and requires good controls.

Nickel (Ni) (and its alloys)

$Z = 28$

$M_a = 58.71$

$d = 8.9$ gm/cm^3

M.P. = 1455°C

$\alpha_T = 0.13 \times 10^{-4}$/K at 12°C

$\rho_V = 7.05 \times 10^{-6}$ ohm-cm

$C_T = 0.9$ watt/cm/°C

$\alpha_R = 60 \times 10^{-4}$/°C

Nickel alloys, in general, are stronger, tougher, and harder than most nonferrous alloys and many steels. Their most important commercial mechanical property, however, is their ability to retain strength and toughness at elevated temperatures.

The range of alloys based on nickel as a major constituent is wide enough to combat a great variety of corrosive environments. Generally, reducing conditions retard, while oxidizing conditions accelerate the corrosion of nickel. Nickel-based alloys often have the excellent corrosion resistance of elemental nickel, as well as that of other elements they contain.

Zinc (Zn) (and its alloys)

$Z = 30$

$M_a = 65.38$

$d = 7.14$ gm/cm^3

M.P. = 419.47°C

$\alpha_T = 0.27 \times 10^{-4}$/K at 12°C

$\rho_V = 5.95 \times 10^{-6}$ ohm-cm

$C_T = 1.2$ watt/cm/°C

$\alpha_R = 40 \times 10^{-4}$/°C

Resistance to corrosion of wrought and cast zinc products is generally retained throughout the product's life. As coating, zinc's protection continues when the underlying material (generally steel) eventually becomes exposed. Corrosive attack is then directed to the zinc rather than the base metal. Low melting point of zinc facilitates low-cost production in cast, wrought, or coating form. While the melting point is in the soldering temperature range, zinc alloys have physical and mechanical properties suitable for many structural applications. Most of the finishes applied to other metal products can also be applied to zinc die castings. These include: (i) Mechanical: buffing, polishing, brushing, and tumbling; (ii) electrodeposited: copper, nickel, chromium, brass, silver, and black nickel; (iii) chemical: chromate, phosphate, molybdate, and black nickel; (iv) organic: enamel, lacquer, and varnish; and (v) plastic.

Precious Metals: The listing and properties of the various precious metals are as follows:

Precious Metals	Z	M_a	d gm/cm^3	M.P. °C	α_T $\times 10^{-4}$ per K at 12°C	ρ_V $\times 10^{-6}$ ohm-cm	C_T watt/cm per °C	α_R
Gold (Au)	79	197.00	19.30	1063.0	0.140	2.21	3.10	34
Platinum (Pt)	78	195.90	21.45	1773.5	0.089	10.42	0.70	38
Palladium (Pd)	46	106.70	12.00	1549.4	0.120	10.55	0.72	_
Iridium (Ir)	77	192.20	22.40	2454.0	0.065	5.05	1.45	_
Rhodium (Rh)	45	102.91	12.44	1966±3	0.084	4.78	1.50	_
Osmium (Os)	76	190.20	22.50	2700	0.048	9.10	0.90	_
Ruthenium (Ru)	44	101.10	12.20	2450	0.080	7.40	1.00	_
Silver (Ag)	47	107.88	10.50	960.8	0.190	1.64	4.20	40

The precious metals are highly resistant to many corrosive environments either in their pure forms or in alloys. Platinum is the most commonly used but the others of the group are also widely used. Silver is tarnished by sulfuric environments. Gold and silver do not oxidize to form a scale even at elevated temperatures, but silver does adsorb considerable oxygen and this must be duly considered in its use. The basic precious-metal alloy system used for structural purposes at elevated temperatures is platinum-rhodium. Pure platinum may be used in cases where the host-strength requirement is not high. Where the strength requirements are higher than can be attained with platinum-rhodium alloys alone, platinum-rhodium alloys are used to sheath higher strength materials such as molybdenum which does not have adequate oxidation resistance. The stability and wide range of high electrical properties make the precious metals useful in a number of areas. Among the useful properties are stable thermoelectric behavior, high resistance to spark erosion, tarnish resistance, and broad ranges of electrical resistivities and temperature coefficients of electrical resistance.

Silver is very malleable and ductile and does not oxidize in the air at room temperature. At temperatures from 200°C and higher, oxidation of silver is, however, vigorous. Like all noble metals, silver is noted for high ductility which permits obtaining foils and wires up to 0.04 mm in diameter. Silver has the highest electric and heat conductivity.

Commercially pure silver has a density of 10,500 kg/m^3, melting point of 960.5°C, and temperature coefficient of expansion of 19.3×10^{-6} per °C, that is, a little higher than that of copper. Basic characteristics of the soft silver are: σ_T = (15-18) $\times 10^7$ N/m^2; e_t = 45-50%; ρ = 0.015 microohm-meter. The characteristics of hard-drawn silver include σ_T = (20-30) $\times 10^7$ N/m^2; e_t = 46%; ρ = 0.0158 microhm-meter; and α_R = 0.00369 per °C.

Silver is a very soft metal and it is not normally used for industrial applications in its pure state, but is alloyed with hardners, usually copper, nickel, or cadmium. Its major uses in industrial applications are in alloy forms as electric contacts of relays, and in other instruments rated for small currents. Silver alloys can also be used as soldering/brazing materials.

Tungsten (W)

Z = 74
M_a = 183.86
d = 19.3 gm/cm^3
M.P. = 3370°C
α_T = 0.046×10^{-4} per K at 12°C
ρ_v = 5.32×10^{-6} ohm-cm
C_T = 1.6 watt/cm/°C

Tungsten is stronger than any other common metal at temperatures over 3500°F. Also, the melting point of tungsten being 6170°F, it is higher than that of any other metal. The electrical conductivity of tungsten is approximately one-third that of copper, much better than the conductivity of nickel, platinum, or iron-based alloys. The resistivity of tungsten in fine wire form has been exploited in making lamp and electronic filaments which serve as a light-emitting or electron-emitting cathodes.

Tungsten is a very heavy grey metal. It is mostly obtained from scheelite, wolframite, and a few other ores as a powder by reduction of the oxide. Tungsten powder may be obtained to a purity of 99.9%. Tungsten parts, rods, and sheets are made by powder metallurgy. In the relevant methods, tungsten powder is molded in steel molds into cores which are then baked at 1300°C. The baked tungsten cores have a coarse grain structure and are brittle. To obtain a mechanically strong version tungsten cores are heated to 3000°C, alternately forged, rolled, and annealed between rollings. Such treatment gives fibrous structure that assures the metal of high mechanical strength and ductility. Tungsten wire drawn from the rolled metal comes in diameters up to 0.01 mm. Tungsten resists oxidation at very high temperatures; it starts corroding at 400°C. Under vacuum, tungsten parts are able to perform without deterioration up to 2000°C.

Commercially available pure tungsten has a density of 19,300 kg/m^3, melting point at 3,380°C; annealed tungsten has a tensile strength of (5-8) $\times$ 10^8 N/m^2 and resistivity of 0.0503 microhm-meter. The tensile strength of hard tungsten is 18 $\times$ 10^8 N/m^2 and ρ = 0.0612 microhm-meter. The temperature coefficient of resistivity is 0.0046 per °C.

Tungsten is widely used in electrical engineering as a wear-resistant material for contacts and parts of vacuum devices, such as filaments of incandescent lamps, and electrodes.

Molybdenum (Mo)

$Z = 42$
$M_a = 95.95$
$d = 10.20$ gm/cm^3
M.P. = 2620 ± 10°C
$\alpha_T = 0.052 \times 10^{-4}$ per K at 12°C
$\rho_V = 5.33 \times 10^{-6}$ ohm-cm
$C_T = 1.45$ watt/cm/°C

On strength basis, pure molybdenum is generally considered as the most suitable of all *refractory metals* for applications at temperatures between 1600 and 3000°F. Small amounts of other refractory metals with molybdenum form alloys that have much greater strength-to-weight ratios at higher temperatures. Thermal conductivity of molybdenum is more than three times that of iron and almost half that of copper. Abrasion resistance of molybdenum is generally outstanding at high temperatures.

Tantalum (Ta)

$Z = 73$
$M_a = 180.95$
$d = 16.6$ gm/cm^3
$\alpha_T = 0.065 \times 10^{-4}$ per K at 12°C
$\rho_V = 13 \times 10^{-6}$ ohm-cm
$C_T = 0.55$ watt/cm/°C
M.P. = 2996 ± 50°C

Tantalum is the most corrosion resistant of the major refractory metals. It closely matches the corrosion resistance of glass. Tantalum is an easy-to-fabricate metal. Spinning, deep drawing, and severe bending can be performed without tears, cracks, or excessive peeling. Ductile, nonporous welds can be made easily by using TIG, resistance, or electron beam welding. Tantalum finds an important use in electrolytic capacitors because it forms tantalum oxide (Ta_2O_2) which has a high dielectric constant and good dielectric strength.

Columbium [Also known as Niobium (Nb)]

At temperatures from 2000 to 3000°F (in vacuum or inert atmosphere), columbium alloys give the best performance on a strength-to-weight basis among metals that exhibit ductile welds. Columbium has excellent corrosion resistance, including resistance to liquid alkali metals. Columbium is the major component in high-field, superconducting alloys for use at cryogenic temperatures. Columbium has good nuclear properties, important in such applications as fuel-element cladding for nuclear propulsion reactors. The material is quite ductile, but is susceptible to air and hydrogen embrittlement at elevated temperatures. While columbium is easily worked when pure, it becomes difficult to work when highly alloyed. At high temperatures, columbium absorbs oxygen, nitrogen, and hydrogen and becomes brittle. It must be alloyed for high-temperature applications. Columbium is a reactive metal and accordingly reacts with coatings at high temperature.

Titanium (Ti)

$Z = 22$
$M_a = 47.90$
$d = 4.5$ gm/cm^3
$\alpha_T = 0.085 \times 10^{-4}$ /K at 12°C
$\rho_v = 53.0 \times 10^{-6}$ ohm-cm
M.P. 1800°C
$\rho_T = 0.2$ watt/cm/°C
$\alpha_R = 38 \times 10^{-4}$/°C

Titanium, with a density of 0.161 lb/in.^{3}, is classed as a light metal, being 60% heavier than aluminum (0.10 lb/in.^{3}), but 45% lighter than alloy steel (0.28 lb/in.^{3}). Titanium-based alloys are extremely strong, with an ultimate stress at about 30,000 lb/in.^{2} and yielding in the neighborhood of 200,000 lb/in.^{2}. Strengths of 160,000 ib/in.^{2} are attainable in the general-purpose (Ti6Al-4V) grade. The strength of titanium alloys is accompanied by excellent ductility. Titanium is virtually immune to corrosion from the atmosphere. Its corrosion resistance is excellent in most oxidizing environments and many mildly reducing environments. Titanium-based alloys offer excellent fatigue properties. Titanium is the only known structural metal with a corrosion-fatigue behavior in saltwater which is practically identical to that in air.

Zirconium (Zr)

$Z = 50$
$M_a = 91.22$
M.P. = 1857
$d = 6.4$ gm/cm^3

$\alpha_T = 0.059 \times 10^{-4}$ /K at 12°C

$\rho_V = 42.4 \times 10^{-6}$ ohm-cm

$C_T = 0.21$ watt/cm/°C

Zirconium displays excellent resistance to many corrosive media. This metal resists both acids and alkalies, in particular, alkali and chloride solutions, some inorganic acids, and chlorine-saturated water. Such resistance covers a wide range of temperature and concentration. Zirconium offers a low absorption cross-section for the slow thermal neutrons necessary to sustain a chain reaction. The material is therefore used in reactor components. Zirconium is lighter than steel and heavier than titanium. When impure, it is very hard and brittle. When pure, it is soft, malleable, and ductile. Its mechanical properties resemble those of mild steel, although they change rather rapidly as temperature is elevated. For example, Zircaloy-2 has a yield strength of 65,000 lb/in.^{2} at room temperature, but drops to 35,000 lb/in.^{2} at 700°F.

9.8 Metals and Alloys Used in Specific Electrical Applications

Iron and Steel

Steel is not very often used as a conducting material because of its low electrical conductivity, in spite of the fact that it has higher mechanical properties and low cost. For making wires, steel with 0.1 to 0.15 percent carbon is used.

Steel is easily corroded by moisture and heat. Overhead steel conductors are galvanized to prevent corrosion. To galvanize, steel is thoroughly cleaned and then dipped in a bath of molten zinc. This layer of zinc protects steel from rusting.

Steel overhead conductors due to less electrical conductivity are used only to transfer small amounts of power. Aluminum conductors used for high voltage systems and long spans are usually reinforced with steel, which gives high tensile strength to overhead lines. Being a ferromagnetic material, steel is often used in electrical engineering as a magnetic core material. (See Chapter 14.)

Bimetallic conductors

In some of the electrical applications, *bimetal conductors* containing copper (high conductivity) and steel (mechanically strong) are used. There are two methods of producing such bimetals. In one method steel ingot is poured into molten copper in a big end-up mold, and then rolled and drawn after cooling. In the other process copper is deposited on a steel wire core electrolytically. Bimetallic conductors are used for high frequency communication lines. They are also used for bus bars, and in switchgears, blades for knife switches, and various current carrying parts in electrical apparatus.

9.9 Soldering and Brazing Materials

Solder material is an alloy used to join two or more pieces of metals. The melting point of a solder should be lower than the materials being joined. The molten solder joins the pieces of metals and the process is known as *soldering*.

Solders used for electrical purposes can be divided into two groups:

(i) *Soft solders* or quick solders (melting point lower than 400°C)

(ii) *Hard solders* (melting point higher than 400°C)

A soft solder is an alloy of tin and lead. The most popular composition is 50% tin and 50% lead. The tin-lead solder serves to join copper, bronze, brass, lead, tinned iron, zinc, etc.

A hard solder is an alloy of copper and zinc. It melts at a very high temperature. It is used for joining brass, copper, iron, and steel. There are two varieties of hard solders, namely, *brazing solder* and *silver solder*.

Brazing is soldering at high temperatures. A non-ferrous filler metal is used for this purpose and has a melting point less than the base metals. The filler metal is distributed in the joint by capillary attraction during brazing.

Tin-lead solders are most widespread among the group of quick solders. They have high fluidity (flowability in liquid state); easily penetrate into thinnest joints; and adhere well to most metals, particularly to copper, brasses, bronzes, steel, and zinc, assuring high firmness and tightness of soldered joints. The solders with less than 15% tin are used to join parts together where high mechanical strength of the joint is not a factor. The tin-lead solders with a high content of bismuth (50-57%) have the lowest melting point (79-95°C). However, the resulting soldered joints are rather brittle.

The basic characteristics of typical quick solders are presented in Table 9.5.

Hard or brazing solders refers to copper-zinc, and copper-silver alloys, and aluminum-copper, aluminum-zinc, and aluminum-silicon alloys. Copper-silver solders are the most common among lead solders. They are noted for low resistivity and so are widely used to join together current-conducting parts of ferrous and non-ferrous metals. These parts when joined are wetted with these solders, forming strong and corrosion-resistant brazed joints.

Aluminum-based solders with addition of copper, silicon and tin are known for their high mechanical strength and resistance to weathering. They are used for brazing together aluminum wires and various parts from aluminum or its alloys.

Copper-zinc brazing solders are brittle and unstable to vibration and impact load, but give joints of a very low electrical resistance. They are commonly used to braze together mating parts made of copper, brasses, bronzes, and steel.

The basic characterisitcs of brazing solders are furnished in Tables 9.6.

Besides the solder, soldering or brazing also requires *fluxes* to remove oxides and other contaminants from the surface of metals being joined and to protect the surfaces and the molten solder from oxidation during brazing.

Fluxes may be solid (powdered substances, such as borax, boric acid, rosin, etc.) or liquid (aqueous solution of zinc chloride, alcohol solution of rosin, etc.). Sometimes, semiliquid flux pastes may be used.

The fluxes used in soldering copper, brasses, and bronzes with quick lead-base, solders should not cause corrosion of the soldered joints. Among such fluxes, called corrosion-resistant fluxes, are rosin, solution of rosin in ethyl alcohol, and other rosin-based compounds. Rosin is a weakly active flux. The surfaces of metals being soldered must therefore be properly cleaned and degreased before the flux of rosin is applied.

Neither rosin, nor other fluxes that readily decompose at elevated temperatures, can be used when brazing with hard solders that melt at temperatures higher than 500°C. For high-temperature brazing of steel, copper, and copper alloys (brasses and bronzes, etc.), the most suitable fluxes are often borax (sodium tetraborate, $Na_2B_4O_7$), or its mixture with boric acid, H_3BO_3, and other salts. For brazing aluminum that readily oxidizes in the air, use is made of especially active fluxes, called *activated fluxes*, able to dissolve the dense film of oxides formed on an aluminum surface. Such fluxes refer to a composition made of lithium chloride, sodium fluoride, zinc chloride, and potassium chloride. The general rule for choosing the proper flux is that the melting point of a solid must be lower than that of the solder, and the soldering temperature be lower than the thermal decomposition temperature of the flux. When using hard or brazing solders, it is essential that the residue of the flux be removed with a brush and hot water to avoid corrosion of the brazed joints.

The composition and basic properties of some fluxes are listed in Tables 9.7 and 9.8.

Table 9.5 Properties of Typical Soft Solders

Composition, % by Weight	Melting Point, °C	Tensile Strength, N/m^2	Application
Pure tin	232	5×10^7	Soldering and tinning of copper, brasses or bronzes to form corrosion-resistant joints and coated plating.
Tin 89-91 Lead 11-9	220	4.9×10^7	Soldering and tinning of copper, brasses or bronzes to form corrosion-resistant joints and coated-plating
Tin 60-61 Lead 40-39	190	4.3×10^7	Soldering and tinning of copper, brasses or bronzes to form corrosion-resistant joints and coated plating
Tin 39-41 Lead 61-59	238	3.8×10^7	Soldering and tinning of copper, brasses or bronzes to form corrosion-resistant joints and coated plating
Tin 9-10 Lead 91-90	299	3.2×10^7	Soldering together of nonvital parts (copper, brasses, and bronzes)
Tin 39-41 Antimony 2 Lead (remaining)	229	4.3×10^7	Soldering and tinning of copper wires and parts made from brasses and bronzes to form impervious joints
Zinc 60 Cadmium 40	300	3.5×10^7	Soldering of wires and parts made from aluminum and aluminum alloys

Table 9.6 Properties of Typical Hard Solders

Composition, % by Weight	Melting Point, °C	Tensile Strength, N/m^2	Application
Silver 72 Copper 28	779	36×10^7	Brazing conducting parts made from copper, brasses, bronzes, and other metals except aluminum
Silver 70 Copper 26 Zinc 4	775	35×10^7	Brazing together conductive parts made from copper, brasses, bronzes, and other metals (except aluminum) to obtain higher resistance to corrosion
Copper 54 Zinc 46	880	26×10^7	Brazing together parts made from copper, brasses, bronzes, and steel to form a brittle joint
Aluminum 66 Copper 28 Silicon 6	525	18×10^7	Brazing together parts made from aluminum or its alloys to obtain joints of high mechanical strength

Table 9.7 Fluxes for Soft Solders

Composition, % by Weight	Application
Esters of tarry acids (light rosin)	Soldering together conductive parts made from copper, brasses, and bronzes
Rosin 25-30 Ethyl alcohol 75-70	Soldering together conductive parts made from copper, brasses, and bronzes
Petrolatum 66 Triethanolamine 6.5 Salicylic acid 6.3 Ethyl alcohol 21.2	Soldering conductive parts made from copper, brasses, aluminum, bronzes, constantan, manganin, silver, platinum, and its alloys
Hydrochloric aniline 1.8 Glycerine 1.5 Rosin 96.7	Soldering conductive parts from copper, brasses, aluminum, bronzes, constantan, manganin, silver, platinum, and its alloys, the flux is more active and cleans well the surfaces of metals being brazed
Rosin 16 Zinc chloride 4 Petrolatum 80	Brazing parts made from ferrous and non-ferrous metals to form joints of higher strength, flux residue must be washed off with water to avoid corrosion of joints

Table 9.8 Fluxes for Hard Solders

Composition, % by Weight	Application
Borax glass	Brazing together the parts of copper, bronzes, platinum, nickel, and steel (using copper and copper-zinc solders), flux residue must be removed to avoid corrosion.
Molten borax 21 Boric acid 70 Calcium fluoride 9	Brazing together the parts of copper, brasses, bronzes, stainless steel, and carbon steel (with copper solders)
Boric anhyride 36 Potassium fluoride 42 Potassium fluoroborate 22	Brazing together the parts of copper, brasses, bronzes, stainless steel, and carbon steel (with silver solders)
Sodium fluoride 10 Zinc chloride 8 Lithium chloride 32 Potassium chloride 50	Brazing together the parts from aluminum and its alloys (with aluminum base solders), flux residue must be removed to avoid corrosion

9.10 Cryogenic Hyperconductors

Cryogenic conductors are essentially metals whose resistivity is rather low at sub-normal temperatures but higher than the critical temperatures of superconductors (see Chapter 11). That is, at the cryogenic temperatures of liquid nitrogen (20.4 K), liquid neon (27.3 K), and liquid nitrogen (77.4 K), the electrical resistance of hyperconductors drops sharply. This is explained by the hypothesis that the vibrating ions in crystal lattice do not impede the flow of electrons constituting the electric current in metal conductors at sufficiently low temperatures and when ions cease to vibrate, the perfect crystal lattice allows the electrons to pass without much scattering. In other words, the ohmic loss is minimized.

The degree of electron scattering in such a case will be determined essentially by the impurities that a cryogenic metal conductor may contain. For this reason, cryogenic hyperconductors are normally conductive metals of a high chemical purity, such as 99.9% copper, 99.999% aluminum, 99.99% silver, and 99.95% beryllium. These metals are usually annealed (to a soft temper) to increase conductivity before they are used for hyperconducting purposes. The coils made from hyperconducting wires for electric machines, transformers, and electrical apparatus destined to operate at cryogenic temperatures permit high current densities being reached without much loss. This allows a considerable reduction of the overall dimensions and mass of electric machines and apparatus (for use at cryogenic temperatures) and markedly increases their efficiency.

9.11 Materials for Electrical Contacts

Electrical contact refers to a temporary, releasable junction established between two conductors for the purpose of making electrical continuity as a "make-break" mechanism. When the contact is "made", the contact points should be able to carry the circuit current. When the contact is released, the gap across the contacts faces a potential across it and therefore offers no breakdown or arc discharge across the gaps.

The materials used for electrical contacts, therefore, operate under severe ampacity and voltage ratings. Such materials should meet several requirements specified by:

- Contact resistance (Low contact resistance is desired.)
- Contact force (High contact force is desired.)
- Current through the closed contacts (High current density is desired.)
- Voltage across the released contacts (High breakdown strength is desired.)
- Make-break cycles (High performance reliability is desired.)

Contacts operating on d.c. circuits are invariably subjected to material transfer; that is, the transfer of metal from the face of one contact to the face of its matching contact exists. In a.c. circuits, this type of material transfer is not usually encountered within the frequency range of about 25 to 400 Hz. The other factors which affect the contact performance are: Frequency of operation (number of operations a pair of contacts may be required to perform in a unit time), speed of contact separation, type of electric load (capacitive, inductive, or resistive), and the medium in which the contact operates.

Circuit-breaking contacts have to withstand arcing or sparkover, whenever they are separated or brought together. They deteriorate with time because of: (1) Mechnical wear; (2) corrosion resulting from oxidation and other chemical reactions due to contact with surrounding media and other factors; and (3) erosion from fusing, evaporation, and wear of the working surfaces during service. Due to corrosion, contact surfaces usually acquire a film of oxide which has low conductivity and reduces the effectiveness of electrical contacts.

Commercially available electrical contact materials are:

- Pure metals acceptable as make and break contact materials: They include copper, molybdenum, nickel, palladium, platinum, silver, tungsten.
- Alloys and heterogeneous mixtures: For contact material applications alloys are produced in the conventional manner by melting, casting, forging, rolling, drawing, stamping, heading, etc. Heterogeneous mixtures such as copper-tungsten, silver-molybdenum, and silver-tungsten carbide, etc. cannot be alloyed by the usual melting practice because of extremely low solubility of the refractory metal in silver or copper. Such heterogeneous mixtures can be formed as contact materials *via* the powder metallurgy technique.
- Copper: Because of low cost and high electrical and thermal conductivities, copper is popular as a contact material. Its drawbacks are poor resistance to oxidation and lending itself in the formation of chemical compounds (for example, sulfides). Its typical applications are in control relays, motor starter switches, and tap changers.
- Silver, silver alloys, and silver heterogeneous mixtures: Silver has high resistance to oxidation and has low contact resistance as well as good current-carrying capacity. Silver contacts are alloys such as silver plus a refractory constituent (for example, tungsten) or silver plus a semi-refractory constituent (for example, cadmium oxide). Some typical compositions of alloys are: Pure silver (99.9%), cadmium silver (Ag plus up to 20% Cd), and palladium-silver (Ag plus upto 10% Pd). Silver and silver alloy contacts are used up to 600 V, 50 amp d.c., and 200 ampere a.c. conditions. They are used for a variety of industrial relays, automatic voltage and current regulators, light switches, generators, cutouts, thermal overhead devices, domestic appliances, limit switches, thermostatic controls, etc.

  An important silver-refractory heterogeneous mixture used as a contact material consists of silver and various percentages of tungsten, tungsten carbide, or molybdenum. Such materials are used for medium and heavy-duty (high current) interrupting apparatus and the contacts operate in air. They are highly resistant to corrosion. Typical compositions are: Molybdenum-silver (Ag plus 40-50% Mo), tungsten-silver (Ag plus 40-75% W), and tungsten carbide-silver (Ag plus 40-60% tungsten carbide). They are operated at voltages up to 300,000 volts, and current up to 10,000 amperes. They are used in arcing tips for air-type multistage circuit breakers, light and medium duty circuit breakers, arcing contact oil circuit breakers, and as contact materials for heavy-duty cut rollers.

Silver-semirefractory heterogeneous mixtures include cadmium oxide-silver (Ag plus upto 15% CdO), nickel-silver (Ag plus 10-15% Ni), and others: They are used at voltages up to 550 V a.c. and 220 V d.c. and up to 300,000 amperes as current-carrying contacts for multistage breaker (maximum normal operating current up to 300 amperes). Such contacts are used for light-duty breakers, heavy-duty motor starters, and commercial relays.

- Copper-tungsten heterogeneous mixtures: These contact materials are similar in properties to silver refractory materials. They are used under severe arcing conditions more often under oil. Copper-tungsten (Cu plus 55-70% W) contacts can operate up to 300,000 volts, and up to 10,000 amperes. Their typical applications are oil-immersed circuit breakers, transformer tap changers, and arcing tips on certain air breakers.
- Precious metal electrical contacts: The most used precious metals and their alloys used for electrical contacts are platinum and palladium.
- Platinum and its alloys: Platinum has a weakness to forming oxides, sulfides, and chlorides, etc. easily. It has high melting point and provides good resistance to electrical erosion. Pure platinum is soft as an electrical contact material. Therefore it is alloyed with other metals in various percentages up to 35%. Common alloying elements are iridium, ruthenium, and osmium. They are used for lightly loaded contacts used up to 500 volts a.c. or d.c. up to 25 amperes. For example, magneto-ignition systems, thermostatic relays, sensitive relays, galvanometer contacts, recording instruments, railway signal equipment, vibrator contacts, relays for the chemical industry, etc. use this version of contact materials.
- Palladium and its alloys: They have similar mechanical and electrical properties of platinum. The melting point is lower than platinum but higher than silver. It is less costly than platinum or silver. Palladium is used for those applications where certain characteristics of silver are inadequate and where the expense of platinum is not justifiable. They are used for voltages up to 500 volts, and up to 15 amperes. Their typical applications are in telephone relays, thermostatic controls, sensitive relays, and relays for the chemical industry.
- Tungsten: The extremely high melting point of tungsten, coupled with high hardness and resistance to abrasion, place tungsten as an outstanding material for specific applications. They operate under not too severe arcing. They are used up to 200 V a.c. and d. c. and up to 15 amperes. Its typical applications are: Battery ignition systems, magneto-ignition systems, vibrators, electric razors, and automotive horn contacts.
- Sliding contact materials: For wire wound variable resistors and rheostats of metal alloy wires with high resistivity, the contact materials used are typically bronze-nickel or platinum alloys.
- Liquid metal contacts: Mercury as a contact material has the merit of being wear free. However, its application in structures open to atmosphere is restricted due to its hazardous vapor. Also, mercury in oxygen ambient may form insulating contacts. Therefore Hg as a contact material is restricted for use in only evacuated bulbs/valves.
- Conductor-included ceramic contacts: *Cermets* are composite materials made of ceramic particles (or grains) dispersed in a metal matrix. Cermet parts are produced by powder metallurgy techniques, that is, by baking the molded powdered units at high temperatures (1000-1400°C). The cermet molding stock is constituted by two powders or more of different metals, one of which is a higher melting point metal compared to others. As the molded stock bakes, the powdered metals that are more fusible melt and fill the pores between the unmelted metal particles, making the molding stock into monolithic cermet molded products. Bonding between the constituents results in this caseis due to the partial solubility. Some systems, however, such as the metal oxides, exhibit poor bonding between phases and warrant additives to serve as bonding agents. Mutual solubility can be used to obtain foamed (porous) cermet products, such as bearings, filters, etc. Hard-phase bonding of cermet parts can be obtained by fusion of the particles of powdered metals that melt at about the same temperature point. Some cermets are

produced by impregnating a porous ceramic structure with a metallic matrix binder. The powdered-metal molding stock is sometimes made with a nonmetal powder, for example, graphite powder, for use in the production of metal-graphitic brushes for electric machines, electric contacts, and other electrical parts. The aforementioned methods and techniques of making molded cermet products from powdered metal compositions by subsequent baking of the metal particles at high temperatures are largely based on powder metallurgical considerations.

Powder metallurgy techniques are adapted where the product cannot be obtained by alloying exceptionally high melting point metals or exceptionally pure metals or where the product requires fusion of metals with nonmetals. Such methods also permit obtaining products to a preset dimension without a need for finishing operations. This significantly reduces losses in the form of scrap metal. Electrical engineering employs powder metallurgy techniques in making electrical carbon elements, certain types of magnetic materials, and high precision electric contacts.

Cermet contacts are distinct from the metal contacts of silver, copper, tungsten, and metal alloys by their high wear resistance, greater contact force and resistance to erosion, or degradation of contact surfaces due to arcing.

Contacts made from an oxide-based cermet (silver and about 15% CdO powders) find extensive use in low-voltage electrical apparatus. They exhibit all the property characteristics of cermets. In addition, they prove reliable in tropical climatic conditions.

Cermet materials produced from powdered silver, tungsten (40-50%) and nickel (2-3%) are valued for high-precision arcing contacts. The cermet parts from this composition are noted for ductility and can also be machined.

Copper-graphite-based cermet contacts are known for minimum welding tendency at break while carrying large currents (30,000-100,000 amperes). High resistance to sticking and welding of contacts can be ascribed not only to the structure and composition of the cermets based on copper and graphite (3-5%) but also to the cermet porosity (10-15%). In the production of cermet parts, the amount of pores can be varied over a wide range. The porosity in cermet contacts rated for small currents should not exceed 2-5 percent. This can be attained by pressing and subsequent sintering of cermet products, which produces a harder and denser structure.

9.12 High Resistivity Alloys

These are specific class of conducting materials usable as:

- Standard resistances (thermostable resistors)
- Rheostats/variable resistor elements
- Furnace (heating) elements
- Filaments for lamps, valves, etc.

Their major requirements are:

- High resistivity
- High mechanical strength
- Noncorrosive at high temperatures
- High melting point
- Malleable and ductile to take wire shapes at requisite sizes
- Low coefficient of resistivity with respect to temperature
- High resistance stability with time
- Low thermoelectromotive effects under bimetallic contact situations

Typical metal alloys used for high electrical resistance applications are tabulated in Table 9.9 along with their characteristics.

Table 9.9 High Electrical Resisitivity Materials and Their Properties

Alloys with Composition	Density g/cm^3	Resistivity ohm-cm ($\times 10^{-4}$) at 12°C	Temperature Resistance Coefficient per ^{o}C	Maximum Working Temperature ^{o}C	Applications
Manganin (85% Cu, 12% Mn, 3% Ni)	8.4	0.42-0.48	5-30 ($\times 10^{-5}$)	60-70	Used in precision instruments, standard resistances, bridge potentiometers, and resistance boxes; exhibits almost temperature independent ohmic resistance and thermo-electromotive force under bimetallic contact with copper is low ($\cong 1 \times 10^{-6}$ volt/^{o}C); annealed (soft) and hard-drawn wire from manganin comes in thickness from 0.02 to 6 mm across; manganin can also be rolled in strips up to 0.08 mm thick and up to 270 mm wide; manganin magnet wire is available with enameled insulation, natural silk insulation, and with a coat of enamel underlying a single layer of natural silk

(continued...)

Alloys with Composition	Density g/cm^3	Resistivity ohm-cm ($\times 10^{-4}$) at 12°C	Temperature Resistance Coefficient per ^{o}C	Maximum Working Temperature ^{o}C	Applications
Constantan (58.8% Cu, 40% Ni, 1.2% Mn)	8.9	0.48-0.52	5-25 ($\times 10^{-5}$)	450-500	Used in rheostats and similar control devices; soft and hard-drawn wire from constantan comes in thickness from 0.03 to 5 mm across; also available is 0.1 mm thick constantan strip; insulated constantan wire is used in combination with a copper wire as bimetallics with thermo-temperatures higher than 500°C.
Nichrome (67.5% Ni, 15% Mn, 16% Fe)	8.4-8.5	1.0-1.1	10-20 ($\times 10^{-5}$)	1100	Used as heating elements in heaters and furnaces, and as filaments; mechanically strong
Fechral (80% Fe, 14% Cr, 6% Al)	_	1.2	10 ($\times 10^{-5}$)	900	Replacement/substitute for nichrome
German silver (65% Cu, 20% Zn, 15% Ni)	_	0.28-0.35	4 ($\times 10^{-5}$)	150-200	_
Nickeline (54% Cu, 20% Zn, 26% Ni)	_	0.39-0.45	2 ($\times 10^{-5}$)	150-200	_
Rhesstan (84% Cu, 12% Mn, 4% Zn)	_	0.45-0.52	40 ($\times 10^{-5}$)	150-200	_
Chromel (26-28% Cr, 0.6% Ni, 5-8% Al)	_	_	5 ($\times 10^{-5}$)	1000-1150	_

In the above class of high resistivity materials, nichrome, fechral, and chromel are known as *heat-resistant conducting materials*. The specific characteristics of such materials are as follows: Heat-resistant conducting materials refer to alloys based on chromium, nickel, and other elements. The heat resistance of these alloys, that is, their ability to resist oxidation at very high temperatures, is due to the formation on their surface an oxide film of high density that excludes access of air/oxygen to the alloy. The chromium oxide (Cr_2O_3) and nickel monoxide (NiO) which do not volatilize from the metal surface even at high temperatures, make the basis of heat-resistant oxide films. Heat-resistant conducting materials based on nickel, chromium, and aluminum are called nichromes, fechrals, and chromals, respectively. They are all solid solutions of metals with a disordered structure. As a result these alloys have high resistivity and low values of temperature coefficient of resistivity.

Besides the principal alloying elements listed, heat-resistant alloys may include impurities (0.06-0.15% carbon; 0.5-0.35% phosphorus; and 0.03% sulfur) which may cause brittleness of the wires and strips made from these alloys.

9.13 Nonmetallic Conductors in Electrical Applications

Electric carbon and graphite elements: Carbon occurs in varied allotropic forms from the transparent diamond with its extreme hardness to graphite, which is grey and relatively soft. Electrical carbon materials are manufactured from graphite and other forms of carbon (coal etc.). Graphite occurs in nature as a mineral with a high content of carbon (up to 90% or more). It is crystalline in structure and has a very high melting point (about 3900°C). Pure carbon is a semiconductor with a negative temperature coefficient of resistivity. Carbon has a conductivity slightly less than that of metals and their alloys. In electrical engineering, carbon elements are extensively used as: (i) Brushes for electrical machines; (ii) carbon electrodes for electric-arc furnaces, electrolytic baths and welding; (iii) non-wire resistors; (iv) arc-light; (v) battery cell elements; and (vi) microphone powders and other components of telecommunication equipment, etc.

The carbon brushes are used for current collection from the rotating parts of electrical machines. Carbon and graphite, which are different forms of the same element, are used as sliding electrical contacts because of their valuable properties as a contact material, prominent among which are:

- They have the ability to withstand high temperatures. Very high instantaneous local temperatures normally exist under all abrading contacts. Carbon retains its properties under these conditions because it remains solid up to temperatures exceeding 3000°C.
- Carbon has a low density and is lighter than most metals; only magnesium approaches its quality for lightweightedness. The low inertia of lightweight carbon brushes makes them easy to follow the contours of the moving surface.
- Carbon does not weld to metals under conditions where metals invariably weld to one another, such as in the heat of an electric arc.
- Carbon brushes provide self-lubrication at the moving contact parts.
- Carbon brushes are fabricated by the agglomeration of fine powders.

Depending upon the requirements of electrical machines (variety of current collecting conditions), a large class of carbon brush grades are manufactured. The most common grades are: (i) Natural graphite – it is natural mineral graphite with the highest degree of lubricating properties. They are greasy and are used in high speed, silent running conditions. (ii) Hard carbon – they are mechanically robust and wear resistant. They are used in moderate speed and current industrial motors due to low thermal and electrical conductivity of amorphous carbons. (iii) Electrographitized – the brush carbon is transformed into crystalline graphite by a thermal process in special electrical ovens (2500°C). In this process, the product retains some of the robustness of carbon while retaining some of the lubricating property of graphite. Electrographite has high thermal and electrical conductivity and is highly heat/flame resistant. It is capable of carrying heavy overload currents. (iv) Metal graphite is used in slip-ring motors, where the brushes have current collection function only. Its comparatively high contact resistance is, however, a disadvantage. The wear of the contact is usually significant with metal graphites because of the absence of lubricating effects. In metal-graphite grades, the advantage is that carbon and graphite mix with many metals in all proportions without alloying. The commonly used combinations are: copper-graphite, bronze-graphite, and silver-graphite.

Electric carbon conductors are made by the following techniques:

Electric carbon elements are made using power metallurgy techniques with carbon in the form of graphite, coke, carbon black, and hard coal (anthracite). Sometimes the carbon is mixed with metal powders (copper, lead, tin, and other powders) to increase electrical conductivity. They may also be made with the impregnating binders. The binder could be bituminous coal tar or a synthetic resin, Bakelite, or silicone resin, etc.

All carbon materials (except the graphite and black carbon) are first calcinated at 1200–1300°C to rid of volatile matters and to reduce the shrinkage. They are then crushed and milled to a finely divided powder, and then thoroughly mixed with the metal powder(s). The mixture is blended with binding agents (resins, coal tar pitch) and mixed again at 110–230°C in a special mixer.

The resulting electric carbon stock is dried, crushed to powder, and sieved. This results in a molding powder which is then pressed in molds into various product shapes or blocks which are then cut and ground into carbon brushes and other carbon parts and products. Depending on the binder, they may be pressed at room temperature or at 180–210°C. Electric carbon elements are pressed, molded or formed at elevated temperatures, if melting or polymerization of the binder warrants high temperatures. Electric carbon products are pressed at 100–300 MPa.

Products with a large linear dimension such as electric light arc electrodes, etc. are made by forcing through a screw-press die a heated fine-powdered stock into pressed compacts. The pressed products are then sintered at a temperature in the range 1200–1300°C to bond the particles and cement them with the carbon material released from the organic binder constituents.

Sintering produces a harder and denser structure; that is, it makes the products mechanically stronger, and reduces resistivity. After sintering, carbon elements can be machined. Electric carbon elements which contain carbon black, coke, and other nongraphitic components are normally carbonized by sintering at 2400–2800°C; thus the process is known as *graphitization*. By this the nongraphitic components in compressed form convert into the soft crystalline structure of graphite and most of the impurities volatilize at the high temperature. Graphitization of carbon brushes (in the electrographitic group) and other similar carbon elements offers a softer structure and has reduced friction coefficient and lower resistivity.

Upon graphitization and machining (cutting, grinding), electric carbon elements have high porosity, up to 30%. To eliminate porosity and minimize hygroscopicity, carbon elements are impregnated with varnishes or waxy materials and, sometimes, with molten metals such as tin or lead, usually at 80 to 200°C when the impregnating substance is still in a molten state. Impregnation increases mechanical strength and electric conductivity.

Electric carbon elements (carbon brushes) are then machined to give them the final shape and surface finish. Compressed blocks of carbon brushes and other elements are cut into small pieces, using special machines, milling cutters, or thin grinding wheels. Holes are drilled to accommodate terminals.

The finished surfaces of some types of carbon brushes and arc electrodes are copper-plated. They are given a layer of copper from 10 to 15 μm thick by electroplating techniques to assure a reliable electric contact between the body of the carbon brush and the holder. The flexible (multi-wire) terminals are expanded, soldered, or press-fitted in the brush body. The finished carbon brushes are then checked for dimensional accuracy, hardness, mechanical strength, resistivity, voltage drop across the brush and the commutator, friction coefficient, contact resistance between the lead and brush, and other characterisitcs.

Among several applications of carbon elements, their uses as carbon brushes and contacts are most widespread. Carbon brushes can be graphitic, carbon-graphitic, metal-graphitic, and electrographitic. Their *in situ* characteristics follow.

Graphitic brushes of natural graphite are softer in structure compared to others, noiseless in operation and used where peripheral velocities range from 20 to 40 meter/second. Their resistivity is 8-30 microohm-meter.

Carbon-graphitic brushes are made of graphite, carbon black, and coke with a binding resin. They are characterized by significant hardness, mechanical strength, and abrasiveness. They are capable of removing oxide films from electric machine commutators and rings and are used at peripheral velocities from 10 to 40 meter/second. Their resistivity is 100-400 microohm-meter.

Metal-graphitic brushes are produced from graphite and copper powders, sometimes blended with powdered tin or silver. They have a low resistivity, on the order of 0.3–0.8 microohm-meter; the resistivity of carbon brushes with a low content of copper is 5–22 microohm-meter. They are used at peripheral velocities 20 to 25 meter/second.

Electrographitic brushes are made of graphite, coke and carbon black with a binding resin. Compressed and sintered, the brushes are subjected to a temperature of 2500°C for graphitization to increase mechanical strength and permit them being used at increased peripheral velocities from 40 to 90 meter/second. Their resistivity is 12–75 microhm-meter. The brushes are used in electric motors and generators with heavy-duty commutation.

Electric light arc electrodes are characterized by arc resistance, very slow oxidation, and noninflammability. They do not burn or melt below 3800°C. Their major applications are in high capacity electric equipment.

Contact elements (slip rings, shoes, etc.) for electric locomotives, trolley buses and other current-collecting machines are made from electrographitic and copper-graphitic carbons. In their finished form, they offer significantly low resistivity, about 0.02–0.05 microhm-meter.

The characteristics of various brush materials and their application potentials are listed in Table 9.10.

Table 9.10 Characteristics and Uses of Various Carbon Brush Materials

Brush Material	Hardness (in Vickers)	Bulk Resistivity ρ_V in ohm-cm ($\times 10^{-3}$)	Contact Potential Drop	Uses
High resistivity carbon with resin bonding	–	5-25	High	Small fractional horse power motors
Low resistivity hard, baked carbon	30	5	Low	Distributors
Medium speed electrographite	15	5	Medium	d.c. machines
High speed electrographite	15	4-6	Medium	Large horse-power machines
Copper or bronze-based graphite	~15	0.5-0.005	Very low	a.c. machines and low voltage d.c. machines

9.14 Fusible Metals/Alloys

These materials are used for circuit interruptions when currents beyond the rated (maximum) values are encountered. When the *fusing current* flows through the fuse material, the resulting joulean heating melts the material and facilitates an open circuit thereby preventing possible damage of the circuit due to heavy currents.

Under nominal operating conditions a fuse carries a *rated carrying current* safely without melting and/or overheating.

Fusing current is defined as the minimum current required in a given time interval to set the fuse at a steady temperature enabling its melting.

The *fusing time* is rather a crucial parameter of these materials. It refers to the minimum time of heating at the flow of fusing current leading the fuse to melt. The fusing current is given by an empirical relation as follows:

$$I_{(fusing)} = A \times (Diameter\ of\ the\ wire)^{n} \tag{9.19}$$

where A and n are empirical constants for a given metal. Typical values of A and n for different metals are given in Table 9.11.

Table 9.11 Values of A and n

Metal	A	n	Melting Point °C
Copper	2.50×10^3	1.48	1084
Aluminum	1.90×10^3	1.50	659
Iron	0.78×10^3	1.49	1535
Tin	0.40×10^3	1.50	232
Lead	0.30×10^3	1.50	327

The low melting point metals and alloys usable as fuse materials are furnished in Table 9.12.

Table 9.12 Fusable Metals and Alloys

Material	Melting Point °C
Lead	327.0
Cadmium	321.0
Bismuth	271.0
Tin	232.0
Indium	156.4

(continued...)

Material	Melting Point °C
20% Bi, 80% Sn	200.0
50% Bi, 50% Pb	160.0
32% Pb, 50% Sn, 18% Cd	145.0
54% Bi, 26% Pb, 20% Cd	103.0
52% Bi, 25% Pb, 13% Sn, 10% Cd	72.0

9.15 Thermoelectric Properties of Metals and Alloys

When the junctions of two dissimilar conductors (metals, alloys or semiconductors) are at differential temperatures, a current flows in the closed circuit constituted by the dissimilar conductors (Figure 9.3). This phenomenon is known as the *Seebeck effect.*

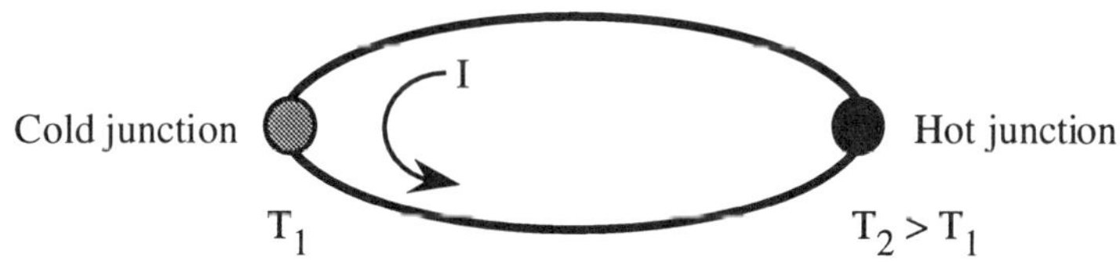

Figure 9.3 Seebeck effect.

Conversly, a reversible change in heat content can be perceived at the junctions when a coulombic charge crosses the junction. This is known as the *Peltier effect.*

The associated power in the above phenomena is known as the *thermoelectric power* and the pair of dissimilar conductors constitute a *thermocouple.* The thermoelectric power of a thermocouple is the algebraic sum of the *absolute thermoelectric powers* (ATP) of its components.

Commonly used alloys as thermocouples are listed in Table 9.13.

Table 9.13 Standard Thermocouples

Thermocouple Designation (ANSI-ASTM)	Positive Component	Negative Component	Temperature Range °C
Type S	$Pt_{90}Rh_{10}$	Pt	-50–1767
Type R	$Pt_{87}Rh_{13}$	Pt	-50–1767
Type B	$Pt_{70}Rh_{30}$	$Pt_{94}Rh_6$	0–1820

(continued...)

Thermocouple Designation (ANSI-ASTM)	Positive Component	Negative Component	Temperature Range °C
Type J	Iron	Constantan	-210–1200
Type K	Ni-Cr Alloy	Ni-Al Alloy	-270–1372
Type E	Ni-Cr	Constantan	-270–1000
Type T	Cu	Constantan	-270–400

Adapted from [4].

Thermocouples are very useful in temperature measurements. Details on thermocouples and their characteristics are available in: ASTM STP 470 B – *Manual on the Use of Thermocouples in Temperature Measurement.*

9.16 Concluding Remarks

Conductive materials constitute the major materials in electromagnetic applications paralleling only the insulating materials. Conductors are also very widely studied materials in electrotechnology perspectives. Since metals and alloys have been known for ages, their physical characteristics have been judiciously chosen to match such materials in electrical engineering practice. Nonmetallic conductors and composite conductors are adjuncts to metals and alloys in modern electrotechnological practice.

References

[1] R. N. Sampson: Materials for electronic packaging, in C. A. Harper (Ed. In-chief): *Electronic Packaging and Interconnection Handbook.* (McGraw-Hill Inc., New York: 1991).

[2] D. D. Pollock: *Physical Properties of Materials for Engineers.* vol. I, II & III (CRC Press, Boca Raton, FL: 1982).

[3] D. D. Pollock: *Electrical Conduction in Solids.* (Americal Society for Metals, Metals Park, OH: 1985).

[4] D. D. Pollock: *Physics of Engineering Materials.* (Prentice Hall, Englewood Cliffs, NJ: 1990).

[5] L. Solimar and D. Walsh: *Lectures on the Electrical Properties of Materials.* (Oxford University Press, Oxford: 1993).

[6] A. J. Dekker: *Electrical Engineering Materials.* (Prentice Hall, Englewood Cliffs, NJ: 1959).

[7] B. Arzamasov: *Materials Science.* (Mir Publishers, Moscow: 1989).

[8] A. Nussbaum: *Electronic and Magnetic Behavior of Materials.* (Prentice Hall, Englewood Cliffs, NJ: 1967).

Defining Terms

Alloys: A mixture or a solid solution of two or more metals.

Bimetallic conductors: A combination of highly electrically conducting metal and a metal with large mechanical strength.

Brazing: Soldering at high temperatures.

Cryogenic hyperconductors: Conductors exhibiting high electrical conductivity at cryogenic temperatures.

Contact materials: Materials used at make-break electrical contacts.

Debye temperature: Temperature above which the resistivity of metals varies linearly with temperature.

Fermi statistics: Statistics governing the probabilistic aspects of electronic distribution across the energy band(s).

Seebeck effect: Generation of electrical potential at the junction of two dissimilar metal/alloys kept at a differential temperature.

Soldering: Joining two pieces of a metal by another metal in a molten state.

Temperature coefficient of resistivity: A coefficient relating the resistivity *versus* temperature.

Thermocouple: A pair of dissimilar metals/alloys used to realize the Seebeck effect.

CHAPTER 10

Semiconducting Materials

10.1 Introduction

Semiconductors are materials which have resistivities between the extremes of metallic conductors and insulators. A typical scale of conductivity of materials is depicted in Figure 10.1.

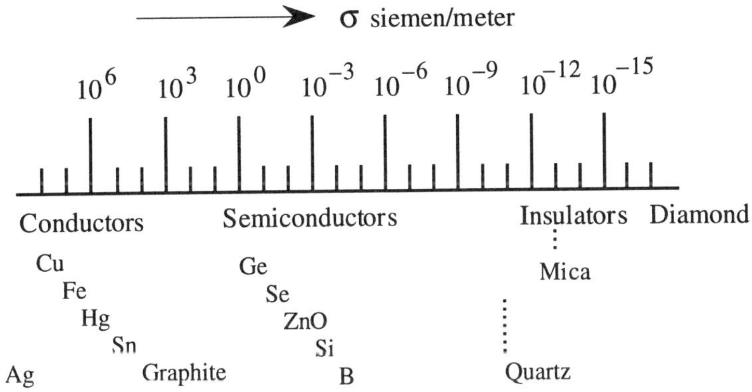

Figure 10.1 Range of conductivities of conducting, semiconducting, and insulating materials.

Semiconductors lie approximately between conductors (with $\sigma > 10^5$ siemen/meter) and insulators (with $\sigma < 10^{-10}$ siemen/meter). Semiconductors are materials with filled valence bands and have a small *forbidden energy gap* between the upper-filled band and the overlapping vacant energy band.

The electric conduction in semiconductors is effected by two types of charge carriers namely, *electrons* (negative charge carriers) and *holes* (positive charge carriers). Descriptions of energy bands and the concepts of free electrons and bound electrons are presented in Chapter 1.

Materials which are intrinsically semiconductors (with no impurities added) are known as *intrinsic semiconductors*. These are group IV elements namely, carbon (C), silicon (Si), germanium (Ge), and tin (Sn). These elements along with their neighbors are presented in Figure 10.2.

II	III	IV	V	VI
Be	B	C	N	O
Mg	Al	Si	P	S
Ca	Ga	Ge	As	Se
Sr	In	Sn	Sb	Te

Figure 10.2 Group IV elements and their neighbors in the periodic table.

Group IV elements have four valence electrons and are chemically similar. They also have similar crystalline structure. An atom of a group IV element shares its four valence electrons with its neighboring atoms forming a *covalent bonding*. The covalent bonding specifies an *octave saturation* of the orbit which corresponds to a saturation of the outermost orbit of the eighth group element with $2n^2$ (n being the orbital number, 1, 2, 3, ...,) law of saturation. As such a covalent bonded pure semiconductor prefers to be chemically inactive (as in the case of eighth group inert gases). The covalent bonding is represented in Figure 10.3, where the connecting parentheses indicate the shared covalence state of the electrons in each atom.

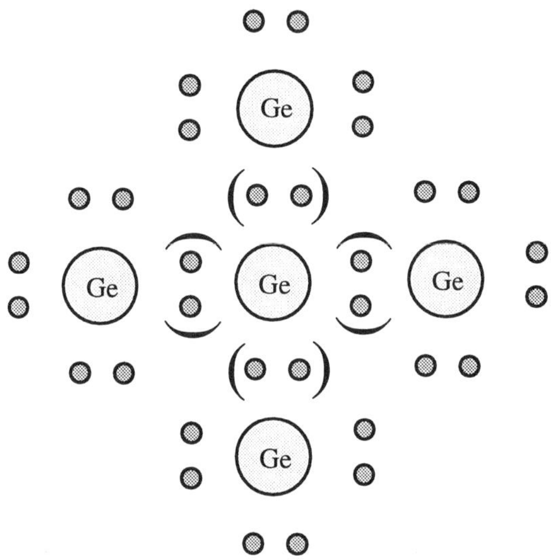

Figure 10.3 Covalent bonding.

In the absence of thermal energy, that is at 0 K, the valence band of a semiconductor is fully occupied and the conduction band has no electrons. As the temperature is raised, a small fraction of electrons in the valence band would acquire enough thermal energy so as to jump the forbidden energy gap E_g. The absence of electrons in the valence band corresponds to "holes". In reference to covalent bonding depicted in Figure 10.3, the movement of a covalently bonded electron away from a parent atom (due to acquiring thermal energy) creates a hole in the covalent structure which would be eventually occupied by an electron. Thus the migration of the electrons from atom-to-atom would necessitate filling and creation of holes. That is, the flow of electrons means a concurrent movement of holes as well. Inasmuch as holes attract the electrons, they can be regarded as positive charges with a value $+e$ ($= 1.6 \times 10^{-19}$) coulombs.

10.2 Properties of Intrinsic Semiconductors

The properties of group IV intrinsic semiconductors are presented in Table 10.1.

Among the four intrinsic semiconductors indicated in Table 10.1, Si and Ge are the materials largely used in the semiconducting devices. Their other physical properties (pertinent to semiconductor technology) are presented in Table 10.2.

Table 10.1 Properties of Group IV Intrinsic Semiconductors

At 20 °C

Elements	Forbidden Energy Gap (E_g) (eV)	Conductivity (σ) siemen/meter	Mobility m^2/volt-sec		Lattice Constant (a) nm
			Electron (μ_n)	Hole (μ_p)	
C (Diamond)	~ 6.0	$<10^{-6}$	0.1700	0.1200	0.3570
Si (Silicon)	1.1	5×10^{-6}	0.1900	0.0425	0.5430
Ge (Germanium)	0.7	2	0.3600	0.2300	0.5660
Sn (Tin)	0.1	10^6	0.2000	0.1000	0.6490

Table 10.2 Properties of Intrinsic Si and Ge

Properties	Ge	Si	Remarks
• **Physical**			
Density (d gm/cm^3)	5.360	2.400	
Atomic number (Z)	32.000	14.000	
Energy gap (Eg), eV	0.700	1.100	at 20°C
Effective mass of electrons (m_e*/m)	0.082	0.190	m: mass of electron (1.6×10^{-31}kg)
Effective mass of holes (m_p*/m)	0.280	0.490	
Electron mobility (μ_e) m^2/volt-sec	0.360	0.190	at 20°C
Hole mobility (μ_p) m^2/volt-sec	0.230	0.043	at 20°C

(continued...)

Properties	Ge	Si	Remarks
Electron diffusion constant D_e (m^2/sec)	9.9×10^{-3}	3.8×10^{-3}	300K
Hole diffusion constant D_p (m^2/sec)	4.9×10^{-3}	1.3×10^{-3}	300K
Effective intrinsic electron concentration (n_i/m^3)	2.5×10^{13}	1.4×10^{10}	300K (measured)
• **Electrical**			
Conductivity (σ) siemen/meter	2.0	5×10^{-6}	at 20°C
Dielectric constant	16.0	11.8	
• **Thermal**			
Melting point °C	958.5	1420	
Thermal conductivity watt/cm/°C	0.06×10^{-4}	0.024×10^{-4}	at 20°C
Coefficient of linear expansion per °C			

10.3 Conductivity of Intrinsic Semiconductors

Since both electrons and holes contribute to the conduction current in a semiconductor, the net conductivity (σ) is decided by the sum of the conductivity components due to electrons and holes. It is given by:

$$\sigma = n_n e \mu_n + n_p e \mu_p \tag{10.1a}$$

where n_n and n_p are the intrinsic electrons and holes per unit volume, respectively, and μ_n and μ_p represent, respectively, the electronic mobility and the hole mobility parameters. The electrons and the hole concentrations in the intrinsic state of the semiconductors at a given temperature are the same. That is, $n_n = n_p \cong 10^{16}$ per meter^3 at 20°C and they increase exponentially with temperature. Theoretical and expermential studies indicate that the conductivity (σ) of an intrinsic semiconductor can be explicitly stipulated by:

$$ln(\sigma) = A - E_g/(1.7 \times 10^{-4} \times T) \tag{10.1b}$$

where E_g is the forbidden gap energy in electron volt (eV), T is the temperature in K, and A is a constant for a material. The change in conductivity σ_1 to σ_2 as the temperature changes from T_1 to T_2 can be written as:

$$\ln(\sigma_1/\sigma_2) = (Eg/1.7 \times 10^{-4})(1/T_1 - 1/T_2) \tag{10.2}$$

10.4 Germanium and Silicon

Among the group IV elements presented in Figure 10.2, Ge and Si became the natural choice for implementing solid-state devices due to their inherent semiconducting properties and their technological match in fabricating such devices.

In the begining of semiconductor technology, Ge was the front candidate inasmuch as it could be produced in large quantites with sufficient purity (with the technological process then available).

Ge has higher mobility than Si which permits Ge devices to operate at high frequencies. However, Ge has a relatively low band gap energy (E_g = 0.67 eV) which makes the Ge devices more sensitive to temperature changes.

Si is one of the most widely used semiconductors in modern times. It has a high gap energy (E_g = 1.12 eV) as well as an abundance of SiO_2 in nature from which Si is extracted, and the superior thermal characteristics of silicon have made this element as the primary choice in semiconductor technology. In addition, SiO_2 allows oxide coating in the planar and IC technology facilitating the photolithographic-based monolithic and miniaturized device fabrication.

10.5 Extrinsic Semiconductors

These are semiconducting materials with deliberately added impurities (*dopants*) so that the electrical conductivity is dominantly dictated by the dopants. The level of impurity concentration ranges from one part per hundred to ten parts per million in typical commercial semiconductors. The maximum level of impurity concentration is so chosen as not to alter the crystalline structure of the host intrinsic medium; that is, a *phase change* is not introduced at this *solid solubility limit* of the impurity-host complex.

The impurity-atom sites in the intrinsic materials are of two types, namely:

- Substitutional impurities.
- Interstitial impurities.

The substitutional impurities occupy the atomic sites in the host lattice and the interstitial impurities fit between regular lattice sites.

Primarily, the doping materials used in semiconductor technology correspond to the adjacent groups, namely, group III and V elements of the intrinsic group IV semiconductors (Figure 10.2). When a group IV semiconductor is so doped, it is known as an *extrinsic semiconductor* since its electrical behavior is more predominantly dictated by the added impurities than by the intrinsic characteristics of the group IV element.

The doping material depending on its origin (group III or V) decides the nature of the extrinsic atoms as *acceptors* or *donors*, respectively. That is, addition of group V impurities, render the extrinsic semiconductor to have a set of easily activatable electrons into the conduction band (as free electrons). Hence, such an extrinsic material has acquired *donor atoms* by virtue of the added group V impurities. Likewise, addition of group III impurities would render the extrinsic material with excess of holes in the valence band, or the material is rich in *acceptor atoms* due to the presence of group III impurity.

The donor type semiconductors are also known as *N-type semiconductors*. As an example, consider a silicon material doped with group V impurity atoms, say P atoms. The pentavalent atom has five valence electrons unlike the tetravalent Si atoms which have only four valence electrons in their outermost orbit. Relevant covalent bonding is illustrated in Figure 10.4. The extra electron at the P atom can be seen independently present unlike the other electron pairs shared to constitute the covalent bonding. This loosely bound electron can be pulled away from the parent P atom with the small energy imparted to the material.

When this electron is detached, the P atom remains in the covalent bonded state in a positively charged ionic form. The detached electron would no longer remain in the already saturated valence band but would be located near the top of the energy gap. From this position, called the *donor level (E_d)*, this extra electron can be activated into the conduction band to become a *charge carrier*. Thus, donor impurities add to the free-electron population in the conduction band facilitating an increased conductivity of the material.

Thus atoms from group V elements used as dopant atoms supply a negative or N-type excess charge carrier to the semiconducting material.

The P-type semiconductors are constituted by the addition of a group III element as the impurity to an intrinsic semiconductor. Say, for example, if Aℓ atoms (with valency three) are added to silicon, the resulting covalent structure is illustrated in Figure 10.5.

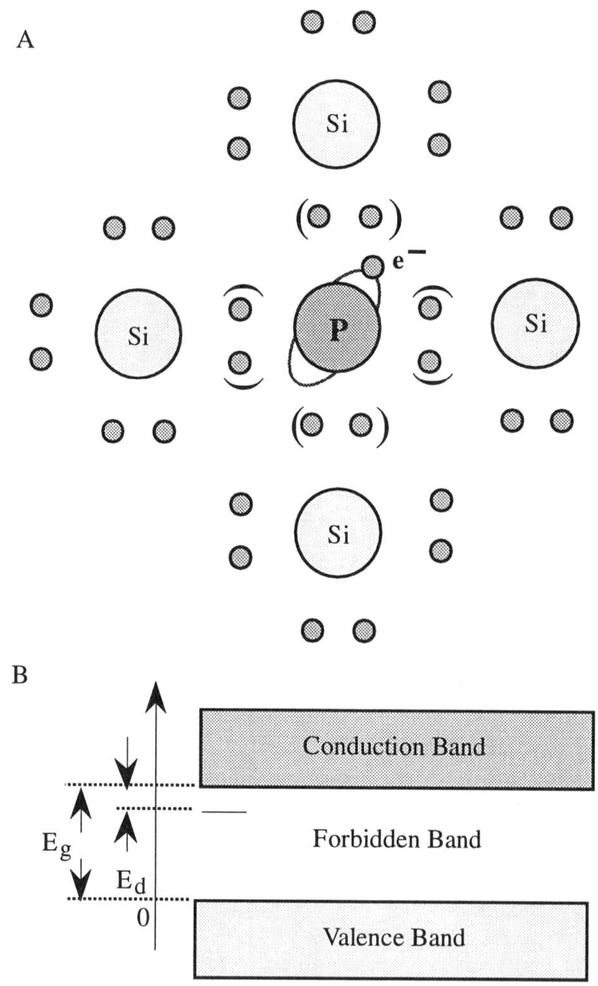

Figure 10.4 N-type semiconducting material.
A. Modified covalent bonding with an excess electron at P atom;
B. Energy-band diagram.

The aluminum atom when forming the covalent sharing with the neighboring Si atoms poses a deficiency in the bonding. This deficient location is ready to accept or attract an electron to form a valence band saturation and therefore could be considered as a positive carrier. Thus the presence of the third group atoms or an acceptor atom presumptuously provides an excess positive carrier at its location. When the hole is filled by an electron, the group III atom becomes negatively ionized, but offers a complete covalent bonding.

The presence of a hole can be attributed with an energy level (known as *acceptor level*) E_a close to the uppermost part of the valence band. E_a depicts the small energy required to accept an electron from the valence band. The resulting deficiency of the electron at the valence band amounts to the inculcation of a positive carrier (hole) in the valence band.

10.6 Conductivity of Extrinsic Semiconductors

Suppose N_d and N_a represent the donor and acceptor impurity concentrations (per meter^3). Then the conductivities of N-type and P-type materials can be defined as follows.

For an N-type material, the charge neutrality condition stipulates that, $n = p + N_D$ where n and p are electron and hole densities, respectively. Further, at equilibrium $np = n_i^2$ where n_i represents the intrinsic state electron (or hole) concentration. Therefore, n can be specified by $[N_D + (N_D^2 + 4n_i^2)^{1/2}]/2$. Normally, $N_D >> n_i$. Hence $n = N_D$.

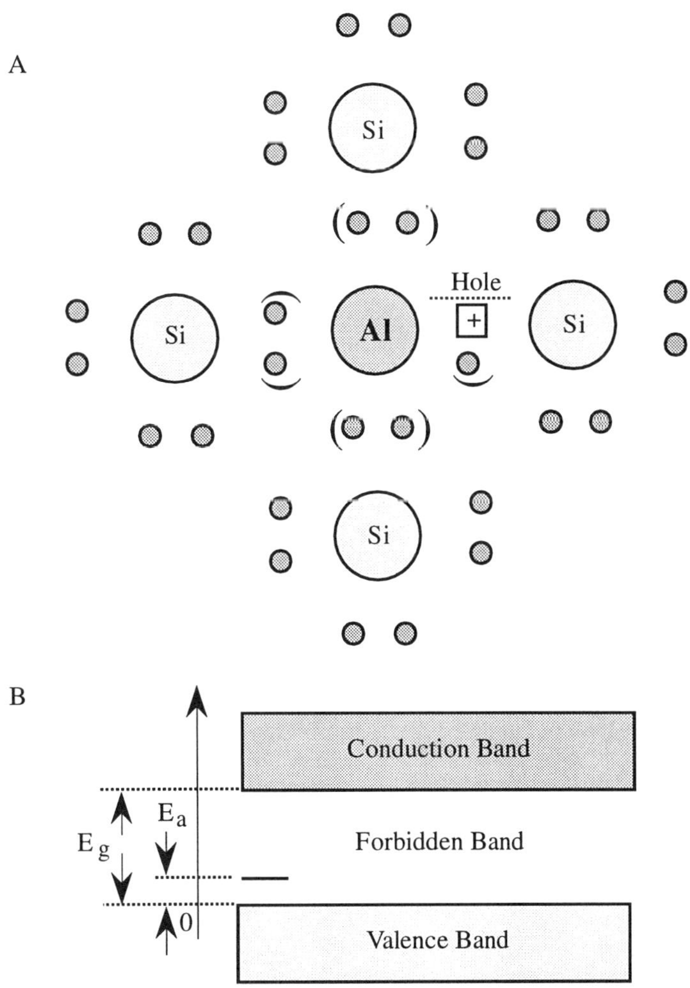

Figure 10.5 P-type semiconducting material.
A. Modified covalent-bonding with a hole created due to the deficiency of electron at Al atom;
B. Energy-band diagram.

Therefore, it follows that:

$$(\sigma)_{N\text{-}type} \cong \mu_e N_D e \qquad \text{siemen/meter} \tag{10.3a}$$

Likewise, for a P-type material,

$$(\sigma)_{P\text{-}type} = \mu_p N_A e \qquad \text{siemen/meter} \tag{10.3b}$$

The donor level energy E_d and the acceptor level energy E_a represent the minimum energy required for the excess electron to be activated into the conduction band and the excess hole to be activated into the valence band, respectively. These are also designated as *ionization energies*. Representative values of E_d and E_a are presented in Table 10.3.

Table 10.3 Typical Impurity Atoms and Their Ionization Energy Levels

Type of Impurity	Element	Ionization Energy			
		E_d (eV)		E_a (eV)	
		Ge	Si	Ge	Si
Donor	P	0.0120	0.0440		
	As	0.0127	0.0490		
	Sb	0.0096	0.0390		
Acceptor	B			0.0104	0.0450
	Al			0.0102	0.0570
	Ga			0.0108	0.0650
	In			0.0112	0.1600

10.7 Majority and Minority Carriers

Table 10.4 Charge Carrier Species in N and P Materials

Type of Extrinsic Semiconductor	Temperature-Induced Charge Carriers (per meter^3)		Dopant Contribution of Charge Carriers (per meter^3)	Remarks
N-type	n_i	p_i	N_D	• $n_i = p_i$. • Majority carrier: $(n_i + N_D)$. • Minority carrier: p_i.
P-type	n_i	p_i	N_A	• $n_i = p_i$. • Majority carrier: $(p_i + N_A)$. • Minority carrier: n_i.

When a semiconductor is doped, this extrinsic state material has the charge carrier at a finite temperature T > 0 K, given in Table 10.4.

10.8 Compound Semiconductors

Semiconducting materials can also be constituted by compounding the elements of groups III–V and II–VI. These materials have the same structures as group IV semiconductors except that alternate atoms are different in their dispositions across the crystal. Typical compound semiconductors and their properties are presented in Table 10.5.

10.9 Alloys of Compound Semiconductors

In certain electronic devices, materials formed by the alloys of compound materials are used. For example, a solid solution of InAs and GaAs provides an alloy designated as (In, Ga)As. Similarly, InP and InAs solid solutions give rise to In(As, P) alloys. Further (In, Ga) (As, P) alloys are also feasible.

Table 10.5 Properties of Compound Semiconductors

Compound	Forbidden Energy Gap (E_g) (eV)	At 20 °C			
		Conductivity (σ) siemen/meter	Mobility m^2/volt-sec		Lattice Constant (a) in nm
			Electron (μ_n)	Hole (μ_p)	
AlSb	1.60	–	0.020	–	0.613
GaP	2.30	–	0.019	0.012	0.545
GaAs	1.40	10^{-6}	0.880	0.040	0.565
GaSb	0.70	–	0.600	0.080	0.612
InP	1.30	500	0.470	0.015	0.587
InAs	0.36	10^4	2.260	0.026	0.604
InSb	0.18	–	8.200	0.170	0.648
ZnS	3.70	–	0.014	0.0005	–
SiC	3.00	–	0.010	0.002	–

10.10 Amorphous Semiconductors

These are noncrystalline (or polycrystalline) materials (usually in thin/thick film forms) which consist of randomly oriented clusters of crystallites. The structure within the crystallites leads to an energy gap. Also the crystallites offer a covalent structure with broken bonds wherever the crystalline orientation changes. This gives rise to *dangling bonds* where unpaired electrons can act as traps for both itinerant electrons and holes. The result is that the mobility is greatly reduced. Also, the doping becomes less effective (in contributing to the conductivity) as the carriers from the dopants are rendered immobile at the traps. However, if hydrogen atoms are introduced in the amorphous material, they neutralize the unpaired electrons, thereby dramatically reducing the dangling bond effects.

Amorphous semiconducting properties are less controllable than those of single crystalline intrinsic and/or extrinsic semiconductors. Still, amorphous structure can be very easily fabricated especially over a large area such as in solar panels or in xerographic applications. Therefore, amorphous semiconductors find potentials in upcoming technological trends.

10.11 Effective Masses of Electrons and Holes in Semiconductors

The charge carrier present in a semiconductor faces interactive force fields; as a result their effective masses are different from their actual masses. Shown in Table 10.6 are the typical values of the effective masses pertinent to various semiconducting materials.

Table 10.6 Effective Masses of Electrons and Holes in Various Semiconductors at Room Temperature

Semiconductor	Effective mass (m_o: free electronic mass)	
	Electrons m_e^*/m_o	Holes m_a^*/m_o
Ge	0.082–0.120	0.280
Si	0.190–0.260	0.500
GaAs	0.067	0.650
GaP	0.350	0.500
InP	0.080	0.200
InSb	0.013	0.180
AlAs	–	–
InAs	0.019	0.410
GaSb	0.047	0.400
CdSe	0.140	0.370
CdS	0.270	0.070

10.12 Thermal Properties of Semiconductors

Much like metals (see Chapter 9), the electrical conductivity (σ) of semiconductors varies in direct proportion to the thermal conductivity (*Wiedemann–Fraz law* for metal). The thermal conductivity of semiconductors is given by:

$$K_{SC} = A_S T\sigma + K_V \tag{10.4}$$

where A_S is a constant, T is the temperature, and K_V is the contribution to thermal conductivity resulting from lattice vibrations.

In the intrinsic state, the electrical conductivity of semiconductors is given by:

$$\sigma_i = e(\mu_n + \mu_p)\,(A_c\,A_v)^{1/2}\,T^{3/2}\,exp(-E_g/2k_BT) \qquad (10.5)$$

where k_B is the Boltzmann constant and (A_c, A_v) are constants independent of temperature given by:

$$A_c = 2(2\pi m_e^* k_B/h^2)^{3/2} \qquad (10.6a)$$

and

$$A_v = 2(2\pi m_n^* k_B/h^2)^{3/2} \qquad (10.6b)$$

where h is the Planck constant and m_e^* and m_n^* are effective masses of electrons and holes, respectively.

10.13 Hall-Effect Properties of Semiconductors

When a semiconductor is subjected to cross-fields of magnetic force $(\boldsymbol{H})$ and electric field $(\boldsymbol{E})$, constituting a conduction current flow, $\boldsymbol{J}_c = \sigma \boldsymbol{E}$, the *Lorentz force* acting on the moving charge carriers would induce the Hall potential field $(\boldsymbol{E}_H)$ orthogonal to $\boldsymbol{H}$ and $\boldsymbol{J}_c$ as shown in Figure 10.6.

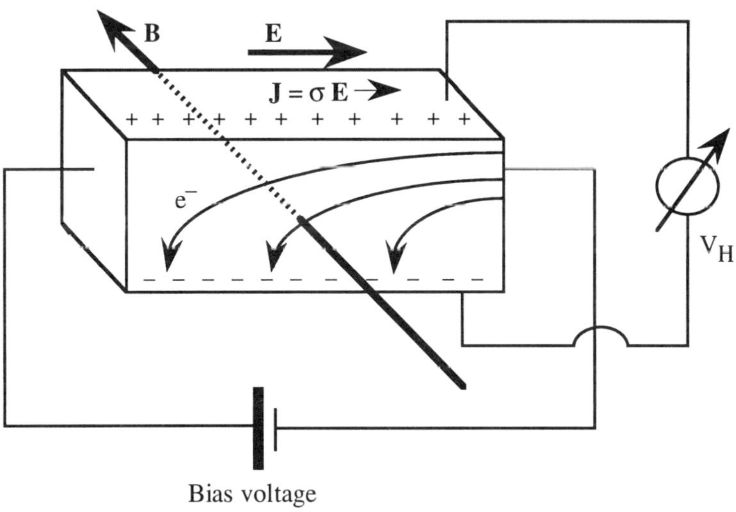

Figure 10.6 Hall-effect in semiconducting materials.
V_H: Hall voltage.

The corresponding *Hall voltage* is given by $|\boldsymbol{E}_H|/|\boldsymbol{B}|$ and *Hall coefficient* (R_H) is defined as $R_H = |\boldsymbol{E}_T/\boldsymbol{J}\boldsymbol{B}|$. Typical Hall coefficients of semiconducting materials are given in Table 10.7.

Table 10.7 Hall Coefficients of Typical Semiconductors

Material	R_H ($cm^3 \times 10^{-5}$/coulomb)
Si	-10^3
Ge	-10^4
InAs	0.5
InSb	10^3

10.14 Optical Properties of Semiconductors

AB-type binary compounds which crystallize in either the cubic zincblende or hexagonal wurtzite structure are typically semiconductors and are usefull as *optical semiconductors, phosphors, photodetectors,* and *injection laser materials.* Typical examples of these materials are: ZnO, ZnS, ZnSe, ZnTe, CuCl, CuBr, GaP, GaAs, and CdS. Properties of these materials are listed in [2].

10.15 Miscellaneous Semiconductors

1. Ternary and quaternary compound semiconductors: These materials, though not in widespread use, are potentially considered in upcoming technologies such as nonlinear optics. They also offer certain distinct luminescence and lasing properties. Typically $CdGeAs_2$, $CdSnP_2$, $ZnGeP_2$, $AgInSe_2$, $AgGaS_2$ and $CuAlS_2$, have been considered as candidate materials.

2. Oxide semiconductors: These are semiconductors with a large forbidden gap. Cu_2O, NiO, and ZnO are typical oxide semiconductors which have been studied. Oxides like VO_2 and V_2O_3 exhibit high conductivity at high temperatures but they behave more like dielectrics at room temperature. Other oxides like $SrTiO_3$ and $BaTiO_3$ which are normally dielectrics could, however, be changed into semiconductors with appropriate doping.

3. Refractory semiconductors: These operate at high temperature and have very large energy gaps. They can be doped to form N- or P-type materials. Examples are: AℓP, AℓAs, InP, AℓN, GaN, InN, BN, Bp, and BAs.

4. Superconducting semiconductors: Several semiconductors go into the superconducting state at temperatures less than 1 K. Typical such materials are: PbS, PbSe, PbTe, GeTe, SnTe, and $SrTiO_2$.

Interestingly, materials like InSb and InTe and the element Te become superconductors under high pressures.

5. Magnetic semiconductors: Some oxide-type and chalcogenide semiconductors exhibit magnetic properties. NiO, CoO, and FeO are antiferromagnetic materials, whereas, europium chalcogenides like EuO, EuS and EuSe, and $CdCr_2Se_4$ are semiconducting ferromagnetic materials. The spinels $M^{2+}OFe_2{}^{3+}O_3$ and garnets $Y_3Fe_5O_{12}$ are also semiconducting ferrimagnetic materials.

6. Organic semiconductors: Organic compounds such as anthracene ($C_{14}H_{10}$) and phthalocyanines have been observed to exhibit semiconducting properties. Single component aromatic compounds such as $C_{10}H_8$, $C_{14}H_{10}$, $C_{18}H_{12}$, and $C_{12}H_{14}$ which are insulators become photoconductive with sufficient photoexcitation. Their energy gap ranges from 2 to 5 eV. Impurities play a dominent role in the electrical conductivity of organic semiconductors.

Two component semiconductors consist of pairs of complementary molecules with large differences in their redox properties.

Low mobility and tendency for carrier trapping are of interest in organic semiconductors studied so far. These limitations hamper the use of these materials in technological applications. Semiconducting p-paraphenylene vinylene (PPV) in layer form has been successfully used in conjunction with CdSe as an electroluminescent structure.

10.16 Nonsemiconducting Materials Used in Semiconductor Technology

Silicon dioxide and silicon nitrite are widely used as *passivation materials* in the fabrication of semiconducting devices. SiO_2 also constitutes the gate oxide part of metal oxide semiconductor devices. The general characteristics of these materials are presented in Table 10.8.

Table 10.8 Properties of SiO_2 and Si_3N_4 at Room Temperature (≅300 K)

Properties	Materials	
	SiO_2	Si_3N_4
Physical properties		
Crystalline structure	Amorphous	Amorphous
Specific gravity	2.2	3.1
Electrical properties		
Relative permittivity	3.9	7.5
Dielectric breakdown strength (volt/meter)	10^9	10^9
DC resistivity (ohm-cm)	10^{14}–10^{16}	~10^{16}
Forbidden gap energy (eV)	9	5
Thermal properties		
Melting point °C	~1600	–
Thermal conductivity Watt/cm-K	0.014	–

Another adjunct material used as a substrate in semiconductor devices is sapphire (such as in silicon-on-sapphire (SOS) devices). Essentially sapphire is $A\ell_2O_3$.

Silicon carbide (SiC) has been tried as an integrated circuit material to survive operating temperatures up to 500°C, three times higher than the failure temperature for conventional silicon-based components.

10.17 Applications of Semiconducting Materials

In modern time, semiconducting materials have facilitated the emergence of innumerable electronic devices for a variety of applications. The descriptions of these devices are abundantly presented in the literature, a few of which are listed as references in this chapter. Essentially, the semiconductor devices can be grouped as follows:

- Junction devices (junction diodes and bipolar transistors)
- Unipolar devices (field effect devices)
- Metal oxide semiconductor devices
- High frequency devices
- Bulk-effect microwave devices
- Optoelectronic devices and semiconductor lasers
- Power rectifiers and thyristor family of devices
- Semiconductor transducers

10.18 Concluding Remarks

Semiconductors are perhaps the most widely studied materials due to their unlimited application potentials in the active device fabrications. The physics of conventional semiconductors have been fairly well comprehended and explained. The exploring trend in finding newer categories of semiconductors is not over. Microminiaturization and high frequency applications of semiconducting devices as well as low power consuming devices are setting the goals in search of better semiconductors in the existing technology and for futuristic trends.

References

[1] K. W. Böer: *Survey of Semiconductor Physics.* (Van Nostrand Reinhold, New York: 1990).

[2] I. P. Kaminow: *An Introduction to Electrooptic Devices.* (Academic Press, New York: 1974).

[3] D. K. Schroder: *Semiconductor Material and Device Characterization.* (John Wiley and Sons Inc., New York: 1990).

[4] M. S. Tyagi: *Introduction to Semiconductor Materials and Devices.* (John Wiley and Sons Inc., New York: 1992).

Defining Terms

Effective mass: Mass of charged carriers modified as a result of prevailing interactive field forces on them.

Extrinsic semiconductors: Semiconducting materials with added dopants (impurities).

Intrinsic semiconductors: Pure semiconducting materials with no addition of impurities.

Mobility: Extent of ease of the movement of charge carriers under the influence of applied electric field force, specified as the velocity per unit electric field applied.

CHAPTER 11

Superconducting Materials

11.1 Introduction

The phenomenon of *superconductivity* came to be known in 1911 by observing that a capillary column of mercury immersed in a liquid helium bath would show an abrupt reduction in resistance, and at 4.2 K it became impossible to measure the low resistance exhibited by the column with the then available measurement techniques. It was concluded that below a *critical temperature* (T_c) the mercury had passed into a new state which on account of its extraordinary electrical properties was called the *superconducting state*. Subsequent studies on superconduction were pursued with tin and lead as candidate materials and it was discovered that a *critical current density* (J_c) is carried by a superconduction sample before it returns to its normal (metallic) state. Similar to the threshold value of current or critical current density, it was also observed that a *critical magnetic field* (H_c) intensity is required to destroy the superconducting effect in a sample; and the following empirical law was established relating to H_c and T_c:

$$H_c(T) \cong H_{co}[1 - (T/T_c)^2] \tag{11.1}$$

where H_{co} refers to the critical magnetic field intensity at zero temperature and T depicts the operating temperature. Hence superconductivity ceases to exist in a sample with the prevalence of the following influences in excess of certain threshold (critical) values:

- Temperature (T)
- Magnetic field $(\boldsymbol{B}$ or $\boldsymbol{H})$
- Current density $(\boldsymbol{J})$

The state of a superconductor can be depicted in terms of $\boldsymbol{B}$ and T as shown in Figure 11.1.

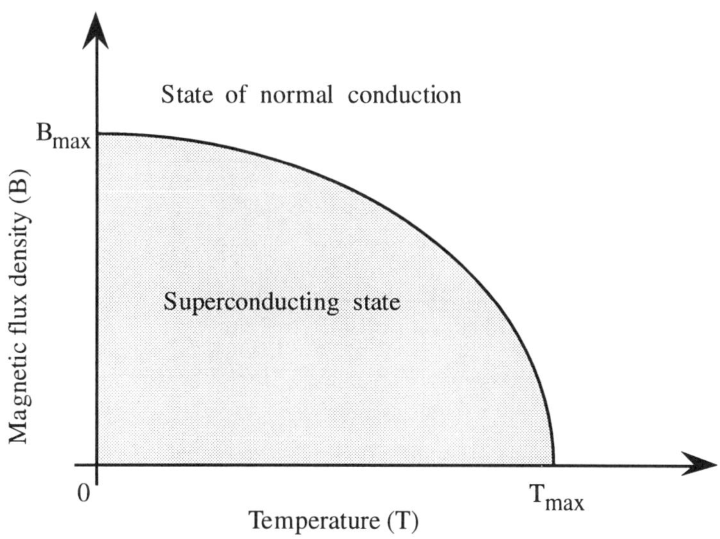

Figure 11.1 Superconducting state as decided by temperature and magnetic flux.

The maximum values of the transitional parameters B_{max} (= μH_{max}, μ: Permeability of the material) and T_{max} are approximately specified by the relation:

$$B_{max} \cong \alpha T_{max} \qquad (11.2)$$

where α is a coefficient equal to 0.02 tesla/K.

A superconductor (unlike the conventional *perfect conductor*) does not *conserve* the magnetic field flux within it. It rather *expels* the flux (assuming that the applied magnetic field is insufficient to destroy the superconduction). A superconductor which seeks to maintain the conduction of magnetic flux density (***B***) equal to zero within itself is known as a *perfect diamagnetic*. The phenomenon of expelling the magnetic field by a superconductor is referred to as the *Meissener effect*. Therefore, it has become known that superconductivity is "more than just perfect conductivity".

It has been established that apart from certain pure elements, a combination of superconductive metals or a superconductive metal plus a nonsuperconductive material may also exhibit superconductivity. Similarly there are compounds whose constituent elements themselves may not be superconductors, but the compounds are superconductors. For example, neither Co nor Si nor O is a superconductor, but $CoSiO_2$ exhibits superconductivity with T_c = 1.4 K. It is also to be noted that the well-known good conductors at ordinary temperature such as Cu, Au, Ag, and Pt have not yet been transferred into a superconducting state even at the lowermost cryogenic temperatures feasible in modern technology.

Table 11.1 Typical Superconductors and Their Critical Temperatures

Elements	T_c K	Compounds	T_c K
Tungsten (W)	0.01	AgF_2	0.07
Iridium (Ir)	0.14	$CuAl_2$	1.00
Zinc (Zn)	0.90	$ReYO_3$	2.00
Aluminum (Al)	1.20	NiBi	4.30
Thallium (Tl)	2.40	La_3S_4	8.30
Indium (In)	3.40	$MoRe_3$	10.00
Tin (Sn)	3.70	MoN	12.00
Mercury (Hg)	4.15	MoC	13.00
Tantulum (Ta)	4.50	V_3Ga	14.00
Vanadium (V)	5.30	V_3In & Nb_3Ga	15.00
Lead (Pb)	7.20	V_3Si & NbN	16.00
Niobium (Nb)	9.40	Nb_3Sn	18.00

Typical superconducting materials and their critical temperatures are listed in Table 11.1. The critical magnetic fields are furnished in Table 11.2.

Table 11.2 Some Superconducting Materials and Their Critical Magnetic Fields

Material	B_c (tesla)
Al	0.0099
Va	0.1370
Ga	0.0051
Ir	0.0020
Cd	0.0030
Nb	0.1944
Sn-α	0.0309
Rh	0.0198
Ti	0.0100
Zn	0.0053
Zr	0.0047

The influence of magnetic field on superconducting properties of materials has led to the classification of superconductors into the following categories:

- *Type I superconductors:* These materials lose superconductivity at exposures even at feeble magnetic fields. Examples are: Pure metals except niobium, vanadium, and technetium. Type I materials do not violate the bulk Meissener effect.
- *Type II superconductors*: These have a higher critical magnetic field than Type I counterparts. Examples are: Superconducting alloys and pure metals like niobium, vanadium and technetium. Doping impurities in these materials improve their critical current densities. The disadvantage of these materials is their inhomogeneity. Say, in a binary alloy constituted by two materials 1 and 2, there is a possibility that the superconduction may vanish in the intermediate range of magnetic field specified by the critical values B_{c1} and B_{c2} corresponding to the constitutive materials 1 and 2, respectively. Type II materials may allow magnetic flux to enter the bulk of their volume.
- *Type III superconductors*: These are similar to Type II except that the aforesaid disadvantages are removed *via* proper compositions. Examples are: $NbSn_2$, NbTi, and NbZrTi.

The resistivity *versus* temperature characteristic of a typical superconductor is illustrated in Figure 11.2 along with the corresponding characteristic of a nonsuperconductor.

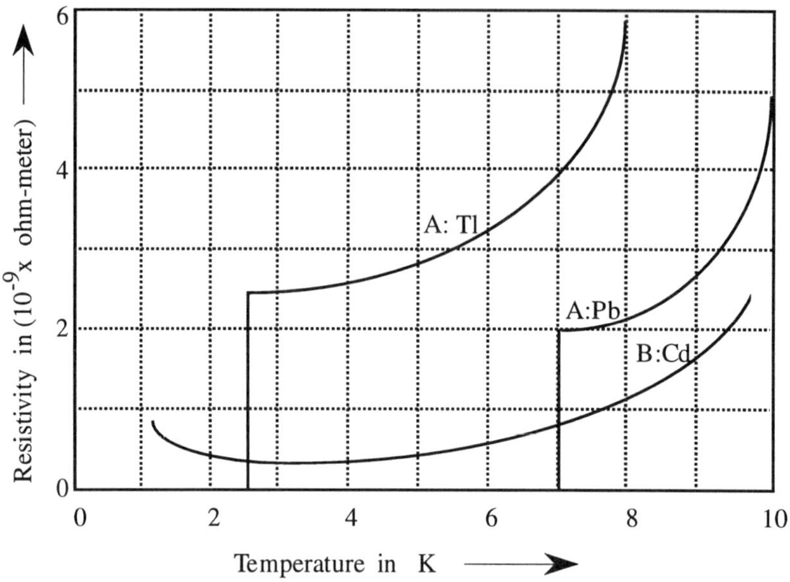

Figure 11.2 Resistivity *versus* temperature.
A: Superconductors;
B: Ordinary conductor which does not pass into a superconducting state.

11.2 Theories and Models of Superconductivity

Classical Model: This model of superconductivity incorporates the concept of zero resistance and perfect diamagnetism into electromagnetic constitutive relations leading to what are known as *London equations*. Relevant expressions due to Fritz and London [1] developed in 1935 are mostly empirical.

This model is based on four fundamental constants which can be associated with the electromagnetic interaction with material or a medium. They are defined as follows:

- *Electromagnetic coupling time* (τ_{em}) delineates what frequencies may be considered low. This *quasistatic regime* is defined by the condition $\omega\tau_{em} << 1$. τ_{em} is equal to $\ell(\mu\varepsilon)^{1/2}$, where ℓ is a geometrical dimension, μ is the permeability, and ε is the permittivity of the material/medium.
- *The charge relaxation time* (τ_e) defines the time over which free charge in the bulk medium relaxes to the surface of the material. It is equal to ε/σ_o where σ_o is the lossy (conductive) attribution to the medium. When $\tau_e > \tau_{em}$, the material/system is considered as *electroquasistatic* corresponding to low frequencies with the energy stored mostly in the electric field.
- *The magnetic diffusion time* (τ_m) is defined as the time required for the magnetic field to distribute itself across a material that is self-consistent with both the applied and induced currents. It is given by $\mu\sigma_o\ell^2$. The *magnetoquasistatic* limit occurs at low frequencies when $\tau_m > \tau_{em}$. Pertinent to superconductors, the magnetoquasistatic limit holds good since it defines the regime of easy conduction of current through the material.
- The average time between successive collisions for a carrier of electric charge in a medium is defined as the *scattering or transport time* (τ_{tr}). For a perfect conductor $\tau_{tr} \rightarrow \infty$.

The interplay between these four *time constants* is considered in defining a *constitutive relation* for a perfect conductor specified by the *first London equation* stated as:

$$\boldsymbol{E} = \partial(\Lambda \boldsymbol{J})/\partial t \tag{11.3}$$

where Λ is an attributable parameter to a perfectly conducting material and $\boldsymbol{E}$ and $\boldsymbol{J}$ are electric field and current density vectors, respectively. (Note: For a normal conductor with a conductivity σ_o, the corresponding constitutive relation is given by the *microscopic Ohm's law*, namely,: $\boldsymbol{J} = \sigma_o \boldsymbol{E}$.) Similar to the electromagnetic *penetration depth* of a normal conductor, the *characteristic length* (λ_s) parameter at the limiting case of perfect conductivity is given by:

$$\lambda_s = (\Lambda/\mu_o)^{1/2} \tag{11.4}$$

On the basis of the above considerations the classical model of superconductivity incorporates the concept of electrodynamics and thermodynamics to explain the superconductivity phenomenology. Corresponding constitutive field equations are as follows:

$$\boldsymbol{E} = \partial[\bar{\Lambda}(T)\ \boldsymbol{J}_s]/\partial t \qquad \text{(\textit{London equation} I)} \tag{11.5a}$$

$$-\boldsymbol{B} = \nabla \times [\bar{\Lambda}(T)\ \boldsymbol{J}_s] \qquad \text{(\textit{London equation} II)} \tag{11.5b}$$

where $\bar{\Lambda}$ is an anisotropic parameter of the superconductor and T is the temperature; and $\boldsymbol{J}_s$ is defined as the current density due to the *superelectrons*. The total current density $(\boldsymbol{J})$ is therefore:

$$\boldsymbol{J} = \boldsymbol{J}_n + \boldsymbol{J}_s \tag{11.6}$$

where $\boldsymbol{J}_n$ is the current density due to normal electrons.

The above model permits the energy associated with the system being partitioned among electric fields, magnetic fields, and *supercurrents*. The *supercurrent* here refers implicitly to the kinetic energy of the superelectrons. The classical model of superconductivity relies on explaining the relevant superconducting properties on the basis of the aforesaid considerations.

Macroscopic Quantum Model: This *MQM model* was developed to demonstrate that superconduction is a manifestation of quantum mechanical phenomenology. This model enclaves the concept of classical model as well as describes self-consistently the various properties of superconductors. This model also explains the anomaly such as why Type II superconductors violate the Meissener effect. The MQM model also explains the *Josephson junction* prevailing in *small-scale superconducting* systems. (Details on the Josephson junction are presented later.)

The MQM theory is based on the unified aspects of electromagnetism, quantum mechanical considerations of superconductivity, the *Ginzburg-Landau theory* and principles of thermodynamics.

In quantum mechanics, the wave-particle duality of nature is explicit. As a consequence, the so-called *Schrödinger wave equation* for a single quantum particle with a scalar potential describes the dynamical evolution of a probability amplitude (wave) function. The physical attribution of this *wave function*, $\Psi(\boldsymbol{r},t)$, is that the square of its magnitude is the probability that the particle will be at a specific place $\boldsymbol{r}$ at a certain time t.

Inasmuch as superconductivity can be envisioned as a coherent phenomenon between all the superelectrons, the entire ensemble of such carriers can be represented by a single macroscopic wave function, namely:

$$\Psi(\mathbf{r},t) = [n^*(\mathbf{r},t)]^{1/2}\ \{exp[j\theta(\mathbf{r},t)]\} \tag{11.7}$$

The above equation is akin to the *Schrödinger wave equation* for a single particle. In short, it is possible to assign a single wave function to depict the entire ensemble of carriers subjected to an electromagnetic excitation in a superconductor wherein the local density of the superelectrons in space and time is specified by $n^*(\mathbf{r},t)$. Equation 11.7 governs the probability of current function in the superconductor. That is, inasmuch as Ψ in Equation 11.7 refers to an ensemble of many particles, the proliferation of probability for the entire ensemble is equivalent to the flow of the macroscopic supercurrent $\boldsymbol{J}_s$; and for an isotropic superconduction the following equation holds good:

$$\Lambda \boldsymbol{J}_S = -\,[\boldsymbol{A}(\mathbf{r},t) - \hbar\ \nabla\theta(\mathbf{r},t)/q^*] \tag{11.8}$$

where $\boldsymbol{A}$ is the *magnetic vector potential*, θ is a real function representing the phase of the complex number in Equation 11.7, $\hbar = h/2\pi$ (h being Planck's constant) and q^* is the charge associated with superelectrons.

Equation 11.8 is known as the *supercurrent equation* and it is of primary importance in the macroscopic quantum model of superconduction. The time derivative of Equation 11.8 yields:

$$\partial(\Lambda \boldsymbol{J})/\partial t = \boldsymbol{E} = \nabla(\Lambda \boldsymbol{J}_s^{\,2}/2)/n^*q^* \tag{11.9}$$

which is the same as the first London equation, which self-consistently includes the effect of magnetic field created by the motion of supercarriers. The curl of supercurrent equation leads to the *second London equation*, namely:

$$\nabla \times (\nabla \boldsymbol{J}_S) = -\,\boldsymbol{B} \tag{11.10}$$

The supercurrent continuity around a closed path is specified by the relation:

$$\oint_C (\Lambda \boldsymbol{J}_S) \cdot d\boldsymbol{\ell} + \iint_S \boldsymbol{B} \cdot ds = n\Phi_O \tag{11.11}$$

which is referred to as the statement of *fluxoid quantization* where Φ_o is the *flux quantum*. Measurement of Φ_o implicitly confirms that the so-called *Cooper pair* of electrons (as will be described in the next section) identically represents the concept of superelectron. The macroscopic quantum model also applies to superconductors which are anisotropic.

Bardeen-Cooper-Schreiffer model: Known as *BCS theory*, it was developed to explain the microscopic aspects of how superconductivity occurs. The central theme of BCS theory is that the electrons that carry lossy currents in the normal (metallic) state pair together in the superconducting state. Such pairs are referred to as *Cooper pairs* or *superelectrons* signifying the lossless supercurrent they carry. The charge of superelectron is equal to twice the electronic charge, as also is the mass of the superelectrons.

The separation of paired electrons under the influence of temperature and/or magnetic field above a certain critical value would destroy the superconducting property of the material. Further, the paired electrons liberate energy in small doses so that the usual *Joule losses* of power observed in metals with normal conductivity do not occur in superconductors.

The BCS theory introduces an energy scale on the basis of the bound energy pertinent to a Cooper pair. This bound energy (2Δ) is typically on the order of 10^{-3} eV for conventional superconductors with $T_c \leq 25$ K. This bound energy is called the *energy gap* of the superconductor. It specifies the minimum amount of energy to split the Cooper pair into two unbound electrons.

The BCS theory is not, however, adequate to explain the high temperature superconductivity.

The evolution of BCS theory is as follows: The macroscopic quantum model discussed earlier can be extended to address two other theories, namely, (i) *Ginzburg-Landau* (GL) theory and (ii) the BCS theory. The GL theory, like the macroscopic quantum model is phenomenological in nature. It specifies two fundamental length scales: The *coherence length* ζ and the *penetration depth* λ_s. It is based on writing the *Gibbs free energy* of the superconductor in terms of an *order parameter* $f(r)$ leading to the following two equations:

$$\zeta^2(\Delta/j + 2\pi\mathbf{A}^2/\Phi_o)^2 f + |f^2|\, f - f = 0 \tag{11.12}$$

(GL equation)

and,

$$J_s = (\Phi_o/2\pi\mu_o\lambda_s^{\,2})\; Re[f^*(\Delta/j + 2\pi\mathbf{A}/\Phi_o)f] \tag{11.13}$$

(Supercurrent equation)

The above two equations are coupled and refer to the minimization of Gibbs energy under equilibrium.

The *Ginzburg-London* (GL) theory is governed essentially by the aforesaid two characteristic lengths, namely, the coherence length (ξ) and the penetration length (λ_s). The coherence length decides the spatial change in the order parameter whereas the penetration length governs the spatial change in the electromagnetic fields and currents. The coupling of these two lengths would lead to (i) the so-called *Josephson behavior* of the current density. That is, when the order parameter is restricted on a length scale smaller than ξ, there exists a probability that a Cooper pair (or its corresponding macroscopic wave function) may *tunnel* from one superconductor to the other as an ordered, coherent process. (ii) The coupling of the energies due to the magnetic field penetration and variations in the order parameter may also cause the observed differences between Type I and Type II superconductors. The interaction between the Gibbs free-energy H_c and the characteristic lengths ξ and λ_s are:

$$H_C = \Phi_o/(2\sqrt{2}\;\pi\mu_o\lambda_s\xi) \tag{11.14}$$

where μ_o is the free-space permeability and Φ_o is the flux quantum.

Apart from λ_s and ξ, there are two more characteristic lengths *vis-a-vis* superconductors. They are:

- Wavelength of the interacting electromagnetic (EM) field $(2\pi c/\omega)$ where c is the velocity of propagation of the EM wave and $\omega = 2\pi \times$ frequency of the wave
- *Mean free path* (λ_{tr}) which refers to the average distance (in the transport of electrons) between successive collisions or scattering events

In superconductors the two paired electrons scatter in a correlated fashion such that the Cooper pair does not feel the drag force. The absence of such scattering is responsible for the perfect conductivity in a superconductor.

11.3 Applications of Superconductivity

The influence of magnetic field on superconductivity refers to both an external magnetic field or to a magnetic field caused by current passing through the superconductor itself. However, there is a limitation on the current which, without destroying the superconductivity, can be passed through a superconductive circuit. This has constrained the practical use of superconductivity in electrical engineering systems in which high currents have to be passed and strong magnetic fields to be realized. Related studies indicate that an alloy of Nb with 25% of Zr with minimum feasible temperature could lead to a magnetic field density (B_c) limited by the critical value of 10.9 tesla. The corresponding values of B_c for Nb_3Sn is 20 tesla and for $V_{2.95}Ga$ is as high as 35 tesla. These studies indicate the feasibility of producing *cryogenic magnets* with cooled superconducting windings with current densities on the order of 10^9–10^{10} ampere/meter^2. Future successes in this area could lead to systems and devices like electric machines, transformers, etc. operating with virtually at no expense of power.

Another feasible electrical engineering application of superconductors refers to *cryotron* computers, the principle of which is illustrated in Figure 11.3. The central conductor A and the winding B are of two different superconducting materials and are kept at a temperature lower than the critical temperatures of both conductors. The change in current (I_B) in winding B would control the current (I_A) through A; and, if I_B reaches to such a value that the corresponding induced magnetic field destroys the superconductivity of A, the value of I_A will reduce instantaneously posing a switching action (binary transition). The collection of such switches can be designed into a *cryotron computer* with miniaturized film structures.

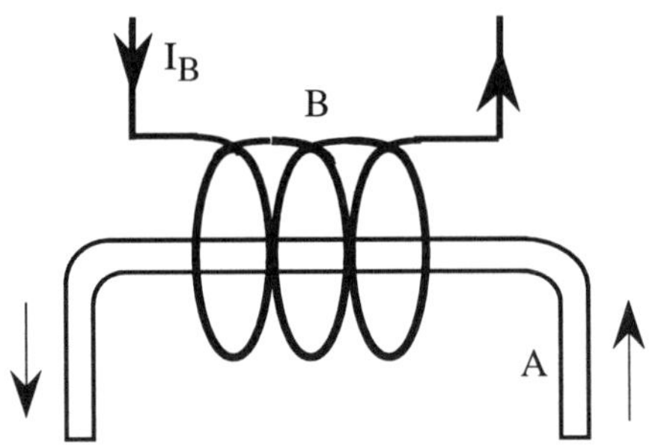

Figure 11.3 Principle of Cryotron computer.

Other uses of superconductors included the following applications:

- Modulators/converters of weak, steady-state currents into audio frequency signals
- Demodulator/rectifier of modulated high-frequency signals making use of the nonlinear behavior of superconductor conductance in the transitional region
- Noncontact commutating switches
- Superconducting memory devices

The following sections describe more specific applications of superconductors.

11.3.1 *Electromagnetic transmission lines*

The transmission lines made of superconducting materials can support a guided wave propagation of EM energy. Such transmission lines can be analyzed *via* electrodynamic principles with appropritate constitutive relations deduced for superconductors beyond the quasistatic approximations. In high frequency, time-varying situations, the parameter Λ that characterizes a superconductor is a function of frequency as well. Pertinent studies include the use of London equations to calculate the EM fields associated with a superconductor and leads to a lumped element model (as illustrated in Figure 11.4) that mimics the behavior of a superconductor.

The feasibility of depicting a superconductor by a conventional set of lumped elements can be extended to represent a transmission line by an electrodynamic structure with the transmission lines being superconductors. Such a model facilitates the elucidation of the associated magnetic field, electric field, normal current density and supercurrent density, components and the *complex propagation constant.* Practical uses of such analysis in the design of waveguide structures and strip lines have been considered and specifically there is a considerable effort directed at microwave transmission line structures of high-T_c and low-T_c superconducting thin films [7,8]. Materials like NbN and $YBa_2Cu_3O_{7-x}$ films have been studied for such applications. Ceramic superconducting wires/transmission lines fabricated from powders of Bi, Sr, Ca, Cu, and O (BSCCO) composition become superconducting above 77 K and sustain magnetic fields above 20 tesla. Such lines are currently in the development stage.

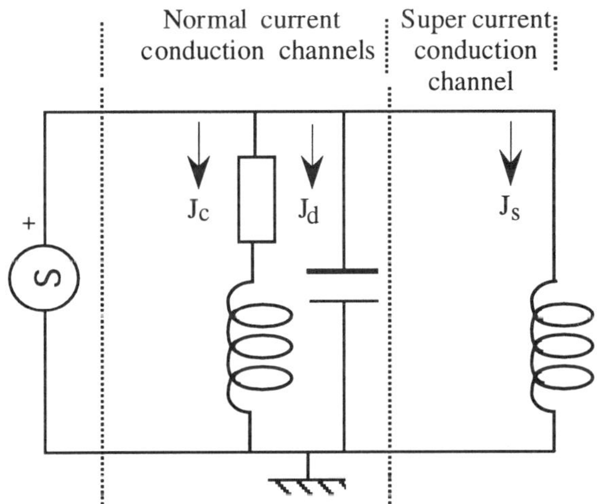

Figure 11.4 A lumped element model of a superconductor.
$\mathbf{J}_c$: Normal conduction current density; $\mathbf{J}_d$: Normal displacement current density;
$\mathbf{J}_s$: Supercurrent density.

11.3.2 *High–T_c superconducting active antennas*

Superconducting materials have also been considered in the development of active antennas. A typical structure refers to using a high-T_c superconducting film (YBCO on MgO substrates) with a corner reflector to detect microwaves.

11.3.3 *Kinetic inductance memory cell*

The Josephson junction refers to a *tunnel junction* between two separated superconductors across which a flow of superconductor can be maintained. Typical current-voltage characteristics of a Josephson junction are illustrated in Figure 11.5.

The current at zero voltage (in Figure 11.5) is a direct result of the Cooper pair tunneling and represents the Josephson current. It flows as a result of Josephson tunneling by the Cooper pair electrons.

When a Josephson junction switches, the voltage across it is typically on the order of millivolts. (On the contrary, conventional semiconductor junctions would require potentional on the order of volts to switch binary states.) Further the superconducting Josephson junction would need three orders of magnitude less power to operate than the standard semiconductor logic devices. Therefore the size of the Josephson memory cells can be extremely small in comparison with semiconductor cells [10].

11.3.4 *Ferroelectric superconductors*

Superconductivity at high temperatures in materials like CuO_2-based perovskite-type structures has indicated a possible relation between superconductivity and ferroelectricity and accordingly relaxor ferroelectric state in CuO-based superconductors (Pb-Bi-Sr-Ca-CuO) has been investigated [11]. Ferroelectric superconductors and viable applications of such materials in practical use provide a niche for upcoming technology.

11.3.5 *Anisotropic superconductors* [1]

In isotropic superconductors, the superconducting properties are not dependent on direction. However, in materials like $NbSe_2$ and $PbMo_6S_2$ and high-T_c superconductors like $Y Ba_2Cu_3O_7$, anisotropic behavior is perceived. That is, the conductivity and the parameter Λ are tensors. The use of anisotropic characteristics of superconductors in practical systems rests in futuristic technology trends.

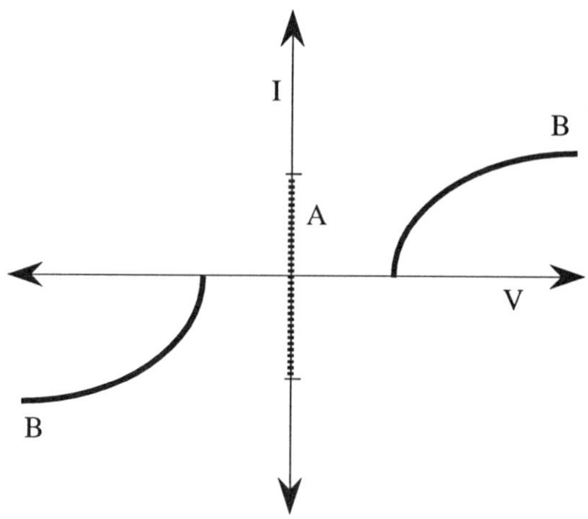

Figure 11.5 Current-voltage characteristics of a Josephson junction. A: Cooper pair tunneling; B: Normal electron tunneling.

11.4 Applications of Superconductors in Electromechanical Systems

The dynamic interaction between a current carrying conductor and a magnetic field is well-known through *Ampere's force law*. In the event of a superconductor being subjected to a magnetic field, an electromechanical *levitation* can be realized. For example, a samarium-cobalt magnet can be levitated by an $YBa_2Cu_3O_7$ superconducting disk, inasmuch as the superconductor (being a perfect diamagnet) prevents the penetration of the magnetic flux through it; and the resulting force of repulsion would levitate the magnet placed in its vicinity.

The concept of magnetic levitation (*maglev*) is being considered for use in levitating a train car above the track to counteract the frictional drag. However, the relevant technology at present is still in its cradle of development.

11.5 The DC SQUID (Superconducting Quantum Interference Device)

SQUID magnetometer is a transducer which produces a voltage signal related to the applied magnetic field. Such SQUIDs make use of superconducting materials and offer extremely high sensitivities [6]. The operation of a SQUID is based on the maximum current through junctions in parallel, being dependent on the magnetic flux enclosed by the loop. Magnetic flux density on the order of 10^{-6} to 10^{-7} weber/meter^2 can be measured with a time constant of 1 sec using SQUIDs. SQUID magnetometers can sense the feeble disturbances in the earth's magnetic field caused nearby submarines, enabling target locations.

11.6 Other Applications

11.6.1 *Radiation detection*

When a superconductor is kept just above its critical temperature (where its resistance varies significantly with temperature), incident radiations can be sensed, inasmuch as such radiations would induce temperature changes (and hence resistance changes) in the superconducting medium.

11.6.2 *Heat valves*

Thermal conductivity of some superconductors would change (increase) by two orders of magnitude when the material is turned into a normal conductor by a magnetic field. This phenomenon can be used to devise a *heat valve* in refrigerating systems operating at cryogenic temperatures.

11.6.3 *Resonant cavities*

The transmission line function of a superconductor can be logically extended to make microwave cavities with very high Q factors, if the cavity walls are coated with a thin film of a superconductor offering low loss characteristics.

11.6.4 *Oxide superconductors*

In an attempt to realize high temperature superconductors, oxide superconductors have emerged in the recent past. Yttrium-barium-copper oxide (YBCO) material has a critical temperature between 90 and 100 K and $Tl_2Ba_2Ca_2Cu_3O_{10}$ (TBCCO) material has yielded T_c = 125 K.

11.6.5 *High-field magnets*

In the present, the most important use of superconductors is in producing high magnetic fields. A magnetic flux density of 20 tesla can be produced by superconductor–based solenoids of about 12 × 20 cm. In comparison, even to realize 5 tesla with conventional conductors, the size of the solenoid would be enormous and would need megawatts of power and an exorbitant cooling system.

Type II superconductors are used to produce high magnetic fields since they are superconducting even under large magnetic fields. Typically Nb-Ti alloy and intermetallic compounds such as Nb_3Sn are used in practice.

11.7 Properties of Typical Superconductors

Table 11.3 Properties of Superconductors

Type I Superconductors

Material	T_c (K)	λ_0(nm)	ξ_0(nm)	Δ_0(meV)	H_{co}(mT)
Al	1.18	50	1600	0.18	10.5
In	3.41	65	360	0.54	23.0
Sn	3.72	50	230	0.59	30.5

Material	T_c (K)	λ_o(nm)	ξ_o(nm)	Δ_o(meV)	H_{co}(mT)
Pb	7.20	40	90	1.35	80.0
Nb	9.25	85	40	1.50	198.0

Type II Superconductors: Conventional Types

Material	T_c (K)	$\lambda_{GL}(0)$(nm)	$\xi_{GL}(0)$(nm)	Δ_o(meV)	$H_{c2,o}$(T)
Pb-In	7.0	150	30	1.2	0.2
Pb-Bi	8.3	200	20	1.7	0.5
Nb-Ti	9.5	300	4	1.5	13.0
Nb-N	16.0	200	5	2.4	15.0
$PbMo_6S_8$	15.0	200	2	2.4	60.0
V_3Ga	15.0	90	2-3	2.3	23.0
V_3Si	16.0	60	3	2.3	20.0
Nb_3Sn	18.0	65	3	3.4	23.0
Nb_3Ge	23.0	90	3	3.7	38.0

Type III Superconductors: High-Temperature Versions

Material	T_c (K)	$\lambda_{a,b}$(nm)	λ_c(nm)	$\xi_{a,b}$(nm)	ξ_c(nm)
$La_{1.85}Sr_{0.5}CuO_4$	40	80	400	4	0.7
$YBa_2Cu_3O_7$	95	30	200	3	0.4
$Bi_2Sr_2CaCu_2O_8$	85	25	500	4	0.2
$Bi_2Sr_2Ca_2Cu_3O_{10}$	110				
$Tl_2Ba_2CaCu_2O_8$	108				
$Tl_2Ba_2Ca_2Cu_3O_{10}$	125				

Type I and II with permission from R.J. Donnelly, "Cryogenics" in Physics Vade Mecum, H.L. Anderson (Ed.) American Institute of Physics: 1981; Type III with permission from T.P. Orlando and K.A. Duelin, Foundations of Applied Superconductivity, Addison-Wesley Publishing Co.: 1991.

Representative values of the parameter of typical superconductors are furnished in Table 11.3.

The values in the above tables are for clean elements. The penetration depth λ_o is given at zero temperature, as are the coherence length ξ_o, the thermodynamic critical field H_{co}, and the energy gap Δ_o.

Further, the values are only representative because the parameter for alloys and compounds depends on how clean or dirty the material is. The penetration depth $\lambda_{GL}(o)$ is given as the coefficient of the Ginzburg-Landau temperature dependence as $\lambda_{GL}(T) = \lambda_{GL}(0)(1 - T/T_o)^{-1/2}$ and likewise for the coherence length where $\xi_{GL}(T) = \xi_{GL}(0)(1 - T/T_c)^{-1/2}$. The upper critical field $H_{c2,o}$ is given at zero temperature as well as the energy gap Δ_o.

The values in Table 11.3 are only approximate because the parameters for high-temperature superconductors have not all been established well enough. The penetration depth is the coefficient of the Ginzburg-Landau temperature dependence $\lambda_{GL}(0)$ as in the table for conventional superconductors; likewise for the coherence length, $\xi_{GL}(0)$. However, since these materials are anisotropic, these lengths are given along the principal axis. The directions $\hat{a}$ and $\hat{b}$ are taken to lie in the plane of the Cu-O planes and $\hat{c}$ is taken to be perpendicular to that plane.

11.8 Concluding Remarks

Superconductors are the most intriguing materials of modern times. Realizing a high temperature superconductor with potentials for technological applications is the target of scientific studies around the world. Though some breakthroughs have been achieved, a comprehensive set of materials for wide-scale applications is yet to be conceived.

References

[1] T. P. Orlando and K. A. Duelin: *Foundations of Applied Superconductivity.* (Addison-Wesley Publishing Co., Reading, MA, 1991).

[2] S. Foner and B. B. Schwartz (Eds.): *Superconductor Material Science.* (Plenum Press, New York: 1981).

[3] A. C. Rose-Innes and E. H. Roderick: *Introduction to Superconductivity.* (Pergamon Press, New York: 1978).

[4] D. J. Quinn and W. B. Ittner III: Resistance in a superconductor. J. Appl. Phys., vol. 33, 1962: 748- 749.

[5] S. Foner and B. B. Schwartz (Eds.): *Superconducting Machines and Devices.* (Plenum Press, New York: 1974).

[6] S. Foner and B. B. Schwartz (Eds.): *Superconducting Applications: SQUIDs and Machines.* (Plenum Press, New York: 1977).

[7] D. E. Oates and A. C. Anderson: Stripline measurements of surface resistance: Relations to HTSC film properties and deposition methods, SPIE, vol. 1187, pp. 326-337, 1989.

[8] D. M. Sheen, S. M. Ali, D. E. Oates, R. S. Withers and J. A. Kong, Current distributation, resistance, and inductance for superconducting strip transmission lines, IEEE Trans. Superconductivity, vol. 1,(2) June 1992, 108-115.

[9] T. Ohnuma and Y. Tanaka, High-Tc superconducting active antenna with reflector, IEEE Trans. Applied Superconductivity, vol. 2,(2) June 1992, pp. 113-115.

[10] G. J. Chen, P. A. Rosenthal and M. A. Beasley, Kinetic inductance memory cell, IEEE Trans. Applied Superconductivity, vol. 2,(2) June 1992, pp. 95-101.

[11] S. C. Mathur, D. C. Dube, S. Sinha, Y. S. Reddy and A. S. Bhalla, Dielectric Properties of (Pb-Bi-Sr-Ca-CuO) Ceramics in normal and superconducting states.

Defining Terms

BCS theory: A theory that suggests the superconduction being instigated by a pair of electrons known as superelectrons.

Cooper pairs: Pair of superelectrons responsible for superconductivity as per BCS theory.

Josephson behavior: The behavior of current density in a superconductor as controlled by coherence length and penetration depth.

CHAPTER 12

Ferroelectric Materials

12.1 Introduction

Ferroelectric materials are typically *nonlinear dielectrics* with the following characteristics:

- They exhibit distinct dependency of permittivity on the intensity of an applied electric field.
- The dielectric displacement *versus* applied electric field follows *hysteresis loop* characteristics.
- The dielectric constant of these materials are invariably very high, on the order of thousands to tens of thousands.
- The dielectric parameters are pronouncedly dependent on temperature and the ferroelectric properities are observed only within a definite range of temperatures.
- There is a presence of spontaneous dielectric polarization in the absence of external electric field and this polarization can be switched in direction by applying an electric field.

The name *ferroelectricity* is derived from its formal similarity with *ferromagnetism*. That is, the spontaneous electric polarization and hysteresis effect in the relationship between dielectric displacement and the applied electric field observed in ferroelectric materials are akin to the spontaneous magnetic polarization and hystersis in the relationship between magnetization and applied magnetic field exhibited by magnetic materials.

Examples of typical ferroelectrics are:

- Alkali metal dihydrogen phosphates such as potassium di-hydrogen phosphate (KDP: KH_2PO_4) and ammonium di-hydrogen phosphate (ADP: $NH_4H_2PO_4$-PO_4) groups linked by hydrogen bonds.
- A number of isomorphous phosphates and arsenates.
- Titanates: barium titanate, calcium titanate, lead titanate and cadmium titanate which contain repeated oxygen octahedral surrounding another type of ion with a general formula, ABO_3.
- Niobates and tantalates : potassium niobate, sodium niobate, sodium tantalate etc. and mixed compounds like cadmium pyroniobate and lead metaniobates.
- Rochelle salt (sodium potassium tartarate tetrahydrate, ($NaKC_4H_4O_6 \cdot 4H_2O$), which are hydrogen bonded crystals.
- Triglycine sulfate (TGS), selenate and fluoroberyllate.
- Certain polycrystalline ceramics.
- Polymeric ferroelectrics.
- Hydrated double sulfate of guanidinium and a trivalent metal, for example, guanidinium aluminum sulfate hexahydrate (GASH)

Ferroelectric materials are subgroups of *piezoelectric* crystals which when subjected to mechanical stress exhibit proportionate electric (surface) charges; or conversely, such crystals are stretched or compressed in an electric field of a given direction. When the field direction is reversed, the sign of the dimensional changes alters. Another property of certain piezoelectrics is *electrostriction* which also denotes a dimensional change caused by an electric field of a given direction but this does not alter when the direction of the field is reversed, and the observed changes are usually several orders smaller in magnitude (in comparison with *converse piezoelectricity*).

In a class of piezoelectric materials electric charges may be formed on the surface as a result of temperature changes when no mechanical stress is present at the same time. This phenomenon is termed as *pyroelectricity*. Every pyroelectric material is piezoelectric, though the converse is not true. There are piezoelectric materials which are nonpyroelectric.

Electects are ferroelectrics which preserve their polarization for a long time after the external electric field (that produced the polarization) is removed. The "permanent" polarization would set up an electrostatic field in the surrounding space similar to a permanent magnet which enables a magnetic field in its surrounding.

12.2 Ferroelectric Characterization

The ferroelectric properties stem from the crystal chemistry of the relevant materials. A simple model representing a two-dimensional crystalline structure of a (hypothetical) ferroelectric material is shown in Figure 12.1. It is comprised of an arrangement of positive and negative ions, each pair of which is located at the lattice points of a simple square net.

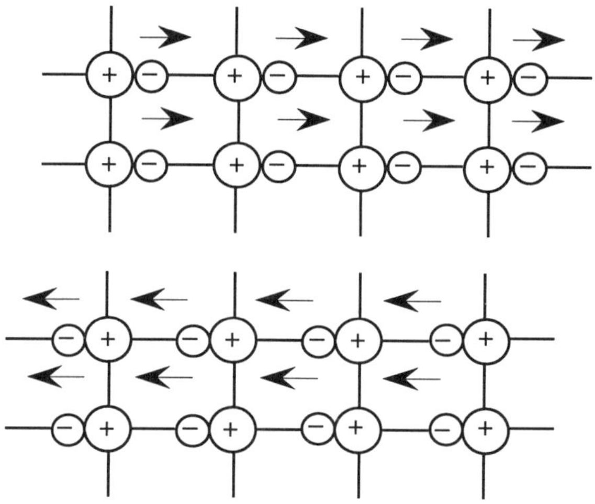

Figure 12.1 Two-dimensional crystalline structure of a hypothetical ferroelectric material.

Under equilibrium conditions, the negative ion with respect to the positive ion can assume two possible positions (left or right) as illustrated. This corresponds to the two (bilateral) symmetric potential minima between a pair of lattice sites as depicted in Figure 12.2.

The transition of location left to right or *vice versa* can occur provided sufficient energy is supplied to overcome the potential barrier, ΔE.

The *dipoles* thus align (at a given temperature,T K) exhibiting a spontaneous polarization measured in terms of number of *dipole moments* per unit volume. The direction of spontaneous polarization is called the *polar axis*. The regions with the set of the spontaneously polarized dipoles are called *domains*. In Figure 12.2, a set of two domains are illustrated.

When positive and negative domains (that is, domains oriented to the right and the domains oriented to the left) are equal in the crystal, the overall polarization is zero. With the application of an external electric field on a virgin ferroelectric crystal as illustrated in Figure 12.3, the domains tend to be polarized along the positive direction; and at a sufficiently large electric field, all the domains are polarized parallel to the applied electric field indicating a *state of saturation* characterizing the crystal as a *single domain*. Upon reducing the applied field, the extent of polarization decreases but does not return to zero when the applied field is brought back to the zero value. In other words, the material is spontaneously polarized and has acquired a *remnant* polarization.

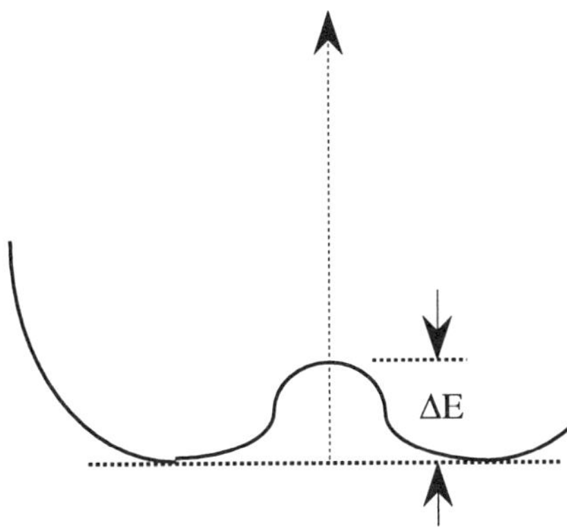

Figure 12.2 Potential well.

That remnant value can be removed by reversing the direction of the applied field to a *coercivity* value. Upon continuing the reverse field, the material again reaches saturation with all its domains polarized along the negative direction, as illustrated in Figure 12.3.

The cyclic polarization depicting *P versus E* curve is known as the *dielectric hystersis loop* and is similar to the B-H curve of a ferromagnetic material.

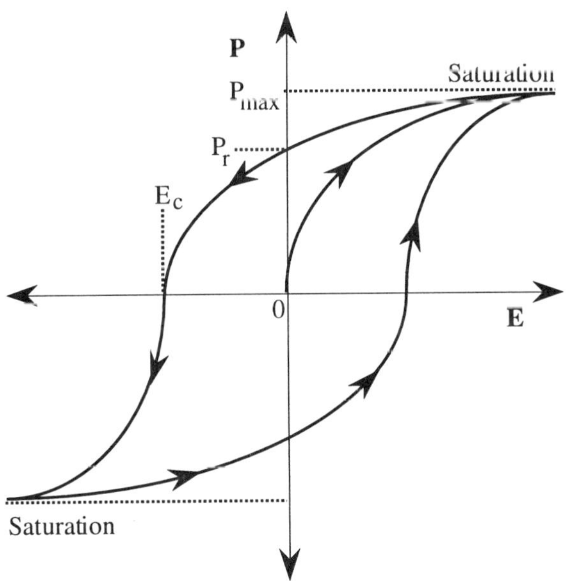

Figure 12.3 **P–E** hysteresis curve.
P_r: Remnant polarization;
E_c: Coersive electric field;
P_{max}: Maximum (saturation) value of the polarization.

With the application of heat energy, the domain switching (from left to right or *vice versa*) is made feasible due to overcoming of the potential barrier ΔE even in the absence of the external electric field. This permits a single equilibrium state to be acquired by the domains and the crystal is no longer polar and would behave like an ordinary dielectric. The critical temperature(T_c) at which the transition from polar to nonpolar state occurs is known as the *Curie temperature* or *Curie point*. The dielectric constant of a ferroelectric when plotted against temperature shows a maximum on transition into the ferroelectric state. In the paraelectric, nonpolar state, the changes in dielectric constant with temperature can be described by the *Curie-Weiss law* stated as :

$$\varepsilon = \varepsilon_{\infty} + A/(T - T_c) \tag{12.1}$$

and the corresponding dielectric susceptibility (χ) is given by :

$$\chi = C_c/(T - T_c) \tag{12.2}$$

where

A = Constant
ε_{∞} = The value of dielectric constant at temperature $T >> T_c$
T = Temperature
T_c = Curie point
C_c = *Curie constant* of the material

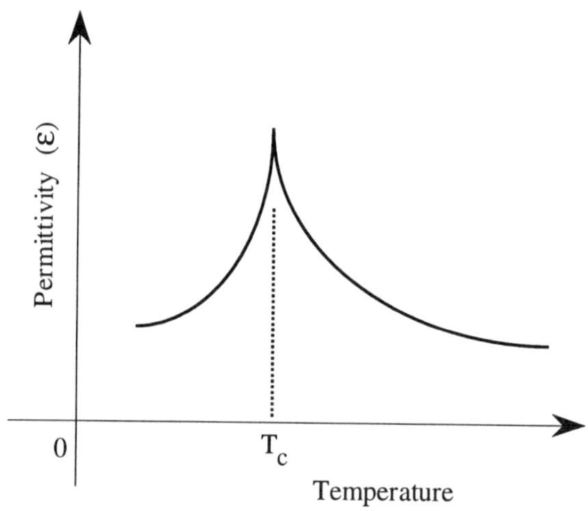

Figure 12.4 Permittivity *versus* temperature characteristics of a ferrielectric material.

12.3 Antiferroelectrics and Ferrielectrics

In certain crystals which are isomorphous with ferroelectrics or have a related structure, a phase transition from a crystal form of higher symmetry into one of lower symmetry can take place with decreasing temperature. This transition is accompanied by a slight deformation of the crystal structure. However, in this case in contrast to ferroelectrics, the phenomenon is not accompanied by spontaneous polarization. In the majority of cases, the crystal structure of lower symmetry can be described as the sum of sublattices which are equivalent to each other and in which the value of polarization is identical but of opposite sign. Crystals of this type are *antiferroelectric*. They have no permanent dipole moment. The neighboring dipole are *antiparallel* to each other.

The ferroelectric and antiferroelectric states are very closely related and are decided by the thermodynamic attributions of the crystals. It has also been observed that under intense electric field, certain antiferroelectrics such as lead zirconate ($PbZrO_3$) convert into ferroelectrics. Antiferroelectric materials also have a transition Curie temperature above which the antiferroelectric property ceases to exist.

Table 12.1 Typical Antiferroelectric Materials and Their Properties

Material	Curie Temperature, T_c°C
Lead zirconate: $PbZrO_3$	+233
Sodium niobate: $NaNbO_3$	+63
ADP: $NH_4H_2PO_4$	–125

Some crystals exhibit over a certain temperature range ferroelectric properties along one axis and antiferroelectric properties along another. These are designated as *ferrielectric* materials.

12.4 Classification of Ferroelectric Materials

Various classifications of ferroelectric materials have been proposed to facilitate the treatment of their properties. The following gives the summary of different classifications of ferroelectric materials:

1. Crystal chemical classification: According to crystal chemical classification, ferroelectrics can be subdivided into two groups, namely:

- Hydrogen-bonded crystals
- Double oxides

Examples of hydrogen-bonded crystals are Rochelle salt, triglycine sulfate, and potassium di hydrogen phosphate.

Examples of double oxides are barium titanate, potassium niobate, and lead niobate.

2. Classification based on the number of directions allowed to the spontaneous polarization: Relevant ferroelectric crystals can be divided into the following two groups:

- Crystals with single axis of spontaneous polarization
- Crystals with multiple axes of spontaneous polarization

Ferroelectric materials are further grouped into two categories. The first one refers to those in which the spontaneous polarization may exist along one crystallographic axis and the second type can be polarized along several axes which are equivalent in the nonpolarized state. The various materials grouped as above are given in the Table 12.2.

3. Classification based on the existence of lack of center of symmetry: This classification is used to study the thermodynamic behavior of ferroelectric crystals. Some ferroelectric crystals are characterized by a nonpolar phase which is noncentrosymmetric. Examples are Rochelle salt, potassium di-hydrogen phosphate and isomorphous compounds. Another group of ferroelectric crystals characterized by a centrosymmetric nonpolar phase are barium titanate, cadmium (pyro) niobate and triglycine sulfate.

4. Classification according to the nature of phase change occurring at Curie point: Some crystals undergo a transition of the order-disorder type during the phase transition. They are:

triglycine sulfate, and potassium di-hydrogen phosphate. Another group of crystals undergo a transition of the displacive type. Examples are titanates and some double-oxide ferroelectrics.

Table 12.2 Classification of Ferroelectric Materials Based on Single or Multiple Axes Spontaneous Polarization

Group	Materials
I (Spontaneous polarization along a single axis)	Segnette salt and related tartrates Ferroelectrics of potassium di-hydrogen phosphate types, $(NH_4)_2\ SO_4$ and $(NH_4)_2\ BF_4$ Colemanite Thiourea, glycine sulfate, glycine selenate
II (Spontaneous polarization along multiple axes)	Ferroelectrics of perovskite type Ferroelectrics of pyrochlore type ($Cd_2Nb_2O_7$) Ferroelectrics of niobate type Alums like $CH_3NH_3\ Al(SO_4)_2 \cdot 12H_2O$ $(NH_4)_2\ Cd_2(SO_4)_3$
Unclassified	Ilemite-type ferroelectrics: $LiNbO_3$ $LiTaO_3$

12.5 Other Properties of Ferroelectric Materials

Optical properties: Optical properties such as the *Kerr effect* are observed in ferroelectric crystals like potassium di-hydrogen phosphate, barium titanate, and Rochelle salt.

Potassium di-hydrogen phosphate crystals are negative uniaxial. The refractive indices for the sodium D lines at 15 °C are $n_1 = n_2 = n_a$ = 1.5095 and $n_3 = n_c$ = 1.4684. The birefringence $(n_3 - n_1)$ increases with decreasing temperature and shows an anomaly at the transition point. This anomaly is due to the *spontaneous Kerr effect* resulting from spontaneous strain and polarization (Chapter 17).

In case of cubic phase barium titanate, the refractive index (n_c) depends on temperature. The temperature dependence of the refractive index is shown in Figure (12.5), but the refractive index (n_a) of the tetragonal phase barium titanate is independent of temperature. The birefringence (Δn) of the tetragonal phase is depicted in Figure (12.6). Thus the birefringence is temperature dependent.

Dielectric properties: Dielectric properties are exhibited by numerous ferroelectric crystals such as triglycine sulfate, potassium di-hydrogen phosphate, barium titanate, potassium niobate, potassium tantallates, and various perovskite-type oxides. For example, in the case of triglycine sulfate, the components of the dielectric constants have the following values at 23°C at a measuring field of 1volt/cm and a frequency of 500 KHz: $\varepsilon_a = 8.6$, ε_b = 4.3, and $\varepsilon_c = 5.7$. These components are dependent on temperature. The temperature dependence of these components is shown in Figure 12.7.

Semiconducting properties: Semiconducting properties are exhibited by ferroelectric crystals of pervoskite structures such as barium titanate, lead titanate, and cadmium titanate.

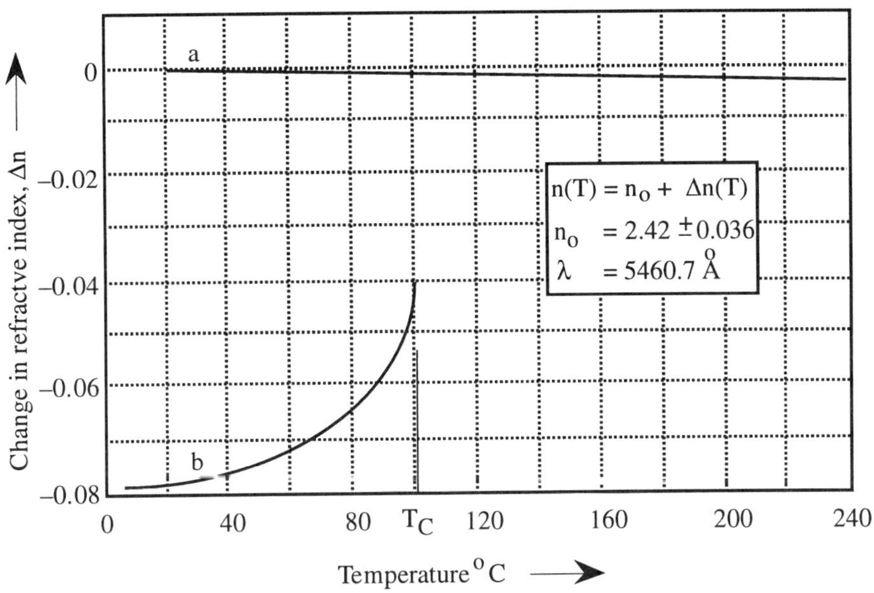

Figure 12.5 Change in refractive index of barium titanate as a function of temperature. a) Refractive index along a-axis; b) refractive index along c-axis.

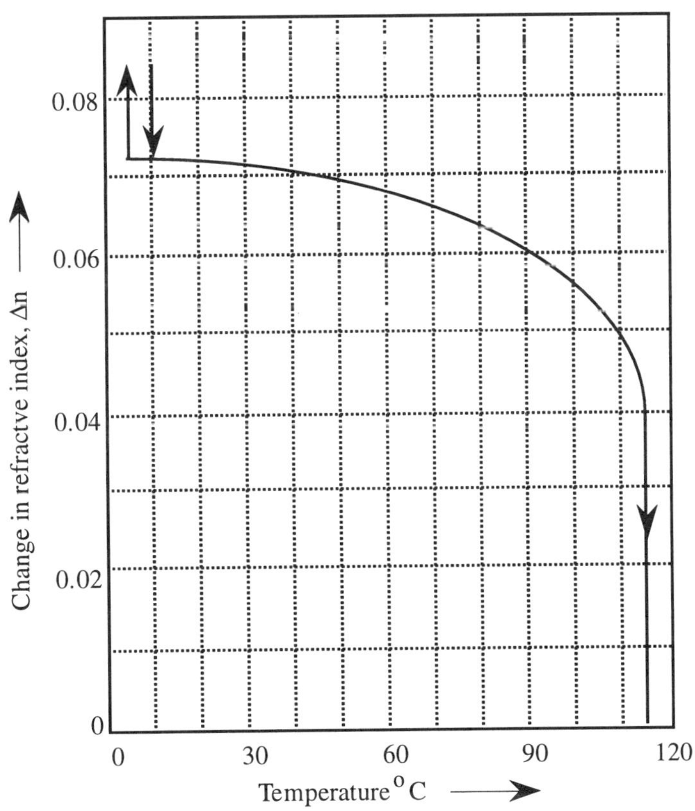

Figure 12.6 Birefringence of a typical tetragonal crystal of barium titanate as a function of temperature.

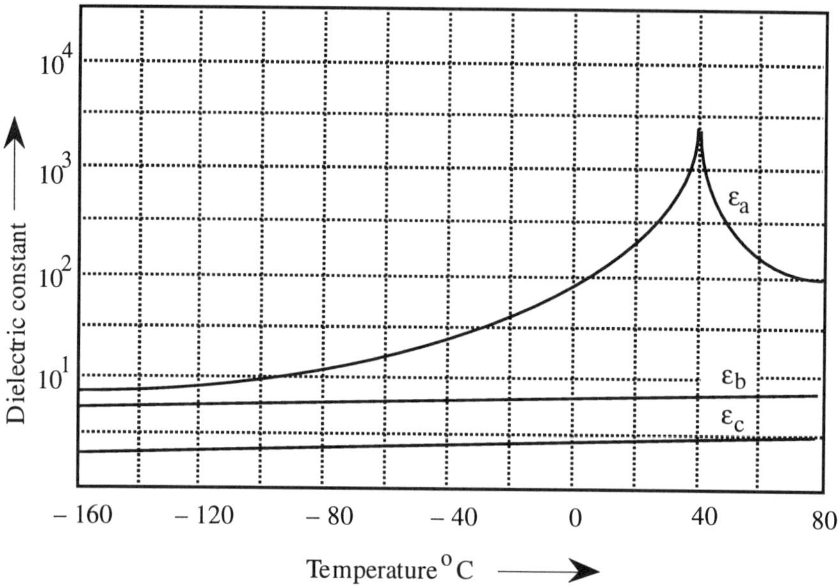

Figure 12.7 Axial components of the dielectric constant of triglycine sulfate as a function of temperature.

Pyroelectric effect: *Pyroelectric effect* is observed in barium titanate, Rochelle salt, and colemanite. Studies on of the field-induced pyroelectric effect at temperatures above the Curie point on barium titanate crystals (with Curie temperature of 10°C) indicate that at temperatures slightly above the Curie point, the pyroelectric current increases rapidly with increasing field and then drops discontinuously to some lower value after which it decreases slowly and smoothly.

With increasing temperatures, the discontinuous drop occurs at increasing fields and also decreases in magnitude until it no longer occurs. Above the latter temperature (T_1) the curve shows smooth peak. These experimental results have been explained by thermodynamic principles.

Electrostrictive properties: *Electrostriction effects* occur usually in all substances whether crystalline or amorphous, solid or fluid. Electrostrictive property is the ability of a crystal to maintain its strain unchanged even though the shape of the crystal changes by reversing the direction of the applied electric field. Thus this effect is quadratic and the strain is proportional to the square of the applied electric field.

Piezoelectric properties: *Piezoelectric property* of a ferroelectric crystal refers to the ability of a crystal to exhibit electric polarity when subjected to stress. The piezoelectric effect is a linear effect. By applying force to a piezoelectric crystal placed between two electrodes a charge flows in the measuring circuit. On the other hand, by changing the direction of application of force, the direction of charge flow in the measuring circuit reverses. Also, if an electric potential is applied between the electrodes containing the piezoelectric crystals between them, the crystal will be stretched. By changing the direction of electric potential, the crystal will be compressed. This is referred to as the *converse piezoelectric effect*. If the applied electric potential is an alternating one, then the crystal will undergo vibrations. Conversely if the crystal experiences vibrations, then an alternating voltage will be introduced between the two electrodes (see Chapter 13).

The piezoelectric coefficient defined as the ratio between piezoelectric strain component to the applied electric field component at a constant mechanical stress or *vice versa* depicts the strain *versus* the electrical phenomenon in the piezoelectric crystals. This coefficient is denoted by d_{mn}. The subscript n (1 – 3) refers to the 3 Elucidean orthogonal axes, and $m = (1-6)$ specifies the mechanical stress-strain components. The unit for d_{mn} is meter/volt.

When an electric field is applied to a piezoelectric material, the ionic core and electron distribution are distorted. This distribution appears as a strain and is proportional to the square of the electric field. Inversely, application of a stress to a crystal results in a net polarization manifesting as surface charges on the crystal.

Denoting the stress tensor by ***T***, and the strain tensor by ***S***, they are interrelated linearly by the following relations:

$$\boldsymbol{T} = c\boldsymbol{S} \tag{12.3a}$$

$$\boldsymbol{S} = s\boldsymbol{T} \tag{12.3b}$$

where c is the elastic stiffness constant and $s = 1/c$. The stress *versus* polarization ***(P)*** is depicted by :

$$\boldsymbol{P} = d\boldsymbol{T} \tag{12.4}$$

with d being the piezoelectric coefficient indicated earlier. The dielectric displacement ***(D)*** in the presence of stress is given by:

$$\boldsymbol{D} = \varepsilon\boldsymbol{E} + d\boldsymbol{T} \tag{12.5}$$

The inverse effect is therefore specified by:

$$\boldsymbol{S} = s\boldsymbol{T} + d\boldsymbol{E} \tag{12.6}$$

As most of the piezoelectric materials are crystalline and anisotropic, the electric and mechanical forces applied along one direction produce effects in the other directions as well.

Table 12.3: Some of the Typical Materials which Exhibit Piezoelectric Properties

Material	Dielectric Constant	Spontaneous Polarization (coulomb/meter^2) $\times 10^{-2}$
Natural quartz	4.5	
Rochelle salt		
ADP		
Lithium sulfate		
Group II-VI compound semiconductors		
Barium titanate	1700	7.5
Lead titanate		
$PbTi_{0.45}Zr_{0.55}O_3$ ceramic	500	30

Five miscellaneous properties are discussed below.
Isotope effects: When hydrogen in KDP and in crystals isomorphous with KDP is replaced by deuterium, the deuterated salt undergoes ferroelectric transitions at higher temperatures than the corresponding hydrogen salts. This effect is known as the *isotope effect*. The magnitude of this isotope effect is an indication of the vital role played by the hydrogen bond in the mechanism of ferroelectricity.

Pressure effect: Displacement of Curie temperature either upwards or downwards due to the application of hydrostatic pressures is referred to as the *pressure effect*. In the case of triglycine sulfate crystals, Curie temperature is displaced linearly upwards by the pressure P according to the relation:

$$T_c = T_{co} + KP \tag{12.7}$$

where T_{∞} = transition temperature at atmospheric pressure and $K = 2.6 \times 10^{-3}$ coulomb/atm.

Radiation effect: By exposing ferroelectric crystals to radiations such as X-ray, gamma radiation, and neutrons, the characteristics of the crystals such as dielectric constants, transition temperature, and elastic constants can be modified. Such effects are known as *radiation effects*.

When a triglycine sulfate crystal is bombarded by X-rays (30 KeV) the hysteresis loop of the crystal becomes distorted. The damages due to radiation result from the effects of ionization rather than from ionic displacements. Radiation effect is the same irrespective of whether the crystal is bombarded above or below the transition point, but application of an alternating electric field during or immediately after the bombardment seems to relieve the distortion temporarily.

Electroluminescent effect: Emission of light from the surface of the crystal due to the application of high radio frequency electric field to the crystal is referred to as *electroluminescent effect*. In case of barium titanate, the intensity of light emitted is given by the expression:

$$I = A\,V^B \tag{12.8}$$

where V is the applied voltage and A and B are constants which depend on electrode material. The intensity of light emitted varies with temperature.

Fatigue effect: *Fatigue effect* or *decay effect* is the gradual reduction of the charges switched or loss of response after a few million cycles of switching. Fatigue effect is observed in barium titanate crystals after repeated pulsing of a given crystal. This effect is of great importance in the field of applications of barium titanate crystals as matrix memories for digital stroage in computer and switching systems. In the case of barium titanate, the reduced switching charge can be restored to its original value by switching the crystal over a few minutes with a 60 Hz sine wave voltage.

12.6 Types of Ferroelectrics

12.6.1 *Ferroelectrics of perovskite structure*: These materials, in general, have pseudocubic rhombic crystalline structure and can be described by a general formula ABX_3 (for example, $CaTiO_3$, $SrTiO_3$, $BaTiO_3$, $PbTiO_3$, etc.). Among these $BaTiO_3$ has been studied exhaustively. This material has considerable importance from the point of view of ceramics.

In perovskite structure, the Ba^{2++} ions occupy the corners of a cube and the centers of the six faces are occupied by O^{2-} ions. The oxygen ions thus form an octahedran at the center of which the Ti^{4+} ions are located.

$BaTiO_3$ is ferroelectric at all temperatures up to 120°C. Below the Curie temperature the structure is no longer cubic. Material becomes spontaneously polarized and the structural changes are as follows:

Temperature (°C)	< –90	> –90 < 5	> 5 < 120	> 120
Structure	Rhombohedral	Orthorhombic	Tetragonal	Cubic

The highest transitional temperatures (Curie temperatures) for the perovskite family of ferroelectrics are as follows:

Table 12.4 Curie Temperature of Perovskite Ferroelectrics

Material	T_c (°C)
$BaTiO_3$	120
$KNbO_3$	415
$PbTiO_3$	490
$KTaO_3$	-260
$NaTaO_3$	475
$SrTiO_3$	-250
$CdTiO_3$	-210

Ferroelectrics of the $BaTiO_3$ type exhibit ferroelectric properties even when polycrystalline because of the high degree of crystal symmetry. General properties of $BaTiO_3$ are:

- Consistent with the temperature ranges and the corresponding crystalline structures indicated above, the dielectric constant changes from about 10^3 at –180°C to 10^4 at 120°C.
- The dielectric constant of $BaTiO_3$ is also frequency-dependent. The ratio between dielectric constant measured at 10^9 Hz and that measured at 10^3 Hz is of the order of 0.7 to 0.8.
- The coersive field (measured at room temperature) shows variations with respect to frequency and amplitude of the applied field and depends on the thickness of the sample.
- The refractive index of the cubic phase is unusually large $n = 22.4$. It also varies with temperature.
- The electrical resistivity of $BaTiO_3$ at room temperature is about 10^{10} ohm-cm. However, with the addition of a small (0.1–0.3%) amount of La_2O_3 or Sm_2O_3, the resistivity drops to $10–10^3$ ohm-cm. Also, this resistivity exhibits a pronounced increase at the Curie point of pure $BaTiO_3$ (around 120°C).
- Elastic compliances and piezoelectric coefficients of $BaTiO_3$ are listed in Table 12.5.

Table 12.5 Elastic Compliances and Piezoelectric Coefficient of Single Crystal Cubic $BaTiO_3$

Elastic Compliances at 150°C 10^{-13} cm²/dyne		Piezoelectric Coefficient at 25°C dik: 10^{-6} Cgs units bik: 10^{-8} Cgs units	
s_{11}	8.33	d_{15}	11.76
s_{12}	–2.68	d_{31}	–1.04
s_{44}	9.24	d_{33}	2.57
		b_{15}	5.07
		b_{31}	–7.67
		b_{33}	19.17

12.6.2 *Perovskite-type oxides*: The family of oxides with perovskite crystalline structure and their solid solutions are also useful ferroelectrics. The oxide family which includes $BaTiO_3$ and the transition temperature to the cubic phase (in each case) are listed in Table 12.6.
The parent member of perovskite-type oxides is the mineral $CaTiO_3$ called *perovskite*. These oxides can be described by a general formula ABO_3 where A is a monovalent, divalent, or trivalent metal and B is pentavalent, tetravalent, or trivalent element, respectively.

Table 12.6 Perovskite-Type Oxides
(C: Cubic, T: Tetrogonal, and O: Orthorhombic Crystalline Structures)

Compound	Structure at 20°C	Transition Temperature to Cubic Phase (°C)
$BaTiO_3$	T	120
$SrTiO_3$	C	-220
$CaTiO_3$	O	1060
$PbTiO_3$	T	490
$CdTiO_3$	O	
$PbZrO_3$	O	230
$PbHfO_3$	O	215
$KNbO_3$	O	435
$NaNbO_3$	O	640
$AgNbO_3$	O	550
$KTaO_3$	C	-260
$NaTaO_3$	O	470
$AgTaO_3$	O	485

Some examples of solid solutions of the perovskite oxides are :
($KNbO_3$ + $KTaO_3$), ($NaNbO_3$ + $NaTO_3$), ($KNbO_3$ + $NaNbO_3$), ($AgNbO_3$ + $AgTaO_3$ + $RbTaO_3$), ($PbZrO_3$ + $PbHO_3$), ($SrTiO_3$ + $CdTiO_3$).

Compounds $KNbO_3$ and $PbTiO_3$ and related binary solid solutions have also been studied extensively for their ferroelectric properties. Ferroelectricity in $KNbO_3$ has been known since 1949. It is the only ferroelectric crystal that exhibits the same phase symmetries and sequence of phase transitions (with respect to temperature) as $BaTiO_3$ [1]. The similarities between $BaTiO_3$ and $KNbO_3$ are furnished in Table 12.7.

Table 12.7 Barium Titanate *versus* Potassium Niobate

Parameter	$KNbO_3$	$BaTiO_3$
Transition temperature (^oC)	425, 225, –10	120, 5, –90
Transition energy (cal/mol)	190, 85, 32	49, 21, 11
Maximum tetrogonal distortion (c/a)	1.017	1.010
Spontaneous polarization at the Curie point (10^{-6} C/cm^2)	26	18
Curie Constant C (K)	2.4×10^5	1.7×10^5
$(T_c – T_o)$ ^oC	58	11

Potassium tantalite ($KTaO_3$) has a cubic symmetry at room temperature phase and its lattice parameters are identical to that of $KNbO_3$ in its centrosymmetric phase, that is, above 435°C. Despite this similarity, $KTaO_3$ has the Curie temperature of 13 K, being one of the two lowest ferroelectric transition temperatures known. (The other compound is lithium thalium tartrate monohydrate with T_c = 10 K)

The dielectric characteristic of $KTaO_3$ follows closely Curie-Weiss law down to 52 K and to a lesser extent below this temperature.

Solid solutions of $KNbO_3$ and $KTaO_3$ have a linear variation of Curie temperature between two end members of the system. Further, the ferroelectric transition which is of the first order in $KNbO_3$ becomes of the second order when the Ta concentration exceeds 55%.

References

[1] F. Jona and G. Shirane: *Ferroelectric Crystals.* (Macmillan Co., New York: 1962).

[2] E. Fatuzzo and W. J. Merz: Ferroelectricity. (North Holland Publishing Co., Amsterdam: 1967).

[3] B. W. Forshberg, Jr.: Piezoelectricity, electrostriction and ferroelectricity in *Handbuch der Physik*, vol. 17 (Springer-Verlag, Berlin: 1956).

[4] E. T. Jaynes: *Ferroelectricity.* (Princeton University Press, Princeton, NJ: 1953).

[5] A. J. Moulson and J. M. Herbert: *Electroceramics: Materials, Properties and Applications.* (Chapman and Hall, New York: 1990).

Defining Terms

Aging effect in ferroelectrics: This refers to the acquiring of a metastable state by the domain configuration of a real crystal.

Antiferroelectrics: These are anti-polar crystals whose free energy is comparable to that of a polar crystal.

Antipolar and polar crystals: Antipolar crystals are characterized by the existence of two oppositely polarized sublattices; and a polar crystal has the same directional polarization of sublattice.

Butterfly loop: In relation to hysteresis effect in a ferroelectric crystal, different portions of the sample may alter orientation at different parts of the hysteresis cycle giving rise to "butterfly-shaped" P-E curve.

Coercive field: It is the extent of electric field required to remove the remnant polarization in the ferroelectric hysteresis.

Curie constant: Temperature dependence of the dielectric constant of a ferroelectric is decided by Curie-Weiss law with a characteristic constant C known as the Curie constant.

Curie point (or temperature): It is the temperature at which a transition from a polar to nonpolar state occurs in a ferroelectric.

Curie-Weiss law and temperature: The dependence of dielectric permittivity (ε) on temperature is specified by the Curie-Weiss law, $\varepsilon = \varepsilon_o + C/(T - T_c)$ where ε_o is the free-space permittivity, C is the Curie constant, and T_c is the Curie-Weiss temperature.

Electrostriction: This refers to the quadratic relation effect between the strain perceived in a ferroelectric material and the electric field applied.

Ferroelectricity, ferroelectrics: Ferroelectricity refers to the phenomenon of spontaneous electric polarization and hysteresis effects in the relation between dielectric displacement and applied electric field. Ferroelectrics are crystals exhibiting ferroelectricity characteristics.

Ferroelectric hysteresis: The relation between the polarization $(\boldsymbol{P})$ in a ferroelectric and the applied electric field $(\boldsymbol{E})$ over a hysteresis cycle.

Piezoelectric effect: It is a linear effect depicting stress *versus* the charge induced (or *vice versa*) in crystals having no center of symmetry (except cubic class 432).

Pyroelectric effect: It refers to the effect that if the temperature of a crystal is altered, a change in the polarization occurs and the electric charge can be on crystal faces perpendicular to the polar axis.

Spontaneous polarization: It defines the orientational assembly of electric dipoles anisotropically pointing in the same direction and measured in terms of dipole moment per unit volume.

Symmetry of crystals: It describes the extent of symmetry elements possessed by a crystal. For example, existence of center of symmetry classifies a crystal as centro-symmetric.

CHAPTER 13

Piezoelectric Composite Materials

13.1 Introduction

A *piezoelectric composite* is a combination of a piezoelectric ceramic and/or a polymer and a nonpiezoelectric polymer constituting a new version of piezoelectric material. In general terms, piezoelectric composite applies to any piezoelectric material resulting from combining a piezoelectric polymer or ceramic with other nonpiezoelectric materials including air-filled voids. Figure 13.1 illustrates different ways of constituting a piezoelectric composite material.

Earlier versions of piezoelectric composites synthesized include barium titanate embedded polymer matrix and lead zirconate-titanate (PZT) ceramic powder dispersed in a polymeric receptacle. Subsequent developments are: (i) Flexible piezoelectric composites using $PbTiO_3$ or PZT plus synthetic rubber; (ii) PVDF-based pyroelectric composites; (iii) woven PZT ceramic/polymer composites; and (iv) calcium modified lead titanate rods embedded in a polymer matrix. The development of piezoelectric composites was motivated by the efforts to find a class of piezoelectric materials which offer substantial improvements over the conventional piezoelectric ceramics and polymers for making ultrasonic transducers used in medical imaging and for hydrophone applications.

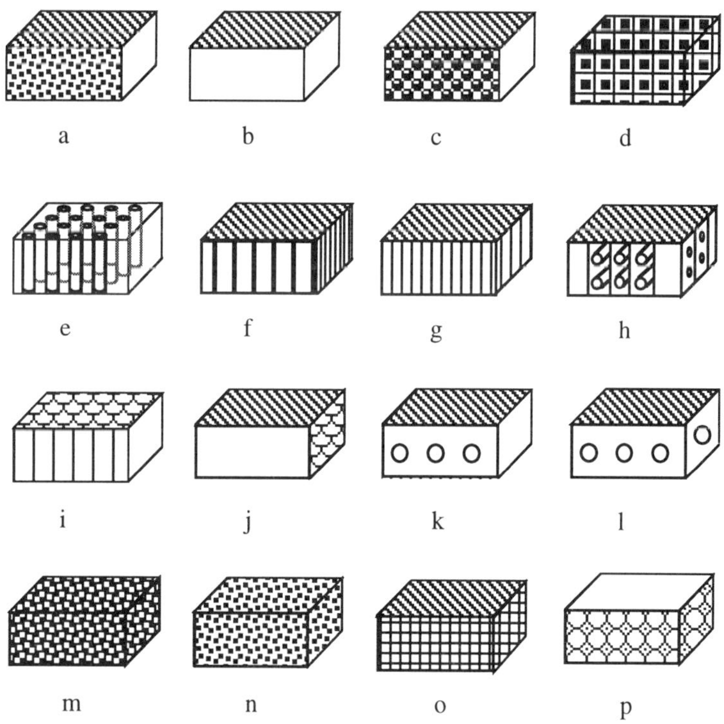

Figure 13.1 Different types of piezoelectric composites [6].
(a) Particles in a polymer: 0-3; (b) PVDF composite: 0-3; (c) PZT spheres in a polymer: 1-3; (d) diced composite: 1-3; (e) PZT rods in a polymer: 1-3; (f) sandwich composite: 1-3; (g) glass-ceramic composite: 1-3; (h) transverse reinforced composite: 1-2-0; (i) honeycomb composite: 3-1P; (j) honeycomb composite: 3-1S; (k) perforated composite: 3-1; (l) perforated composite 3-2; (m) replamine composite: 3-3; (n) burps composite: 3-3; (o) sandwich composite: 3-3; (p) ladder-structured composite: 3-3.

Combining a piezoelectric ceramic and a passive polymer to form a piezoelectric composite facilitates transducer designs that offer several advantages over the use of conventional piezoelectric ceramics or polymers. For example, the rod composite geometry (Figure 13.2) allows enhanced electroelastic coupling and permits designs of transducers with adequate impedance matching feasibilities. Likewise, the dice-and-fill technique (Figure 13.3) adopted in constituting a piezoelectric composite allows shaped geometries of the transducers facilitating focused ultrasonic beams. Further, judicious rod spacing in the composite yields materials with low cross-talk between array elements.

Figure 13.2 Rod piezoelectric composite (Example: PZT rods in a polymer).

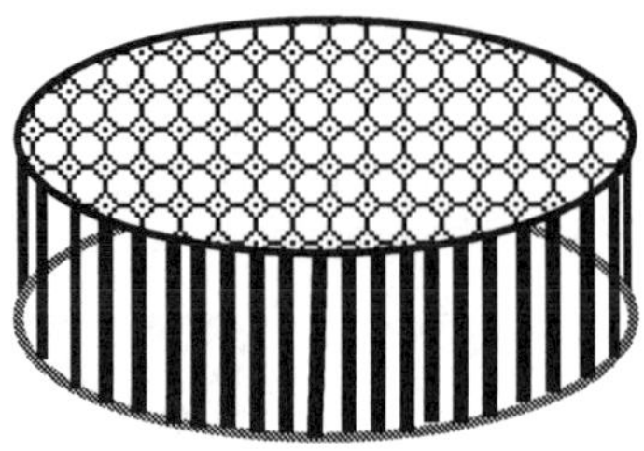

Figure 13.3 Dice-and-fill composite.

Existing studies [1-12] address the various trial-based synthesis of piezoelectric composites with different constituent materials, measurement of the electroelastic properties of the test composites, application of the piezoelectric composites and analytical endeavors to formulate the effective electroelastic parameters of the composite. In the following sections descriptions of various composites are developed; their fabricational aspects, functional characteristics, application potentials and a summary of the theoretical considerations are presented.

13.2 Connectivity-Based Structured Piezoelectric Composites

To understand how different versions of piezoelectric composites are constituted, it is necessary to define a critical parameter, namely, the *connectivity* which refers to the manner or the pattern in which the diphasic or multiphasic constituents are self-connected in zero, one, two or three dimensions. Denoting the connectivity as AB, ($A = B \in 0, 1, 2, 3$), the zero depicts the total absence of linkage between the particulates. That is, the particles of a given constituent remain discrete and isolated totally delinked from the other particles. When the particles are connected across one, two, or three dimensionally, the corresponding designations are 1, 2, and 3, respectively. Thus, for diphasic constituents, the designation 02, for example, means that one of the materials has particles which are discretely isolated and delinked from each other; and the other material has particulate dispersion with the formation of chain links in two dimensions. Sketched in Figure 13.4 are cubical representations of self-connected systems.

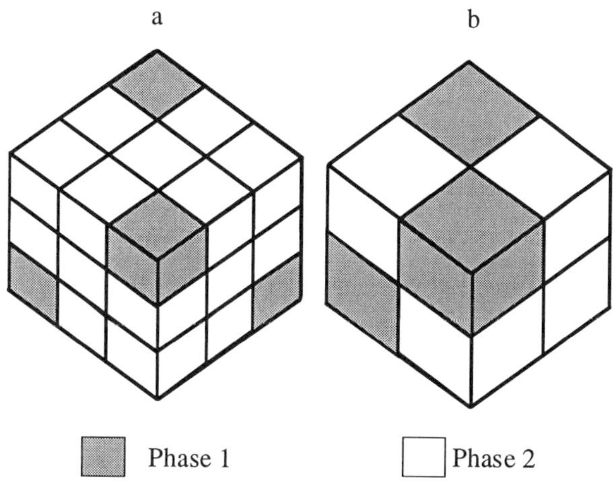

Figure 13.4 Cubic representation of a self-connected diaphasic system. (a) Volume fraction of phase 1 is smaller than the volume fraction of phase 2; (b) Both phases have equal volume fractions.

In view of the self-connecting arrangements of the diphasic constituents, practical structuring of piezoelectric composites is manifold and the schematic diagrams of such various types of structured piezoelectric composites are depicted in Figure 13.5.

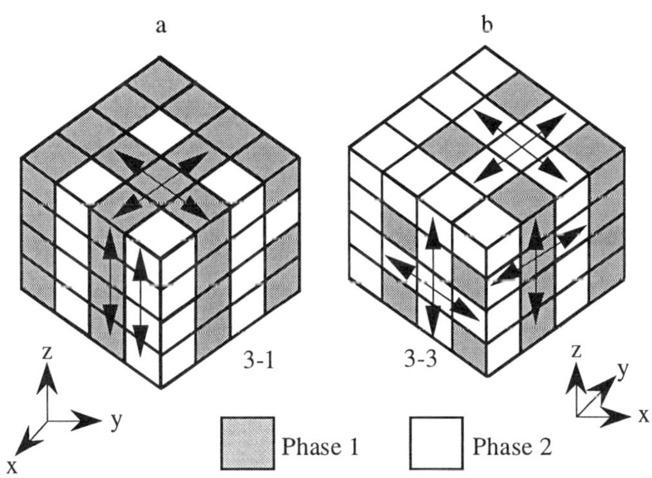

Figure 13.5 Two typical connectivity patterns for a diphasic system. Examples: (a) In the 3-1 composite the shaded phase is three-dimensionally connected and the unshaded phase is one-dimensionally connected. (b) Composite 3-3 shows connectivity pattern. (A set of ten connectivity patterns are illustrated in [11].)

13.3 Fabricational Considerations

13.3.1 *Rod composites*: One of the most popular piezoelectric composites refers to "PZT rods in a polymer" with 1-3 connectivity. It has been identified as a promising element for medical ultrasonic transducer applications. The rod composites are fabricated as follows:

- Slender rods of a piezoelectric ceramic (such as PZT) are aligned in a parallel stack, a polymer is cast between them, and the desired composite link is sliced off (Figure

13.2). The above method is effective for making samples with rod diameters about 200 microns or more. Finer spatial scales are difficult to achieve due to the handling of delicate ceramic rods.

- For spatial scales below 50 microns, a large number of carbon fibers are woven into the desired structure by textile methods and the carbon structure is replicated with piezoelectric ceramic.

- Alternatively, a complementary structure is formed in plastic, and a ceramic slip is injected into this mold and fixed. The plastic mold burns away during the firing, and a polymer is cast back into its place (*lost wax method*). This method yields large area, low cost composites.

- *Dice-and-fill technique*: Figure 13.3 illustrates a widely spread fabrication method for piezoelectric composites. Deep grooves are cut into a solid ceramic and a polymer is cast into these grooves. The resulting composite disk is then sliced off the ceramic base. The dicing operation is feasible for rod dimensions down to 50 microns. Finer spatial scales are possible *via* laser machining to cut the grooves. Laser-induced etching and/or laser ablation may permit scale sizes as low as 10 microns. The dice-and-fill technique is devoid of rod-fragility problems. However, machining and polishing of brittle ceramic and soft polymer combinations may pose engineering problems. Temperature problems and polymer shrinkage effects are other technology factors associated with this method of fabrication.

- *Lamination technique:* Alternate plates of piezoelectric ceramic and a passive material are glued to form a layered stack. Slicing perpendicular to this stack yields a composite thin loaf with 1-3 connectivity. The passive material could be nonpolymeric as well. (Figure 13.6)

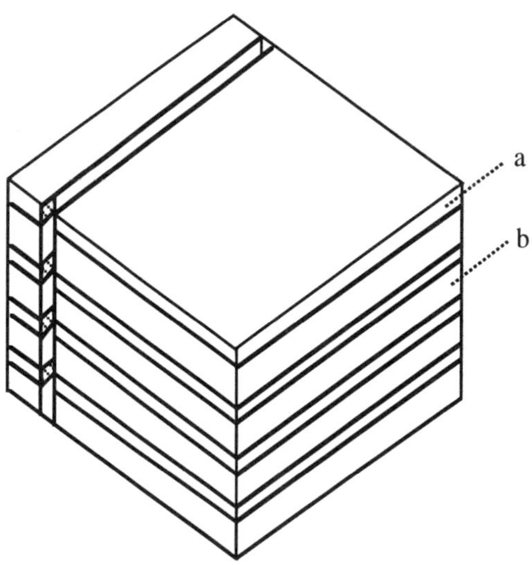

Figure 13.6 Lamination-type composite.
(a) Piezoelectric plate; (b) passive material; (c) a sliced part.

13.4 Flexible Composite Fabrication

Typical flexible composites are :

(a) Gould flexible composite: 5 to 10 μm piezoelectric particles embedded in a polyurethane matrix

(b) Honeywell (T-flex)™ composite: 120 μm piezoelectric composites embedded in a silicone rubber matrix

(c) Honeywell's large-sized piezoelectric embed in a host silicone rubber matrix

In polymer-based flexible composites, the low permittivity polymer layer normally interleaves the piezoelectric particles preventing saturation poling, after the composite is formed. When the piezoelectric particulate bridges the electrodes, it eliminates the poling problem. However, the performance of the composite is controlled by particulate position.

The fabrication procedure involves mixing the piezoelectric ceramic particles (typically Bi_3O_3-modified $PbTiO_3$ or WO_3-modified $Pb(Zr,Ti)O_3$ and synthetic rubber (or chloroprene rubber) and rolling down about 0.5 mm thick sheets at 40°C using a hot roller; and then it is heated at 190°C for 20 minutes under pressure of 30 kg/cm^2. The conductive paste (Fujikura Chemical, Dotite D500™) is attached on both sides of the sheet as electrodes. The specimens are polarized at 60°C in silicone oil by applying a DC field of 100 kV/cm for 1 hour.

Another example of flexible piezoelectric composites is: NTK piezo-rubber™ (NTK Technical Ceramic Division, Japan). It consists of $PbTiO_3$ ceramic powder in chloroprene rubber matrix. It is available in piezo-rubber sheet form and/or as piezo-rubber wire form [12].

Woven PZT ceramic/polymer composite: A *replication process (replamineform process)* is used to fabricate woven PZT/polymer composites. Fabric templates consisting of woven carbon fibers are impregnated with PZT by soaking it in a solution containing stoichiometric amounts of dissolved lead, zirconium, titanium, and niobium. Subsequent heat treatment burns out the carbon, leavning a PZT replica with the same form as the initial carbon weave. To form a composite, replicas are sintered in a controlled atmosphere and backfilled with epoxy polymer. This process has been attempted with as-received activated and non-activated carbon fabrics, as well as those fabrics pretreated with hydrogen peroxide.

Constituting a 3-3 composite by replication process should characterize the end product ideally with the following features: (i) A narrow pore-size distribution; (ii) pore volume closely equal to solid-phase volume; and, (iii) complete pore interconnectivity.

Table 13.1 Characteristics of Piezoelectric Materials

Poled Piezoelectric Materials	Piezoelectric Coefficient 10^{-12} coulomb/newton	Piezoelectric Voltage Coefficient 10^{-3} m^2/coulomb	Remarks
• **Ceramics**			
Barium titanate, $BaTiO_3$	34	2	

(continued...)

Poled Piezoelectric Materials	Piezoelectric Coefficient 10^{-12} coulomb/newton	Piezoelectric Voltage Coefficient 10^{-3} m^2/coulomb	Remarks
Lead niobate, $PbNb_2O_6$	67	34	
Lead zirconate titanate, Pb(Ti, Zn)O_3, PZT	20-50	2-9	
Sodium potassium niobate, (Na,K)NbO_3	40	10	
• **Composites**			
Stycast composite with 25% PZT, by volume	32	25	Under hydrostatic pressures
Spurr composite with 25% PZT, by volume	66	52	
Flexible 33 composite (PZT/$PbTiO_3$)	100	300	
Piezo-rubber	14-48	39-111	

13.5 Application Aspects of Piezoelectric Composites

Merits of piezoelectric composites as electroelastic transducers are:

- Large piezoelectric coefficient (*d* or *g*)
- High electromechanical coupling
- Controllable acoustic impedance
- High sensitivity and compact impulse response
- Complex shape(s) facilitating focused (ultrasonic) excitation reception
- Low cross-talk between electrode arrays
- Low density and mechanical flexibility
- Trade-off optimization in property coefficients of constituent materials
- Fabrication with desirable connectivity (such as thickness node devices)

Typically piezoelectric composite are used in:

- Medical ultrasonic imaging
- Hydrophones

13.6 Theoretical (Design) Considerations

Designing a biphase piezoelectric composite involves the selection of appropriate constituent materials in a proportion such that the end product yields desirable electroelastic characteristics.

The physical properties of the composite material are determined by: (i) The physical properties of the constituent phases; (ii) the volume fractions of the constituents; and, (iii) the structural aspects as decided by the type of connectivity.

Series and parallel arrangements: In a simple diphasic arrangement involving lamellar disposition of the constituents, two types of models, namely, series connection and parallel connection are considered normally (Figure 13.7). These are illustrated in Figures 13.7 corresponding to 3-3 connectivity.

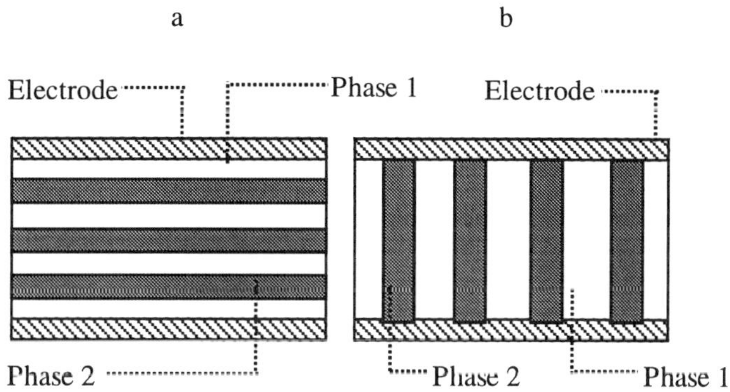

Figure 13.7 Lamellar diphase composite structures interposed between a set of electrodes. a. Series layups of alternate phase 1 (piezoelectric) and phase 2 (nonpiezoelectric) materials; b. Parallel stacking of phase 1 and phase 2 materials.

The effective piezoelastic coefficient *(d_{33})* and effective piezoelectric voltage coefficients of series arrangement are given below:

Effective piezoelectric coefficient:

$$(\bar{d}_{33})_{eff} = \frac{v_1(d_{33})_1(\varepsilon_{33})_2 + v_2(d_{33})_2(\varepsilon_{33})_1}{v_1(\varepsilon_{33})_2 + v_2(\varepsilon_{33})_1} \tag{13.1}$$

Piezoelectric voltage coefficient:

$$(\bar{g}_{33})_{eff} = \left(\frac{d_{33}}{\varepsilon_{33}}\right)_{eff} = \frac{v_1(d_{33})_1}{(\varepsilon_{33})_1} + \frac{v_2(d_{33})_2}{(\varepsilon_{33})_2}$$

$$= v_1(g_{33})_1 + v_2(g_{33})_2 \tag{13.2}$$

Parallel arrangements follow.

Piezoelectric coefficient:

$$(\bar{d}_{33})_{eff} = \frac{v_1(d_{33})_1(s_{33})_2 + v_2(d_{33})_2(s_{33})_1}{v_1(s_{33})_2 + v_2(s_{33})_1} \tag{13.3}$$

Piezoelectric voltage coefficient:

$$(\bar{g}_{33})_{eff} = \frac{v_1(d_{33})_1(s_{33})_2 + v_2(d_{33})_2(s_{33})_1}{[v_1(s_{33})_2 + v_2(s_{33})_1][v_1(\varepsilon_{33})_1 + v_2(\varepsilon_{33})_2]} \tag{13.4}$$

Hydrostatic sensitivity is shown below.

For the parallel connection as specified in Figure 13.7b:

$$(d_h)_{eff} \cong [\frac{v_1(d_{33})_1(s_{33})_2 + v_2(d_{33})_2(s_{11})_1}{v_1(s_{33})_2 + v_2(s_{33})_1}] + 2[v_1(d_{31})_1 + v_2(d_{31})_2] \tag{13.5}$$

where subscripts 1 and 2 indicate phases 1 and 2, respectively; and subscript 33 refers to 3-3 connectivity. Further, the other entities indicated represent:

v : volume fraction
d : piezoelectric coefficient
ε : dielectric permittivity
s : elastic compliance

Generalized formulations to calculate the properties of a 0-3 piezoelectric composite (modified cube model due to Banno and Saito [7]) follow.

Dielectric constants ($\bar{\varepsilon}_{33}$, $\bar{\varepsilon}_{22}$, $\bar{\varepsilon}_{11}$):

$$\bar{\varepsilon}_{33} = a^2 [a + (1-a)n]^2 (\varepsilon_{33})_1(\varepsilon_{33})_2/\{a(\varepsilon_{33})_2 + (1-a)n\,(\varepsilon_{33})_1\} + \{1 - a^2 [a + (1-a)n]\}(\varepsilon_{33})_2 \tag{13.6}$$

Equations for $\bar{\varepsilon}_{11}$ and $\bar{\varepsilon}_{22}$ can be obtained by substituting ε_{11} and ε_{22} for ε_{33}, ℓ and m for n, respectively, in Equation 13.6.

Elastic constants ($\bar{s}_{33}$, $\bar{s}_{22}$, $\bar{s}_{11}$):

$$1/\bar{s}_{33} = a^2 [a + (1-a)n]^2/\{a(s_{33})_1 + (1-a)n\,(s_{33})_2\} + \{1 - a^2 [a + (1-a)n]\}/(s_{33})_2 \tag{13.7}$$

Equations for $\bar{s}_{11}$ and $\bar{s}_{22}$ can be can be obtained by substituting s_{11} and s_{22} for s_{33}, ℓ and m for n, respectively, in Equation 13.7.

Poisson's ratio ($\bar{\sigma}_{31}$, $\bar{\sigma}_{32}$, $\bar{\sigma}_{12}$, $\bar{\sigma}_{13}$):

$$\bar{\sigma}_{31} = a^2 [a + (1-a)n]\{a\,(s_{33})_1(\sigma_{31})_1 + (1-a)n\,\,(s_{33})_2(\sigma_{31})_2\} [\{a\,\,(s_{33})_1 + (1-a)n\,\,(s_{33})_2\} + \{1 - a^2\,\,[a + (1-a)n]\}(\sigma_{31})_2]^{-1} \tag{13.8}$$

An equation for $\bar{\sigma}_{32}$ can be obtained by substituting σ_{32} for σ_{31} in Equation 13.8.

$$\overline{\sigma}_{12} = a^2\,[a + (1-a)\ell]\,\{a\ (s_{11})_1(\sigma_{12})_1 + (1-a)\ell\ \ (s_{11})_2(\sigma_{12})_2\}$$
$$[\{a\,(s_{11})_1 + (1-a)\ell\,(s_{11})_2\} + \{1 - a^2[a + (1-a)\ell]\}\,(\sigma_{12})_2]^{-1} \quad (13.9)$$

An equation for $\overline{\sigma}_{13}$ can be obtained by substituting σ_{13} for σ_{12} in Equation 13.9.

Dielectric loss tangent ($tan\ \overline{\delta}_{33}$, $tan\ \overline{\delta}_{22}$, $tan\ \overline{\delta}_{11}$):

$$tan\ \overline{\delta}_{33} = A_1/B_1 \quad (13.10)$$

where

$$A_1 = a^2\,(\varepsilon_{33})_2\{a(\varepsilon_{33})_1(tan\delta_{33})_1 + (1-a)n\,(\varepsilon_{33})_2(tan\delta_{33})_2\}$$
$$\times\{a/(\varepsilon_{33})_1 + (1-a)n/(\varepsilon_{33})_2\} + \{1 - a^2[a + (1-a)n]\}$$
$$\times\{a\,(\varepsilon_{33})_1 + (1-a)n(\varepsilon_{33})_2\}(tan\delta_{33})_2 \quad (13.11a)$$

$$B_1 = a^2\,(\varepsilon_{33})_2\{a(\varepsilon_{33})_1 + (1-a)n(\varepsilon_{33})_2\}\{a/(\varepsilon_{33})_1 + (1-a)n/(\varepsilon_{33})_2\}$$
$$+ \{1 - a^2[a + (1-a)\,n]\}\{a(\varepsilon_{33})_1 + (1-a)n(\varepsilon_{33})_2\} \quad (13.11b)$$

Equations for $tan\ \overline{\delta}_{11}$ and $tan\ \overline{\delta}_{22}$ can be obtained by substituting $tan\ \delta_{11}$ and $tan\ \delta_{22}$ for $tan\ \delta_{33}$, ε_{11} and ε_{22}, for ε_{33}, ℓ and m for n, respectively, in Equation 13.11.

Mechanical loss tangent ($tan\ \overline{\delta}_{m\,33}$, $tan\ \overline{\delta}_{m\,22}$, $tan\ \overline{\delta}_{m\,11}$):

Equations for mechanical loss tangent can be obtained by substituting s for ε, $tan\ \delta_m$ for $tan\ \delta$ in Equation 13.11.

Piezoelectric constants $\overline{d}_{31}$, $\overline{d}_{32}$, $\overline{d}_{12}$, $\overline{d}_{13}$:

$$\overline{d}_{33} = A_2/B_2$$

where

$$A_2 = a^2[a + (1-a)n]^2(S_{33})_2\{a(d_{33})_1(\varepsilon_{33})_2 + (1-a)n(d_{33})_2(\varepsilon_{33})_1\}$$
$$+ \{1 - a^2[a + (1-a)n]\}\{a(S_{33})_1 + (1-a)n\,(S_{33})_2\}$$
$$\times\{a(\varepsilon_{33})_2 + (1-a)n\,(\varepsilon_{33})_1\}(d_{33})_2 \quad (13.12a)$$

$$B_2 = a^2[a + (1-a)n]^2(S_{33})_2\{a(\varepsilon_{33})_2 + (1-a)n(\varepsilon_{33})_1\}$$
$$+ \{1 - a^2[a + (1-a)n]\}\{a(S_{33})_1 + (1-a)n(S_{33})_2\}$$
$$\times\{a(\varepsilon_{33})_2 + (1-a)n(\varepsilon_{33})_1\} \quad (13.12b)$$

$$\overline{d}_{31} = A_3 + B_3 + C_3 \quad (13.13)$$

where

$$A_3 = \{a^2[a + (1-a)n]/[a(\varepsilon_{33})_2 + (1-a)n\,(\varepsilon_{33})_1]\}$$
$$\times\{a(d_{31})_1(\varepsilon_{33})_2[a + (1-a)\ell]/[a((S_{11})_1/\ \overline{S}_{11})$$
$$+ (1-a)\ell(S_{11})_2/\ \overline{S}_{11})]\} + (1-a)n(d_{31})_2(\varepsilon_{33})_1/(S_{11})_2/\ \overline{S}_{11})\} \quad (13.14a)$$

$$B_3 = a^2(1-a)\ell[a + (1-a)\ell](d_{31})_2/\{a((S_{11})_1/\ \overline{S}_{11}))$$

$$+ (1-a)\ell((S_{11})_2/\bar{S}_{11})\} \quad (13.14b)$$

$$C_3 = \{(1-a)m/[a+(1-a)m] + a(1-a)\ell\,(1-a)n\,\} \times (d_{31})_2/((S_{11})_2/\bar{S}_{11}) \quad (13.14c)$$

An equation for $\bar{d}_{32}$ can be obtained by substituting d_{32} for d_{31}, S_{22} for S_{11}, ℓ for m, m for ℓ in Equation 13.13, where volume fraction of phase 1, namely, $(v)_1$ is given by:

$$v_1 = a^3 \quad (13.15)$$

Ceramic piezoelectrics with pores (modified cube model due to Banno and Saito [7]):

$$\bar{d}_{33} = (d_{33})_2 \quad (13.16a)$$

$$\bar{d}_{31} = (d_{31})_2\{1 - a/K_s^{1/3} + (a/K_s^{1/3})(1 - a/K_s^{1/3})/(1 + a^2 K_s^{1/3})\} \quad (13.16b)$$

$$\bar{d}_h = d_{33} + 2d_{31} \quad (13.16c)$$

$$\bar{d}_h/\bar{d}_{33} = 1 - (-2(d_{31})_2/(d_{33})_2)\{1 - a/K_s^{1/3} + (a/K_s^{1/3})(1 - a/K_s^{1/3})/(1 + a^2 K_s^{1/3})\} \quad (13.16d)$$

$$\bar{d}_h/(d_h)_2 = 1 + (-2(d_{31})_2/(d_{33})_2)\{(a/K_s^{1/3}) - (a/K_s^{1/3})(1 - a/K_s^{1/3})\} \times [(1 - (-2(d_{31})_2/(d_{33})_2)]^{-1} \quad (13.16e)$$

$$\bar{\varepsilon}_{33}/(\varepsilon_{33})_2 = 1 - (a/K_s^{1/3})^2\{1 - 1/(a((\varepsilon_{33})_2/\varepsilon_0 - 1) + 1)\} \quad (13.16f)$$

where the volume fraction of pore v_1 is a^3 and the notation K_s is a parameter attributed to pore shape of phase 1. When phase 1 pores are cubic or flat tetragonal, K_s becomes unity or less than unity, respectively (Figure 13.8).

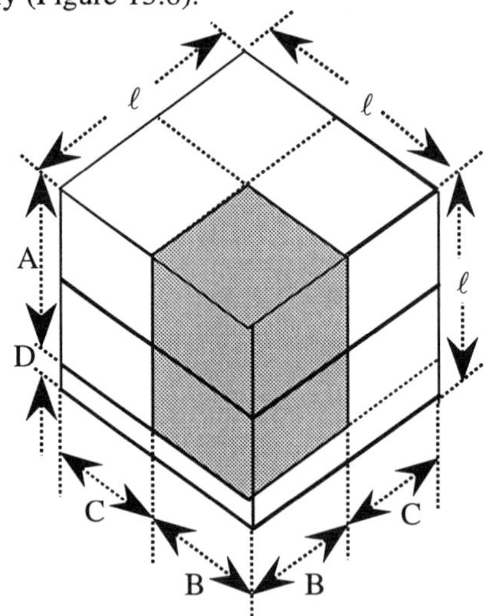

Figure 13.8 Shaped factor of the phase 1 in modified cubes model.
A: a/[a + (1 – a)n].
B: a/[a + (1 – a)m].
C: (1 – a)m/[a + (1 – a)m].
D: (1 – a)n/[a + (1 – a)n].
Shape factor of phase 1: K_s = A/B.

Using K_s and assuming that the relative dielectric constant of the pore is unity, theoretical equations of the dielectric and piezoelectric $\bar{\varepsilon}_{31}$ constants of the porous ceramics are obtained as follows:

$$\bar{\varepsilon}_{33} = (\varepsilon_{33})_2 \left\{ 1 + \frac{1}{a((\varepsilon_{33})_2 - 1)K_s^{2/3} + 1K_s^{2/3}} \frac{a^2}{} - \frac{a^2}{K_s^{2/3}} \right\} \tag{13.17a}$$

$$\bar{d}_{31} = (d_{31})_2 \left\{ 1 - \frac{a}{K_s^{1/3}} + \frac{\frac{a}{K_s^{1/3}}\left(1 - \frac{a}{K_s^{1/3}}\right)}{1 - a^2 K_s^{1/3}} \right\} \tag{13.17b}$$

where $(\varepsilon_{33})_2$ and $(d_{31})_2$ are the relative dielectric and piezoelectric constants of the bulk ceramic material, respectively, and a^3 $(= v_1)$ is the volume fraction of the pore.

Binary piezoelectric composite with a continuous dielectric host medium and piezoelectric ellipsoidal inclusions: An example of this system refers to a mixture of lead zirconate titanate (PZT) particles dispensed in a poly-vinylidene fluoride (PVDF) receptacle. Relevant formulations to calculate the effective mixture parameters follow.

Dielectric constant:

$$\varepsilon_{eff} = \varepsilon \left[1 + [n\theta(\varepsilon_2 - \varepsilon_1)]/[n\varepsilon_1 + (\varepsilon_2 - \varepsilon_1)(1 - \theta)] \right] \tag{13.18}$$

where $n = 4\pi/\text{m}$ is a parameter attributed to the shape of the ellipsoidal particles and θ is the volume fraction of the ellipsoidal particles; and

$$m = \int_0^{\infty} du/[(a^2 + u)\, \beta_u]$$

$$\beta_u = \{[(a^2 + u)\,(b^2 + u)\,(c^2 + u)]^{1/2}\}/2\pi abc$$

a,b,c : Semiaxial lengths of the ellipsoidal particles.
ε_1 : Dielectric permittivity of the host medium.
ε_2 : Dielectric permittivity of the ellipsoidal inclusions.

Piezoelectric constant:

$$d_{eff} = \theta\, \alpha\, G\, d_2 \tag{13.19}$$

where

$$G = n\left\{(\varepsilon_2/\varepsilon_1) - 1 + n + (n-1)[(\varepsilon_2/\varepsilon_1) - 1]\theta\right\} / \left\{[(\varepsilon_2/\varepsilon_1) - 1 + n]^2 + [(\varepsilon_2/\varepsilon_1) - 1][(n-1)^2 - (\varepsilon_2/\varepsilon_1)]\theta\right\}$$

α = Poling ratio
d_2 = Piezoelectric constants of the piezoelectric particles.

Elastic constant:

$$E_{eff} = E_1 \{1 + [\theta (E_2 - E_1)]/[E_1 + n'(E_2 - E_1)(1 - \theta)]\} \quad (13.20)$$

E : Young's modulus
$1,2$: Subscripts to denote the host medium and the dispensed particles
n' : $(1/3)(1 + \sigma)/(1 - \sigma)$
σ = Poisson's ratio of the host medium

Neuromimetic model [8]: In [8], the author and Park had developed a neuromimetic model to describe the electroelastic synergism in piezoelectric composites. That is, the electroelastic response of a piezoelectric composite material (constituted by piezoelectric particulate dispersions in a nonpiezoelectric receptacle) is emulated analogous to the collective response of randomly interconnected neurons. By comparing the stochastic aspects of both systems, the effective parameters of the piezoelectric composite are deduced. Theoretical results on the effective piezoelectric coefficients of two types of test composites are compared with the relevant available experimental data. The neuromimetic concept envisaged facilitates a understanding of the behavior of advanced piezoelectric composites *vis-a-vis* the constituent materials of such composites. Relevant study also refers to the effects of the size, shape, volume fraction, and orientation of the inclusions and the characteristics of the host medium in deciding the net electroelastic response of the composite. The strategy presented in [8] indicates a neural network approach in studying such composites constituted by randomly dispersed interacting inclusions.

On the basis of the above considerations, the theoretic formulations obtained are summarized below.

For dispension of shaped piezoelectric particles with arbitrary orientational dispositions, the effective piezoelectric coefficient is given by :

$$d_{eff} = \{(d_{eff})_p\}^u / \{(d_{eff})_s\}^{u-1} \quad (13.21)$$

where

$$(d_{eff})_p = \{d_2/[1 + \frac{s_1(1-\theta)}{s_2 \ \theta}] + d_2/[1 + \frac{s_2 \ \theta}{s_1(1-\theta)}]\}$$

and 1, 2 are subscripts denoting the media 1 and 2, respectively. Further,

d : Piezoelectric coefficient
ε : Dielectric permittivity
θ : Volume fraction of medium 1
s_1 : Elastic compliance
u $= (u_S + u_G)^{1/2}$
u_G : Order parameter decided by the particulate shape; $1/3 \leq u_G(a/b) \leq 1$
u_S : Order parameter decided by the particulate (an isotropic) spatial orientation $1/3 \leq u_S \leq 1$
a/b : Ratio of semiaxial lengths of the spheroidal inclusions

$$u_G = [1 - \frac{m^R}{\sum_{\ell=0}^{R} m_o^{\ell}}]/[1 - \frac{m_\infty^R}{\sum_{\ell=0}^{R} m_o^{\ell}}]$$

R = Nearest integer of [0.5 + 0.5 (*a/b*)], if $a/b > 1$
= Nearest integer of [0.5 + 0.5 (*b/a*)], if $a/b < 1$

m $= e^2[(1 - \sqrt{1 - e^2}\, arcsin(e)/e)^{-1}$; (Sillars' shape parameter)

e : Particulate eccentricity

$$= \begin{cases} [1 - (b/a)];\ a > b \\ [(a/b) - 1];\ a < b \end{cases}$$

m_o = 3

m_∞ = 1

u_S $= <cos^2\alpha>$, where α refers to the angle of preferential orientation (or degree of anisotropicity) in the spatial dispositions of the particles.

α = 1/3: Totally isotropic random dispension of the particulate inclusions.
= 1: Totally anisotropic (parallel or antiparallel arrangement of the particles).

13.7 Experimental Data on Piezoelectric Composites

Figures 13.9–13.16 depict typically the variations of the characteristics of piezoelectric composites with respect to temperature, volume fraction of piezoelectric material, poling conditions and adding of another piezoelectric materials.

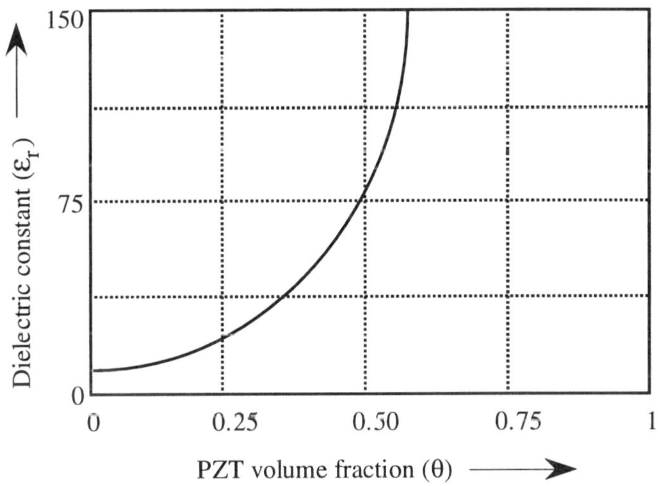

Figure 13.9 Dielectric constant (ε_r) *versus* volume fraction (θ) of PZT content in a piezoelectric composite at room temperature.

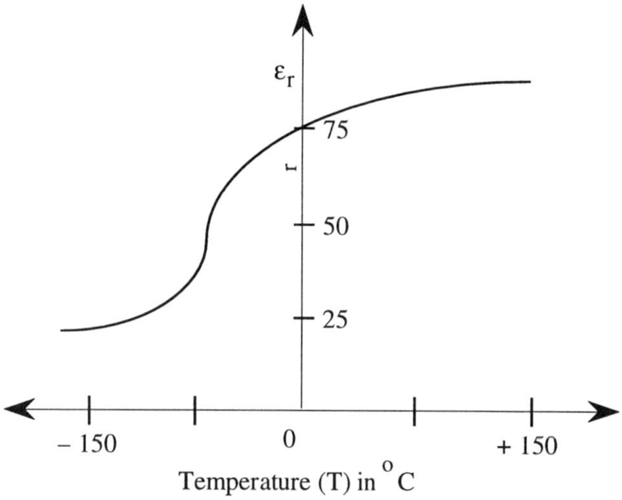

Figure 13.10 Temperature (T) *versus* dielectric constant (ε_r) of a PZT-included piezoelectric composite. PZT volume fraction (θ) ≈ 0.5.

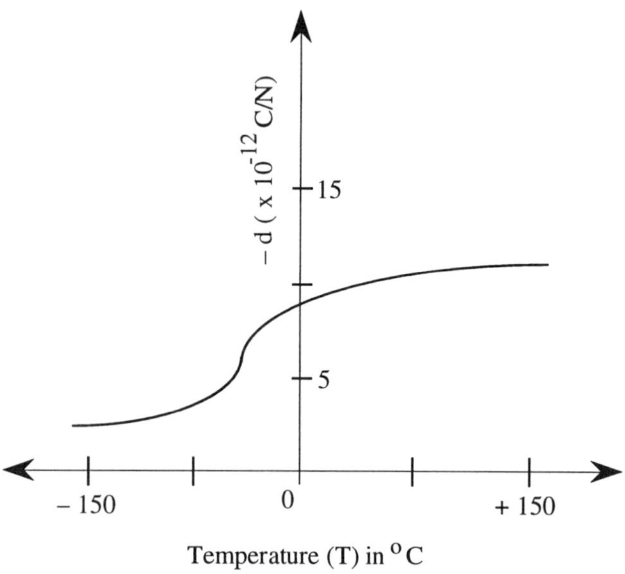

Figure 13.11 Temperature (T) *versus* piezoelectric coefficient (d) of a PZT-included piezoelectric composite. Volume fraction of PZT (θ) ≈ 0.5.

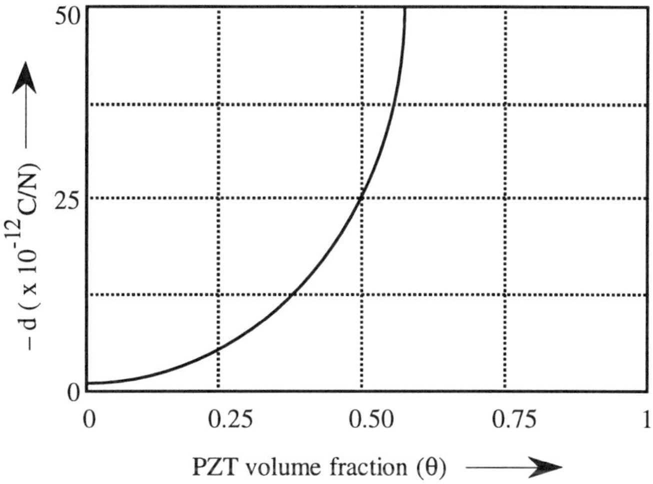

Figure 13.12 PZT volume fraction (θ) *versus* piezoelectric coefficient (d) of a PZT-included piezoelectric composite at room temperature.

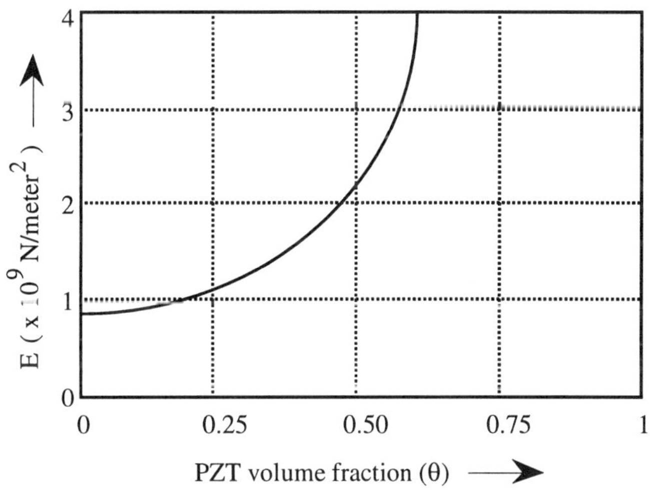

Figure 13.13 PZT volume fraction (θ) *versus* elastic modulus (Young's modulus E) of a PZT-included piezoelectric composite at room temperature.

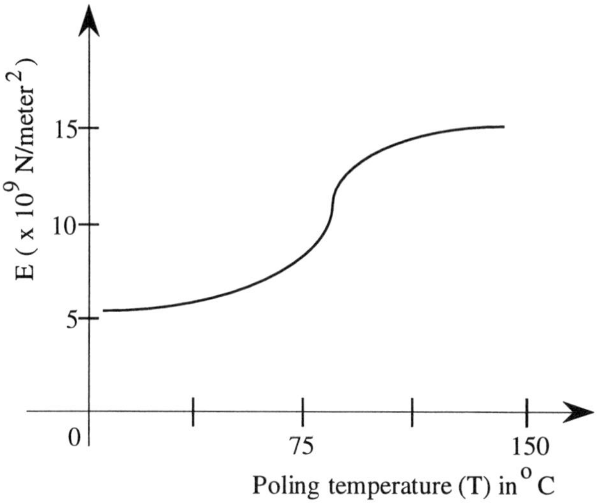

Figure 13.14 Poling temperature (T) *versus* piezoelectric coefficient (d) of PZT-included piezoelectric composite. (PZT volume fraction $\theta \approx 0.5$; poling electric field (ξ): 10×10^9 volt/meter; Duration of poling: 120 minutes.)

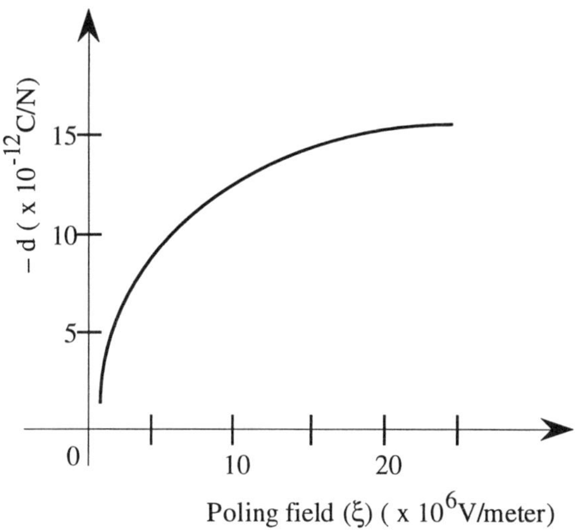

Figure 13.15 Poling field (ξ) *versus* piezoelectric coefficient (d) of a PZT-included piezoelectric composite. ($\theta \approx 0.5$; Poling temperature: 100°C; and poling time: 120 minutes.)

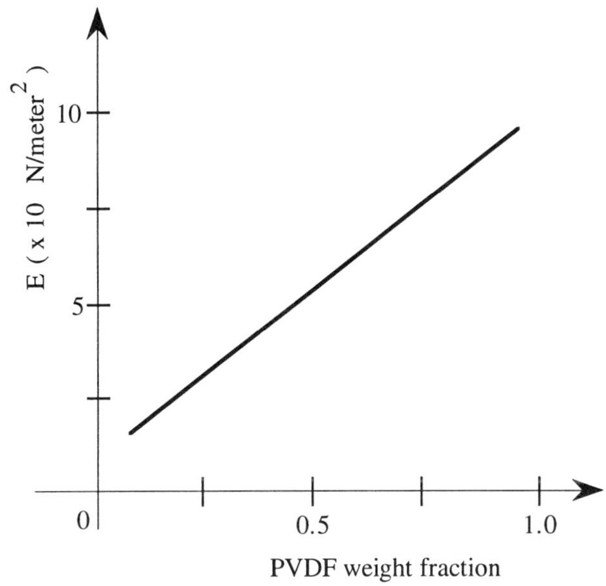

Figure 13.16 Weight fraction of PVDF *versus* Young's modulus (E) of a PVDF-based piezoelectric composite at room temperature. (Volume fraction of PVDF (θ) ≈ 0.7.)

References

[1] J. Wolak: Dielectric behavior of 03-type piezoelectric composites. IEEE Trans. Elec. Insulation, vol. 28(1), 1993: 116-121.

[2] H. Zewdie and F. Brouers: Theory of ferroelectric polymer-ceramic composites. J. Appl. Phys., vol. 68, 1990: 713-718.

[3] M. Chino et al.: Microwave absorbers using ferroelectric/rubber composite structure and their evaluation. Ferroelectrics, vol. 93, 1989: 67-71.

[4] H. Banno: Theoretical equations for dielectric, piezoelectric and elastic properties of flexible composite consisting of polymer and ceramic powder of two different materials. Ferroelectrics, vol. 95, 1989: 111-115.

[5] H. Banno: Recent progress in science and technology of flexible piezoelectric composite in Japan, Proc.7th International Symp. on Application of Ferroelectrics, pp. 67-72.

[6] W. A. Smith: The role of piezocomposites in ultrasonic transducers, Proc. 1989 Ultrasonics Symp., pp. 755-766.

[7] H. Banno and S. Saito: Piezoelectric and dielectric properties of composites of synthetic rubber and $PbTiO_3$ or PZT. Japanese J. Appl. Phys., vol. 22 (Supplement 22-2), 1983: 67-69.

[8] P. S. Neelakanta and J. C. Park: Neuromimetic model of electroelastic synergism in piezoelectric composites. Biomimetics, vol. 2(1), 1993: 33-56.

[9] A. A. Shanlov, W. A. Smith and R. Y. Ting: Modified lead-titanate/polymer composites for hydrophone applications. Ferroelectrics, vol. 93, 1989: 177-182.

[10] D. P. Skinner, R. F. Newham and L. E. Cross: Flexible composite transducers. Mat. Res. Bull., vol. 13, 1978: 599-607.

[11] R. E. Newham, D. P. Skinner and L. E. Cross: Connectivity and piezoelectric-pyloelectric composites. Mat. Res. Bull., vol. 13, 1978: 525-536.

[12] Data sheet: NTK Piezoelectric Rubber. (NTK Technical Ceramic Division), NGK Spark Plug Co. Ltd., Mihuho, Nagoya, 467 Japan).

Defining Terms

Connectivity: Refers to the pattern in which the diphasic or multiphasic constituents in a anisotropic composite are self-connected in zero, one, two, or three dimensions.

Dielectric constant: Relative permittivity of the medium; in the anisotropic case as in piezoelectric composites, is a tensor parameter.

Dielectric loss tangent: Depicts the lossy nature of a dielectric (monolithic or composite) material.

Dice and fill process: Process in which deep grooves are cut in a piezoelectric ceramic and grooves are filled with a polymer to realize a piezoelectric composite.

Elastic constants: Young's bulk or shear modulii of elasticity tensor parameters; in anisotropic materials such as piezoelectric composites.

Hydrophone: An underwater acoustical transducer.

Laminated piezoelectric composite: Alternate stacking of piezoelectric and nonpiezoelectric materials.

Neuromimetic model: Behavior of a material mimicking the neuronal state transitional characteristics.

Piezoelectric composites: A combination of a piezoelectric-ceramic and/or a polymer and a nonpiezoelectric polymer.

Piezoelectric constant: Expresses the polarization along a particular direction produced by an elastic strain in a monolithic or composite piezoelectric material.

Piezoelectric coefficient (d or g): The piezoelectric stress per charge induced in a monolithic and/or composite piezoelectric material.

Poling field: Minimum electric field (kV/m) required to polarize a monolithic or composite piezoelectric material.

Poling temperature: Temperature at which piezoelectric poling is done.

Replication process (replamineform process): Fabric templates consisting of woven carbon fibers impregnated with PZT and subsequent heat-treatment burns out carbon, leaving behind PZT.

Rod composites: Rod-like piezoelectric material embedded in a polymer or ceramic with 1-3 connectivity.

Rubber piezoelectric composites: Flexible piezoelectric composites constituted by piezoelectric inclusions in a rubber material.

Ultrasonic imaging: Using ultrasonic transducers, reflection of ultrasonic energy from a body which is processed to "image" the body irradiated with ultrasonics.

Woven ceramic/polymer composites: Piezoelectric composites with interwoven dispersion of piezoelectric inclusions obtained by refraction or replamineform process.

Table 13.2 Comparison of Physical Properties of Typical Piezoelectric Composites

Materials	Parameters									
	K_{33}	d_{33} (pc/N)	d_h (pc/N)	g_h ($\times 10^{-3}$ Vm/N)	$d_h g_h$ ($\times 10^{-5}$ m^2)	d_{31} (pc/N)	ε_r	tanδ	PF* KV/m	Density
PZT single phase	1760	450	42	2.7	113	–204				
PZT 1-3 composite	22	217	10	52	50-520	–83				
PZT 1-3-0 composite 20% void	110	270	220	220	228	–25				
PZT 1-3-0 composite 30% void	24 - 25	225	60-100	295-446						
PZT 1-3-0 composite 40% void	110	310	284	285	94	–13				
PZT plus polymer (Spurr epoxy)		25	6	18	108		38	0.017	66	
PZT plus Eccogel		90	30	108	2100		48	0.25	40	
Piezo-rubber (NTK)		34-56	17-44	45-124	765-5084	–2.5- –18.5	43-45 (ε_{33})	2 - 6		5.3-5.9
PZT 3-3 composite with silicone rubber		100		300 (g_{33})			40			3.3
Calcium modified lead titanate plus polymer (Spurr epoxy or Stycast)		49-59	32-25	66-52	2100- 1300	–8.5- –17.0			54-55	

*PF: Poling Field.

CHAPTER 14

Ferromagnetic Materials

14.1 Introduction

Ferromagnetic materials have the ability to carry *high magnetic flux*. *Ferromagnetism* refers to the property of increasing the magnetic flux associated with the material when a *magnetizing force* is applied, but there exists a *saturation point* for most of the magnetic materials beyond which the associated magnetic flux does not increase. This condition is referred to as *magnetic saturation*.

Ferromagnetic materials can be distinctly classified into two groups, the *magnetically soft* and *magnetically hard*. The distinguishing properties are having high permeability and having easy to magnetize for soft materials and hard materials having high *coercivity* so that once magnetized the materials must be able to resist demagnetizing forces due to any applied or stray magnetic fields. Soft magnetic materials are used commonly in applications such as computer memory cores, television receivers, communication and radio components, microwave components, and magneto-optic devices. Hard magnetic materials find use in audio/video recording, energy conversion, electron flow control, and similar applications.

14.2 Classification of Magnetism

In modern terms, magnetic materials can be classified broadly on the basis of the atomic structure of the material and values of relative permeability as follows:

- *Diamagnetic materials* with relative permeability slightly less than unity.
- *Paramgnetic and antiferromagnetic materials* (like MnO, FeO) with relative permeabilities slightly greater than unity.
- *Ferromagnetic and ferrimagnetic materials* (like Fe_3O_4) with relative permeability appreciably greater than unity.

Para- and diamagnetic materials automatically come under nonferromagnetic materials and these two magnetic properties are substantially independent of the applied magnetizing force.

Tables (14.1 and 14.2) show a broad comparison between the aforesaid classes of magnetic materials and Table 3 provides descriptions of these classes.

Table 14.1 Comparison Chart between Dia-, Para- and Ferromagnetic Materials

Properties	Diamagnetic	Paramagnetic	Ferromagnetic
Susceptibility	Small and negative	Small and positive	Large and positive
Permeability	< 1	> 1	>> 1
Magnetic spin alignment	Opposite	Spins too far apart	Same direction

(continued...)

Properties	Diamagnetic	Paramagnetic	Ferromagnetic
Susceptibility *versus* temperature	Independent	Inversely proportional (Curie's law)	Above Curie point similar to paramagnetics
Examples	Cu, He	Na, Al	Fe, Co, Ni

Table 14.2 Classification of Magnetic Materials Based on Arrangements of Dipole Moments or Spins

Type	Susceptibility χ_m	χ_m *versus* Temp. (T)	Examples	Arrangment of Dipole Moments or Spins
Diamagnetic	~ -10^{-6} (Negative)	Independent	Cu,Ag,Au, Ge etc.	
Paramagnetic	~ $+10^{-3}$ (Positive)	$\chi_m = C/T$ (Curie law) $\chi_m = C/(T - T_c)$ (Curie-Weiss law) C: Curie constant T_c = Paramagnetic Curie Temp.	Materials having atoms with odd numbers of electrons; ionic crystals etc.	
Ferromagnetic	Extremely large and positive	$\chi_m \rightarrow \infty$	Fe, Co, Ni, Gd	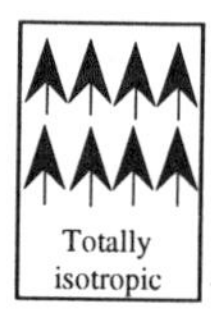
Antiferromagnetic	Small and positive	$\chi_m \propto 1/T$	Salts and oxides of transition metals (e.g. NiO, MnF_2)	
Ferrimagnetic	Large and positive	$\chi_m \rightarrow \infty$	Ferrites (e.g. Fe_2O_3)	

(continued...)

Type	Susceptibility χ_m	χ_m *versus* Temp. (T)	Examples	Arrangment of Dipole Moments or Spins

Note: Types of interaction between magnetic dipoles:
Paramagnetism : No long range order of interaction.
Ferromagnetism : Positive exchange interaction.
Antiferromagnetism : Negative exchange interaction.
Ferrimagnetism : Negative interaction between unequal moments.

Table 14.3 Descrption of Magnetic Behavior of Materials

Type	Description
Diamagnetism	A diamagnetic material exhibits a negative magnetism. Its magnetic susceptibility is independent of external magnetic field and temperature. Its relative permeability $\mu_r \approx 1$. Its magnetic susceptance is given by : $\lambda_m = (-Ne^2/4m)r^2\mu_0$ where N = number of dipoles induced per unit volume; (e, m) are electronic charge and mass, respectively; r is the atomic radius; and μ_0 : permeability of free space. Diamagnetic materials being nonmagnetic are of no use as magnetic materials.
Paramagnetism	When the permanent magnetic moments (resulting from the electric current due to electronic spin, orbital motion, and nuclear spin angular momentum) in the atoms or ions do not mutually interact and are dispersed with isotropic randomness, the corresponding materials are known as paramagnetic materials. Though an external magnetic field may tend to polarize the random moments, the thermal agitation would oppose this causing only a very small (partial) magnetization (quantified in terms of number of dipoles per unit volume aligned in the direction of external field). Therefore, temperature plays a dominant role in deciding the magnetization as per Curie or Curie-Weiss law. Curie law : $\chi_m = (NP_m^2 \mu_0 H)/ k_B T$ where N : Number of dipoles per unit volume p_m : Magnetic moment per spin μ_0 : Free-space permeability H : Applied magnetic field intensity k_B : Boltzmann constant T : Temperature C : $(NP_m^2 \mu_0 H/ k_B) \cong$ Curie constant Some materials do not follow Curie law but obey a modified law (Curie-Weiss law):

$\chi_m = C/(T - T_{pc})$

T_{pc} = Paramagnetic Curie temperature

	Materials		
	Fe	Co	Ni
T_{pc} °K	1093	1428	650

Ferromagnetism

This is characterized by the presence of parallel orientation of magnetic dipole moments. It occurs in materials with partly filled inner electron shells. In ferromagnetic materials, magnetization is spontaneous; that is, below a certain critical temperature (T_c, known as Curie temperature), the material exhibits spontaneous magnetization as per Curie-Weiss law, $\chi_m = C/(T - T_i)$

	Materials				
	Fe	Co	Ni	Gd	Dy
T_c °K	1043	1393	631	289	105

Below T_c, ferromagnetic materials follow the B-H hysteresis over magnetization cycle. In single crystal magnetic materials (such as iron), the magnetic properties depend on the direction in which they are measured. That is, iron exhibits preferred directions of magnetization in *'easy'* directions with reference to the crystal axis. This is known as the *magnetocrystalline anistropy.*

When a ferromagnetic material is magnetized, changes in physical dimensions, in general, occur. This is known as *magnetostriction.* It could be a longitudinal, transverse, or volumetric dimensional change. The dimensional change occurring along the direction of induced magnetic field is called *Joule effect magnetostriction.* The converse of magnetostriction is known as the *Villari effect.*

Antiferromagnetism

Typically, antiferromagnetic materials show a small positive dependence on temperature as illustrated

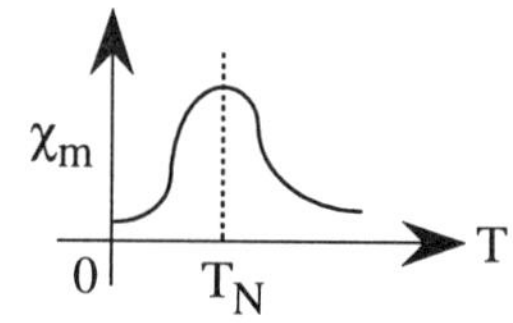

T_N : Neel temperature

	Materials			
	MnO_2	MnO	FeO	CoO
T_N °K	84	122	198	292
T_{pc} °K	316	610	570	280

Ferrimagnetism

Also known as ferrites, ferrimagnetic materials have nonzero net magnetization of magnetic sublattices. Hence they possess a net magnetic moment which disappears at $T > T_c$ due to thermal energy forcing (randomizing) individual magnetic moments.

Ferrites have high electrical resistivity. They are in general complex oxide compounds of various metals. They are derived from *magnetite* by replacing the divalent iron atom by an atom of another metal ($Fe^{++}O$, $Fe_2^{+++}O_3$). Many ferrites are isomorphic with the mineral spinel structure $MOFe_2O_3$ with M being a divalent metal. (e.g. stoichiometric formula: $\alpha NiO \cdot \beta ZnOFe_2O_3 \cdot \gamma H_2O$). Rare-earth ferrites called *garnets* are: $xM_2O_3 \cdot yFe_2O_3$, M being a rare-earth material like yttrium. The dielectric constant of ferrites is 10~12 at microwave frequencies with low loss tangent. Permeability of ferrites is normally on the order of several tens. Mechanically, ferrites are hard brittle and difficult to machine.

14.3 Magnetization Curves and Hysteresis Loop

Figure 14.1 shows the typical magnetization curves for the above classes of magnetic materials.

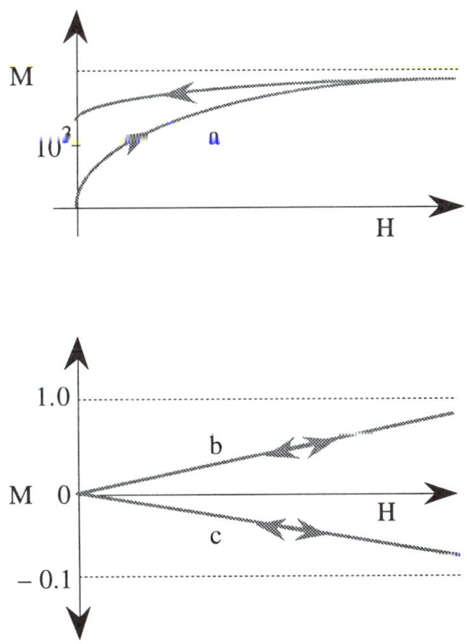

Figure 14.1 Magnetization curves of (a) diamagnetic, (b) paramagnetic, and (c) ferromagnetic materials.

In magnetic applications, ferro- and ferrimagnetic materials are of prime importance. Figure 14.2 shows the magnetization curves for mild steel, cobalt and nickel iron. Cobalt and nickel are typical ferromagnetic materials widely used in EM applications.

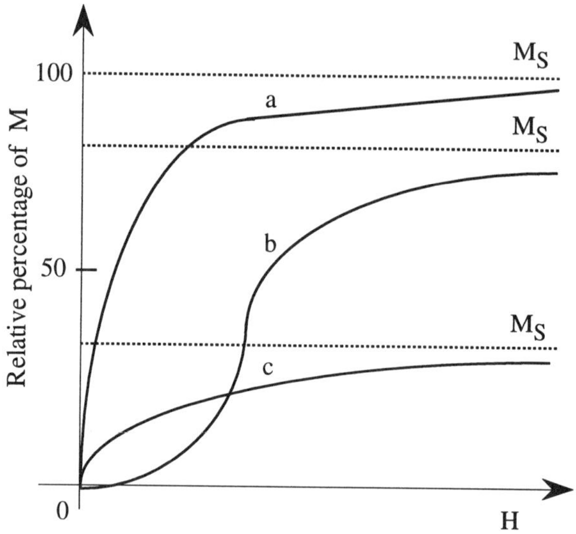

Figure 14.2 Magnetization curves of ferromagnetic materials. (a) Mild steel; (b) Cobalt; (c) Nickel.

The flux density *(B)* increases linearly with magnetizing force *(H)* in the case of air, but for other materials, the *B-H curve* is nonlinear. A comparison of the contrasting behavior between ferromagnetic materials is best seen in their magnetizing curves which also indicate how much magnetizing force would be needed to reach the magnetic saturation.

Both ferro- and ferrimagnetic materials differ widely in the ease with which they can be magnetized. Soft magnetic materials require a small applied field to produce saturation whereas hard magnetic materials require large field strengths. Figure 14.3 shows the magnetization loop in terms of B. Flux density continues to increase with magnetizing force, because the force forms a part of the parameter B. Continued increase of H beyond saturation will cause μ ($= B/H$) to tend to unity as H tends to infinity. If H is reduced to zero after saturation ($B = B_s$) has been attained in the positive direction, the flux density will reduce (but not to zero) to a finite value, $B_r < B_s$. The corresponding flux is called *remanence* or *residual flux*. As the magnetizing force is further reduced (by reversing the applied field), the flux density will reduce to zero and this negative magnetizing force is referred to as *coercivity* (H_c). If the reversed field is increased further, saturation in the negative direction is attained ($-B_s$). If the field is further reduced to zero and applied in the positive direction, the flux density will follow $-B_s$, $-B_r$, $+B_s$ [1]. The loop that is finally traced out is known as the *hysteresis loop.*

The material remains magnetized in one direction or the other (when $H = 0$) as long as it is cycled in a symmetric loop. Demagnetization of the specimen is accomplished by interrupting this cycling at some point. Following this the cycling continues in a field reduced a little during each cycle until the flux linking is nearly zero. The energy expended over a hysteresis cycle is proportional to the area under the hysteresis loop.

The other alternative to demagnetize a ferromagnetic material is to raise its temperature beyond the *Curie point.* Above this temperature the thermal energy will overcome the force that aligns the magnetic spins and the material behaves like a paramagnetic. If at this stage they are cooled in the absence of the applied field, demagnetization is accomplished at room temperature. Mechanical stresses/shocks applied to a magnet may also cause a demagnetization.

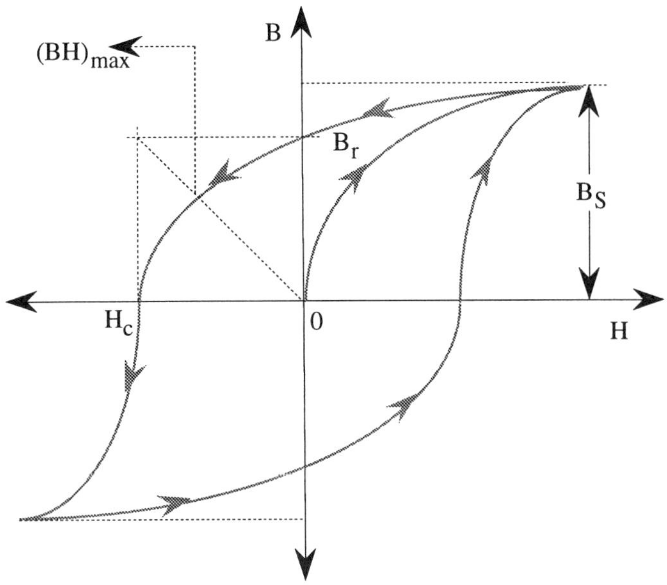

Figure 14.3 Hysteresis loop: BH curve.

14.4 Theories of Ferromagnetism

Instead of considering two separate classes of ferromagnetic materials, namely, hard and soft, it is more appropriate to treat them as one single group and study the conditions under which magnetic losses can be minimized and coercivity maximized as would be required in practical applications. This brings in the whole subject of magnetism in line with the developments in *atomic energy band theory, domain theory, and crystal structure analysis,* as briefed below.

14.5 Domain Theory

All ferromagnetic atoms and molecules can be recognized as elementary tiny magnets capable of rotating on their own axes in an applied magnetic field. A model made from these individual tiny magnets provides a way to predict the magnetization curves but not the magnitude of magnetizing forces required. This mystery was resolved by Weiss who proposed his *domain theory*. Weiss's theory is also referred to as the *molecular field theory*. This hypothesis can be elaborated based on the so-called Curie-Weiss law that relates mass susceptibility *(c)* and absolute temperature *(T)* as follows:

$$c = C(T - q) \tag{14.1}$$

where C is the Curie constant per gram; and q is a factor directly related to the molecular field (H_m). Weiss surmised that the molecular field H_m (assumed to be caused by the magnetization of the surrounding material) is directly proportional to the magnetization.

Qualitatively Curie-Weiss law (also referred to as the *classical theory of paramagnetism*) portrays a simple paradigm: Paramagnetic materials consist of atoms and molecules, each of which possesses a net magnetic moment due to the *spin* and *orbital* moments of electrons. In the absence of the applied field, these atomic (spin) moments nullify each other and the material's magnetization is zero. When a field is applied, the atomic moments tend to align towards the direction of the field which is opposed only by the *thermal agitation* of the electrons. This results in only partial alignment of the atomic moments and therefore a small positive susceptibility is perceived in the material.

When ferromagnetics are heated above the Curie point (T_c), they become paramagnetic due to the increased randomness of atomic moments. This randomness is due to greater thermal agitation. The susceptibilities then follow the Curie-Weiss law with the value of q nearly equal to T_c. This larger value of q (on the order of 1000 K for iron) directly affects the molecular field, H_m. This fact was used by Weiss to make an assumption that molecular field acting on a ferromagnetic material (below and above the Curie point) is sufficient to magnetize the material to saturation even in the absence of applied magnetic field. The material is then said to be *spontaneously magnetized.* Simultaneously, Weiss also assumed that a ferromagnetic material consists of a larger number of tiny regions called *domains* in the demagnetized state. Here, each domain is spontaneously magnetized, but the directions of magnetization of the domains act in such a way that the net magnetization effect is zero.

The process of magnetization can thus be explained as the transformation of the material from randomly arranged multi-domain state into a single-domain state magnetized in the direction of the applied field. The four stages are shown in Figure 14.4. Figure 14.4 indicates a portion of a ferromagnetic crystal in which there are parts of two domains. The boundary separating these regions are called *domain walls.* These domain walls can be "stretched" by applying a magnetic field so that they return to the non-magnetized state on removal of the applied field. On increasing the applied field one domain "grows" in size at the expense of the other until the domain wall is out of the region considered. Still greater applied fields will "rotate" the domain's magnetization in *parallel* with the applied field and finally giving the effect of magnetic saturation.

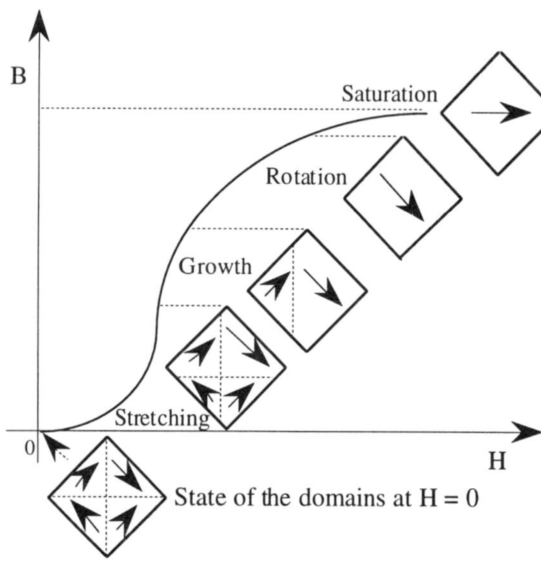

Figure 14.4 Magnetization process: Stepwise transformations.

Weiss's domain theory therefore presents two important postulates that formed the basis for research in the following decades. The two postulates were:

1. Spontaneous magnetization
2. Division into domains

14.6 Band Theory

In every atom the electrons revolve around the central, positively charged nucleus in an orbital fashion and at varying distances from the nucleus. Apart from this, the electrons spin on their own axes in the positive or negative direction. Hence the magnetic field will either be similar or opposed depending on the spin directions.

Thus the magnetic effect of a host of spinning electrons can give a strong or negligible magnetic field according to their pattern and direction. The orbital movement is relatively unimportant. Elements with pronounced electron spin unbalance owing to their atomic structure can exhibit ferromagnetic properties. Iron is an outstanding example.

The atomic structure of an element is made up of spherical orbits of electrons grouped in *shells* or *levels* around the central nucleus. All elements are divided into six shells denoted by the corresponding initial letters of their spectroscopy terms (*s* – sharp, *p* – principal, *d* – diffuse, *f* – fine). In each element the shell patterns are distinctive and different.

Returning to the problem of ferromagnetism, it is clear that completely filled levels contribute zero magnetic moment, because pairs of electrons in each level have opposite spin and cancel out each other. Consider an atom with just one electron in a particular energy level when the atom is free. Suppose n such atoms are grouped to form a crystal; then the single level in the free atom is divided into n levels with lower $n/2$ containing 2 electrons each. If one electron reverses its spin, then a spin unbalance is created. The force creating this spin unbalance in a ferromagnetic is the *exchange force*. The maximum magnetic effect is experienced in a half-filled energy level. In atoms of iron the *1s, 2s, 3s, 4s* energy levels contain two electrons each and *2p* and *3p* are filled with six electrons each. Now ten electrons are required to fill the *3d* level and iron has only six left, which gives the reason as to why it is strongly ferromagnetic. The outer *4s* level contains two electrons which explains the electrical conducting property of the metal.

Although the number of ferromagnetic elements is small, the possible combinations of these elements in both metallic and oxide form are quite large in number. Explanations for magnetic performance of these "alloys" require some knowledge of crystallography, which is dealt with in the following section.

14.7 Crystal Structures and Analysis

Materials exhibit a definite crystalline form when found in their natural state. Crystallography is important for a clear understanding of magnetic performance which depends on crystal size, shape, arrangement and lattice distortion.

Iron, cobalt, and nickel crystallize into three different geometrical forms which influence their magnetic behavior. Iron has a *body-centered* cubic crystal lattice structure (Figure 14.5a) which can be most readily magnetized along the edge of a face. Cobalt has a *hexagonal* crystal lattice structure (Figure 14.5b) and can be most readily magnetized in its axial direction. However, with nickel this lies along a body diagonal of its *face-centered* cubic structure (Figure 14.5c).

This property of a crystal to be more readily magnetized in one direction than others is of utmost importance in magnetic materials and is referred to as *crystal anisotropy*. Conversely, predetermined critical proportions of iron and nickel in specially heat-treated alloys give the characteristic high permeability and low hysteresis of the permalloys, but the crystal structures of such materials are found to be strain sensitive and the hysteresis loss is greatly increased by the mechanical stress required to manufacture them. Also crystals of ferromagnetic materials exhibit small changes in permeability and linear dimensions along the direction of magnetization as manifestations of the effect of stress due to mechanical strain. This effect, which is known as *magnetostriction* is widely used in magnetic applications.

Actually there is no single comprehensive theory for explaining ferromagnetism. There are at least two distinct theories that exist –

1. Localized moment theory
2. Band theory

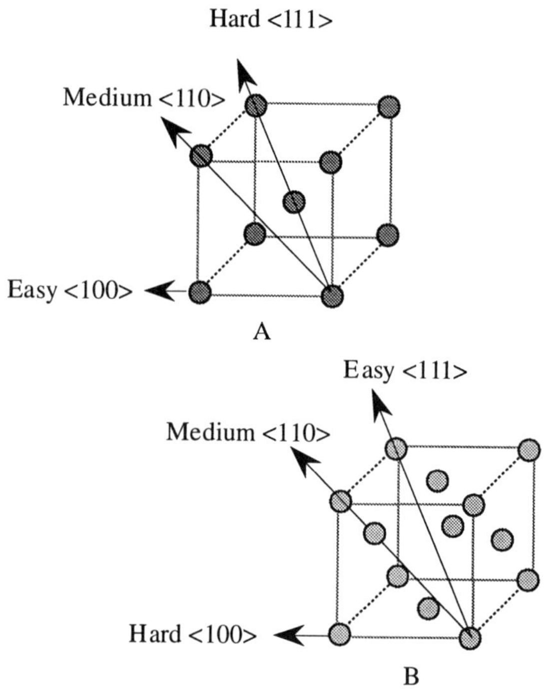

Figure 14.5 Crystal structures of ferromagnetic materials with specified directions of "easy" magnetization. (A) Iron. (B) Nickel.

In localized moment theory, the electrons responsible for ferromagnetism are not free. These electrons contribute a certain finite, localized magnetic moment to each atom. This is implicitly stated in the molecular field theory (domain theory); but this theory fails to account for the non-integral values of observed magnetic moments per atom. Usually, the moment per atom, if due to localized electrons, should be an integral value.

The band theory on the other hand is often referred to as the "collective-electron" theory when applied to magnetic properties because the localization of outer free electrons of the atom is abandoned. The electrons responsible for ferromagnetism are considered as entities belonging to the crystal as a whole. They are capable of moving from one atom to another and are not localized in an atom. The single important accomplishment of this theory is that it accounts for the non-integral values of the moment per atom.

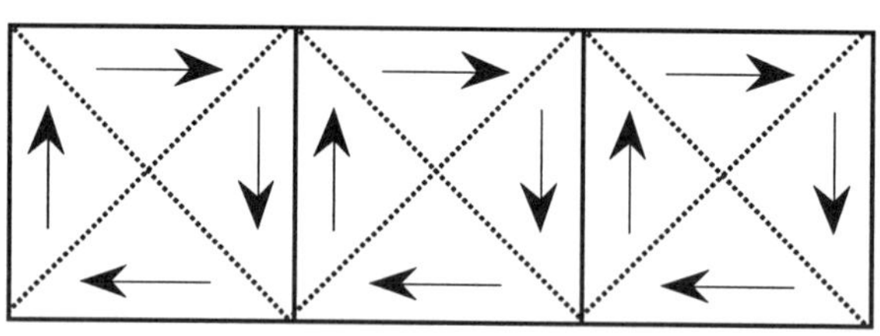

Figure 14.6: Domains in a demagnetized ferromagnetic material. Triangles represent the domains. Each domain has parallel-oriented dipoles.

The general conclusion is that the molecular field theory (domain theory) with its assumption of localized moments does not explain ferromagnetism in metals. In comparison, the band theory is regarded as correct though problems like understanding the precise form of various energy levels, how they are occupied by electrons, and how exchange forces operate are still unresolved.

An example of domains in a demagnetized ferromagnetic material is illustrated in Figure 14.6.

14.8 Magnetic Anisotropies

Magnetic anisotropies are important in elucidating the properties of permanant magnet materials. High coercive force indicates the existence of high anisotropic forces. Such anisotropies can be classified as *magnetocrystalline, shape, strain, and pair-ordering (or deformation-induced).*

Magnetocrystalline anisotropy is due to the existence of preferred crystalline axes for magnetization as dictated by the preferred direction of magnetic dipoles. It is an intrinsic property of the material. The energy required to rotate the magnetization vector from the easy to hard direction of magnetization is an implicit measure of magnetocrystalline anisotropy.

The crystal anisotropy *(K)* is a measure of theoretical maximum coercive force. The relevant relation is:

$$H_c = 2K/B_s \qquad (14.2)$$

where B_s is the saturation magnetization.

14.9 Shape Anisotropy

It is the preferential alignment of atomic moments in a given direction due to the shape of the magnetic particle. For an elongated particle, it is easy to magnetize parallel to the long direction. The implicit measure of shape anisotropy is *via* coercive force, given by:

$$H_c = (N_t - N_o)B_s \qquad (14.3)$$

where N_t is the demagnetizing factor in the narrow direction and N_o is the demagnetizing factor in the long direction.

14.10 Strain or Magnetostriction Anisotropy

As the name implies, it refers to the anisotropy resulting from the combination of strain and magnetostriction from the crystal. Application of stress along a given crystal axis may result in an inverse or a decrease in the magnetization for a given applied field along that direction. (This effect has a converse counterpart.) In terms of magnetostriction constant (λ_m) and the Curi-axial stress, *(T)*, the coercive force (as a measure of anisotropy) is given by:

$$H_c = 3\lambda T/B_s \qquad (14.4)$$

Table 14.4 Magnetic Units and Their Conversions

Magnetic Parameter	CGS Units	MKS Units	CGS - MKS Conversion	From - To Multiplier
Flux, Φ	Maxwell (line)	Weber	1 Wb = 10^8 Mx (lines)	
Flux Density, B	Gauss 1 line/cm^2	Tesla weber/meter^2	1 T = 10^4 G	Gauss-line/in.^{2} 6.4516

(continued...)

Magnetic Parameter	CGS Units	MKS Units	CGS - MKS Conversion	From - To Multiplier
Magneto - motive force, mmf	Gilbert	Ampere - turn	1 Gilbert = 0.796 ampere - turn	
Magnetizing field, H	Oersted	Ampere-turn per meter	1 Oerstead = 79.57 ampere - turn per meter	Oersted - ampere turn/in 2.0213

14.11 Types of Magnetic Materials

The magnetic materials can be classified on two basic considerations, namely, compositional and application-based. The first type refers to the inherent basic ferromagnetic nature of materials and the second version depicts the materials constituted for specific applications. The composition-based materials are:

- Metals and alloys
- Ferrites/ceramics
- Rare-earth inter metallics

The magnetic materials for specific applications are:

- Magnetic core materials
- Permanent magnetic materials
- Materials for laminations
- Materials for high-permeability requirements
- Magnetic memory core materials
- Magnetic core materials for high frequency applications
- Magnetic shielding materials
- Magnetic composite materials
- Magnetic superconductor materials
- Magnetic bubble materials
- Ferromagnetic liquids (ferrofluids)
- Ferromagnetic amorphous film materials
- Ferromagnetic insulating materials (garnets)

The aforesaid application-specific materials are subsets of composition-based set of materials indicated above.

14.12 Magnetic Materials

Basically, the following are the metals alloys which exhibit ferromagnetic properties (in different extents). Specific types of these materials and their magnetic properties are listed in Table 14.6.

A. **Metals/alloys**

Iron, steels (e.g. silicon steel, carbon steel, Tungsten carbon steel, chromium steel, cobalt steel)
Nickel-iron alloys
Cobalt-iron alloys
Vanadium-cobalt-iron alloys
Aluminum-nickel-iron alloys
Cobalt-nickel-copper-iron alloys

Vanadium-silver-manganese-aluminum-iron alloys
Carbon-tungsten-chromium-molybdenum-platinum-iron alloys
Grain-oriented silicon steel
Non-oriented silicon steel

B. **Ferrites (ferrimagnetic materials)**

(General formula: $Me_2^{+}Fe_2^{3+}O_4^{2-}$ with Me^{2+} represents a variety of divalent metallic ions)
Fe_2O_4
$CoFe_2O_4$
$MgFe_2O_4$
$NiFe_2O_4$
$MnFe_2O_4$
$CnFe_2O_4$

C. **Rare-earth intermetallic compounds**

Rare-earth metals are the fifteen elements which range from lanthanum (La, atomic number 57) to lutetium (Lu, atomic number 71). These are usually outside the regular array of the atomic table because they all exhibit similar chemical characteristics. The reason is that these materials have an outer shell electron structure being the same for all of them. Inasmuch as the outer shell electronic structure determines the chemical properties, all rare-earth metals are alike in their participation in chemical reactions.

The magnetic behavior of the rare-earth elements is almost entirely due to the existence of unpaired electrons in the 4f orbitals. The magnetic momentum is based on the net angular momenta of these unpaired 4f electrons and their orbital angular momenta. (When all 4f electrons are paired as in Lu, the material is diamagnetic.) In an unpaired situation, the net moment is localized about the atomic core.

The rare-earth metals exhibit large magnetic moment at low temperatures. Therefore for high temperature applications alloys of rare-earth and 3d-transition metals were developed. In alloy formation between a rare-earth (R) and 3d-transition elements (such as Mn, Fe, Co, or Ni), the terminal solid solubility is poor due to a large difference between the atomic radii of rare-earth and 3d-transition metals. As such, the alloys formed are invariably not ductile. This limits realizing only a few alloys and intermetallic compounds useful as practical magnetic materials.

Of particular interest for use as permanent magnet materials are those rare-earth Co and Cu hexagonal phases having the Cu_5Ca structure. Such Cu_5R compositions are shown in Table 14.5.

The other alloy phases are constituted by Fe_5R such as Fe_5Y, Fe_5Ce, Fe_5Sm and Fe_5Gd.

Multicomponent magnetic alloys with rare-earth elements are essentially Co-Cu-Ce, CO-Cu-Sm, Co-Cu-Fe-Ce, Co-Cu-Fe-Sm, and Co-Cu-Fe-Ce-Sm systems. For example, stoichiometric formulas of typical multiphase alloys are: $Y(Cu_xCo_{1-x})_5$ and $Sm(Cu_xCo_{1-x})_5$

There are also $A_{17}R_2$ phases of stoichiometry with A $\Rightarrow$ Co, Fe, or Ni which are closely related to Cu_5Ca structure.

Table 14.5 Cu_5R-Type Hexagonal Phases

Cobalt-based	Copper-based	Nickel-based
Co_5La	Cu_5La	Ni_5La
Co_5Ce	Cu_5Ce	Ni_5Ce
Co_5Pr	Cu_5Pr	Ni_5Pr
Co_5Nd	Cu_5Nd	Ni_5Nd
Co_5Sm	Cu_5Sm	Ni_5Sm
Co_5Gd	Cu_5Gd	Ni_5Gd
Co_5Tb	Cu_5Tb	Ni_5Tb
Co_5Dy	Cu_5Ho	Ni_5Dy
Co_5Ho	Cu_5Y	Ni_5Ho
Co_5Er		Ni_5Er
Co_5Y		Ni_5Yb
		Ni_5Y

14.13 Properties of Ferromagnetic Materials

From an engineering design point of view, the essential properties of ferromagnetic materials in general are specified by their B-H hysteresis characteristics. Referring to Figure 14.7, the following magnetic parameters can be enumerated:

μ :	B to H ratio
B_m :	Maximum magnetic induction
H_m :	Maximum magnetizing force
B_r :	Residual induction (remanence induction)
H_c :	Coercive magnetizing force
T_c :	Curie temperature
(H_d, B_d) :	Any operating point on the demagnetization curve
B_dH_d :	Energy product

Associated auxiliary parameters of engineering importance are :

s :	Material specific gravity
ρ :	Electrical resistivity
α:	Coefficient of thermal expansion

Pertinent to various ferromagnetic metals and alloys the parameters indicated are presented in Table 14.6.

In general, the metallic or alloy magnetic materials can be divided as *hard* and *soft* types. It depends on the ease with which the direction of magnetization can be altered by an applied magnetic field.

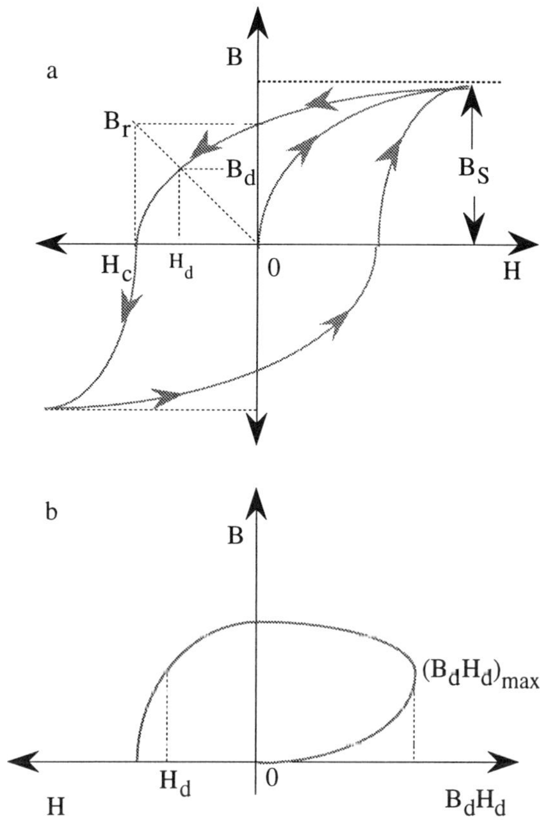

Figure 14.7 (a) Hysteresis loop. (b) Demagnetization curve.

Soft magnetic materials have high permeability and low coercive force. Both low and high remanent flux can be realized with these materials. The range of materials of this category are:

- Iron-based alloys
- Nickel-iron alloys with 30-80% Ni (ferrites, which nonmetallic materials are also "soft" in nature)

Application of soft magnetic materials include transformers and machine cases. In such applications, high permeability ensuring a tight coupling between windings, low eddy and hysteresis losses, ease of fabrication as laminations, and low cost are the engineering considerations in selecting a magnetic material. High permeability also allows saturation with a small applied field so that primary ampere-turns are kept low. Further, high permeability ensures a narrow hysteresis loop and hence the hysteresis power loss (proportional to the area of the loop) is small. Typical values of magnetic parameters of soft materials are : $\mu_r = 10^5$, $B_m \cong 1$ Tesla, $H_c \sim 0.3$ ampere/meter and hysteresis loss is a few hundred times less than that of soft iron.

14.14 Hard Magnetic Materials

These are intended for permanent magnets. Hence, they possess extremely high retentivity. That is, they are "hard" to be demagnetized with a coercive force. Permanent magnets find use in small electrical machines, meters, transducers, electron tubes (magnetrons), focusing magnets in TV tubes, etc. Characteristically, the hard magnetic materials have:

- High coercive force $(H_c) \cong 10^4$ ampere / meter
- High retentivity $\cong 1 - 1.5$ tesla
- High permeability
- High Curie temperature

Permanent magnets are designed to yield maximum B-H products with minimum volume of the material:

$$(BH)_{max} = \frac{kR_a\Phi^2}{(volume)_{min}}$$

where k is the flux leakage factor due to pole separation geometry, Φ is the total flux across the air gap between the poles, and R_a is the selectance of the air gap. $(BH)_{max}$ specifies the figure of merit of the magnet's quality.

14.15 Types of Hard Magnetic Materials

Hardened steels: Carbon steel, chrome steel, cobalt steel, and tungsten steel can be hardened *via* heat treatment and hot rolled to form bars, rods, strips, etc. They possess good physical characteristics such as high strength, machinability, and largely low-budget materials. However, they gave relatively low BH products. Relative performance of different hardened steels are presented in Table 14.6.

Table 14.6 Magnetic Properties of Hardened Steel

Material	Parameters				BH Product Applications
	Coercivity (H_c) ampere-turn/meter	Retentivity (B_r) tesla	$(BH)_{max}$ Product joules/meter^2	Physical Qualities	
Carbon steel 98% Fe, 1% Mn 0.9% C	Low (4000)	High (1.0)	Low (1400)	Casting material	Low cost applications in toys, latch relays, meters
Tungsten steel (5% Wo + 0.7 C+Fe)	Medium (5000)	High (1.0)	Medium (2500)	Casting material	High-cost applications such as magnetrons, d.c. meters

(continued..)

Material	Parameters				BH Product Applications
	Coercivity (H_c) ampere-turn/meter	Retentivity (B_r) tesla	$(BH)_{max}$ Product joules/meter^2	Physical Qualities	
Chromium steel (0.5 Cr+0.8 C+Fe)	Medium	Medium	Low	Casting material, no cracking on annealing	Low cost
Cobalt steel (3-38% Co + 2-8% Cr and W+Fe)	High (7000)	High (3.4)	High (8500)	Casting material, high tensile strength, poor machinability	Popular magnet material
Alnico family (carbon free) (8% Al, 14% Ni, 24% Co, 3% Cu + Fe)	High (50000)	High (1.2)	High (4500)	Low cost. Sintered and pressed, non-machinable, stable magnetic properties with temp. and shock	
Remalloy (12% Co, 17% Mo) Cunife (60% Cu, 20% Ni)	High (2500 - 5000)	High (0.97 - 0.54)	High (10000 - 13500)	Casting material, machinable stable ductile more expensive	
Iron	Low (75)	Low (1.5)	Low (100)	Cast magnet	

14.16 Magnetic Stainless Steel

These are Si-Fe and Ni-Fe magnetic steels. They are corrosion resistant and used in solenoidal values as solid sections (e.g. Ludlum Stainless Steel Type 416™). Another category of this steel is straight Cr-Fe steels which provide improved magnetic perfomance *via* heat treatment after fabrication. A nonhardening Cr-Fe steel has modified chemical structure to give high electrical resistivity and resists oxidation. It is used in motor starting rheostats and other motor control equipment.

14.17 Silicon Steel

Typical silicon steel perfomance can be ascertained from the following comparisons in Table 14.7.

Table 14.7 Relative Magnetic Perfomance of Silicon-steel

Material	μ_r	ρ in micro-ohm-cm at room temp.	Hysteresis Loss joule/kg/Hz
99.9% Fe	5000	10	0.03
97% + 3% Si	7500	35	0.02
96% Fe + 4% Si (Grain - oriented steel)	30000	55	0.005

Typical engineering requirements for silicon steel in their applications in electrical machines are specified in Table 14.8.

Table 14.8 Silicon Steel and Their Engineering Requirements

Applications	Power loss (watts) at 1&1.5 tesla, 50/60Hz	DC Magnetization tesla	Magnetic Induction ampere-turn per meter	Stacking Factor*	Aging**
Transformers	0.9-1.1 at 1.5 T	1.8-1.9	1600	98%	0%
		1.9-2.0	10000		
Small generators and motors	2.0-4.5 at 1T	1.35-1.45	1600	95%	5%
	4.5-10.0 at 1.5 T	1.70-1.80	10000		
Heavy duty generators and motors	1.0-2.0 at 1T	1.35-1.40	1600	96%	3%
	2.25-4.5 at 1.5 T	1.70-1.72	10000		

* Stacking factor : Effective material content due to lamination stacking

** Aging : Acceleration change in specific core loss due to continuous heating at 100°C for 600 hours.

14.18 Iron-Cobalt Steel

Typically these alloys are made of 49% Fe + 49% Co + 2% Va to realize high saturation density. They are used widely in air/space-borne electrical machines, servos and synchros, loudspeakers, telephone receiver diaphragms and magnetic pole shoes. Vanadium (or alternatively chromium) makes the material malleable and ductile. The composition makes these materials expensive.

14.19 Nickel-Iron Steel

There are three versions of these steels depending on the content of Ni:

36%Ni alloy: High electrical resistivity and low permeability characteristics. Used in high frequency devices, wideband transformers, inductors, and high speed relays.

50% Ni alloy: High flux density and maximum permeability. These alloys replace silicon steel where low-loss and small-size requirements are posed. (e.g. small relays, transformers, inductors, synchros, and minimotors). Other applications include magnetostrictive transduction due to almost rectangular hysteresis loop. They are also used in switching magnetic systems such as pulse transformers, magnetic amplifiers, and invertors. Small-sized cores can be used for magnetic memory core purposes.

70% Ni alloy: High permeability and low losses. These are used in precision current-voltage transformers, inductive bridges, recording heads, sensitive relays and wideband transformers. Also they are used extensively in magnetic shielding applications.

Typical Ni alloy soft magnetic materials and their magnetic performance are presented in Table 14.9.

Table 14.9 Magnetic Performance of Ni Alloy Soft Magnetic Materials

Material	Parameters		
	μ_r	ρ in micro-ohm-cm	Hysteresis Loss joules/Kg/Cycle
Hypernik™ 50% Fe,50% Ni	6×10^4	45	0.003
Supermalloy™ 79% Ni,15% Fe, 5% Mo,0.5% Mn	10^6	65	0.0001
Mumetal™ 77% Ni,14% Fe, 5% Cu,4% Mo	9×10^4	62	
Permalloy 78™ 78% Ni,22% Fe	10^5	16	0.0005

14.20 Conclusions

In the postwar years, the developmental activities pertinent to magnetic materials have been stupendously high *vis-a-vis* the understanding of magnetic properties and synthesizing new materials. With the advent of superconductivity, realization of high-gauss permanent magnets for relevant applications has set new goals and strategies towards the science of magnetic materials. Thus, as conceived today, magnetic materials form the basics of simple technology of electrical materials as well as high technology of superconduction, magnetic bubble memory, etc. They constitute a vital set of modern electromagnetic materials.

References (General Reading)

[1] J. E. Thompson: *The Magnetic Properties of Materials.* (CRC Press, Cleveland, OH: 1968).

[2] B. D. Cullity: *Introduction to Magnetic Materials.* (Addison-Wesley Publishing Co., Reading, MA: 1972).

[3] G. R. Polgreen: *New Applications of Modern Magnets.* (Boston Technical Publishers, Inc., Cambridge, MA: 1966).

[4] J. Smit (Ed.): *Magnetic Properties of Materials.* (McGraw-Hill Book Co., New York: 1971).

[5] E. A. Nesbitt and J. H. Wernick: *Rare Earth Permanent Magnets.* (Academic Press, New York: 1973).

[6] A. J. Dekker: *Electrical Engineering Materials.* (Prentice-Hall of India Pvt. Ltd., N. Delhi: 1988), Chapter 4.

[7] L. F. Bates: *Modern Magnetism.* (Cambridge University Press, London: 1951).

[8] J. L. Snoek: *New Developments in Ferromagnetic Materials.* (Elsevier, New York: 1947).

[9] L. Solymar and D. Walsh: *Lectures in the Electrical Properties of Materials.* (Oxford University Press, Oxford: 1993), Chapter 11.

Defining Terms

Anisotropy (magnetic): Directional or orientational effects in crystal structure of materials which can provide better magnetic performance along certain (preferred) axial directions.

Coercivity (H_C): The value of "negative" magnetizing force required to reduce the flux density *(B)* due to magnetic induction to a zero value after a material has been magnetized to saturation and demagnetized to a remnant state.

Curie point (T_C): The characteristic temperature above or below which the properties of ferromagnetic materials differ widely. Above this Curie point, ferromagnetic materials become paramagnetic.

Curie-Weiss law: Describes the quantitative realtionship between susceptibility and temperature for paramagnetic materials. For paramagnetics, susceptibility is inversely proportional to the absolute temperature.

Demagnetizing curve: Refers to the portion of the hysteresis loop when the reverse field conditions are applied and also represents practical operating circumstances.

Domain: The smallest zone of a magnetic material which retains its magnetic identity. The domain theory assumes domains are spontaneously magnetized to saturation, but can be moved and oriented by externally applied fields.

Ferromagnetism: Magnetic materials with permeability much greater than that of free space which are classified as ferromagnetic. High-coercivity materials with permeability nearly unity are also categorized as ferromagnetic materials.

Hysteresis: When a ferromagnetic material is placed in an alternating magnetic field, the flux density *(**B**)* lags behind the magnetizing force *(**H**)* that causes it. This effect is known as hysteresis. The area under the hysteresis loop is the hysteresis loss per cycle, and is maximum for permanent magnets and minimum for high permeability, low-loss magnetic materials.

Intensity of magnetization (M): This entity determines the extent to which a body is magnetized. Quantitatively, it is defined as the magnetic moment *(**m**)* per unit volume *(v)* or the pole strength per unit cross-sectional area *[**M** = m/v]*. It is also referred to as *magnetization*.

Magnetic dipole: It is the shortest magnet of finite moment. Any magnet can be visualized to be made up of several magnetic dipoles.

Magnetic flux density (B): Flux lines passing through a unit area of the uniform material surface.

Magnetic field intensity (H): Measure of magnetic field strength or magnetomotive force.

Magnetic moment (m): It is the couple acting upon a magnet of length ℓ, having poles of strength p placed at an angle q in a magnetic field of strength H.

Magnetic poles: Magnets experience forces of attraction and repulsion that originate at hypothetical locales near the ends of the magnet, called poles. These poles always occur in pairs in magnetized bodies.

Magnetic saturation: The upper limit to the capability of carrying flux by ferromagnetic materials called magnetic saturation.

Magnetization curve: A magnetization curve (***B*** *versus* ***H***) represents the level of magnetizing force needed for a magnetic material to attain magnetic saturation.

Magnetostriction: Dimensional changes in ferromagnetic materials exposed to magnetic fields. Conversely, magnetic property changes experienced due to mechanical stress applied on the materials.

Permeability (μ): The ratio of flux density *(**B**)* and magnetizing force *(**H**)* producing it is called permeability. It physically refers to the extent of magnetic flux allowed by a material to permeate through it.

Remanence or retentivity (B_r): The value of flux density *(**B**)* of a ferromagnetic material retained with the magnetizing force removed after magnetizing the material to saturation.

Susceptibility (χ): It is the ratio of magnetization (***M***) to field intensity. Susceptibility describes the way in which ***M*** varies with ***H*** and indicates the response of a magnetic material to an applied magnetic field.

CHAPTER 15

Ferrite Materials

15.1 Introduction

Ferrites are mixed crystals of various metallic oxides and may be considered to consist of oxygen ions in a closed-packed structure with *cations* filling the interstices. They have a general formula: $MOFe_2O_3 \cdot xH_2O$ where M is a divalent metallic ion such as Mn^{2+}, Fe^{2+}, Co^{2+}, Ni^{2+}, Zn^{2+}, Mg^{2+}, and Cd^{2+}. In general, the magnetic saturation intensity of ferrites is lower than that of various metallic magnetic alloys, but the ferrites have very high resistivity and as such are insulator-like. Therefore, the power loss in ferrites is very small especially at high frequencies. Further, ferrites have a narrow hysteresis (BH) loop and can be made with square-loop characteristics. These properties permit their applications in high frequency devices and for switching (memory) cores.

Ferromagnetic materials such as steel which have wide applications in electrical engineering have a disadvantage in that they have low electrical resistivity. The laminations used for electrical machines, for example, have a resistivity of about 14×10^{-6} ohm-cm and the highest value obtainable in ferromagnetic alloys is less than 10^{-4} ohm-cm. This disadvantage of ferromagnetic materials limits their application in the high frequency alternating current applications. High *eddy current losses* occur in metallic sheets even at low frequencies. Ferrites, on the other hand, with useful magnetic properties have d.c. resistivity of many orders higher than in iron and are used at frequencies up to microwaves.

Ferrites are essentially *ferrimagnets*. That is, if the net magnetization of magnetic sublattices is not zero, the material exhibits ferrimagnetism and possesses a net *magnetic moment*. (This moment disappears above the Curie temperature T_c analogous to Néel temperature at which thermal energy randomizes the individual magnetic moments and the material becomes *paramagnetic*.)

Ferrites (ferrimagnetic materials) in general, as mentioned earlier, are complex oxide compounds of various metals and oxygen. Ferrites are derived from *magnetite* by replacing the divalent iron atom by an atom of another metal. The formula for magnetite is $(Fe^{2+}O, Fe_2{}^{3+}O_3)$.

The general properties of ferrites as compared to metallic soft-alloy materials are listed in Table 15.1.

Many of the useful ferrites are isomorphic (similar structure) with the mineral *spinel* with the formula $(MO\ Fe_2O_3)$, M being the replacement metal as mentioned earlier. Usable replacement metals as mentioned earlier are divalent metals such as manganese, magnesium, nickel, copper, cobalt, zinc, and cadmium. The properties of final ferrite depend on the nature of the replacement metal and on its properties. When Zn or Cd are the replacement for Fe, the ferrite obtained is nonmagnetic. When it is entirely replaced by one of the other metals mentioned above, the material is magnetic with a high permeability but large hysteresis losses. Complex ferrites are also possible when the iron atoms are replaced by two divalent atoms at the same time. Typical examples are : Manganese-magnesium, nickel-zinc, nickel-cobalt, nickel-aluminum ferrites. The chemical formula for nickel-zinc ferrite is $(\alpha NiO\ \beta ZnO\ Fe_2O_3)$, $(\alpha + \beta) = 1$. In some cases the trivalent atoms of iron in Fe_2O_3 are replaced by atoms of another trivalent metal such as $A\ell$.

The intense intrinsic magnetization of ferrites is due to the ion distribution of the inverse spinel type and also an anti-parallel alignment of the spins on the A and B sites as shown in Figure 15.1.

Table 15.1 Ferrite *versus* Metallic Alloys: General Magnetic Properties

Material	μ_r (Initial)	μ_r (Max)	H_c (oersted)	B_s (gauss)	Saturation BH loss (erg/cm^2)	ρ (ohm-cm)
Grain-oriented Si-Fe 3% Si, 97% Fe	1500	40000	0.1	20000	700	47
Supermalloy 5% Mo, 79% Ni	100000	10^6	0.002	7900	8	60
Ferroxcube3 Mn-Zn Ferrite	1000	1500	0.01	2500	130	10^8

Remarks: μ_r = Relative permeability.
H_c = Coercive magnetic field intensity.
B_s = Saturation value of magnetic flux density.
ρ = Resistivity.

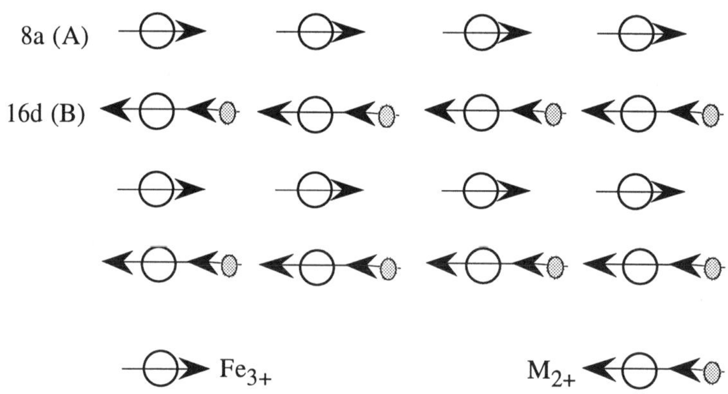

Figure 15.1 Spin-arrangement in inverse spinel structure.

The opposing Fe^{3+} at A and B sites cancel each other being M^{2+} ions at the B sites to yield an effective moment. The extent of this effective moment depends on the maximum spin magnetic moment attained by the alignment of 3d spins in the transition metal. The absolute saturation moment of a simple ferrite per unit cell is denoted by M/M_B.

Ferrites have largely spinel or inverse spinel structures. (However, garnet and perovskite structures are also not excluded as will be detailed later.) The spinel structure is illustrated in Figure 15.2.

As an example, considering $MOFe_2O_3$ ferrite the divalent metal M^{2+} ions reside at the interstitial 8a-sites (A sites) and Fe^{3+} ions reside at 16d sites (B sites). This is a *simple* or *normal ferrite*. The oxygen ions in this lattice touch each other and form a close-packed face-centered cubic lattice.

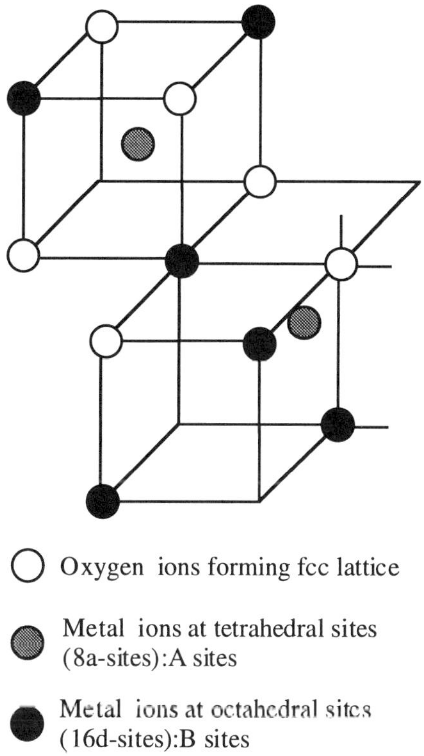

Figure 15.2 Spinel structure.

In another crystalline structure known as *inverse spinel structure*, half of the Fe^{3+} ions reside at A sites and the remaining half as well as all the M^{2+} ions are located at B sites.

In normal spinel-type structure, the spinel unit cell of a ferrite consists of a close packed cubic array of 32 oxygen anions between which there are 96 spaces or interstices: 24 are filled with cations, and the remaining 72 are empty. The sites occupied by the cations are of two kinds, tetrahedral or A sites and octahedral or B sites. The A sites (of which eight are occupied), are surrounded by four oxygen anions, and B-sites (of which sixteen are occupied) are surrounded by six oxygen anions.

General magnetic and electric properties of ferrites can be enumerated as follows:

- A very high resistivity, generally, in excess of 10^5 ohm-cm
- Dielectric constant of the order of 10–12 at high frequencies (microwaves) with extremely low dielectric loss
- Permeability of several tens
- Saturation magnetization is appreciable, but significantly smaller than that of ferromagnetic materials
- Low coercive force
- Curie temperature varies from 100°C to several hundred °C
- Mechanically hard, brittle, and not easily machinable

Ferrimagnetic domains in certain materials can be altered in their shape by the application of an external field. These domains can be used to store information. The materials of interest in such a category are the rare-earth *orthoferrites* and the rare-earth *iron garnets*.

The orthoferrites are not ferrimagnets but antiferromagnets in which the direction of the sublattice magnetizations make a small angle with each other giving rise to a magnetic state called *weak ferromagnetism.* In such materials, domain pattern can be altered by applying a field normal to the crystal face to become a set of individual cylindrical domains known as *magnetic bubbles.*

Rare-earth iron garnets, for example, $Er_3Fe_5O_{12}$ (erbium-iron-garnet), holmium-iron-garnets, etc. also show bubble domain characteristics; and hexagonal ferrites $BaFe_{12}O_{19}$ and $PbFe_{12}O_{19}$ have also been used as domain materials.

Anisotropic properties in ferrimagnetic materials made by pressing and sintering *ceramic powder* can be obtained by applying a magnetic field appropriately during the firing and cooling processes. These are cost-effective materials and can be used *in lieu* of metallic magnets to realize high resistivity and low eddy current losses.

Maximum magnetic moment of a ferrite is $5\ M_B$, (where M_B is the Bohr magneton moment) with inverse spinel structure. This value can be increased by adding some amount of normal spinel ferrite, for example, $ZnFe_2O_4$. The Zn^{2+} ions would then occupy the A sites and force the same number of Fe^{3+} ions from A to B sites; and this will result in additional magnetization of B sites. By adding a fraction of $ZnFe_2O_3$ to the fraction (1 – x) of $(Fe^{3+})O \cdot (Fe^{3+} + M^{2+})O_3$, the resulting structure $(Fe^{3+}_{1-x}Zn^{2+}_{x})O \cdot (Fe^{3+}_{1+x}\ M^{2+}_{1-x})O_3$ will have $M = [n + (10 - n)x]\ M_B$ where n is the number of Bohr magnetons of M^{2+} moment. Thus with Zn^{2+} ions, M_B increases to $10\ M_B$. Experimental evidence indicates such an increase with a small concentration of the added ferrite. At higher concentrations deviations from the increase are observed (Figure 15.3) [1].

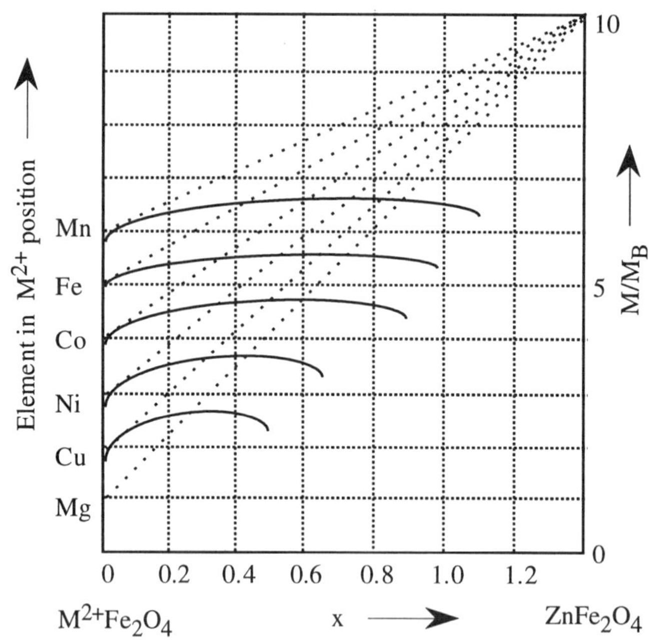

Figure 15.3 Fractional addition of normal spinel ferrite *versus* enhancement of magnetic moment of $M^{2+}Fe_2O_4$.

15.2 Temperature Dependency of Intrinsic Magnetization of Ferrites

This is described by the *Néel theory.* Considering the proportion of Fe^{3+} ions on A sites as α and on B sites is β (so that $\alpha + \beta = 1$), then $\alpha = 0$, $\beta = 1$ specifies a *normal spinel*; and $\alpha = \beta = 0.5$ refers to *inverse spinel.* Since M^{2+} ions occupy the remaining sites, the distributation of ions over A and B sites can be indicated by:

$$(Fe^{3+}_{2a}M^{2+}_{1-2a})O \cdot (Fe^{3+}_{2b}M^{2+}_{2-2b})\ O_3 \tag{15.1}$$

The total intensity (I) of magnetization is $\alpha I_A + \beta I_B$ where I_A and I_B are contributions from sites A and B, respectively, at thermal equilibrium. These depend on exchange interactions between both atoms of different sites and atoms on the same site. Hence above the Curie temperature, by consideration of *Langevin's theory of paramagnetism**, the magnetic susceptibility χ_m, *versus* temperature *(T)* can be written as:

$$(1/\chi_m) = [(T/C) + (1/\chi_o)] - [\sigma/(T - \theta)] \tag{15.2}$$

where C, σ, and θ are constants for a material and $(1/\chi_o)$ is the y-intercept of $(1/\chi_m)$ *versus* T curve shown in Figure 15.4.

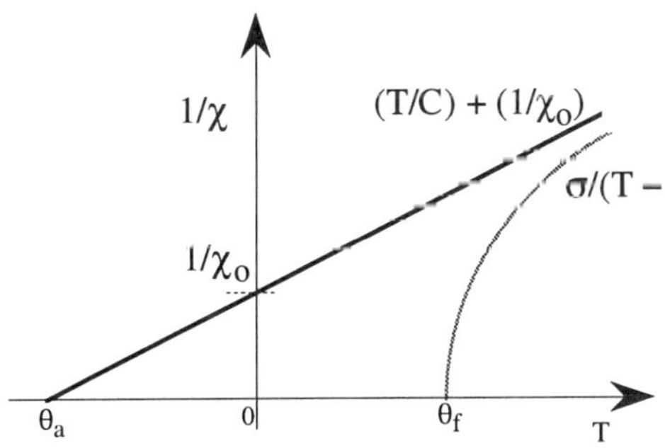

Figure 15.4 Temperature *versus* magnetic susceptibility of a ferrite.

Referring to Figure 15.4, the constant C and θ_a are related by the equation $\theta_a = (-\ C/\chi_o)$ and θ_a is designated as the *asymptotic Curie point.* Further, θ_f is the temperature at which $1/\chi_m$ becomes zero and at the same time a spontaneous magnetizationn appears and is referred to as *ferrimagnetic Curie point* or *Néels temperature.* If $\theta_f < 0$, there is *paramagnetism* for the whole temperature range whereas if $\theta_f > 0$, there is *ferrimagnetism* below this temperature.

---

* *Langevin's theory of paramagnetism* refers to a simple model specifyiing the net atomic moments being zero in the absence of an applied field due to their mutual cancellation as a result of isotropic random orientation. With an applied field, the atomic moments tend to align along the field and the resulting orientation of the moments (with the counteraction due to thermal agitation) is partially anisotropic. This partial anisotropic orientation of the atomic moments yields a small susceptibility. With increase in temperature, the thermal agitation upsetting this partial anisotropicity leads to a decrease in susceptibility. Quantitatively, the relative magnetization (M/M_o) is governed by the *Langevin function* L(a) equal to Coth(a) – 1/a where $a = \mu H/k_B T$. Here, μ is the permeability, H is the magnetizing external field, k_B is the Boltzman constant, and T is the temperature.

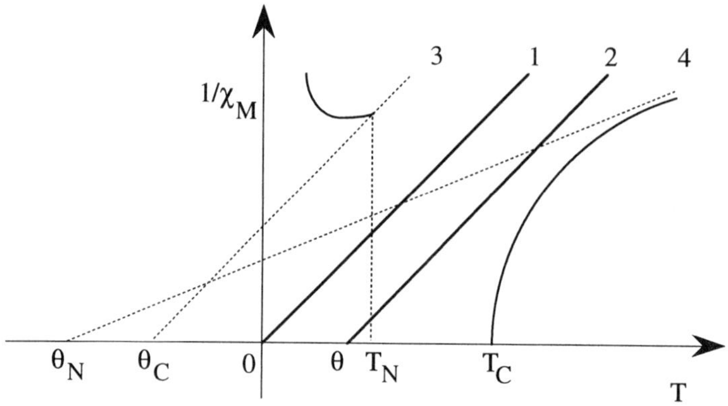

Figure 15.5 Reciprocal of magnetic susceptibility as a function of temperature for various types of magnetic materials.

As a summary, Figure 15.5 depicts the variation of reciprocal magnetic susceptibility (χ_m) as a function of temperature for various magnetic materials. The curves shown are:

1. Paramagnet-Curie law: $\chi_m = C/T$
2. Ferromagnet-Curie-Weiss law: $\chi_m = C/T - \theta$; $T > \theta$ where θ is Curie temperature.
3. Antiferromagnet-Curie-Weiss law: $\chi_m = C/(T + \theta_N)$; at $T > T_N$

   θ_N- Extrapolated Néel temperature

   T_N- Néel temperature
4. Ferrimagnet-Curie-Weiss law: $\chi_m = C/(T + \theta_o)$; $T > T_o$

   T_o- Ferrimagnetic Curie temperature

   θ_o- Extrapolated Curie temperature

15.3 Magnetization Characteristics of Ferrites

When ferromagnetic or ferrimagnetic materials are magnetized, the direction of magnetization will be rotated from its preferential direction. This corresponds to an anisotropic behavior. On removing the magnetizing force, the total magnetization will in general have a non-zero value. The above characteristics are due to crystal stress and shape anisotropies, respectively.

Crystal anisotropy: Crystal anisotropy does not yield a variation of permeability in a ferrite with the direction of applied magnetizing force because the bulk material consists of an aggregate of crystals whose axes are randomly oriented.

Stress anisotropy: Simple ferrites (with the exception of Fe_3O_4) have a negative *magnetostriction constant*; that is, upon magnetization the ferrites contract in the direction of magnetization. Therefore under a compressive force they can be magnetized most easily in the direction of magnetization and under tension they magnetize easily in a direction perpendicular to the direction of magnetization. Thus useful properties can be infused into ferrites under controlled stresses. However, occurrence of arbitrary stresses is undesirable and is avoided. As such, while preparing the ferrites, care has to be exercised to reduce any unspecified stresses to a minimum.

Shape anisotropy: A ferromagnetic body is magnetized most easily in the direction of its largest dimension. Non-magnetic inclusions and voids in the body have a similar influence producing a local preferential direction of magnetization parallel to its largest dimension. Different degrees of porosity exist in ferrites depending upon the method of preparation. Ferrites are therefore especially prone to the internal shape anisotropy. Hence, it is often endeavored to make these materials less porous because shape anisotropy impairs their usefulness for many applications. The effect of porosity on the magnetization curve and the hysteresis loop of a ferrite is shown in Figure 15.6.

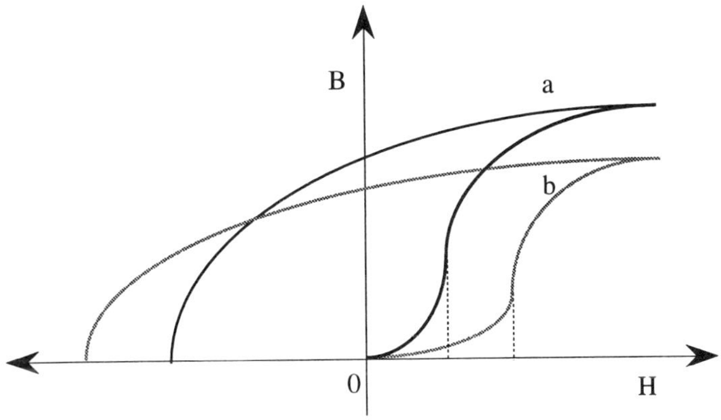

Figure 15.6 Effect of porosity on the magnetization curve of a ferrite. (a) Non-porous material; (b) Same ferrite with substantial porosity.

The porous material has less remnant magnetism than the non-porous material but the demagnetization field required to reduce the magnetism to zero is larger for a porous material than for the non-porous one.

Ferrites are hard and brittle materials and as mentioned earlier they cannot be shaped by ordinary machining processes. In general, a diamond impregnated slitting wheel is necessary for cutting the ferrite rods.

The mechanical and thermal properties of ferrites are compared with those of iron in Table 15.2.

Table 15.2 Properties of Ferrite and Iron

	Ferrite (Ferrox-cube)™	Iron
Specific gravity	4.30	7.80
Young's modulus lb/sq.in.	21×10^6	31×10^6
Tensile strength lb/sq.in.	2,600	43,000
Crushing strength lb/sq. in.	10,400	43,000
Coefficient of linear expansion $^oC^{-1}$	11×10^{-6}	12×10^{-6}
Specific heat	0.17	0.11
Thermal conductivity Cal/cm-sec-^oC	8×10^{-3}	0.18

Table 15.2 shows that ferrites have a lower tensile strength and a lower thermal conductivity as compared to iron. Ferrite component should not, therefore, be subjected to tension. Low thermal conductivity may result in high local temperature and large temperature gradients in the ferrite subjected to rapid reversals of magnetic fields. Apart from mechanical fracture which may take place due to high temperature gradients, the magnetic performance may also be seriously impaired if the Curie temperature is approached at any of the hot spots within the material.

15.4 Magnetic Resonance in Ferrites and Faraday Rotation

A rotating body will produce a *kinetic moment* along the axis of rotation. This kinetic moment is proportional to the speed of rotation. The rate of change of the kinetic moment with time is equal to the couple of the applied force.

An electron having a mass and spin will give rise to both magnetic moment and kinetic moment. If $\boldsymbol{M}$ is the magnetic moment and $\boldsymbol{P}$ is the kinetic moment, then the *gyromagnetic ratio* is given by:

$$\gamma = |\boldsymbol{M}/\boldsymbol{P}| \tag{15.3}$$

The electron rotates in a fixed direction along the axis OC (Figure 15.7) when an external field is applied. If an external field $\boldsymbol{H}$ is applied along the OZ axis, this magnetic field would exert a couple $\boldsymbol{C}$ on moment $\boldsymbol{M}$; that is, $\boldsymbol{C} = \boldsymbol{M} \times \boldsymbol{H} = d\boldsymbol{P}/dt$. Further, since $d\boldsymbol{P}/dt = (-1/\gamma)\, d\boldsymbol{M}/dt$, it follows that $\boldsymbol{M} \times \boldsymbol{H} = (-1/\gamma)\, d\boldsymbol{M}/dt$. This is the differential equation of motion of the moment $\boldsymbol{M}$. The tip of the $\boldsymbol{M}$ vector will rotate in a circle lying in a plane perpendicular to OZ. The corresponding angular velocity is given by: $\omega = 2\pi f$ where $f = (\gamma/2\pi)H = 2.8 \times 10^6\, H$ for electrons.

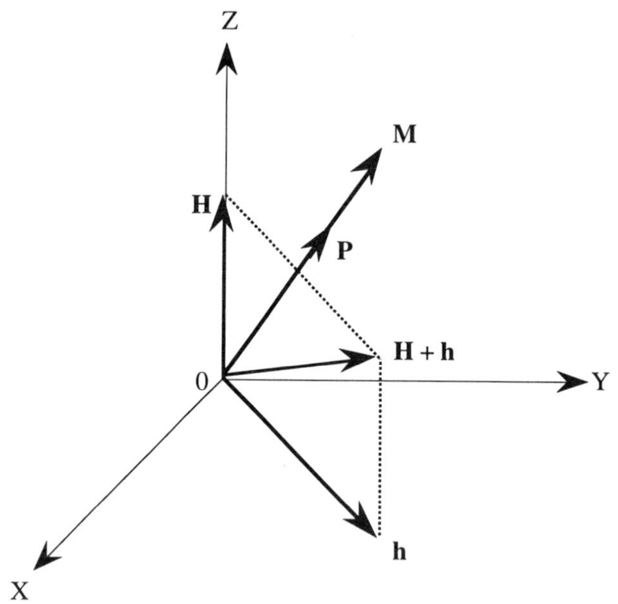

Figure 15.7 Illustration of the axes pertinent to magnetic resonance phenomenon.

When a circularly polarized high frequency field $\boldsymbol{h}$ is applied, this field will rotate in the plane of magnetic polarization which is the plane XOY in Figure 15.7. Under this condition, the vector $(\boldsymbol{H} + \boldsymbol{h})$ describes a cone carrying the moment $\boldsymbol{M}$ with which it acquires a *precessional motion* around it. If the sense of rotation of $\boldsymbol{h}$ is such that the resultant vector

$(\boldsymbol{H} + \boldsymbol{h})$ rotates in the same sense as that of precession and if the frequency of $\boldsymbol{h}$ is the same as that of the precession, then there will be a synchronization of the two phenomena resulting in the *magnetic resonance*. This will cause absorption of energy from the high frequency field. This phenomenon is known as *gyromagnetic resonance*. On the other hand, if the sense of rotation of the field $\boldsymbol{h}$ is such that the resultant vector rotates in the opposite sense to that of the precession, then there is no synchronization and hence no magnetic resonance will take place.

For a circularly polarized high frequency wave the sense of rotation depends on the direction of propagation. Resonance occurs only in one direction of propagation. In the other direction there is no resonance and hence no significant absorption.

The effective permeability of the ferrite in respect to the above two cases (that is, with a circularly polarized field in the positive sense and that in the negative sense) can be designated by μ^+ and μ^-, respectively. Further, these permeabilities are assumed as complex quantities to represent the losses. Hence it follows that:

$$\mu^+ = (\mu')^+ - j\,(\mu'')^+ \qquad (15.4a)$$
$$\mu^- = (\mu')^- - j\,(\mu'')^- \qquad (15.4a)$$

Typical variations of the real and imaginary parts of the permeabilities with the applied magnetic field for the positive and negative cases are shown in Figure 15.8. The application of ferrites at microwave frequencies is to make use of their nonreciprocal behavior so that the wave transmission properties depend on the polarization and the direction of incident electromagnetic wave. The most widely used such devices are *isolators* and *circulators* constructed in waveguide, coaxial or stripline geometries. In the forward direction these devices pose a low loss (usually 3.0 dB or less) and in the reverse direction the loss is 20-30 dB or more. The choice of ferrite material for a resonance isolator depends upon the factors like dielectric loss, resonance line width, saturation magnetization, and Curie temperature.

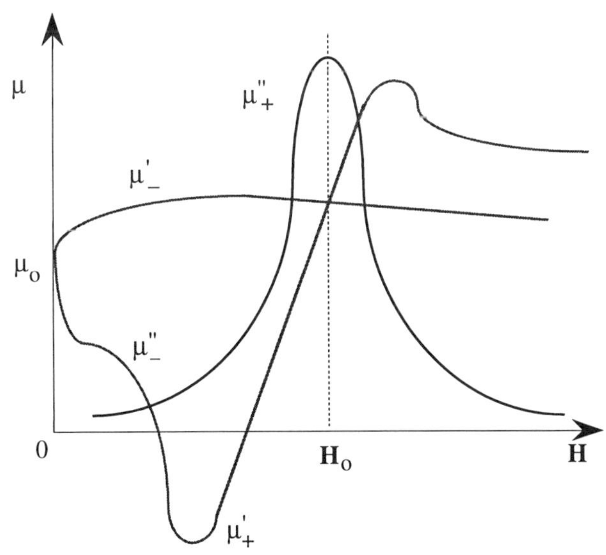

Figure 15.8 Magnetic permeability components as functions of applied field H.

15.5 Classification of Ferrites and Their Subclasses

15.5.1 Spinel structure ferrites

As mentioned earlier, the general formula for a ferrite can be expressed as $MO \cdot xFe_2O_3$ where M is a divalent metal ion. Various constants of simple ferrites are presented in Table 15.3. Unit cells of these ferrites have spinel structure.

Table 15.3 Constants of Simple Ferrites [1]

Ferrite	M/M_B	θ_f (°C)	Density (gm/cc)	Saturation Magnetization (kilogauss)	Initial Permeability
$MnFe_2O_4$	4.6-5.0	300	5.00	5.20	250
$FeFe_2O_4$	4.1	585	5.24	6.20	70
$CoFe_2O_4$	3.7	520	5.29	5.00	
$NiFe_2O_4$	2.3	585	5.38	3.40	10
$CuFe_2O_4$	1.3	455	5.38	1.70	
$MgFe_2O_4$	1.1	440	4.52	1.40	10
$Li_{0.5} \cdot Fe_{2.5}O_4$	2.5-2.6	670	4.75	3.90	

Practical ferrites of this category are made by two constituent mixtures. For example, Zn-Fe ferrite has a stoichiometric formulation $Zn_\alpha Fe_{1-\alpha}(Ni_{1-\alpha}Fe_{1+\alpha})\ O_4$ where α and $(1 - \alpha)$ are the molecular fractions of Zn and Fe ferrites, respectively, and the symbols inside the parenthesis indicate ions occupying octahedral positions (*inverted type*). (Normally, the divalent ions otherwise occupy the tetrahedral positions.)

15.5.2 γFe_2O_3 ferrites

These refer to an oxide called *maghemite* (a natural oxide) which also forms a spinal lattice. The chemical formula of this oxide is:

$$4Fe_2O_3 \rightarrow 3[(Fe^{3+})O \cdot (Fe^{3+}_{5/3}\ V_{1/3})\ O_3] \tag{15.5}$$

where *V* is a vacancy in the lattice.

15.5.3 Ferrites of corundum-type oxides

Hematite (αFe_2O_3) and ilmenite ($FeTiO_3$) are natural oxides like magnetite (Fe_2O_3) which exhibits rhombohedral lattice symmetry with metal ions occupying various sites as illustrated in Figure 15.9.

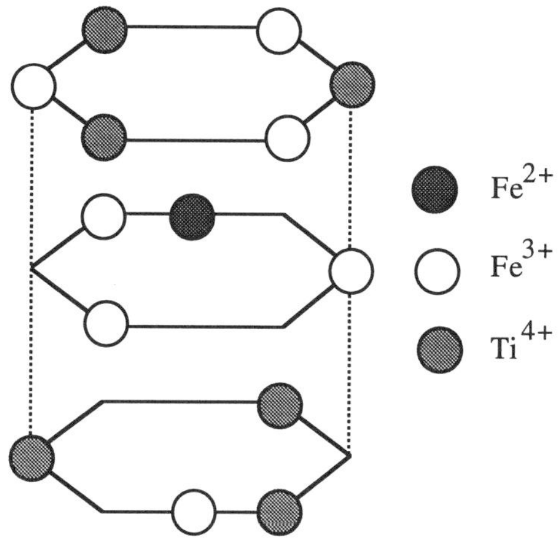

Figure 15.9 Ilmenite lattice structure.

15.6 Ferrites of Magnetoplumbite-Structured Oxides

These have a lead-based composition given by $PbFe_{11}AlO_{19}$ or with a generic formula $MO \cdot 6Fe_2O_3$ where M represents a divalent ions such as Ba^{2+}, Sr^{2+}, or Pb^{2+}. This type of oxide has hexagonal structure composed of stacked spinel ionic layers with interspaced ionic layers of M^{2+}, O^{2-}, and Fe. These types of oxide ferrites have extensive magnetocrystalline anisotropy due to their low crystal symmetry. A typical example of this ferrite is the barium ferrite which is popularly used in making permanent magnets.

Another class of oxides which resemble magnetoplumbite with hexagonal crystal structure are known as *W-type*, *Y-type,* and *Z-type* ferrites as indicated below:

W-type:	$1BaO \cdot 2MO \cdot 8Fe_2O_3$
Y-type:	$2BaO \cdot 2MO \cdot 6Fe_2O_3$
Z-type:	$3BaO \cdot 2MO \cdot 12Fe_2O_3$

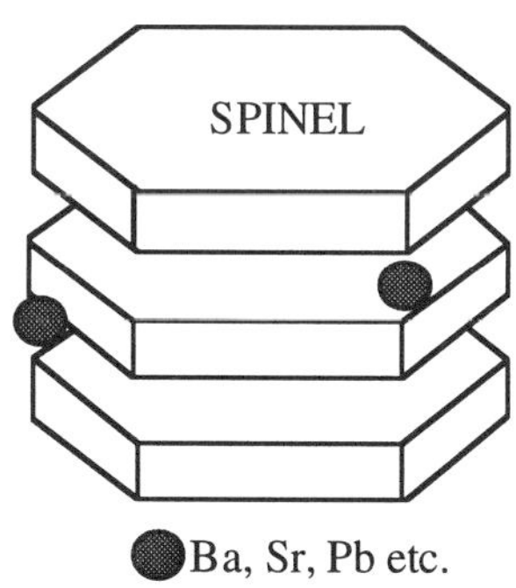

Figure 15.10 Magnetoplumbite.

where again M denotes a divalent metal ion such as Mn^{2+}, Fe^{2+}, Co^{2+}, Ni^{2+}, Zn^{2+}, or Mg^{2+}. Among these, Co_2Z and Mg_2Y exhibit negative anisotropy constant (meaning a basal plane with an easy plane of magnetization). Hence, they are useful as excellent high frequency magnetic materials. Saturation magnetization range of magnetoplumbite oxides is similar to spinel ferrites and the Curie points range from 400 to 500°C.

15.7 Ferrites of Perovskite-Type Oxides

Perovskite is a Calcium titanate ($CaTiO_3$)-based mineral. If Ti is replaced by Fe, a magnetic perovskite-type oxide with a formula $MFeO_3$ is realized where M represents a large metal ion such as La^{3+}, Ca^{2+}, Ba^{2+}, or Sr^{2+}. These oxides have cubic structure as illustrated in Figure 15.11.

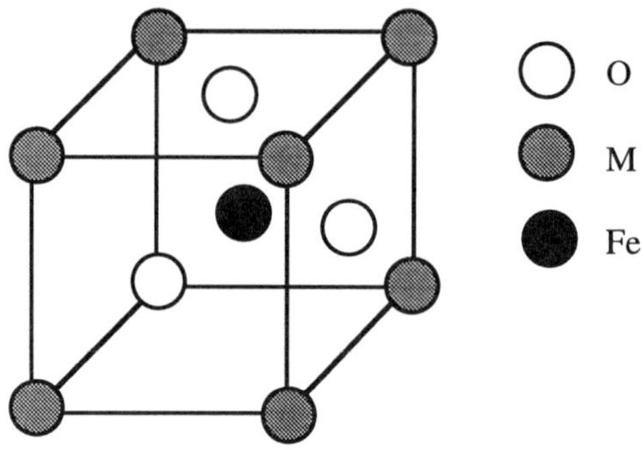

Figure 15.11 Perovskite-type metal oxide.

Another class of ferromagnetic materials of this type are obtained by solutions of (M_1^{3+} $Mn^{3+}O_3$ + M_2 $Mn^{4+}O_3$) where (M_1 and M_2) are (La, Ca, Sr, or Ba), (although single oxides are invariably antiferromagnetic).

15.8 Garnet-Structured Ferrimagnets

Garnets are minerals with pyrope composition, namely, Mg_3Al_2 $(SiO_4)_3$. If Si is replaced by Fe, a family of ferrimagnetic garnets with stoichiometric formulation $3M_2O_3 \cdot 5Fe_2O_3$ are obtained where M represents a rare-earth element, namely, Sm, Eu, Gd, Tb, Dy, Ho, Er, Tm, Yb, Lu, or Y.

Ferrimagnetic garnets have complex cubic crystalline structure. Properties of these materials are listed in Table 15.4.

Table 15.4 Properties of Typical Garnets

Rare-earth Element (M) in the Garnet	Curie Point (K)	M/M_B	Density (gram/cm^3)
Sm	560	9.30	6.235
Eu	570	5.00	6.276
Gd	564	30.00	6.436
Tb	568	31.40	6.533
Dy	563	32.50	6.653
Ho	567	27.50	6.760
Er	556	23.10	6.859
Tm	549	2.00	6.946
Yb	548	0	7.082
Lu	539	8.30	7.128
Y	560	9.44	5.169

Garnet-type ferrites exhibit high resistivity due to the absence of divalent metal ions. Therefore, they are low-loss materials at microwave frequencies.

15.9 NiAs-Type Compounds

Pyrotite compound (Fe_xS) is a typical example of this material which is ferrimagnetic. The index x may range from 0.90 to 0.875. Another example is Cr_yS (with y = 1.17) which is ferrimagnetic between -114^oC and its Curie point, 40^oC. The (MnSb-CrSb) system also shows ferrimagnetic properties with Néel points varying with the volume fractions of the composition.

15.10 Hard and Soft Ferrites

On the basis of hysteresis performance of a ferrite it can be classified as *hard* or *soft* as follows:

Soft ferrite: $H_c \rightarrow 0$; $B_r \rightarrow 0$ and $B_m \rightarrow$ large value
Hard ferrite: $H_c \rightarrow$ high value; $B_r/B_m \rightarrow 1$ (squareness) and $B_m \rightarrow$ large value.

Typical characteristic parameters of hard and soft ferrites are listed in Table 15.5.

Table 15.5 Properties of Soft and Hard Ferrites

Type	Ferrite	B_m	H_c	B_r	μ_r
Soft	0.8Ni + 0.2 Fe	0.38	0.10	~0.60	50,000
	Mn Zn				
	$Co_{10}Fe_5Si_{15}B_{10}$	0.24	0.02	~0.40	5,000
		0.31	0.10	~0.50	10,000
Hard	γ-Fe_2O_3	0.24	250-450	~0.25	~120
	CrO_2	0.24	450-600	~0.60	120
		0.18	800-3,000	~0.25	320
	$BaO \cdot 6Fe_2O_3$	0.50	40,000	~1.40	720
	Sm CO_3	0.60	17,000	~1.75	920
	Sm_2CO_{17}	~0.60	12,000	~2.00	310
	$Fe_{14}B\ Nd_2$				

All values relative to those of Fe.

15.11 Applications of Ferrites

Ferrimagnetic materials are used widely in storing data (as memories) in computers and in high frequency applications. Based on their, applications ferrites can be classified into the following categories:

Ferrites for permanent magnets: Hard ferrites
Ferrites for transformers and inductors: Soft ferrites
Data storage: Rectangular loop ferrites
Microwave applications: Ferrites and garnets

As permanent magnets, barium or strontium ferrite materials are mostly used ($BaFe_{12}O_{19}$ and $SrFe_{12}O_{19}$, respectively). These materials have high value of uniaxial anisotropy field, high coercive force, and high resistivity. They are used as focusing magnets in television tubes. Typical commercial versions of barrium ferrite permanent magnets (ceramic materials) are Ferroxdure™, Magnadur™, and Feroba™. For the $BaFe_{12}O_{19}$ ferrite permanent magnet, the value of $(BH)_{max}$, remanence, and coercivity are 8 Joule/meter^3, 0.21 weber/meter^2, and 14000 ampere/meter, respectively.

Another major application of ferrites refers to the category of soft ferrites. They are used as inductor cores, transformer cores, television transformers and rod aerials. For these applications, ferrites should have high permeability, low coercive force, low eddy current losses, and ability to operate up to frequencies of 10 MHz and with special requirements extending up to 1000 MHz. The ferrites used for this purpose are manganese zinc ferrites and nickel zinc ferrites.

Some ferrites have hysteresis loops which are almost rectangular in shape. This property makes them suitable for use in a *magnetic memory core* in computers. Figure 15.12 shows an ideal rectangular hysteresis loop of a ferrite. The two points –P and +P represent two stable states of magnetization of a ferrite material which is usually in the form of a small toroid. These states can represent a "zero and one" in digital storage of binary information. The loop can only be traversed in anticlockwise direction and the state –P can be changed rapidly to +P by the application of a field greater than or equal to H_a. In the same way +P can be changed to –P by the application of a field less than or equal to $-H_d$. The switching time is on the order of 1 microsecond.

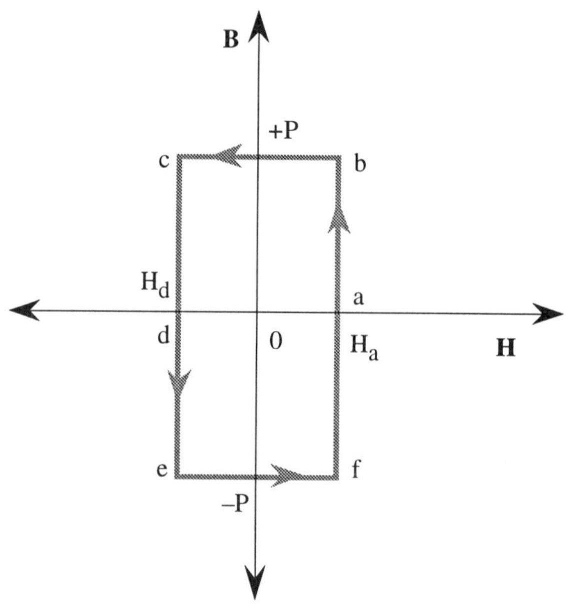

Figure 15.12 Ideal hysteresis loop (BH curve) of a ferrite core.

Figure 15.13 shows a two-dimensional matrix array of toroidal ferrite core loops used to store binary information. If current corresponding to a field $H_A/2$ is passed through an X and a Y wire, only where they intersect will the field be of sufficient magnitude to switch that core, should it be in state –P(o); otherwise, if it is in state "1" already, it will remain in that state. A negative current corresponding to a field $-H_d/2$ in both wires will cause the core at the intersection to change from "1" to "0", if state "1" prevails at the intersection, but will have no effect if it is in state "0". In actual practice, a three-wire system is used for a three-dimensional matrix array of ferrite cores. Normally a fourth wire is used to inhibit the reading of states which are not required. Ferrites commonly used for this application are manganese-magnesium type, manganese copper ferrite, and lithium nickel ferrite. In practice the loops look more like Figure 15.14.

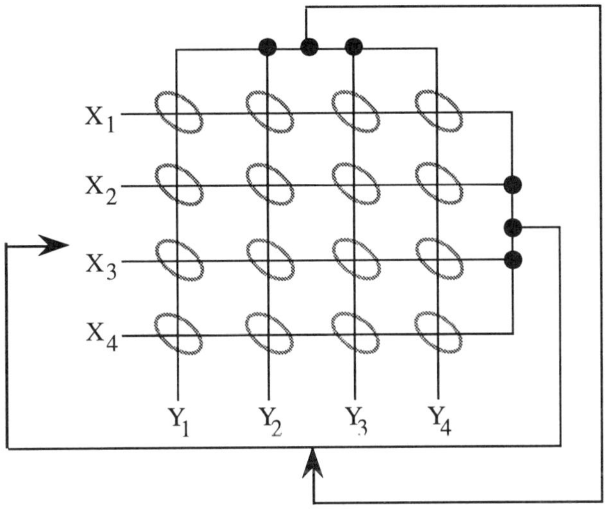

Figure 15.13 A matrix array of ferrite cores.

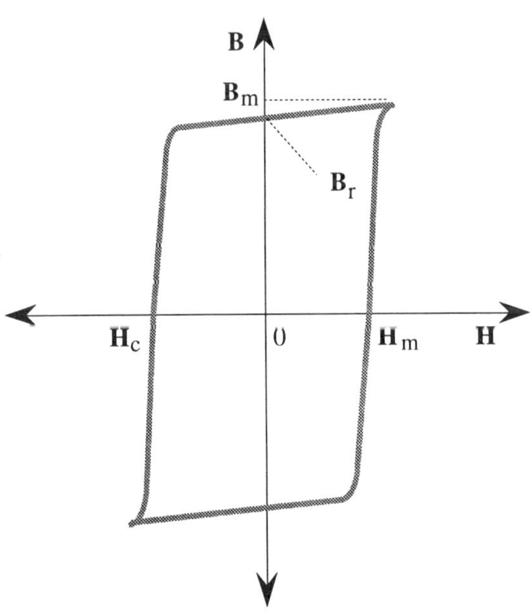

Figure 15.14 Actual hysteresis loop of a ferrite core.

A Mn|Mg|Zn ferrite has a maximum flux density, 0.25 weber/meter^2; remanent flux density, 0.21 weber/meter^2; coercive force, 23.8 ampere/meter; H_m, 43.7 ampere/meter, T_c, 160°C; and permeability μ_r, 260 (initial) and 4500 (maximum).

Microwave ferrites are used in the frequency range of 1-100 GHz. In this range of frequency, electromagnetic waves interact with the spin magnetic moments in the ferrite. As discussed in Section 15.4, the process which takes place is called *Faraday rotation*. This is the rotation of the plane of polarization of a plane electromagnetic wave as it travels through a ferrite in the direction of an applied magnetic field.

Application of Faraday rotation is in waveguides to accept or reject polarized microwaves. These are known as *nonreciprocal microwave devices.* Thus (unwanted) reflected signals are prevented in the so-called *isolators* and rotation of the plane of polarization is achieved in the devices known as *gyrators.* Ferrites are also used in *phase shifters* and *circulators.* Some of the microwave ferrites are: Single crystalline; $MnFe_2O_4$, $NiFe_2O_4$, $CoFe_2O_4$, $5Fe_2O_4$, $3Y_2O_3$, etc.; commercial ceramic ferrites-nickel ferrite, magnesium ferrite, etc.; and commercial garnets -YIG, YI(AI)G, YI(AI, Gd)G.

Ferrites can also be used in the fabrication of EMI shields (Chapter 21) and EM absorbing materials (Chapter 22).

A detailed description of microwave applications of ferrites is available in [5].

15.12 Semiconductor Ferrites

Ferrites with semiconducting properties may exhibit the *Hall effect.* For example, Mn-Zn ferrite, under controlled (isothermic) conditions has been shown to produce a Hall electromotive force (EMF). Semiconducting ferrites (which are distinct from insulating or dielectric ferrites) are characterized by a magnetic arrangement which yields a Hall EMF (E_H) given by:

$$E_H = R_o H + R_A M \tag{15.6}$$

where R_o is the *Hall coefficient* and H is the intensity of external magnetic field. Further, R_A is the analogous *Hall coefficient* and M is the magnetization. Typically $(R_A M/R_o H)$ ratio is about 6 in Mn-Zn ferrites. The flow of charge carriers due to the Hall effect is controlled by the *Hall mobility* μ_H.

The temperature dependency of the electrical conductivity (σ) in semiconductor ferrites is given by:

$$\sigma = \sigma_o \, exp[-\Delta E_\sigma / k_B T] \tag{15.7}$$

where ΔE_σ is the activation energy for electrical conduction as dictated by hopping of electrons and/or holes. The mobility (μ) of charge carriers in semiconductor ferrites is given by:

$$\mu = (C/k_B T) \, exp(-\Delta E_h / k_B T) \tag{15.8}$$

where C is a constant characteristic of the type of crystal and ΔE_h is the activation energy associated with the hopping process. The mobility that corresponds to charge conduction mechanism is labeled as the *drift mobility* (μ_D) which is on the order of 10^{-3}-10^{-7} cm^2/volt-sec.

Typical electrical parameters of semiconducting ferrites (with spinel structure are tabulated in Table 15.6 [2].

Table 15.6 Electrical Parameters of Semiconducting Ferrites

Ferrite	ΔE_σ (ev)	ΔE_h (ev)	Gap Energy E_g (ev)	μ_H at 300° (cm^2/volt-sec)
Fe_2O_4	0.3-0.05	0.03-0.95		1
$CoFe_2O_4$	0.2	0.2	0.6	1

(continued...)

Ferrite	ΔE_σ (ev)	ΔE_h (ev)	Gap Energy E_g (ev)	μ_H at 300° (cm^2/volt-sec)
$MgFe_2O_4$	0.35			
$NiFe_2O_4$		0.1	1.1	
$MnFe_2O_4$	0.05-0.08	0.05-0.08	0.3	10^{-1}
$ZnFe_2O_4$	0.17	0.12		10^{-1}

15.13 Ferrite Dielectrics

The dielectric properties of ferrites depend on: Preparation of the mixed medium, chemical composition, grain structure and/or size of the constituent particulates, and type of sintering. In general, the dielectric response of a ferrite is characterized by a complex permittivity $(\varepsilon' - j\varepsilon'')$. The real part is mainly decided by the average grain size of the specimens in the ferrite composition and ε'' refers to the frequency dispersion. The corresponding relaxation occurs at a low frequency due to the conducting grains in the medium being specified by the insulating layers. (This corresponds to a model of a heterogeneous dielectric structure.) There are also high frequency relaxations due to the presence of low conducting surface layers on the grains of the ferrite. Like any dielectric, as presented in Chapter 2, the relaxational attributes of a ferrite can be described by its conductivity (σ) and the dielectric constant (ε') as follows:

$$\sigma = \sigma_0 + (\sigma_0 - \sigma_\infty)/(1 + \omega^2\tau^2) \tag{15.9a}$$

$$\varepsilon' = \varepsilon_\infty + (\varepsilon_0 + \varepsilon_\infty)/(1 + \omega^2\tau^2) \tag{15.9b}$$

where the subscript o indicates the static values and ∞ refers to the high frequency (optical) limit. Further, τ depicts the relaxational time constant.

15.14 Conclusions

Ferrites as a class constitute a major subset of magnetic materials and are of vital engineering utility in modern high-tech applications. Especially at high frequencies, ferrites offer unique application potentials. Generically being a composite material, there is an abundant niche to search for newer ferrite compositions to yield better performance characteristics. Ferrites being ceramic in nature constitute the nonconducting class of magnetic materials distinctly different from conventional metallic/alloy-type magnetic materials, and they offer great potentials for further studies and technological utilities.

References

[1] S. Chikamuzi and S. H. Charap: *Physics of Magnetisms.* (John Wiley and Sons Inc. , New York: 1959).

[2] B. Viswanathan and V. R. K. Murthy: *Ferrite Materials.* (Springer-Verlag/Narosa Publishing House, New Delhi: 1990).

[3] A. Nussbaum: *Electric and Magnetic Behavior of Materials.* (Prentice-Hall Inc. , Englewood Cliffs, NJ: 1967).

[4] R. C. Dorf (Ed.): *The Electrical Engineering Handbook.* (CRC Press Inc., Boca Raton, FL: 1993), pp. 816-818.

[5] R. F. Soohoo: *Theory and Application of Microwave Ferrites.* (Prentice-Hall Inc., Englewood Cliffs, NJ: 1971).

Defining Terms

Antiferromagnetism: Clustering of oppositely oriented ions with parallel spins in a lattice structure having random spins, leading to nil magnetic susceptibility of the material.

Faraday rotation/effect: Change of plane of polarization of EM wave being transmitted upon reflection by the direction of magnetization of the surface of the material.

Ferrimagnetism: Refers to antiferromagnetism with incompletely cancelled spins.

Ferrites: Mixed metal oxides exhibiting ferrimagnetic properties and are antiferromagnets with incompletely cancelled spin system.

Garnets: Ferrimagnetic materials composed of oxides of iron and a rare-earth element.

Inverse spinel-structured ferrites: Ferrites with a spinel structure with half of Fe-ion at A sites and the remaining Fe-ion as well as divalent metal ions at B sites.

Néels temperature: Critical temperature above which thermal agitation destroys antiferromagnetic alignment.

Spinel structure: A lattice structure in which each oxygen ion surrounded by one tetrahedral ion (A-ion) and 3 octahedral ions (B-ions) constitutes a basic unit. Spinel-structured ferrites have divalent metal ions at A site and iron ions at B sites.

CHAPTER 16

Solid Electrolytic Materials

16.1 Introduction

Also known as *superionic conductors*, *solid electrolytes* are ionic materials which exhibit high electrical conductivity (in comparison with liquid electrolytes) above a certain critical temperature. They are also termed as *fast-ion conductors*. A solid electrolyte has the following characteristics:

- It is crystalline with ionic bonding.
- Its electrical conductivity (over a specified temperature) is high ($10\text{-}10^{-2}$ siemen/meter).
- Its principal charge carriers are ions. Hence, the fractional contribution of the ionic conductivity to the total conductivity (termed as *ionic transference number*) is almost equal to one.
- Its electronic conductivity is small. The corresponding *electronic transference number* is less than 10^{-4}.

Examples of typical superionic conductors and their temperature *versus* electrical conductivities are presented in Figure 16.1.

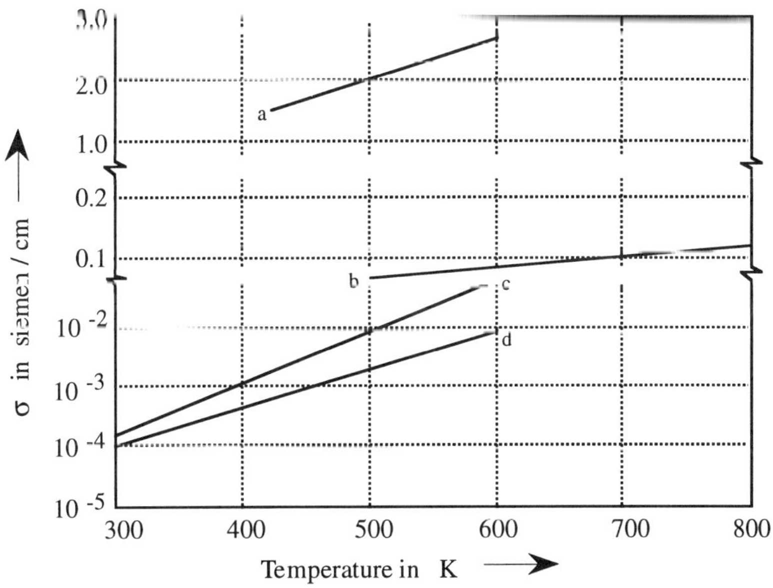

Figure 16.1 Electrical conductivity *versus* temperature of typical solid electrolytes: (a) β-AgI; (b) $LiZr_2(PO_4)$; (c) Li_3N; and (d) $Li_3B_7O_{12}S$.

16.2 β-Phase to α-Phase Transition

Superionic materials exhibit high electrical conductivity above a certain critical temperature. This temperature is well defined in some cases and in some other materials the transitions are not abrupt (Figure 16.1). The temperature-dependent increase in the electrical conductivity is known as the *β-phase to α-phase transition.*

16.3 Structure of Fast-Ion Crystals

The ionic crystalline materials which permit fast-ion transport in general may have:

- Disordered structures
- Channeled structures
- Layered structures

The crystalline structure determines the dominance of electronic or ionic conduction. It is directly related to the *band-gap energy* of the materials. For solid electrolytes, the electronic band-gap energy is always in excess of (T/300) eV (*Heyne's condition*) where T is the temperature in Kelvin. This is only a necessary condition, but need not be sufficient. The high conductivity of solid electrolytes is attributed to ion-ion interactions. Further, defects and disorders are eventually needed in the crystalline structure to sustain a significant ionic transport.

16.4 Types of Defects and Disorders in Solid Electrolytes

Essentially, there are two possible *crystalline defects* pertinent to solid electrolytes. They are:

- Point defects
- Molten sublattice defects

Point defects allow ionic transport through *Frankel (or Schottky) defect pairs*, induced thermally. Hence, the number of defects and the proportionate charge carriers are functions of temperature with the *activation energy* in excess of 1 eV. Examples of solid electrolytes with point defects are:

- Solid electrolytes with dilute defect density: AgCl, B-AgI, NaCl, KCl, etc.
- Solid electrolytes with concentrated defect density: Stabilized zirconia, (*hafia*), CaF_2, etc.

Molten sublattice defects correspond to insufficient sites in the sublattice region for the available ions to occupy them. This results in *ionic hopping* on *free-ion movement* from one available site to the other. This process allows all the ions to participate in the conduction phenomenon with the result the activation energy is rather low. In view of the fact that the defect infestation and the conduction of ions are decided by a statistical average, the solid electrolytes are termed as *average structures* rather than as *rigid structures*.

16.5 Free-Ion-Like Theory

The movement of ions seeking the defects has been modeled as free-ionic motion. Accordingly, fast-ion conductors are classified into three categories:

- Type I: Ionic solids with mobile defects of low concentration ($\approx 10^{18}/cm^3$). These are the same as dilute point-defect versions indicated before.
- Type II: Conglomerated high-density defects extending over the substructure at microscopic dimensions. The corresponding defect concentration is on the order of $10^{20}/cm^3$. These solids are known as *concentrated point-defect versions*.
- Type III: This refers to the participation of all the ions in the conduction process. The mobile ions as charge carriers amount to nearly $10^{23}/cm^3$. The material is *liquid-like molten sublattice* often realized *via* channeled or layered structures.

In practice, Types II and III are commonly useful as fast-ion conductors.

16.6 Ionic Bonding

In some materials, the atoms are bound such that the transference of electrons from one atom to another is feasible and renders the atoms as ions. These ions have closed-shell electronic structure as illustrated in Figure 16.2. The formation of the pair of ions or the *ionic bond* is sustained due to the binding coulombic force of attraction between the *anion* (the atom that accessed extra electrons) and the *cation* (the atom that lost electrons becoming positively charged).

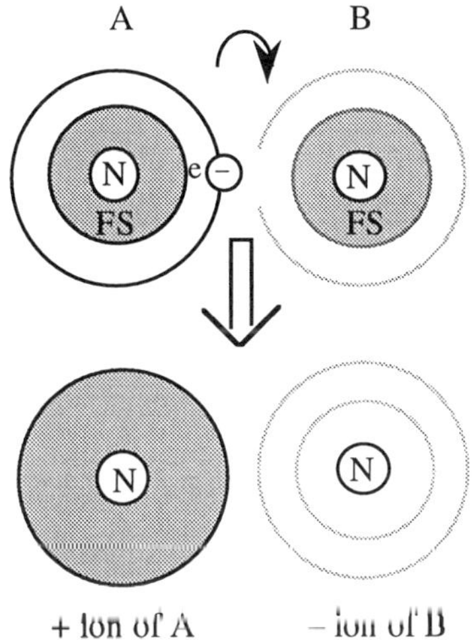

Figure 16.2 Ionic bond formation between two atoms (A and B).
N: Nucleus. FS: Filled shells. e: Electrons.
Atom A: Almost unfilled outershell with loosely bound one electron.
Atom B: Almost totally filled outershell with one deficiency to accommodate an electron.

16.7 Ionic Conductivity Based Classification of Solid Electrolytes

Both cations and anions participate in the electric conduction process in a solid lattice. However, the extent of mobility of these ions could be quite different. Alkali metal ions contribute a high conductivity in solid phase, as first observed in sodium β-alumina. Since then different alkali ion compositions have been synthesized and evaluated as superionic conductors. Typically, a large number of lithium ion conductors have emerged as such superionic materials. This is mainly due to the small ionic radii of L_i^+ offering greater carrier mobility. A few examples of good lithium ionic conductors are lithium β-alumina, LiI doped with CaF_2 or Al_2O_3, $LiAlCl_4$, $Li_4B_7O_{12}$, Li_4SiO_4 + 43 mol% Li_3PO_4, $Li_{0.80}Z_{11.80}Ta_{0.20}(PO_4)_3$, $LiHF_2(PO_4)_3$, and lithium-enriched Li_3N.

Structurally such compounds possess a high degree of disorder or channelled arrangement constituting Type III superionic conductors.

Further, the lithium compounds exhibiting superionic conduction contain tetrahedral anions like SO_4, SiO_4, PO_4, GeO_4, AlO_4, ZnO_4, TiO_4, and GaO_4.

Apart from lithium-based monophase compounds, binary solid solutions of $LiSO_4$ also show high conductivity. Examples are ($LiSO_4 + M_2^{mo}SO_4$) and ($LiSO_4 + M^{di}SO_4$) where M_2^{mo} are monovalent atoms like Na, K, Rb, Cs, Ag, and M^{di} are divalent atoms like Be, Mg, Zn, Mn, Cd, Ca, and Sr.

Ternary solid solutions of LiSO4 which are superionic are: ($Li_2SO_4 + ZnSO_4 + Na_2SO_4$) and ($Li_2SO_4 + Ag_2SO_4 + AgI$).

Other compounds which are of interest as solid electrolytes are:

- Lithium orthosilicate (Li_4SiO_4) (polyhedra structure)
- Li_4SiO_4 with partial substitution of Si with Ge, P, Ti
- Li_4SiO_4 with partial substitution of Li with Zn and Al
- (Li_4SiO_4) modified by substituting SiO_4 with PO_4, GeO_4
- $LiZr_2(PO_4)_3$, $LiHf_2(PO_4)_3$, $LiZr_2(PO_4)_3$ (Distorted rhombohedral symmetry)
- Li_5AlO_4, Li_3GaO_4, Li_6ZnO_4, Li_2S, Li_2O, Li_3AlN_2, Li_{8+4x} $Sn_{1-x}P_4$ anion polyhedra structure)
- Lithium nitride (Li_3N)
- Boracites of lithium (Li_4, $B_7O_{12}X$, X: Cl, Br, S)
- Sodium and potassium based compounds:
  $NaTaWO_6$, $NaTa_2O_5F$, $NaSbO_3$, $2M_2O \cdot 3Nb_2O_5$ (M: Na, K)
  $K_{0.72}L_{0.72}M_{0.28}O_2$ (L: Se, In, M: Hf, Zr, Sn)
  $Na_{0.5}\,In_{0.5}\,Zr_{0.5}\,S_2$
  $Na_{0.8}\,Zr_{0.2}\,S_2$
- β-alumina-type gallates

16.8 β-Alumina

This represents a class of materials having structures similar to the following composition: $M_2O \cdot xAl_2O_3$ (M: Na, K, Rb, NH_4, Tl, Ag and x = 5 to 11). Aluminum could also be replaced by Fe or Ga. Also, Al_2O_3 could be substituted with $Al(OH)_3$, $NaAlO_2$, or $Al(NO_3)_3$ in the sodium β-alumina ($Na_2O \cdot xAl_2O_3$). Again, the Na_2O radical can also be substituted with sodium nitrate, oxalate, carbonate or hydroxide.

16.9 Silver-Ion Conductors

The following silver-ion compounds are also usable as solid electrolytes:

- Ag_2S, Ag_2Se, Ag_2T_e
- AgI
- Solid solution of AgI: MI + 4AgI (M: K, Rb, NH_4)
- MAg_4I_5

16.10 Copper-Ion Conductors

C_u^+ superionic conductors commonly used are: CuI, KCu_4I_5, $RbCu_xAg_{4-x}I_5$, CuRbClI, Cu_2S, and Cu_2Se.

16.11 Oxygen-Ion Conductors

ZRO_2, HFO_2 or CeO_2 forming various solid solutions with other oxides of Ca, Sr, Nd, Sm, Eu, Gd, Dy, Ho, Yb, or Tb.

16.12 Halide-Ion Conductors

- MF_2 (M: Pb, Ca, Ba, Sx)
- MF_3 (M: La, Y, Lu, Ce)
- $La_xSr_yF_z$, $Ca_xY_yF_z$
- $(CeF_3)_x$ $(MF)_y$, (M: Ca, Sr, Ba, Th)
- $Pb_{1-x}Bi_2F_{2+x}$

- $MBiF_4$(M: K, Rb, Ti)
- BiO_xF_y
- $BaCl_2$
- SrBr

16.13 Activation Energy of Superionic Compounds

Ionic movement in a crystal refers, in general, to (1) normal ions into an adjacent interstitial site; (2) interstitial site into an empty normal site; (3) normal ions into adjacent empty normal sites; and (4) interstitial ions into adjacent empty interstitial sites. Accordingly, the free energy required for the transfer of an ion into an interstitial site and the free energy of activation for the mobility out of an interstitial site govern the fast and high ionic conduction. Both these free-energies should be low for such a conduction.

In addition to ionic conduction mode solid electrolytes, there are also fast-ion conductors based on *proton conduction* and *electronic conduction* with ionic transport.

16.14 Engineering Applications of Solid Electrolytes

Fuel-cells: *Fuel-cells* refer to devices in which the embedded electrochemical species are constantly consumed and a continuous electrical energy is made available as the output. The type of solid electrolytes that the fuel cells use are oxide electrolytes.

The solid electrolyte is interposed between a gaseous phase (oxygen) on either side. Depending on the pressure difference of the oxygen in the sides (or the concentration gradient of the oxygen), the O_2^- ions will flow through the solid electrolyte generating an electromotive force given by.

$$V = [RT/|Z|F] \ln(P_2/P_1) \qquad \textit{(Gibbs-Duhem relation)} \qquad (16.1)$$

where, P_1 and P_2 are the gas pressures, F is the *Faraday constant*, $|Z|$ is the absolute value of the valency, R is the gas constant, and T is the temperature.

This type of a cell is known as the *concentration cell.* Typically, the material dispositions in a fuel-cell arrangement are as follows:

$$\text{Pt/(}O_2\text{ or Air)/Oxide electrolyte/ (}H_2, H_2O\text{) or CO or }CH_4\text{/Pt} \qquad (16.2)$$

The transported oxygen across the electrolyte is consumed continuously by burning it with a combustible fuel such as H_2, CO or CH_4. Thus, a continuous electromotive force (EMF) is generated.

Use of manganese-doped covalent cobalt chromite, nickel or cobalt-zirconia cermets as the solid electrolyte in fuel cells has also been recommended. Thin film fuel cell structures with stabilized zirconia electrolyte have been developed. Limitations of solid electrolyte-based fuel cells (as compared to molten salt fuel cells) are: (1) Higher temperature operation; (2) lower power delivery; and (3) use of expensive electrodes such as platinum. However, use of solid electrolyte fuel cells in conjunction with fluidized bed coal reactors have a promising future.

Solid-state batteries: Conventional batteries such as Volta or Leclanche cells use aqueous electrolytes. They have limited performance at excessive or high temperatures and limited shelf-life. They are also bulky and less rugged.

Substitution of aqueous electrolytes with solid electrolytes has led to *solid-state batteries*. Electrochemical EMF develops in a solid-state battery across a pair of electrodes between which a solid electrolyte is sandwiched and if a continuous flow of charge is maintained through the electrolyte, it would constitute an electrical current between the electrodes externally. For example, the cell M/MX/X can be used as an electrochemical cell with M ions moving across the electrolyte with the following set of reactions.

At the anode: $M - e = M^{+}$ (16.3a)

At the cathode: $M + e = X^{-}$ (16.3b)

$$M^{+} + X^{-} + MX \rightarrow M + X \quad (16.3c)$$

Corresponding EMF developed is given by:

$$V = \Delta G/(|Z_x|F) \quad (16.4)$$

where ΔG is the *Gibbs free energy* involved in the reaction, $|Z_x|$ is the valency of the moving ion and ΔG depends on temperature. It should be negative and large for a conceivable EMF; and the ionic conductivity due to the ions participating in the electrochemical reaction must be high in the electrolyte to obtain a low internal cell resistance and a high current delivery.

Further, electronic conductivity must be negligible lest internal battery "short circuit" may prevail, reducing the shelf-life of the battery. Electrode-electrolyte combination, must be physically and chemically compatible in terms of interfacial effects, interdiffusion, pitting, dendritic growth, etc. Typical electrode-electrolyte assemblies for solid-state batteries and their characteristics are presented in Table 16.1.

Coulometer: A solid-electrolyte coulometer is useful in estimating the electric charge and in RC time-constant networks (known as electrochemic timers marketed by Sanyo, Japan and Gould Ionics, USA).

Constructionwise, a coulometer has a solid electrolyte sandwiched between two electrodes. One electrode is made of the same material as the mobile ion species of the solid electrolyte. The other electrode is nonreactive (or insoluble to the mobile ion). For example, referring to Figure 16.3, the electrolyte is a silver ion conductor such as AgBr or $RbAg_4I_5$. One of the electrodes is, therefore, made of silver and the other electrode is either gold or graphite.

When the device is charged across the terminals on the electrodes (of gold or graphite being at a negative or lower potential), the mobile silver ions flowing through the electrolyte accumulate at the gold (or graphite) electrode. When the device is discharged, these accumulated silver ions flow back to the silver electrode. As long as the silver is present on the gold (or graphite) electrode, the voltage across the device is a simple ohmic drop between the silver-to-silver couple. As the accumulated silver ions on the gold (or graphite) electrode deplete, the voltage across the device surges suddenly to an open-circuit voltage. The step or impulsive change of the voltage across the device can be used as a trigger in timing circuits. The Gould Ionics (USA) product is termed as the Coulister Timer™. A typical structure of the device is illustrated in Figure 16.3.

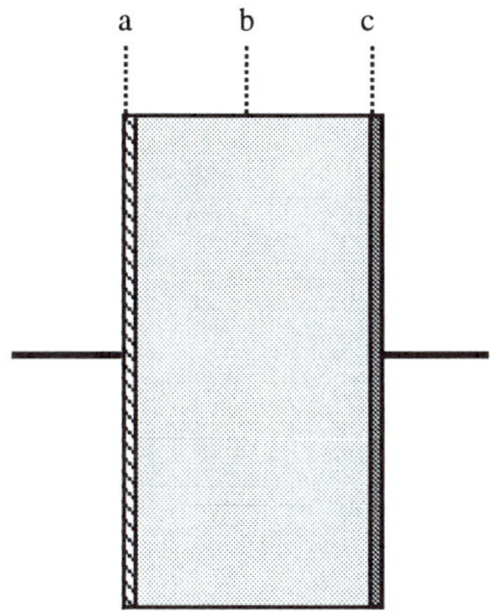

Figure 16.3 A typical solid-electrolyte-based coulometer.
(a) Silver film; (b) Solid electrolyte such as $RbAg_4I_5$; (c) Gold or graphite film.

Electrochemical capacitor: The commonly used liquid/gel-based electrolytic capacitors have limited operating temperature range (0–100°C). They have also limited shelf-life due to the possibility of ionic leakage-current arising from aging and drying of the electrolyte. Further, to prevent electrolyte oozing, hermetical sealing is required which hampers any miniaturization feasibilities. These capacitors require the formation of nonconduction anodic film to function as a capacitor.

Solid-electrolyte-based capacitors have *double-layer capacity* at the ionic-solid-inert-metal electrode interface and could offer capacitance in the range 10–10^3 microfarad/cm^2. Referring to Figure 16.4, the capacitor arrangement has a solid electrolyte with M^+ mobile ions, a reversible electrode of metal M and a polarization electrode of metal M' to which a negative voltage is applied. For example, if the electrolyte is AgBr in which Ag^+ interstitials and vacancies are mobile, the electrode M is Ag while M' could be platinum or graphite.

With the negative voltage applied to M', the defect concentration near the electrode surfaces is illustrated in Figure 16.4. The concentration of positive interstitials, n_i increases towards the electrode while the vacancies (effective negative charges) decrease as per Poisson-Boltzmann distribution of charges in an electric field. This gives rise to a *diffuse space-charge layer* and a *diffuse layer capacity*. The resultant capacitance is decided by the series combination of two capacitors termed as the diffuse-layer capacitor (C_d) and an inner-layer capacitor (C_i). C_d is decided by the Faraday electrolytic chemistry and is a function of temperature. C_i is essentially a parallel-plate capacitance given by $\varepsilon_o\varepsilon_r/d_c$, where ε_o is the absolute permittivity of free space, ε_r is the dielectric constant of the electrolyte, and d_c is a close separation from M' electrode as decided by the finite size of the ions. Effectively, the total capacitance $C = (C_i + C_d)$ depends on mobile ion or defect concentration. Existing developments largely refer to capacitors with Pt/AgBr, Pt/AgI, C/AgI, and Pt/$RbAg_4I_5$ electrode-electrolyte combinations.

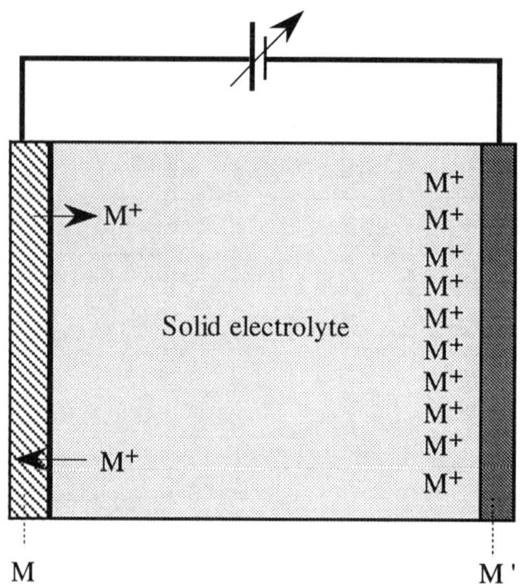

Figure 16.4 Solid-electrolyte-based capacitor. M: Reversible ion electrode (positive); M': Polarized metallic electrode (negative).

Solid-electrolyte electrochromic devices: *Electrochromic devices* are based on materials in which a color change is induced due to an applied electric field or current (see Chapter 18). The mechanisms considered as responsible for the electrochromic effects are:

- Creation of color centers to electron trapping induced by the electric field
- Electric field induced charge transference to an impurity center causing a growth of absorption band at that center
- Shift in absorption band (and hence color) due to tunneling process induced by the applied electric field (*Franz-Keldysh effect*)
- Electrochemical redox reactions in which ions or molecules can be reduced or oxidized (redox) electrochemically with a change in color

Solid-electrolyte-based electronic devices have, for example, the following cell geometry (Figure 16.5):

$$Ag(1)\,|RbAg_4I_5|\,WO_3,\,SnO_2,\,Ag(2) \qquad (16.5)$$

with WO_3 as the cathode. With the application of a potential (less than about 1 volt) across the cell, a blue coloration appears on the $RbAg_4I_5$ film. Upon voltage reversal, the coloration disappears. Other possible cell geometries are structured as follows:

$$Au/WO_3\,|RbAg_4I_5|\,SnO_3 \qquad (16.6a)$$

$$Au/|RbAg_4I_5|\,WO_3/SnO_3 \qquad (16.6b)$$

The observed blue coloration is due to the formation of tungsten bronze A_xWO_3, (A: H^+, Li^+, Ag^+, etc.).

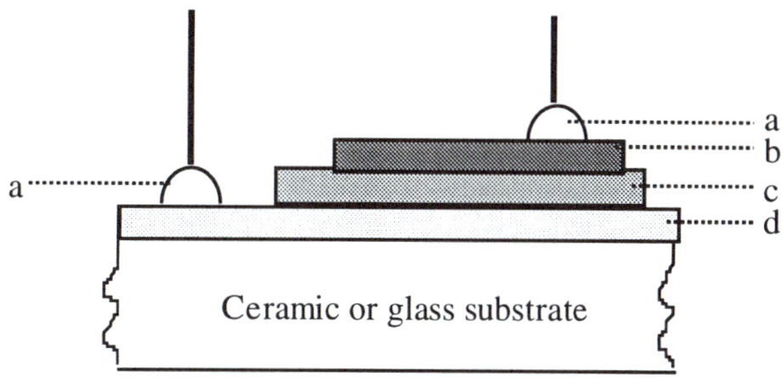

Figure 16.5 Construction of an electrochromic cell. (a) Silver-dag contact material. (b) $RbAg_4I_5$. (c) WO_3. (d) Transparent tin oxide electrode.

Oxygen sensor (Concentration Type): An oxide electrolyte (for example, calcium stabilized zirconia, CSZ) with oxygen on both sides with a pressure difference would allow oxygen ions to flow through the electrolyte developing a potential across it.

$$Pt\ |\underset{at\ p_1}{O_2}|\ CSZ\ |\underset{at\ p_2}{O_2}|\ Pt \tag{16.7}$$

The potential difference so developed is a direct measure of the pressure difference $(p_2 - p_1)$ of the oxygen. In other words, the cell voltage could implicitly measure the oxygen content on one side if the other side is kept at a standard oxygen pressure. Measurement of oxygen content is feasible down to 10^{-7} atm in oxygen containing inert gases, down to 10^{-15} atm in pure O_2, and down to 10^{-25} atm in reducing atmospheres like CO/O_2, CO/CO_2, H_2/H_2O, etc. For oxygen ion mobility across the CSZ cell, the electrolyte warrants a temperature of operation as 1000°C. Lower temperature (700°C) measurements for oxygen pressures on the order of 10^{-10} atm is feasible with Cu-Cu_2O cells. Variations of oxygen measurement cells include oxygen or sulfur sensors and are known as *formation-type cells.*

In concentration, cells with the electrolyte replaced by those carrying other types of mobile ions are useful as sensors for other ionic species. For example, using a sodium-ion conductor (sodium β-alumina), the relevant cell can be used for measuring sodium content in sodium-mercury amalgam.

Solid-electrolyte thermometer: This in principle, is similar to CSZ concentration cell (Pt $|O_2$ at $p_1|$ CSZ $|O_2$ at $p_2|$ Pt). The corresponding EMF developed across the cell is: $V = (RT/4F)\ln(p_1/p_2)$. Hence, the measurement of V directly yields the system temperature. This thermometer is useful for temperatures higher than 1000°C at which gas thermometers are limited due to lack of compatible materials. With modifications, such thermometers could also be deployed for the measurement of thermal gradients.

16.15 Solid-Electrolyte-Based Thermoelectric Generation

A solid electrolyte develops an EMF when its two faces are either at different temperatures or at different pressures due to the electrochemical activity of mobile ionic species in the conductor. In principle, this refers to *thermoelectric power generation.*

16.16 Solid-Electrolyte-Based Active Electromagnetic Surfaces

Controllable electromagnetic absorption and/or reflection by materials are of importance in the development of radar absorbing surfaces and in certain EMI/EMC problems [4,5]. Conventionally, microwave materials composed of a combination of metallic and/or nonmetallic (dielectric) absorbing constituents are used for this purpose (see Chapter 22). For discrete-tuned frequency applications magnetically and dielectrically lossy materials could be blended to obtain moderate performance on absorption/reflection characteristics. The base materials for such applications include: Graphite/iron/aluminum particles (spherical/fibrous/flaky) dispersed in a host medium such as natural rubber-latex, polyisoprene, neoprene, silicone, urethane, etc. However, for better absorption the frequency-tuning is done by the principle of quarter-wave window(s) *via* multiple layers of lossy dielectrics.

An alternative approach suggested by Meyer et al. [7] consists of distributing a large number of magnetic dipoles on a conducting surface to achieve pronounced reflection/absorption characteristics depending on the orientation and distribution of the dipoles. A successful application of this principle has been reported by Chatterjee et al. [8]. Typically, a reflectivity reduction on the order of –20 to –30 dB could be accomplished at selective resonance frequencies by these passive surfaces. Uses of such single-frequency tuned-absorbers include narrowband RCS reduction, minimizing unwanted reflections inside aircraft romes and reducing reflections from shipborne structures, etc.

Modified versions known as graded absorbers designed for broadband applications have been developed for the purpose of broadband RCS reduction, EMI shielding, sidelobe absorption in antennas, and test screens used to prevent personnel radiation hazards in high power radar range applications.

The aforesaid materials are, in general, known as *passive absorbers* and are detailed in Chapter 22. In contrast, recently a class of electromagnetic materials/surfaces have been studied which can be manipulated electrically/electronically to alter their reflection characteristics, scattering pattern and frequency selectiveness. Such materials are known as active media with the surface made "actively" sensitive to incident microwaves. The design principle of such surfaces is the logical extension of smearing the surface with the dipoles (as described by Meyer et al. [7] and Chatterjee et al. [8]) except that the included dipoles at the resonant structures should be made electronically "active" or "tunable" so as to yield desired reflection/absorption characteristics.

The use of pyrosensitive solid electrolytes (such as AgI) has been suggested by the author [4] as active elements at the nodes of synaptic arrangement. On thermally energizing these nodes, the solid-electrolyte material would exhibit superionic electric conduction at elevated temperature(s). With the result the surface (at the nodal points) which is dielectric at cold conditions becomes conducting at hot conditions. Thus, the microwave reflection at this test surface can be effectively altered by the electrothermal synergism. The heating of the nodes can be performed by conductive line segments of heating elements joined (synapsed) at the nodes where the solid electrolytes are planted.

To demonstrate the feasibility of realizing an active surface of the type under discussion, a test surface shown in Figure 16.6 has been developed and tested by the author as reported in [4]. It consists of a heat-resistant dielectric such as ceramic plate with two-dimensional array of holes made to accommodate the pellets of a solid electrolyte. There are a number of solid electrolytes which exhibit high electric conductivity (on the order of 10^{-1} to 10^{-4}siemen/cm) at characteristic temperatures. For example, $RbAg_4I_5$ has a high conductivity (0.27 siemen/cm) even at room temperature; other materials like β-alumina and β-AgI show increasing conductivity with increasing temperature. The compound β-AgI exhibits superionic conductivity with an abrupt transition at a temperature close to 147°C. This transition as mentioned earlier is known as the β - to α-phase transition.

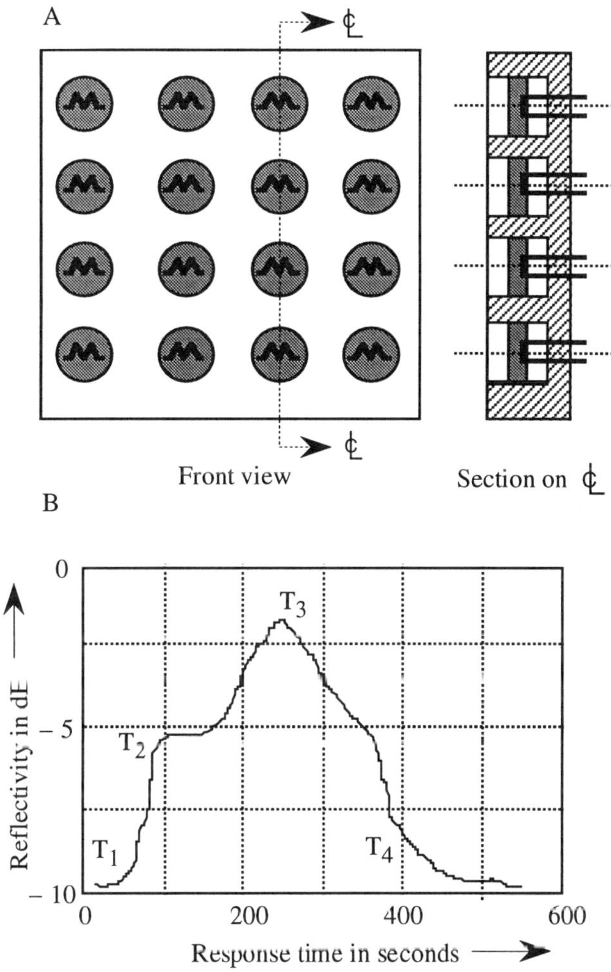

Figure 16.6 (A) An active electromagnetic reflective surface constituted by pellets of AgI embedded in a passive substrate. (B) Reflectivity response of the active surface. (a) Heat resistant substrate. (b) AgI pellet. (c) Nichrome filament. $T_1 = 45^oC$; $T_2 = 150^oC$; $T_3 = 180^oC$, and $T_4 = 80^oC$. (Dimensional details of a typical active test surface is given in [4].)

In the construction of the active test surface of Figure 16.6, the silver iodide (AgI) powder was pelleted as tables and used in the nodes of the ceramic plate. These nodal points were connected by constantan wire on the rear side of the ceramic plate. By properly energizing the matrix of heating elements from a direct current source, the solid-electrolyte pellets in the two-dimensional array could be chosen for heating selectively.

Upon energization those pellets which receive the heat energy would switch to α-phase posing thereby a high conductivity zone. Microwave energy falling on these zones would therefore suffer intense reflection. Hence, with the proper choice of nodes selected for heating, the reflection/transmission pattern of the overall surface can be controlled and configured as desired.

16.17 Kondo Insulators

A class of materials which can be regarded as the duals of solid electrolytes refers to *Kondo insulators* [9]. They exhibit Kondo lattice behavior near room temperature, but step into semiconducting properties with very small energy gaps as the temperature is lowererd. In other words, such materials exhibit higher conductivity at low temperatures and become insulators at higher temperatures.

Certain metallic rare-earth and actinide compounds exhibit Kondo insulation characteristics. Specifically, cerium compounds such as $Ce_3Di_4Pt_4$ have a resistance rising some two orders of magnitude on cooling from 300 to 4 K.

16.18 Conclusions

Fast-ion conductors have creditable performance characteristics which set them at the brink of massive industrial application potentials in the near future. Their applications involve their use mostly in bulk form. This offers their usability with large power handling capabilities.

References

[1] S. Chandra: *Super-ionic Solids*. (North-Holland Publishing Co., Amsterdam: 1981).

[2] B. V. R. Chowdari, P. S. Neelakantaswamy and S. K. Akther: Application of logarithmic law of mixing for estimation of complex permittivity and electrical conductivity of fast-ion conductors at microwave frequencies. Solid State Ions, vol. 18/19, 1986: 122-126.

[3] J. Hladik: Transport processes in solid electrolytes and in electrodes. in *Physics of Electrolytes*, vol. I, (Academic Press, London: 1972), pp. 35-39.

[4] P. S. Neelakanta, J. Abello and C. Gu: Mocrowave reflection at an active surface imbedded with fast-ion conductors, IEEE Trans. Microwave Theory Tech., vol. 40(5), 1992: 1020-1030.

[5] C. Gu, P. S. Neelakanta, V. Ungvichian and P. F. Wahid: A microwave gaussian beam launcher with an active aperture-blockage to control the spot-size of the beam. IEEE Trans. Microwave Theory Tech., vol. 42(3), 1994: 520-522.

[6] P. Vashishta et al. (Eds.), *Fast Ion Transport in Solids*. (North-Holland Publishing Co., Amsterdam: 1979).

[7] M. Meyer, H. Severin and G. Umlauff: Resonanzabsorber fur elektromagnetische Wellen. Z. Physik, vol. 134, 1954: 465-477.

[8] S. K. Chatterjee, H. Kaushal and R. Chatterjee: A two dimensional array absorber for microwaves. J. Ind. Inst. Sci., vol. 51(1), 1969: 103-113.

[9] G. Aeppli, and Z. Fisk: Kondo insulators. Comments on Modern Physics: Part B, vol. 16(3), 1992: 125-190.

[10] M. J. Rice and W. L. Roth: Ionic transport in super ionic conductors: a theoretical model. J. Solid State Chem., vol. 4, 1972, 294-310.

Defining Terms

Active electromagnetic surfaces: The surface which exhibits electromagnetic absorption or reflection of electromagnetic waves controllable by an external application of voltage or current.

Coulometer: A device to estimate electric charge in RC-type networks.

Electrochemical capacitor: A capacitor with an electrochemical material as the dielectric medium between the electrodes.

Electrochromic device: A device exhibiting a color change due to an applied electric field or current.

Electronic conduction: Electrical conductivity dictated by the flow of electrons.

Fuel cells: Devices which consume embedded electrochemical species and deliver electrical energy.

Fast-ionic conductor: Same as solid electrolyte.

Ionic bonding: The bonding between the atoms which permits transfer of electrons from one atom to the other, rendering the atoms as ions.

Ionic conduction: Electrical conductivity dictated by the flow of electrons *via* ionic hopping or free ion movement .

Ionic transport: Transport of ions in solid electrolytes constituting ionic conduction process. (*Note*: An interesting theory on ionic transport in solid electrolytes has been portrayed by Rice and Roth [10], a summary of which is presented in Appendix 16A.)

Table 16.1 Details on Typical Solid-Electrolyte-Based Batteries [1]

Materials*	Solid Electrolyte	Cathode Materials	Electrochemical Reaction	Approximate Open cell Voltage (V)	Current Density per cm^2	Approximate Energy Density ($W\ h\ kg^{-1}$) Theoretical	Practical	Remarks
(1) Silver or (Ag-Hg amalgam)	AgI-based solid electrolytes	(a) I_2 solid, RbI_3, $(C_4H_9)_4$, NI-I_2 complex + C	$Ag + 1/2I_2 \rightarrow AgI$	0.69	~ 1mA	49	1-10	A(R) B(M) V(L)
		(b) Ag_2 Te-Te, C	$2Ag + Te \rightarrow Ag_2Te$	0.22	~ 1mA	29	–	C(H)
		(c) Ag Se-Se, C	$2Ag + Se \rightarrow Ag_2Se$	0.27	~ 1mA	–	–	R(L)
(2) Lithium	LiI (pure & doped)	(a) I_2, C	$Li + 1/2I_2 \rightarrow LiI$	3.0	~ 10 μA	–	–	A(R)
		(b) AgI, Ag	$Li + AgI \rightarrow Li + Ag$	2.11	~ 10 μA	560	–	B(VL)
		(c) CuI, Cu	$Li + CuI \rightarrow LiI + Cu$	2.08	~ 10 μA	–	–	V(H)
		(d) PbI_2, Pb	$2Li + PbI_2 \rightarrow 2LiI + Pb$	1.9	~ 10 μA	–	–	A(H)
	Lithium β-alumina or molten eutectics like LiCl-KCl	Molten sulfur impregnated graphite or (polysulfides)	$2Li + xS \rightarrow LiI_2S_x$	2.3	~ 1 A	1400	250	B(H) R(R)
(3) Sodium	Sodium β-alumina	Sulfur-impregnated graphite or polysulfides	$2Na + xS \rightarrow Na_2S_x$	2.0	~ 1 A	760	250	A(H) B(H) R(R)
		C_8CrO_3		3.5	~ 1 A	1100	–	

* Mobile cation type.

A(R): Room temperature operation; A(H): High temperature operation (~ 300°); B(M): Moderate current drain; B(L): Low current drain; B(VL): Very low current drain; B(H): High current drain; V(L): Low voltage; V(H): High voltage; C(H): High cost; R(L): Limited rechargeability; R(R): Rechargeable.

Appendix 16A: Theory of Ion Transport in Solid Electrolytes [10]

Ionic solids exhibiting exceptionally high levels of ionic conductivity are found among the cation disordered ionic compounds of the silver halide-chalcogenide type, the various substituted beta aluminas, and certain defect-stabilized ceramic oxides. A theoretical model for ionic transport phenomena in such "super" ionic conductors can be based on the hypothesis that there exists in the ionic conductor an energy gap E_o above which ions of mass M, belonging to the conducting species, can be thermally excited from localized ionic states to free-ion like states in which an ion propagates throughout the solid with a velocity v_m and energy $E_m = 1/2Mv_m^2$. On account of the interaction with the rest of the solid such an excited free-ion like state is supposed to have a finite life-time τ_m. On the basis of a postulated Boltzmann transport equation for the thermal occupations of the various free-ion like states, simple expressions can be derived for the ionic conductivity σ, thermal conductivity K_I, and thermoelectric power Q. The theoretical result for Q is well substantiated in [10] by available experimental data. The result for σ may be used to deduce empirical values for the characteristic "mean-free path", $\ell_o = v_o\tau_o$, of the free-ion like state excited at the gap entry E_o. The characteristic life-time τ_o could be deduced in principal from measurements of the frequency dependent ionic conductivity $\sigma(\omega)$ which, according to the model of [10], should be of the Drude type.

CHAPTER 17

Electrooptic Materials

17.1 Introduction

The term *electrooptic effect* refers to a change in the refractive index of a transparent substance induced by an applied electric field, usually at a frequency below the *optic vibrational resonance* of the lattice or the molecules involved. The linear and quadratic electrooptic effects are known as the *Pockel's* and *Kerr effects,* respectively.

The *linear electrooptic (or Pockel's) effect* refers to a change in relative optical dielectric *impermeability* (B_{ij}) proportional to an applied electric field E_K, whose highest frequencies are below the *lattice resonance* of a crystal. The *impermeability* parameter is defined as:

$$B_{ij} = \varepsilon_o \partial E_i / \partial D_j = (1/\varepsilon)_{ij} \tag{17.1}$$

where ε_o is the absolute permittivity of free space, D is the electric flux density, and ε is the relative permittivity (dielectric constant) of the crystal.

In reference to the materials which exhibit linear electrooptic effect, the following parameters are defined to characterize the electrooptic phenomenon:

1. *Refractive index* of a crystal: Considering light propagation through a crystal, the electromagnetic energy stored in a unit volume of dielectric is represented by an ellipsoid with a geometry that reflects the symmetry of the crystal. This geometrical construction is termed as *optical indicatrix, index ellipsoid,* or *ellipsoid of wave normals* (Figure 17.1) in *X* space with coordinate $s(X_1, X_2, X_3)$, constituting a *principal axis* system. In terms of the optical dielectric impermeability, the refractive index of a crystal is described by indicatrix:

$$B_{ij} X_i X_j = 1 = B_{11}X_1^2 + B_{22}X_2^2 + B_{33}X_3^2 + 2B_{23}X_2X_3 + 2B_{13}X_1 X_3 + 2B_{12}X_1X_2 \tag{17.2}$$

in which summation over repeated indices is implicit and $B_{ij} = B_{ji}$.

2. *Electrooptic coefficient* ($r_{i,j,k}$): This relates to a change in the optical dielectric impermeability and the applied electric field. That is,

$$\Delta B_{ij} = r_{i,j,k} E_k \tag{17.3a}$$

or alternatively written as:

$$\Delta B_{\ell} = r_{\ell,k} E_k , \tag{17.3b}$$

where *i, j, k* represent the Cartesian coordinate axes 1, 2, 3; and ℓ = (ij) refers to the reduced six permuted combinations, namely, 1 ⇒ (11), 2 ⇒ (22), 3 ⇒ (33), 4 ⇒ (23), 5 ⇒ (13), and 6 ⇒ (12).

The other electrooptic parameters of interest are :

(a) *Clamped* (S) and *free* (T) *electrooptic coefficients*: These depict, respectively, the electrooptic coefficients of a crystal at frequencies well above the acoustical resonances (clamped crystal) and at frequencies well below the acoustical resonances (free crystal) at a given constant strain. The parameters $r^S_{i,j,k}$ and $r^T_{i,j,k}$ are related by an expression:

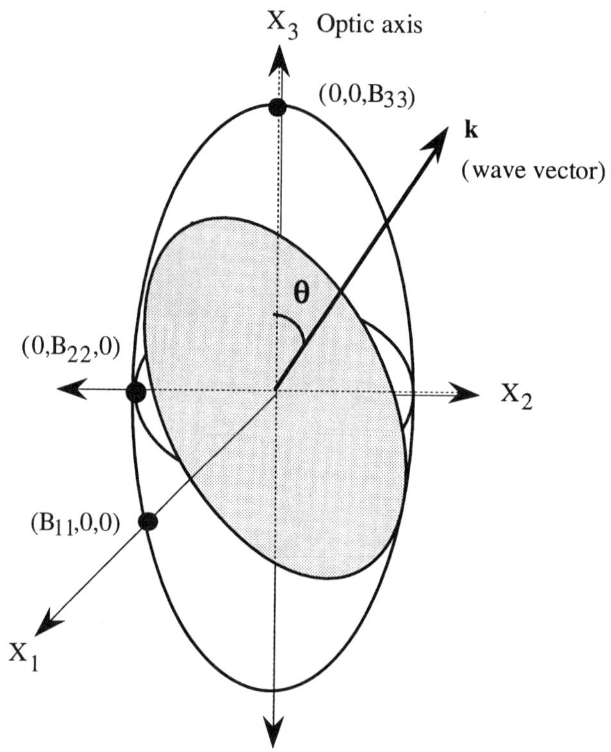

Figure 17.1 Optical indicatrix.

$$r^S_{i,j,k} = r^T_{i,j,k} + p_{i,j,rs}\, d_{i,j,rs} \tag{17.4}$$

where $p_{i,j,rs}$ and $d_{i,j,rs}$ are *elastooptic* and *piezoelectric coefficients,* respectively.

(b) *Nonlinear susceptibility* ($\chi_{i,j,k}$): This is defined as:

$$\chi_{i,j,k} = P_i/\varepsilon_o E_j E_k \tag{17.5}$$

where P_i is the dielectric polarization.

(c) *Electrooptic polarization coefficient* ($f_{i,j,k}$): This defines the electrooptic effect in terms of the optical dielectric parameters rather than electric field. That is:

$$f_{i,j,k} = r_{i,j,k}/[\varepsilon_i(\varepsilon_k - 1)] \tag{17.6}$$

The corresponding *Miller delta* is defined by:

$$\delta_{i,j,k} = \chi_{i,j,k}/[2(\varepsilon_i - 1)(\varepsilon_j - 1)(\varepsilon_k - 1)]$$

$$= [-\varepsilon_i\varepsilon_j r_{i,j,k}]/[4(\varepsilon_i - 1)(\varepsilon_j - 1)(\varepsilon_k - 1)] \tag{17.7}$$

where ($\varepsilon_i, \varepsilon_j, \varepsilon_k$) are *optical dielectric parameters* of the crystal.

(d) *Figure of merit (F)* for a substance in relation to its applications of the electrooptic effect to modulation of light, is defined by:

$$F = (2\varepsilon_o/\pi^2)(\lambda^2 \eta_\alpha \Delta f/P) \tag{17.8}$$

where λ is the optical wavelength, η_α is the *phase modulation index*, P is the *modulating power,* and Δf is the *modulating bandwidth.*

(e) *Half-wave voltage* of an electrooptic crystal $[E_k \ell]_{\lambda/2}$ is the product of applied electric field strength E_k and the propagation distance ℓ, required to produce a phase difference of π radians between orthogonal polarizations (that is, half-wave retardation). For an uniaxial crystal with the optical axis being the z-axis (or 3-axis),

$$[E_3 \ell]_{\lambda/2} = \lambda / n^2{}_3\, r_c \tag{17.9}$$

where $r_c = (r_{33} - n^3{}_1/n^3{}_3)\ r_{13}$ with $n_1 = \sqrt{\varepsilon_1}$ and $n_3 = \sqrt{\varepsilon_3}$.

17.2 Types of Electrooptic Materials

The linear electrooptic materials can be grouped according to their general structures as follows :

- Isomorphs of ferroelectric KH_2PO_4 (KDP) and antiferroelectric $NH_4H_4PO_4$ (ADP)
- ABO_3- type crystals which ferroelectric or pyroelectric similar to perovskites
- Coordinated binary AB compounds which are semiconductors with cubic or hexagonal ZnS structure
- Miscellaneous types

Table 17.1 contains the list of these materials and their characteristic parameters.

KDP-ADP-type materials: Potassium di-hydrogen phosphate (**KDP**) and ammonium di-hydrogen phosphate (**ADP**) are the most widely known electrooptic materials. They are normally grown at room temperature from a water solution and are free of the strains (which are often found in crystals grown at high temperatures). Crystals as large as 5 cm in any dimension can be made at a nominal cost. The crystals are water soluble and fragile. However, they can be handled, cut and polished without difficulty. The electrical resistivity of these materials is typically 10^{10} ohm-cm with a change in the refractive index being a quadratic function of applied field or dielectric polarization.

For effective electrooptic application, use of these materials at T_C (antiferroelectric transition temperature, Curie temperature) is recommended. A material with T_C near the room temperature can be synthesized by forming a solid solution of two materials, one with T_C above and the other with T_C below the room temperature. Such a material, for example, is : $KTa_{0.65}\ Nb_{1.35}\ O_3$ (KTN). However, these mixed crystals are not consistently uniform in their structural, optical, and electrical characteristics.

Table 17.1 Linear Electrooptic Materials and Their Characteristics

A: KDP-ADP Type: 42 Symmetry

	T_C	r_{63}	r_{41}	n_3	n_1	ε_3	ε_1	$\tan\delta_2$	$\tan\delta_1$
KH_2PO_4	123	(T) -10.5	+8.6	1.47	1.51	(T) 21	42		
(KDP)		(S) 9.7				(S) 21	44	(S) 7.5×10^{-3}	4.5×10^{-3}
KD_2PO_4	222	(T) 26.4	8.8	1.47	1.51	(T) 50			
(DKDP)						(S) 48	58	(S) 1.0×10^{-1}	2.5×10^{-2}
KH_2AsO_4	97	(T) 10.9	12.5	1.52	1.57	(T) 21	54		
(KDA)						(S) 19	53	(S) 8.0×10^{-3}	7.5×10^{-3}
RbH_2AsO_4	110	(T) 13.0		1.52	1.56	(T) 27	41		
(RDA)						(S) 24	39	(S) 5.0×10^{-2}	3.0×10^{-2}
$NH_4H_2PO_4$	148*	(T) -8.5	+24.5	1.48	1.53	(T) 15	56		
(ADP)		(S) 5.5				(S) 14	58	(S) 6.0×10^{-3}	7.0×10^{-3}

*Antiferroelectric transition temperature; T_C in K; r_{mi} in 10^{-12}m/V.
(T) = Constant stress; (S) = Constant strain; refractive index at 0.546 μ; tanδ(S) at ~10^{10} Hz.

B: Cubic Perovskites: m3m Symmetry

	T_c	g_{11}	g_{12}	$g_{11} - g_{12}$	g_{44}	n
$BaTiO_3$	401	+0.120	- $\lvert<0.01\rvert$	+0.130		2.40
				0.100		
				(T) 0.088		
				(S) 0.031		
$SrTiO_3$	Low			+0.140		2.38
$KTaO_3$	4			+0.160	+0.120	2.24
$KTa_{0.65}Nb_{0.35}O_3$	~283	+0.136	-0.038	0.174	+0.147	2.29

T_C in K; g_{mn} in m^4/C^2; all measurements at constant stress except where specified; (T) = Constant stress; (S) = Constant strain.

Table 17.1 (*Continued...*) C: Ferroelectric Perovskites

	Symmetry	T_C	r_{13}	r_{32}	$r_{51} = r_{42}$	r_{22}	r_c	$n_1 = n_2$	n_3	$\varepsilon_1 = \varepsilon_2$	ε_3
$BaTiO_3$	4 mm	3930			(T) 1640		(T) 108	(a) 2.440	(a) 2.37	(T) 3000	170
					(S) 820		(S) 23			(S) 2000	100
			(c) (S) 8.0	(c) (S) 28			(S) 19	(b) 2.390	(b) 2.33		
$LiNbO_3$	3 m	1470	(c) (S) 8.6	(c) (S) 30.8	(S) 28	(S) 3.4	(S) 21	(b) 2.286	(c) 2.20	(S) 28	(S) 28
						(T) 7	(T) 19			(T) 78	(T) 32
$LiTaO_3$	3 m	8900	(c) (S) 7.0	(c) (S) 30.3			(S) 24				(T) 47
			(S) 7.9	(S) 35.8	(S) 20	(S) ≈ 1	(S) 28	(b) 2.176	(b) 2.18		(S) 43
							(T) 22				

(Continued...)

(a) at 546 μ, (b) at 633 μ, (c) $r_{13}/r_{33} > 0$; T_C in K, r_{mi} in 10^{-12} m/V, (T) = Constant stress, (S) = Constant strain.

Table 17.1 (*Continued...*) D: AB-Type Semiconductors

	Symmetry	r_{mi}	λ	n_i	λ	ε_i
ZnO	6 mm	(S) $r_{33} = 2.6$	0.63	$n_3 = 2.123$	0.450	$\varepsilon \approx 8.15$
		(S) $r_{13} = 1.4$	0.63	$n_2 = n_1 = 2.106$	0.450	
		$r_{33}/r_{13} < 0$		$n_3 = 2.015$	0.600	
				$n_3 = n_1 = 1.999$	0.600	
ZnS	$\bar{4}3$ m	(T) $r_{41} = 1.2$	0.40	$n_0 = 2.471$	0.450	(T) 16.0
		2.0	0.546	2.364	0.600	(S) 12.5
		2.1	0.65	2.315	0.800	
	6 mm	(S) $r_{33} = 1.85$	0.63	$n_3 = 2.709$	0.360	
		(S) $r_{13} = 0.92$	0.63	$n_2 = n_1 = 2.705$	0.360	
		$r_{33}/r_{13} < 0$		$n_3 = 2.368$	0.600	
				$n_2 = n_1 = 2.363$	0.600	
ZnSe	$\bar{4}3$ m	(T) $r_{41} = 2.0$	0.546	$n_0 = 2.660$	0.546	9.1
						8.1
ZnTe	$\bar{4}3$ m	(T) $r_{41} = 4.55$	0.59	$n_0 = 3.1$	0.570	10.1
		3.95	0.69	2.910	0.700	
		(S) $r_{41} = 4.3$	0.63			
CuCl	$\bar{4}3$ m	(T) $r_{41} = 6.1$		$n_0 = 1.996$	0.535	(T) 10.0
		(T) $r_{41} = 1.6$		1.933	0.671	(S) 8.3
						(S) 7.7

(continued...)

	Symmetry	r_{mi}	λ	n_i	λ	ε_i
CuBr	$\bar{4}3$ m	(T) $r_{41} = 0.85$		$n_0 = 2.16$	0.535	
				2.09	0.656	
GaP	$\bar{4}3$ m	(S) $r_{41} = 0.5$		$n_0 = 3.4595$	0.540	10.0
		(S) $r_{41} = 1.06$	0.63	3.315	0.600	12.0
GaAs	$\bar{4}3$ m	(T) $r_{41} = 0.27$ to 1.2	1.0 to 1.8	$n_0 = 3.60$	0.900	(T) 12.5
			1.0 to 1.8	3.50	1.020	(S) 10.9
		(S-T) $r_{41} = 1.3$ to 1.5	0.9 to 1.08	3.42	1.250	(S) 11.7
			3.39 & 10.6	3.30	5.000	
		(S) $r_{41} = 1.2$				
		(T) $r_{41} = 1.6$				
CdS	6 mm	(T) $r_{51} = 3.7$	0.589	$n_3 = 2.726$	0.515	(T) $\varepsilon_1 = 10.6$
		(T) $r_c = 4$	0.589	$n_2 = n_1 = 2.743$	0.515	(T) $\varepsilon_3 = 7.8$
		(S) $r_{33} = 2.4$	0.630	$n_2 = n_1 = 2.493$	0.600	(S) $\varepsilon_1 = 8.0$
		(S) $r_{13} = 1.1$	0.630			(S) $\varepsilon_3 = 7.7$
		$r_{33}/r_{13} < 0$				

(continued...)

r_{mi} in 10^{-12} m/V; λ in microns; (T) = Constant stress, (S) = Constant strain.

Table 17.1 (*Continued...*) E: Miscellaneous Crystals

	Symmetry	r_{mi}	λ	n_i	λ
$Bi_4(GeO_4)_3$	$\bar{4}3$ m	(T) $r_{41} = 1.03$	0.450 to 0.620	$n_0 = 2.07$	
$C_6H_{12}N_4$–(HMT)	$\bar{4}3$ m	(T) $r_{41} = 4.18$	0.365 to 0.600	$n_0 = 1.591$	0.589
		(T) $r_{41} = 0.80$	0.546		
		(T) $r_{41} = 7.3$	0.547		
Hauynite (mineral)	$\bar{4}3$ m	(T) $r_{41} = 0.04$		$n_0 = 1.496$	
Langbeinites:					
$K_2Mg_2(SO_4)$	23	$r_{41} < 0.04$		$n_0 = 1.535$	
$(NH_4)_2Cd_2(SO_4)_3$	23	(T) $r_{41} = 0.80$	0.546	$n_0 = 1.57$	
$(NH_4)_2Mn_2(SO_4)_3$	23	(T) $r_{41} = 0.60$	0.546	$n_0 = 1.57$	
$NaClO_3$	23	(T) $r_{41} = 0.4$	0.589	$n_0 = 1.515$	
$Na_3SbS_4 \cdot 9H_2O$	23	(T) $r_{41} = 5.66/n_o^3$	0.42		
		(T) $r_{41} = 5.62/n_o^3$	1.08		
Sodium uranyl					
Acetate	23	(T) $r_{41} = 0.87$	0.546	$n_0 = 1.507$	0.546
$LiKSO_4$	6	(T) $r_c = 1.6$	0.546	$n_0 \approx n_1 = n_2 = 1.474$	0.546
$LiNaSO_4$	3 m	(T) $r_{22} < 0.02$	0.546	$n_3 = 1.495$	
				$n_1 = n_2 = 1.490$	
Tourmaline	3 m	(T) $r_{22} = 0.3$	0.589	$n_3 = 1.65$	
				$n_1 = n_2 = 1.63$	

(continued...)

	Symmetry	r_{mi}	λ	n_i	λ
$K_2S_2O_6$	32	(T) $r_{11} = 0.26$	0.546	$n_3 = 1.1518$	
				$n_1 = n_2 = 1.456$	
$Cs_2C_4H_4O_6$	32	(T) $r_{11} = 1.0$	0.546	$n_3 = 1.546$	
				$n_1 = n_2 = 1.564$	
$SrS_2O_6 \cdot 4H_2O$	32	(T) $r_{11} = 0.1$	0.546	$n_3 = 1.528$	
				$n_1 = n_2 = 1.532$	
SiO_2 (Quartz)	32	(T) $r_{11} = -0.47$	0.409 to 0.605	$n_3 = 1.555$	0.546
		(T) $r_{41} = 0.20$		$n_1 = n_2 = 1.546$	0.546
		(S) $r_{11} = 0.23$			
		(calculated)			
		(S) $r_{11} = 0.1$			
$(C_6H_{12}O_6)_2NaBr \cdot H_2O$	32	(T) $r_{11} = 0.1$	0.546	$n_3 = 1.560$	0.546
				$n_1 = n_2 = 1.528$	
Rochelle salt	222	(T) $r_{41} = -2.0$	0.589	$n_1 = 1.491$	0.589
		(T) $r_{52} = -1.7$	0.589	$n_2 = 1.493$	0.589
		(T) $r_{63} = 0.3$	0.589	$n_3 = 1.497$	0.589
$C(CH_2OH)_4$	2	(T) $r_{52} = 1.45$	0.46 to 0.70	$n_1 = 1.528$	
		(T) $\mid r_{12} - r_{32} \mid = 0.7$	0.46 to 0.70	$n_2 \approx n_3 = 1.560$	
$Ca_2Nb_2O_7$	2	(T) $\mid r_{22} - (n_3^3/n_2^3) r_{32} \mid = 14$	0.63	$n_1 = 1.970$	
		(T) $\mid r_{22} - (n_1^3/n_2^3) r_{12} \mid = 12$	0.63	$n_2 = 2.160$	
		(S) $\mid r_{22} - (n_3^3/n_2^3) r_{32} \mid = 13$	0.63	$n_3 = 2.170$	

Adapted from A. Yariv: *Optical Electronics.* (Saunders College Publishing. Philadelphia, PA: 1991), Chapter 9.

Among the ferroelectric perovskites, $BaTiO_3$ possesses good optical and electrical properties and is available as platelets (1/2 mm × 10 mm × 10 mm). Its crystals are in tetragonal phase between 0–120°C.

Rhombohedral crystal of $LiNbO_3$ and $LiTaO_3$ also have good optical and electrical properties. They can be pulled from a melt into a single domain and poled near T_C.

Both KDP and ADP belong to piezoelectric group at room temperature. Below Curie temperature (T_C), KDP crystals become ferroelectric and ADP-type crystals become antiferroelectric below the transition temperature.

The atoms K, H, P in KDP can be replaced by some of the other atoms from corresponding columns in the periodic table without altering the crystalline structure. For example, when H is replaced by deuterium, a significant change in the dielectric properties are observed. A partially deuterated KDP is designated stoichiometrically as: $KD_{2x}\,H_{2(1-x)}\,PO_4$. This family of crystals are transparent for wavelengths as long as 0.2 micron. The dielectric behavior of these crystals changes rapidly near the Curie temperature. It is also accompanied by an increase in loss tangent. The electrooptic coefficient is practically the same for all isomorphs and is independent of wavelength in the transparent region for KDP and ADP.

In the absence of electric field, KDP-type crystals are uniaxial. That is, light polarized parallel or normal to the Z_r-axis travels as a *principal wave* with a refractive index n_3 or n_1, respectively. A field applied along X_1 rotates the indicatrix through a small angle about X_1.

Perovskite family materials: These refer to a group of crystals with structure resembling that of the mineral perovskite, $CaTiO_3$. Particularly $A^{2+}\,B^{4+}\,O_3$ and $A^{1+}\,B^{5+}\,O_3$ oxides which often exhibit ferroelectric properties are of interest in electrooptic applications. These crystals are water insoluble and more rugged and have higher refractive index (and dielectric constant) than KDP. They are transparent between 0.4 and 0.6 micron. Perovskite crystals exist in several forms with varying point symmetries, are derived from the ideal cubic perovskite structure by continuous lattice distortions. In cubic form, they are nonpiezoelectric and nonferroelectric (paraelectric). With structures like tetrahedral or rhombohedral, they assume ferroelectric and /or piezoelectric characteristics.

In cubic structure, $BaTiO_3$, $SrTiO_3$, $KTaO_3$, and $K(Ta,Nb)O_3$ are *centrosymmetric*. Unlike $BaTiO_3$, these crystals can be cut and polished without creating additional domains. These materials are preferable to $BaTiO_3$ due to their ease in handling and availability in crystals. They have lower piezoelectric resonance effect and higher dielectric Q.

AB-type semiconductors: These are binary compounds which are crystallite in either the cubic zinc-blende structure or in the hexagonal wurtzite structure. Examples of these materials are: ZnS, CuCl, CdS, and GaAs. All these materials have large refractive indices.

Miscellaneous versions: As tabulated in Table 17.1 a group of materials which are optically transparent piezoelectrics with various symmetries have been studied as candidates for electrooptic applications.

17.3 Nonlinear Optical Effects (Kerr Effect)

In electrooptical interaction, some materials exhibit nonlinear characteristics. The following parameters are specified to quantify these effects:

Second-order nonlinear susceptibility: It is defined as by the following relation:

$$P_i(t) = \varepsilon_o \chi_{i,j,k} E'_j(t) E'_k(t) \tag{17.10}$$

where $E'_j(t)$ and $E'_k(t)$ represent total fields polarized along x_j and x_k, respectively. The products of these field components lead to *second-order generation* (SHG) effects, quantified by $P_i(t)$.

Dispersion and classification of nonlinear coefficients: The nonlinear coefficients exhibit dispersion (strong function of frequency) near the lattice and electro-resonances. The nonlinear optical properties are derived from two sources: First, from the perturbation of optical polarizability by an electromagnetic field acting through a lattice displacement (deformation potential interaction); second, from the perturbation of optical polarizability produced by the direct action of the field on the electronic energy levels (non-lattice electronic interaction).

The Kerr effect in essence is the birefringence induced in the material by a strong electric field. In terms of a *Kerr constant* ΔB, the birefringence is therefore a quadratic function of the field (E) to a first approximation. That is,

$$\Delta n = \Delta B\, E^2 \tag{17.11}$$

where Δn is the difference between the refractive index $n_{||}$ for a polarized light wave of wavelength λ and electric vector parallel to the applied field, and the refractive index $n_{\perp}$ for the same polarized light wave with its electric vector is perpendicular to the applied field. Approximately Δn is specified by:

$$\Delta n = [N\Delta\alpha/30k_BT\varepsilon_o]\,(\Delta\alpha + (m_e^2/k_BT))\,E^2 \tag{17.12}$$

where $\Delta\alpha$ is a parameter depicting the *polarization anisotropy* which is the difference between the polarizability of the dipoles under parallel and antiparallel arrangements with respect to the applied field. The factor m_e refers to the dipole moment, N is number of dipoles per unit volume, and k_BT is the Boltzmann energy corresponding to temperature T.

The above expression indicates that the Kerr constant is essentially determined by the polarization anisotropy. If the polarizability is isotropic, Kerr birefringence will be absent regardless of strength of the permanent dipole moment. Materials can exhibit both positive or negative Kerr constants.

17.4 Physical Origin of Electrooptic Effects

The linear electrooptic effect results from two distinct types of microscopic interactions: In the first type, the applied field modifies the electronic polarizability of the crystal directly with no elastic changes in the crystal structure. In the second version, the applied field causes a lattice displacement which in turn modifies the electronic polarizability. Both these effects are observed at frequencies far above the acoustic resonances of the crystal.

There is also a third mechanism pertinent to electrooptic effect observed at frequencies of the applied electric field close to acoustical resonances. In this case, the applied field strains the crystal *via* piezoelectric or electrostrictive coupling; and the indices of refraction change due to the strain-optic effect.

17.5 Characteristics of Electrooptic Materials for Practical Applications

- Small half-wave voltage (V_π)
- Small stored-energy parameters $(U_\pi = \varepsilon V^2_\pi)$
- Low dielectric dissipation (loss)
- Good thermal conductivity
- Good optical grade in proper size

- Compatibility for ohmic contacts
- Retention of poled status (in ferroelectric materials)
- Free from optically induced refractive index changes (optical damage)

17.6 Applications of Electrooptic Materials

Electrooptic materials in effect exhibit a change in their indices of refraction proportional to an applied electric field. This property affords a convenient and widely used means of controlling the intensity or phase of a propagating electromagnetic wave. This *modulation* is useful in a number of applications including:

- Impression of information onto optical beams
- Q switching of lasers for generation of intense optical pulses
- Mode locking
- Optical beam deflection

17.7 Electrooptic Amplitude Modulation

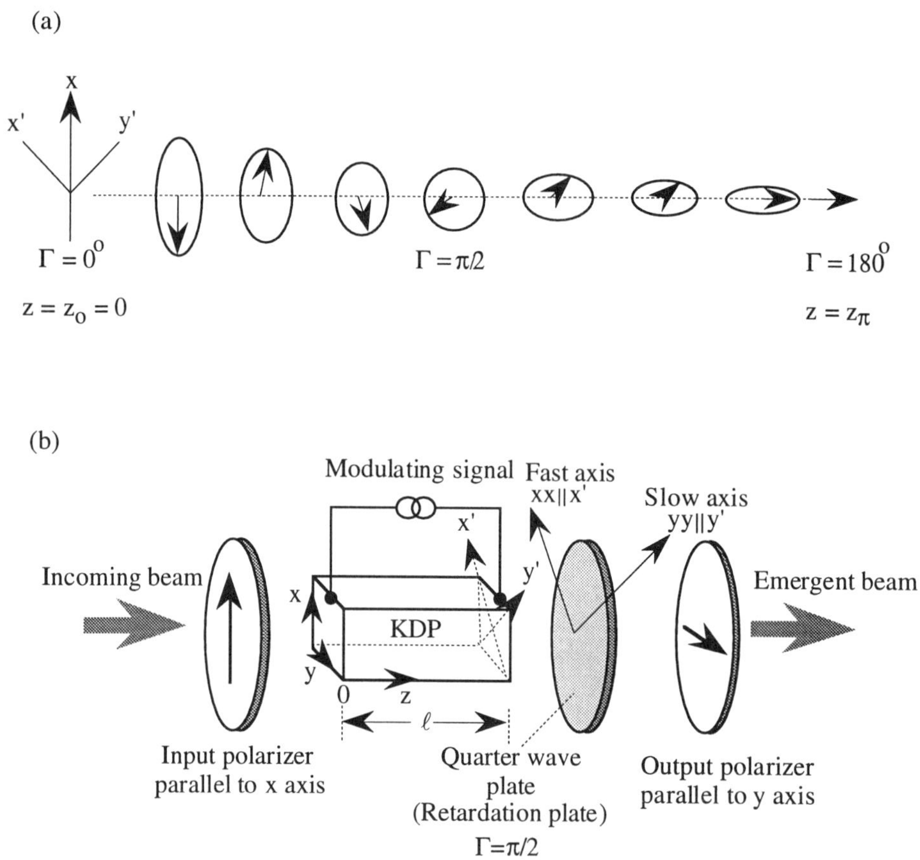

Figure 17.2 (a) Electrical birefringence of a polarized wave. (x-polarized wave acquires a y polarization as it propagates from z = 0); (b) Electrooptic amplitude modulator.

This refers to the electrically induced *birefringence* causing a wave launched at $Z = 0$ (Figure 17.2) with its polarization along x to acquire a y polarization. That is, given a direction in a crystal in general, two possible linearly polarized modes exist, termed as *rays of propagation*. Each mode possesses a unique direction of polarization (direction of electric displacement vector $\boldsymbol{D}$) and a corresponding index of refraction (or the velocity of propagation). When there exist two rays, namely, *ordinary* and *extraordinary* (in the mutually orthogonal directions) with different indices of refraction, it refers to the birefringence; and the electrooptic effect is the change in the indices of these ordinary and extraordinary rays proportional to the applied field.

Referring to Figure 17.2a, the wave which acquires y polarization grows with distance at the expense of the x component until a point z_{π} at which the phase difference between the two components (*retardation*), namely, Γ, becomes equal to π. That is, the polarization becomes parallel to y axis.

If point z_{π} corresponds to the output plane of the crystal and if one inserts at this point a *polarizer* at right angles to the input polarization - that is, one that allows only E_y to pass - then with the field on, the optical beam passes through unattenuated; whereas with the field off ($\pi = 0$), the output beam is blocked off completely by the crossed output polarizer. This control of optical flow of energy serves as the basis of the electrooptic modulation of light. A typical arrangement of the modulator is depicted in Figure 17.2b which includes a KDP crystal. The total retardation Γ, pertinent to this arrangement is the sum of the fixed retardation bias ($\Gamma_B = \pi/2$) introduced by the quarter-wave plate and that attributed to the electrooptic crystal. The ratio of the output intensity to the input is $I_o/I_i = \sin^2(\Gamma/2)$. The process of amplitude modulation is illustrated in Figure 17.3. The applied voltage (modulating voltage) controls the factor Γ. Assuming $\Gamma = (\pi/2 + \Gamma_m \sin \omega_m t)$, and $\Gamma_m << 1$ (low modulation index), the corresponding intensity ratio is $I_o/I_i \cong [1 + \Gamma_m \sin(\omega_m t)]/2$ which is a linear replica of the modulating voltage. If the modulating voltage, for example, represents a signal from a phonograph stylus, the corresponding amplitude modulation on a laser beam could allow the recovery of the signal at an optical detector.

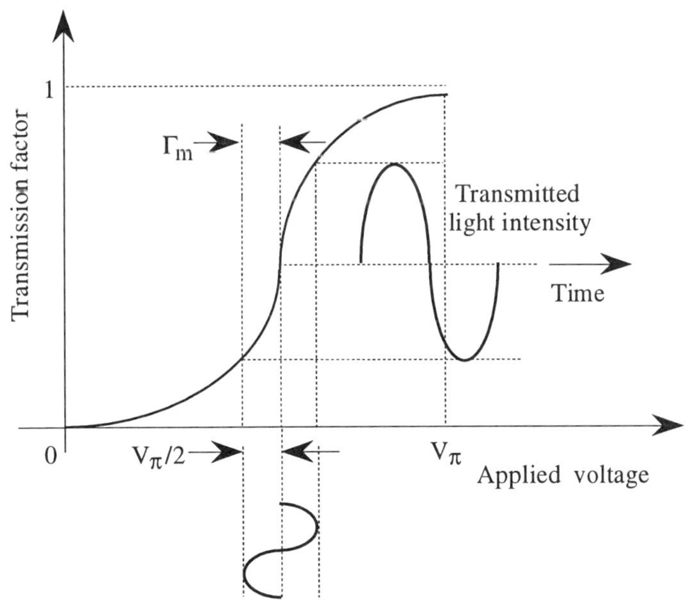

Figure 17.3 Transmission factor *versus* applied voltage in a cross-polarized electrooptic amplitude modulator.

17.8 Phase Modulation of Light

When an arrangement as in Figure 17.4 is made such that instead of there being equal components along the induced birefringent axis *(x' and y')*, the incident beam is polarized parallel to one of them say, *x'*. In this case the application of the electric field does not change the state of the polarization, but changes only the output phase by $\Delta\varphi'_x = (-\omega\ell/c)\Delta n'_x$ where $\omega = 2\pi \times$ frequency, ℓ is the distance transversed by the beam in the crystal, and *c* is velocity of propagation of light. The corresponding output signal can therefore be represented by a phase-modulated waveform given by $e_o = A\cos(\omega t + \delta \sin \omega_m t)$ where δ is termed as the *phase modulation index* decided by the crystal parameters and the wavelength of excitation.

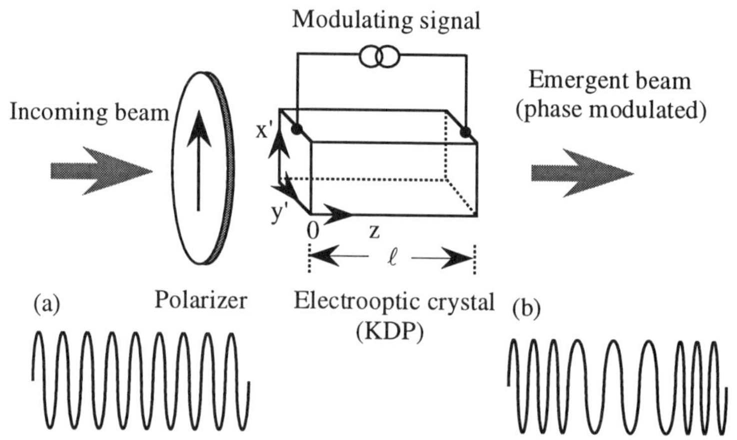

Figure 17.4 An electrooptic phase modulator with optical polarization parallel to electrically induced polarization along x'; (a) Carrier waveform; (b) Phase modulated carrier waveform.

17.9 Transverse Electrooptic Modulation

In the previous modulation techniques, the electric field is applied along the direction of propagation of light. This is called *longitudinal modulation.* Suppose the electric field is applied transverse to the optical beam as illustrated in Figure 17.5. The light propagates along *y'* with a polarization in the *x'-z* plane at 45° from the *z*-axis. The retardation with a field applied along z is given by

$$\Gamma = \frac{\omega\ell}{c}\left[\Delta n + \frac{n_o^3}{2} r_{63} \frac{V}{d}\right] \qquad (17.13)$$

where Δn is the birefringence due to the indices of refraction of ordinary (n_o) and extraordinary rays (n_e); that is, $\Delta n = (n_o - n_e)$. Further, *d* is the crystal dimension along the applied electric field $(E=V/d)$ and r_{63} is a nonvanishing element of the electrooptic tensor of KDP.

The advantages of transverse modulation are that electrodes that apply *E* field do not interfere with the beam and the retardation can be increased by long crystals.

At the high frequency operations, the capacitive effect of the crystal decides:

- Effective modulation bandwidth
- Finite transit time

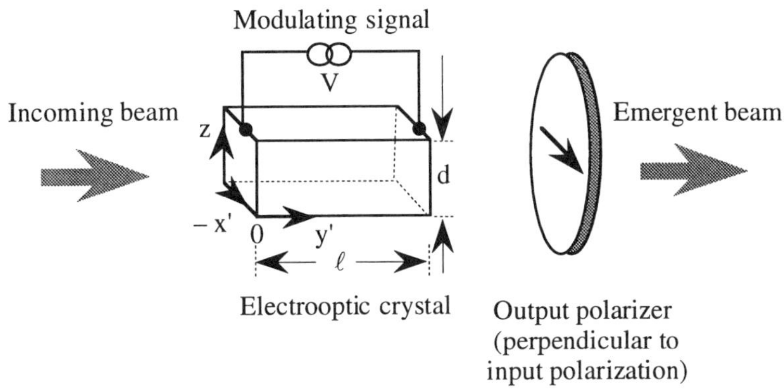

Figure 17.5 A transverse electrooptic amplitude modulator
(z: Direction of optical polarization normal to direction of propagation of the beam).

The *modulation bandwidth* specified as $\Delta\omega/2\pi$ refers to the frequency spectrum occupied by the modulation signal. For a given power P to realize a (peak) retardation (Γ_m), the corresponding modulation bandwidth is given by

$$\Delta\omega/2\pi = [4\pi\ell\, n_o^6 r_{63}^2\, P]/[\Gamma_m^2 \lambda^2 A\, \varepsilon] \tag{17.14}$$

where $n_o\ell$ is the length of optical path in the crystal of area of cross-section A (transverse to ℓ), ε is the dielectric constant at the modulation frequency, and λ is the wavelength of the beam.

Due to the capacitive effect, the peak retardation is reduced by a factor r_c given by:

$$r_c = [1 - \exp(-j\omega_m \tau_d)]/[j\omega_m \tau_d] \tag{17.15}$$

where τ_d is the transit time of the beam through the crystal $(n\ell/c)$ and ω_m is the modulation frequency.

The transit-time limitation can be overcome in principle, by applying the modulation signal in the form of a traveling wave with a transverse modulation arrangement. If the optical and modulation field phase velocities are equal to each other, then a position of an optical waveform will exercise the same instantaneous electric field which corresponds to the field it encounters at the entrance face, as it propagates through the crystal, and the transit-time problem is therefore eliminated.

17.10 Electrooptic Beam Deflection

Light beams can be deflected *via* electrooptic effects. The principle of operation is illustrated in Figure 17.6. An optical wave incident on a crystal in which the optical path depends on the transverse position x, if the index of refraction n depends on x, say a linear function of x (that is $n(x) = n + (\Delta n/D)\, x$), the difference in transit-times experienced by the rays in the upper section and those in the lower section yields a deflected output beam with an angle of deflection $\theta = -\ell\,(dn/dx)$. The deflection phenomenon can be practically obtained by a double-prim arrangement as shown in Figure 17.7. This offers distinct refraction of indices to rays at upper and lower sections at the input leading to a deflection angle $\theta = (\ell/D)\, n_o^6 r_{63}^2\, P\, E_z$, where E_z is the z-directed E field.

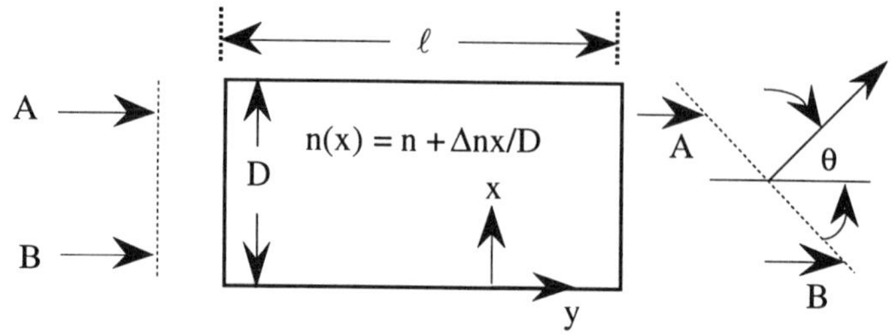

Figure 17.6 Principle of electrooptic beam deflection.

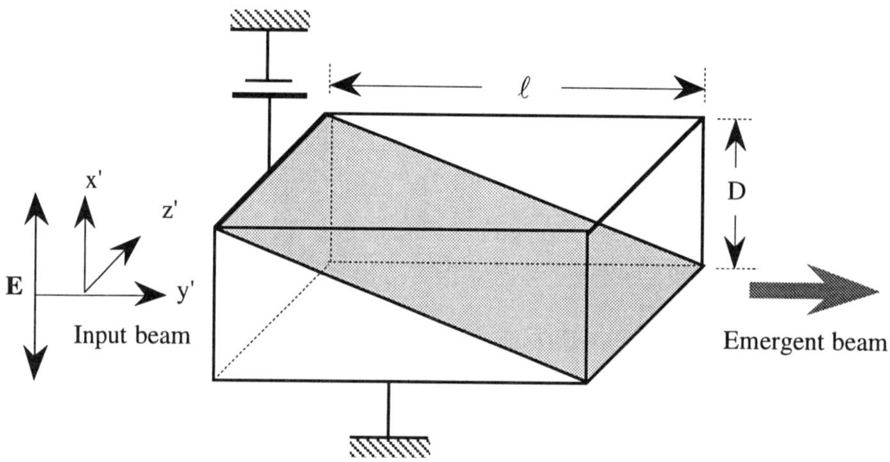

Figure 17.7 Light beam deflection using an electrooptic double prism.

17.11 Magnetooptical Effect

An effect of the magnetic field analogous to Kerr's electrooptical phenomenon was discovered by Kerr and Majorona for certain colloidal iron and iron-oxide solutions. The latter are doubly refracting and pleochroic when observed in a direction perpendicular to the magnetic field lines. Since these liquids are not homogeneous solutions but rather suspensions, the observed phenomenon is attributed to the orientation of suspended particles in the magnetic field.

The phenomenon of rotation of plane of polarization of simple materials such as lead glass in the direction of the lines of force in a strong magnetic field was first observed by Faraday. The rotation direction is reversed with the direction of the field and the extent of rotation per unit length increases with field intensity. The magnetic rotation is also observed in crystals of quartz, beryllium, and tourmaline in the direction of the optical axis.

Magnetooptic modulation of practical value can be achieved with a high quality single crystal of yttrium iron garnet (YIG) with absorption loss less than 0.3 dB/cm at room temperature in the wavelength (λ) range of 1.15 micron to 4.5 micron. High saturation optical rotation (172°/cm at λ = 1.52 micron) together with low absorption loss characteristics permits the use of YIG for efficient modulators in the near infrared band.

Figure 17.8 shows a typical *magnetooptic modulator*. A d.c. magnetic field (H_{dc}) applied along z-direction saturates the YIG rod in the direction normal to the longitudinal axis of the rod. The modulating (RF) voltage (of small signal level) is applied along the rod-axis to "wobble" the magnetization within a small angle to produce a small component of RF magnetization M_x along the x-axis. Since the magnetic rotation is proportional to the component of magnetization along the direction of light propagation (persumed to be along the rod axis), the emergent light beam will have its plane of polarization rotation by an amount proportional to $m_x \ell$ where ℓ is the length of the rod. A linear intensity modulation is obtained after the light beam passes through an analyzer set to 45° with respect to the input polarizer. The modulation power required is given by:

$$P = (\pi/64)\ \Delta f M H_i [\nu/\phi_{s(rod)}]^2 [d^2/l] 10^{-7} \text{ watt} \tag{17.16}$$

where Δf = modulation bandwidth, M = saturation magnetization of the sample, H_i = d.c. effective magnetic field, ν = modulation index (depth of modulation), $\phi_{s(rod)}$ = saturation magnetic rotation per unit length (in radius), d = diameter of the rod and ℓ = length of the rod.

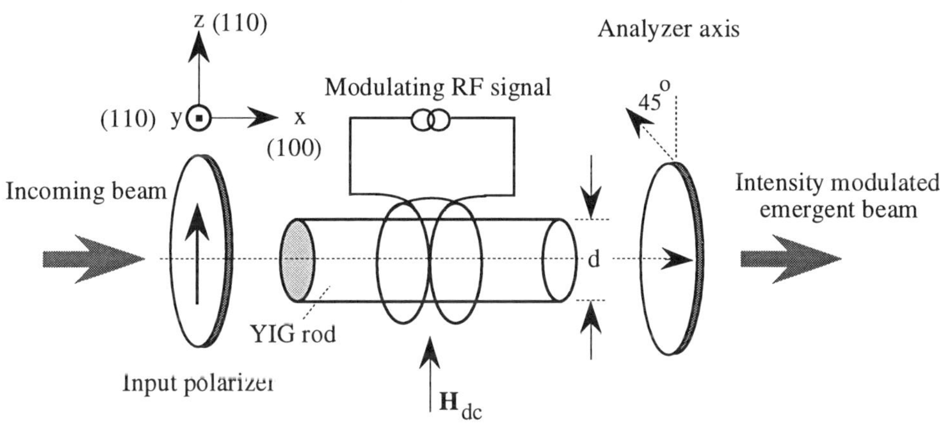

Figure 17.8 Magnetooptic light intensity modulator.

Doping YIG with materials like gallium would reduce the saturation magnetization. Proper choice of crystal orientation helps in increasing the efficiency of the system in tilting the magnetization along the rod axis. Typically gallium-doped YIG modulators provide a bandwidth of a few hundred megahertz in the near infrared band with low absorption of about 0.25 dB/cm.

References

[1] A. Yariv: *Optical Electronics.* (Saunders College Publishing, Philadelphia, PA: 1991), Chapter 9.

[2] I. P. Kaminow: *An Introduction to Electrooptic Devices.* (Academic Press, New York: 1974).

Defining Terms

Electrooptic effect: Change in the refractive index of a transparent material with the application of an electric field below a critical frequency.

Magnetooptic effect: Similar to electrooptic effect with the excitation field being a magnetic field.

Pockel's effect: Linear electrooptic effect.

Kerr effect: Quadratic (nonlinear) electrooptic effect.

CHAPTER 18

Electrochromic Materials

18.1 Introduction

A set of materials exhibit *electrochromic activity* when constituted as a thin film structure whose coloration can be changed reversibly by electrical charging and discharging. There are two types of such *electrochromic materials*. The first type is uncolored when unpowered and changes to a colored state when a voltage is applied and returns to its uncolored state when the voltage is removed. In the second type, an opposite voltage is required to be applied for *bleaching* (or *decoloration*). The *degree of coloration* is a direct visual indication of the state of charging present in the material.

Electrochromic devices are constituted by multilayered structures consisting of an electrochromic electrode, a transparent electrolyte, and a transparent counter or complementary electrochromic electrode.

The phenomenon of coloration (or decoloration) in an electrochromic material is a result of two possible reactions set forth by electrical energization, namely, the *reduction* of an electrochromic electrode and the *oxidation* of a counter/complementary electrode. This phenomenon warrants the flow of compensating ions (typically alkali ions or protons) across the electrolytic layer in order to maintain the charge neutrality of the system.

A well-known electrochromic material is Prussian blue (PB) which in conjunction with WO_3 can be used as a primary or as a complementary electrochromic window [1].

There are two types of PB, namely, soluble PB $[KFeFe(CN)_6]$ and insoluble PB, $Fe_4[Fe(CN)_6]_3$. Both versions have cyanide-bridge iron atoms forming a nominally cubic molecular framework with distinct iron sites. The soluble or insoluble property refers to the relative peptization of each type.

The soluble type is an N-bounded Fe^{3+}-type iron site and in an electrochromic reaction, the reduction of N-bounded Fe^{3+} yields the transparent Everitt's salt (ES). The relevant reaction is as follows:

$$M\,Fe^{III}[Fe^{II}(CN)_6] + e^- + M^+ \leftrightarrow \underset{\text{(Everitt's salt)}}{M_2Fe^{II}[Fe^{II}(CN)_6]} \tag{18.1}$$

where M^+ is an alkaline ion (such as potassium ion). Thus the reduction of soluble PB leads to a (reversible) bleached condition, inasmuch as ES is optically transparent.

The oxidation of soluble PB is also an electrochromic reaction:

$$\underset{\text{(Prussian blue)}}{K\,Fe^{III}[Fe^{II}(CN)_6]} \leftrightarrow \underset{\text{(Prussian yellow)}}{M_2Fe^{III}[Fe^{III}(CN)_6]} + e^- + K^+ \tag{18.2}$$

Since Prussian yellow (PY) absorbs light much less than PB, PB is bleached (reversibly) by this reaction. Thus soluble PB offers reversible bleaching both by reduction as well as by oxidation.

A single film electrochromic device is a simple way to get dark and bright effects by application of external voltage. Relevant structure consists of a PB kept between two transparent conducting plates as shown in Figure 18.1a. When voltage is applied, oxidation occurs near the positive electrode and reduction near the negative electrode to yield PY and ES, respectively, as indicated in Figure 18.1b. This conversion of the outer portions of the film results in the net bleaching of the device. A single film of PB is thus made to function as both electrochromic and complementary electrodes as well as an electrolyte. Materials like PB are known as *ion insertion electrochromic* materials.

Water plays a major role in the electrochromic mechanism. For example, tungsten oxide (WO_3) colors cathodically in aqueous solution. When nickel oxide (NiO) is used as a positive electrode, it colors anodically only in its hydrated form in aqueous electrolytes [2,3]. Under appropriate conditions, NiO can undergo a reversal lithium intercalation-deintercalation process in aprotic nonaqueous electrolytes. This type of electrochromic effect may be described by the following reaction:

$$\underset{\text{(Colored)}}{xLi + NiO} \leftrightarrow \underset{\text{(Bleached)}}{Li_x\,NiO} \qquad (18.3)$$

Under this circumstance, the oxide is electrochromic in a complementary mode with respect to Tungsten oxide whose reaction is given by:

$$\underset{\text{(Colored)}}{xLi + WO_3} \leftrightarrow \underset{\text{(Bleached)}}{Li_x\,WO_3} \qquad (18.4)$$

Hence, WO_3 and LiO can be used in order to create efficient electrochromic windows provided that the two materials are adjoined by a suitable electrolyte.

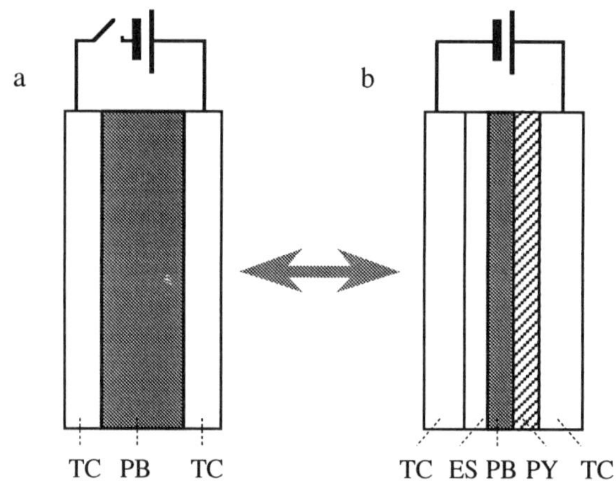

Figure 18.1 SFPB Cell. A: With no bias voltage. B: With bias voltage applied. (TC: Transparent conductor; PB: Prussian blue; ES: Everitt's salt; PY: Prussian yellow).

The chemical reactions depicted by Equations 18.3 and 18.4 show that the NiO film becomes transparent in the cathodic cycle (Li insertion) and dark brown in the anodic cycle (Li extraction). The WO_3 film becomes transparent in the anodic cycle (Li extraction) and blue in the cathodic cycle (Li insertion).

Figures 18.2a and 18.2b illustrate the electrochromic properties of individual (electrochromic) layers by depicting the cyclic voltametric curves for NiO and WO_3 immersed in a typical nonaqueous electrolyte.

The coloring and bleaching mechanism for WO_3 and NiO can be specified by the following reactions:

$$\underset{\text{(Clear)}}{WO_3 + x\,H^- + x\,e^-} \leftrightarrow \underset{\text{(Colored)}}{H_x\,WO_3} \qquad (18.5)$$

$$\underset{\text{(Clear)}}{NiO + OH^-} \leftrightarrow NiOOH + e^- \qquad (18.6)$$

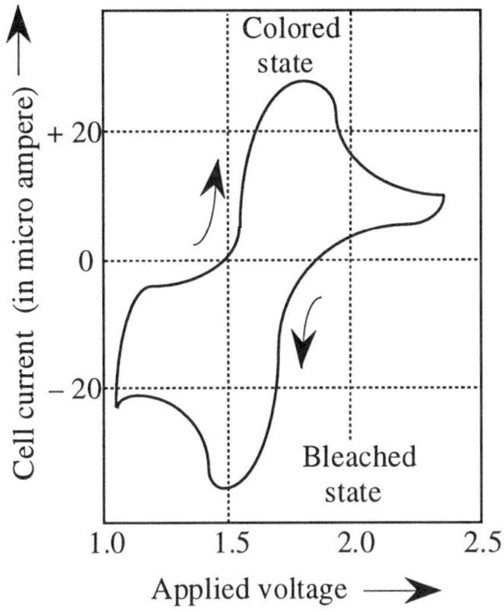

Figure 18.2a Cyclic volt-ammogram of a typical NiO-based electrochromic layer.

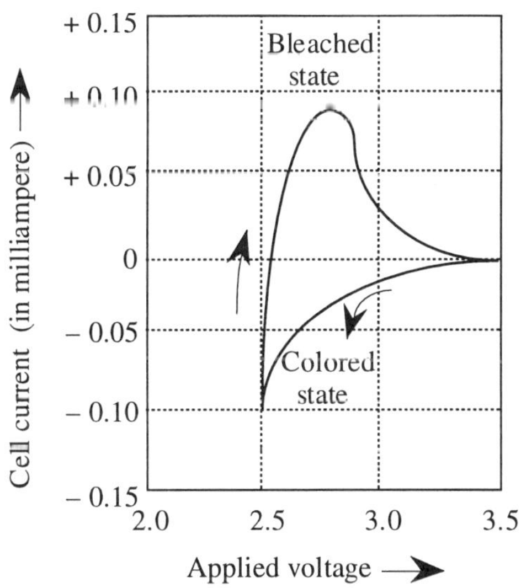

Figure 18.2b Cyclic volt-ammogram of a typical WO_3-base electrochromic layer.

Thus, the observed coloration is due to the formation of hydrogen-tungsten bronze with the inclusion of protons into the tungsten oxide layer with electrons from an external voltage source. Simultaneously, the nickel oxide absorbs an hydroxide ion to form a nickel hydroxide compound with the release of an electron into the external maintaining a charge neutrality.

18.2 Electrochromic Mirror Systems

An application of an EC device is to control the passage of the incoming light. That is accomplished by an EC window which is transparent in one state and colored in another state with appropriate application of electrical voltage. This switching phenomenon can be effected by an EC mirror whose reflectivity can be changed by introducing or removing an electric field. This effect is prevalent in two types of materials, namely:

- Liquid crystal (LC) materials
- Electrochromic (EC) materials

The most popular LC mirror is a *guest-host liquid-crystal mirror* shown in Figure 18.3. In this type of material, specialized dyes (the *guest*) are combined with the liquid crystal material (the *host*) which acts as a matrix to orient the dye molecules. The host material reorients the dye molecules when an electric field is applied to it and causes a change in the amount of light that can be absorbed. The darkness depends on the dye efficiency and dye concentration. In the dimmed mode, higher concentrations lead to a darker mirror and also it results in higher residual coloration due to incomplete ordering of the dye molecules.

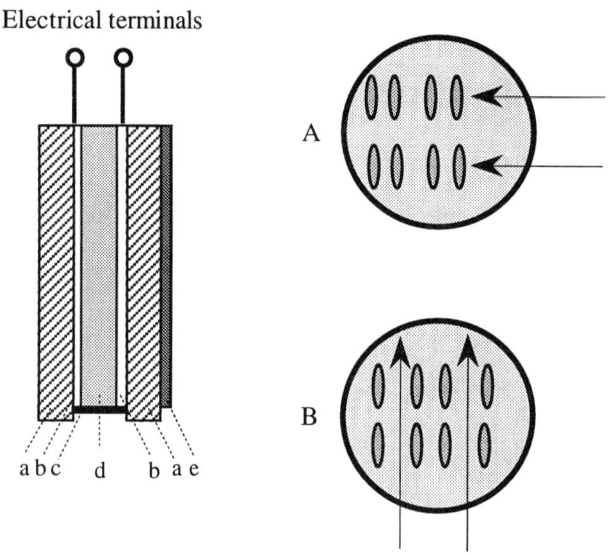

Figure 18.3 A guest-host liquid-crystal mirror.
A: Dark-mode wherein the dye absorbs the light.
B: Bright-mode wherein the passage of light is not inhibited.
(a: Glass substrates; b: Transparent conductors; c: Seal; d: Liquid-crystal/dye solution; e: Silver reflector.)

Hence, *guest-host mirror* can be classified as negative or positive guest-host mirror. The negative one is bright when powered and dark when unpowered and the positive-type mirror is bright when unpowered and dark when the electric field is present. Thus by applying an electric field, the reflectivity of the mirror can be changed and it acts as a switching device yielding a chopped square wave corresponding to the incoming light falling on the mirror.

Another type of mirror is the electrochromic mirror [4-7] which changes its color in the presence of an electric field typically through the injection of ions and electrons or through electrochemical reactions.

EC mirrors are typically constructed in two ways. In one type, the reaction takes place in a chemical solution. In the other construction, the reaction takes place within thin solid films. The first type of mirror uses a solution which contains two EC materials; one colors cathodically and the other colors anodically. The solution is sealed between two sheets of conductive coated glass as shown in Figure 18.4.

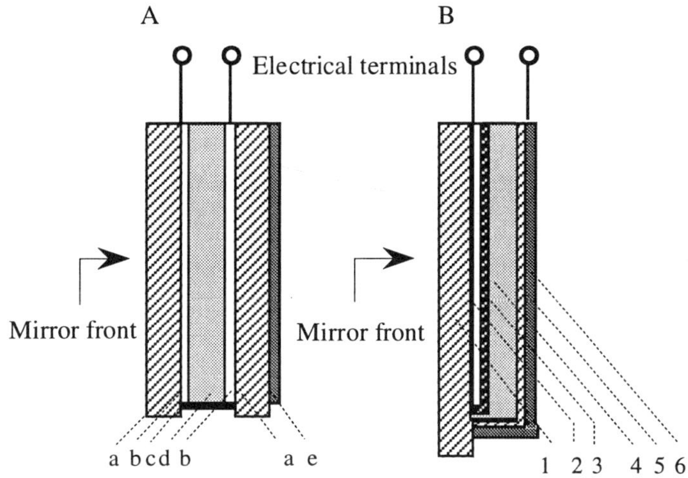

Figure 18.4 Electrochromic mirrors.
A: Using electrochromic solution.
B: Solid-state type.
(a: Glass substrates; b: Transparent conductors; c: Seal; d: Liquid-crystal/dye solution; e: Silver reflector.)
(1: Glass substrate; 2: Transparent conductors; 3: Anodic electrochromic layer; 4: Solid-state electrolyte; 5: Cathodic electrochromic layer; 6: Silver reflector.)

A piece of glass is coated with a reflective film to form the mirror. The EC material is uncolored when unpowered. When a voltage is applied between the two glass sheets, one material is cathodically reduced and the other is anodically oxidized to convert them to the colored state. On removing the voltage, the system spontaneously returns to the uncolored state.

The second type of mirror refers to a multilayer structure which consists of a stack of thin films deposited on a single piece of glass as shown in Figure 18.4. An electric field is set up between the transparent conductive material and the metallic layer by an external bias. Sandwiched between these layers, is a layer of anodic and a layer of cathodic electrochromic materials. When a voltage is applied, both the anodic and cathodic layers get colored and the mirror darkens. When the voltage is removed, the thin film electrochromic layers retain their color. An opposite voltage is required to reverse the electrochromic reactions and return the mirror to full brightness.

18.3 Optical Switching Applications of Electrochromic Materials

Inasmuch as an electrochromic mirror or a window offers optical transparency or opacity to the incoming light, it is possible to produce a square wave corresponding to the light absorbed or transmitted by it. The rate of optical switching depends on the dynamic/transient (switching) characteristics of the ER material used.

Figure 18.5 illustrates the dynamic performance of the current and photonic reflectance of an EC mirror for a coloring voltage of 1.8 volt d.c. followed by a bleach voltage of – 0.8 volt (d.c.). The reflectance varies between 75 to 10 percent with a switching time of < 20 second.

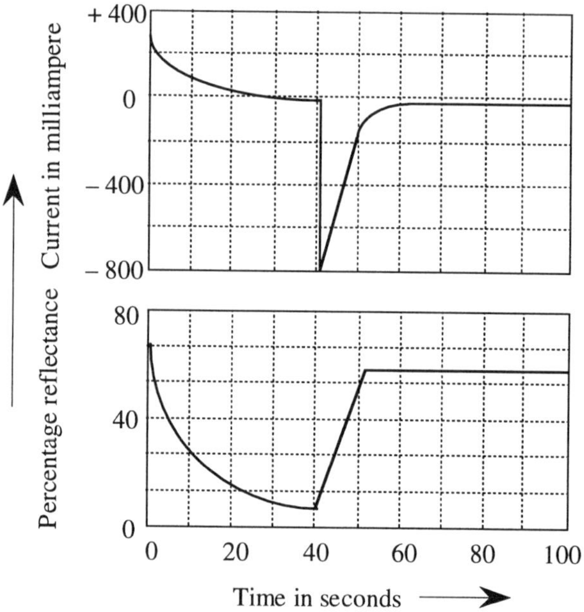

Figure 18.5 Time-response characteristics of an electrochromic mirror.

In the colored state the mirror becomes deep blue; therefore reflectance drops. The spectral response of an EC mirror is shown in Figure 18.6 in both clear and colored state. The EC material adopted is same as the tungsten oxide (WO_3) and nickel oxide (NiO) combination as the active cathodic and active anodic part, respectively, as discussed earlier.

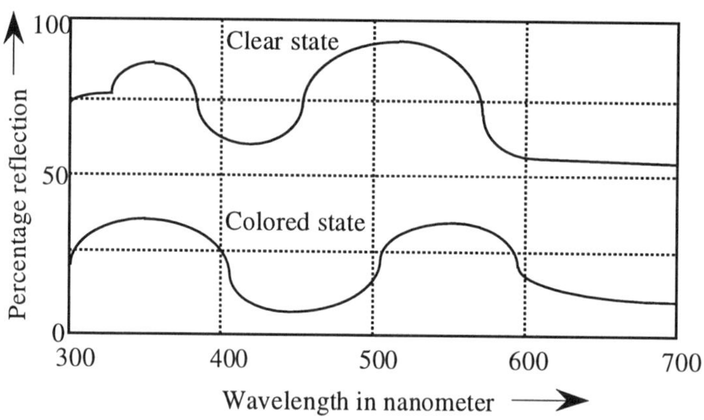

Figure 18.6 Visible reflectance at the electrochromic mirror under clear and colored states.

18.4 Characteristics of Constituent Materials in an EC System

The electrolyte in a conventional cell is an electrochromic insulator which prevents electron flow between the two electrodes whereas the electrolyte in an EC device such as the single film Prussian blue (SFPB) is a mixed conductor with enough electronic conductivity to allow the system to equilibrate rapidly after removal of the driving voltage. Therefore an electrochromic device repeatedly bleached by the application of a voltage across it would result in partial oxidation near the anode and partial reduction near the cathode. The applied voltage determines the extent of bleaching of the device. In addition, the response of the device is also sensitive to the amount of pressure applied to it.

The optical absorbance spectra of a single film Prussian blue (SFPB) device for different voltage is shown in Figure 18.7. Significant absorbance changes in the SFPB cell occur over a wide spectral range (550-900 nm). It is seen that the open-circuit state, when no voltage is applied, corresponds to maximum absorbance and the absorbance decreases with an increase in the applied voltage. Hence the more voltage is applied, the more it is bleached.

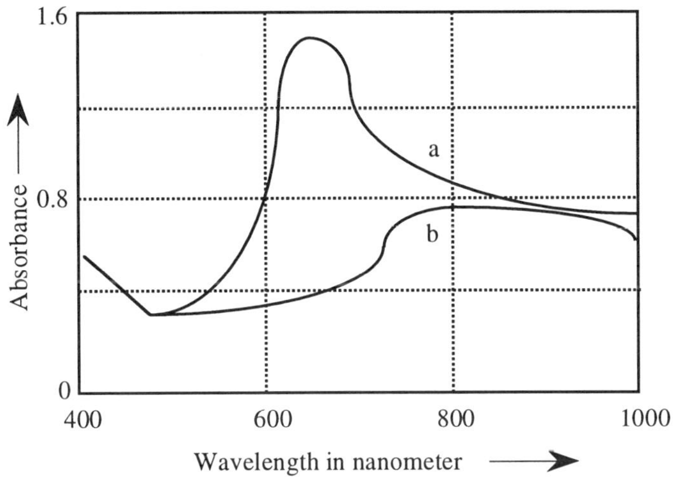

Figure 18.7 Optical absorbance of a typical SFPB device. (a: Zero bias voltage; b: Bias voltage ~ 8 volts.)

Figure 18.8 shows the typical absorbance (at 690 nm) of a cell during repeated switching. The cell is alternately powered over 15 seconds at 12 volts and turned off with open-circuit state for 15 seconds. The absorbance changes between 0.92 (when voltage is 12 volt) and 1.85 (when no voltage is applied). It could be seen from the figure (Figure 18.8) that both coloration and bleaching occur within the first 5 seconds of the transient state of the voltage.

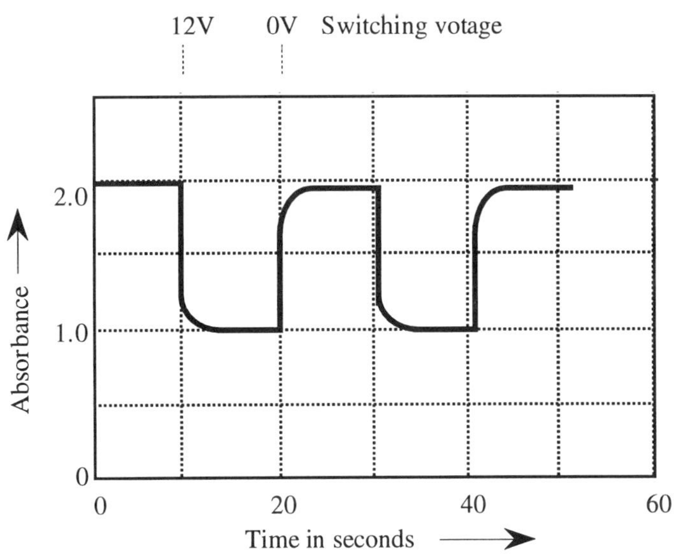

Figure 18.8 Optical transients of a SFPB device in response to an applied switching voltage.

The various aspects of electrochromic materials discussed so far indicate that an EC device can be used as an optical switch whose switching state can be alternated by changing the voltage applied to it. That is, by changing the voltage applied to an EC material, the incoming light can be alternately blocked or allowed to pass. This principle can be used in the *reticle* design to chop an incoming beam of light. In the conventional reticles a rotating disk is used to modulate the light at a desired rate and modulation profile (such as amplitude modulation, frequency modulation etc.). Such reticles (or *choppers*) find useful applications in passive homing systems in pursuing a target *via* tracking the heat (infrared) emissions from the target. The infrared emission intercepted at the seeker is chopped (modulated) by a rotating disk reticle, to acquire position information of the target [8]. The rotating disk systems, however, is rather cumbersome *vis-a-vis* considering the vibrations involved, mounting of the motor, wobbling of the disk, etc.

In order to overcome the shortcomings of rotating reticles, a *stationary chopper/reticle* can be used as an alternative. This stationary reticle can be designed [9] with EC materials making use of the electronically controllable optical transparency/opacity characteristics of the EC device as discussed earlier.

Other applications of the EC principle include the design of rear view mirrors [10-13] with glare control abilities and high resolution flat panel display technology for cockpit applications.

18.5 Concluding Remarks

Electrochromic materials refer to one of newest breeds of electromagnetic media. Though successful studies have been made and some potential applications have been put into practice, still the material science aspects and possible technological innovations are many to be explored *vis-a-vis* electrochromic materials. In the applications of EC materials, several technological desirabilities such as contrast (in color changing display), grey shades involved, switching speed, lifetime, resolution of the displays etc. pose persistent problems yet to be addressed in detail. Nevertheless, EC materials constitute a set of the most promising items of material science and technology of the future.

References

[1] M. K. Carpenter and R. S. Conell: A single film electrochromic device. J. Electrochem. Soc., vol. 137(8), 1990: 2464-2467.

[2] M. A. Hansen, I. A. Macabe, B. P. Hichwa, J. Gordon and H. Matheu: EC technology, no longer a late curiosity. Photonics Spectra, Jan. 1992.

[3] S. I. Cordova-Torresi et al.: Electrochromic behavior of nickel oxide electrodes. J. Electrochem. Soc., vol. 138(6), 1991: 1548-1553.

[4] N. Lynam and K. Seah: *Electrochromic Mirror*, U.S. Patent #4,712,879 dated December 15, 1987.

[5] C. M. Lampert: Electrochromic materials and devices for energy efficient windows. Solar Energy Materials, vol. 17, 1984: 1-27.

[6] K. C. Ho, D. E. Singleton and C. B. Greenberg: The influence of terminal effect on the performance of electrochromic windows. J. Electrochem. Soc., vol. 137(12), 1990: 3858-3864.

[7] R. B. Golder et al.: Recent research related to the development of electrochromic windows. Solar Energy Materials, vol. 14, 1986: 195-203.

[8] P. R. Mahapatra, S. Ramakrishna and P. S. Neelakantaswamy: A pulse-modulated eccentric chopper optical tracking system. Electro-Tech., vol. XI(6), 1971: 199-211.

[9] Md. Hoque: Studies on Reticle Performance in Passive Homing Systems. M.S.E Thesis, Department of Electrical Engineering, Florida Atlantic University, Boca Raton, FL, 1995.

[10] S. E. Selkowitz and C. M. Lampert: Applications of large-area chromogenics to architectural glazings, in C. M. Lampert and C. G. Granqvist (Eds.): *Large Area Chromogeneics: Materials and Devices for Transmittance Control.* vol. IS4 (SPIE Optical Engineering Press, Bellingham, WA: 1988), pp. 22-45.

[11] N. R. Lynam and A. Agarwal: Automotive applications of chromogenic materials, in C. M. Lampert and C. G. Granqvist (Eds.): *Large Area Chromogenics Materials and Devices for Transmittance Control.* vol. IS4 (SPIE Optical Engineering Press, Bellingham, WA: 1988), pp. 45-84.

[12] D. J. Helder: Large-area variable reflectance mirror for trucks and buses. SAE Technical Paper Series #912705 (Presented at International Truck and Bus Meeting and Exposition, Chicago, IL, Nov. 18-21, 1991).

[13] When you think electrochromics, Think Donnelly: Pamphlet issued by Donnelly Corp., 114 E. Fortieth St., Holland, MI 49423.

Defining Terms

Electrochromic material: An electrochromic material changing its color, when subjected to an external electric field.

Electrochromic mirror: A mirror system with variable transmittance/reflective characteristics constituted by elastrochromic materials.

Static reticles: An optically, nonmoving chopper system using electrochromic materials.

CHAPTER 19

Electronic Packaging Materials

19.1 Introduction

Electronic packaging (EP) refers to the integrated efforts of combining engineering and manufacturing methods to translate an electronic cricuit/subassembly/system into a manufactured assembly. The associated electronic packaging materials are those which maintain the proper functioning and expected lifetime of the packaged assembly.

Essentially in a packaged electronics, the associated materials could be conductors, semiconductors and insulators. Conductors may exist in bulk form (as wires, solders, contacts and sheaths) or as films/coatings/platings. The insulating systems usually prevail as solids, semisolids (gels), liquids or as gases. While the metallic materials provide the pathways to the signal, insulators confine their routings restricted and maintain the lifetime of the constituent parts.

Apart from conductors and insulators which essentially play the role as electrical materials, an electronic packaging may also include materials which provide mechanical support and structural form or offer protective shield against moisture, heat, contamination, radiations, chemicals, etc.

19.2 Classification of Electronic Packaging Materials

On the basis of electrical characteristics (represented by the volume resistivity) the EP materials can be classified as.

- Insulators – 10^{19} to 10^{6} ohm-cm
- Semiconductors – 10^{6} to 10^{-3} ohm-cm
- Conductors – 10^{-3} to 10^{-6} ohm-cm

19.3 Insulating-Type EP Materials

Electronic packaging uses two types of dielectric (insulating) materials, namely:

- *Organic* materials
- *Inorganic* materials

Organic compounds are based on carbon-to-carbon (C-C-C) molecular chains with the insertions of O, S, N, Cl, H and a variety of other atoms modifying their characteristics. Straight chains are designated as *aliphatic* and ring structures are known as *aromatic* compounds. The complex forms of such molecular chains lead to polymeric compounds known as *plastics* which could be either *thermosetting or thermoplastics.*

Thermosetting plastics chemically react while being formed so that further heating renders them infusible and not liquefiable. *Epoxy glass* is a typical packaging material of this class.

Thermoplastic polymers lend themselves to a liquefying–solidifying cycle *via* heating and cooling repeatedly. A typical example of this category is paraffin wax.

Another class of organic compound usable as a packaging material is *elastomeric rubber* which can be either thermosetting or thermoplastic. Characteristically, these materials structurally yield extensive elongation under applied stress.

Inorganic compositions are stable ionic salts like NaCl or fluids like H_2O, HCl, NaOH etc. Also, inorganic ceramics are not precluded from being electronic packaging materials. Carbides, nitrides, silica (SiO_2), alumina (Al_2O_3), beryllia (BeO) and several glass compositions could be classified as ceramic–like and are popular electronic packaging materials.

Electronic systems may use gaseous media such as sulfur hexafluoride (SF_6), dry nitrogen (N_2), and CO_2 as compatible insulating materials in certain applications. Gases in the liquified state are also used in EP strategies to achieve the cryogenic state.

Semiconductors both intrinsic form and/or added with impurities (dopants) have been successful candidates in electronic packaging needs. Si, Ge and GaAs are typical examples of semiconductors, a detailed description of which are presented in Chapter 10.

Among the gamut of conductors, metals and metal-alloys yielding a resistivity of 10^{-6} to 10^{-3} ohm-cm are very commonly used in electronic packaging applications. Examples of such conducting materials are: Cu, Al, Au, Pt, and solder alloys (Sn + Pb). There are also high resistance alloys which find certain exclusive applications in electronics and electrotechnological packaging techniques.

19.4 Insulating Materials for EP Applications

Dielectric media or insulators take different manifestations as bulk media, films, plates, coatings, rods, foams, adhesives, laminations, gels, liquids, and gases in their implementation as EP materials. Characteristics and properties of typical polymeric/plastic materials, inorganic insulators and elastomers are presented in Chapter 3. Typical examples of these materials are summarized in Tables 19.1 and 19.2.

Table 19.1 Insulating EP Materials: Organic Types

Thermosetting Plastics	Thermoplastic Plastics	Elastomers	Solvents	Other
Alkyd	ABS- acrylonitrile-butadiene-styrene	ABR- acrylate butadiene	Alcohols	Oils
Allyl	Acetal	BR- polybutadiene	Aromatics	Paper
Epoxy	Acrylic	CO- epichlorohydrin	Benzene	Rayon
Melamine	Cellulose	COX- butadiene-acrylonitrite	Hydrocarbons	Silk
Phenolic	Fluoropolymer	CR- chloprene- neoprene	Ketones	Wood
Polyester	Ionomer	CSM- chlorosulfonated polyethylene		
Polyimide	LCP- liquid crystal polymers	EPDM		
Polyurethane	Nylon	EPM- ethylene-propylene copolymer		
Silicone	Parelene	FPM- florinated copolymer		
	PEEK- polyether-ether-ketone	IIR- isobutylene-isoprene		

(continued...)

Thermosetting Plastics	Thermoplastic Plastics	Elastomers	Solvents	Other
	Polyarylate	IR		
	Polycarbonate	NBR- butadiene-acrylonitrite		
	Polyester	NR		
	Polyether imide	PVC/NBR- polyvinyl chloride/nitrile Buna N		
	Polyethylene	SBR-styrene:butadiene		
	Polyimide	SI-silicone copolymer		
	Poly methylpentene	T-polysulfide		
	Polystyrene	U-polyurethane		
	Polysulfone			
	PPO- poly-phenylene oxide			
	PPS- poly-phenylene sulfide			
	PVCI- polyvinyl chloride			

Table 19.2 Insulating EP Materials: Inorganic Type

Ceramics	Glasses	Chemicals	Gases
Alumina	Aluminosilicate	HCl	Air
Beryllia	Borosilicate	H_2O	Carbon dioxide
Carbides	Glass ceramics	H_2SO_4	Nitrogen
Lava	Lead	NaCl	Sulfur hexafluoride
Magnesia	Silica	NaOH	
Nitrates	Soda lime		
Titanates			

19.5 EP Applications of Insulating Materials

19.5.1 *Thermosetting plastics*

Alkyds: These are molding compounds formulated from polyester resins and diallyl phthalate monomers and used in mineral-filled or glass-fiber-filled forms up to 350°C. Typical electrical properties are listed in Table 19.3.

Table 19.3 Electrical Properties of Alkyds

Type	Dielectric Constant (ε_r)	Dissipation Factor ($\tan\delta$)	Dielectric Strength (volt/mil)	Bulk Resistivity (ohm-cm)	Arc Resistance (seconds)
Mineral-filled	6.3 at 60 Hz 4.7 at 10^6 Hz	0.04 at 60 Hz 0.02 at 10^6 Hz	400	10^{14}	>180
Glass- filled	5.6 at 60 Hz 4.7 at 10^6 Hz	0.10 at 60 Hz 0.02 at 10^6 Hz	375	10^3	~180

Allyls: These are molding materials based on monomers (diallyl phthalate or diallyl isothalate) reinforced with glass, mineral or synthetic fibers. They are used in making connectors, chip carriers, terminal boards, and switches. Also in fiber laminated form, they are used in radome structures.

Epoxies: These have a characteristic epoxide ring $CH\,/\overset{O}{-\,-}\backslash\,CH_2$ structure and combined with glass fabrics are widely used to produce laminated printed circuit boards (PCBs) and also conformal coatings, varnishes, and adhesives. By reacting with phenolic resins, molding compounds are achieved to make coil bobbins, connectors, and chip carriers.

Table 19.4 Electrical Properties of Epoxies

Type	Dielectric Constant (ε_r)	Dissipation Factor ($\tan\delta$)	Dielectric Strength (volt/mil)	Bulk Resistivity (ohm-cm)	Arc Resistance (second)
Mineral-filled	5.0 at 60 Hz 4.6 at 10^6 Hz	0.01 0.01	360	3.8×10^{15}	140
Glass- filled	4.0 at 60 Hz 5.0 at 10^6 Hz	0.01 0.01	400	9×10^{15}	180

Phenolics: These resins are products of chemical reaction between phenol and formaldehyde. These can be used as molded parts or laminated structures (as in printed circuit boards), as chip carriers, connectors, or bobbins.

Unsaturated polyesters: These are used as bulk molding compounds, sheet molding compounds, hand lay-ups, stray-ups, resin-transfer molding, laminates, filament windings, and *pultrusion*. Laminated versions with random glass-fiber mats, fillers (aluminum oxide trihydrate) and polyester resins are very popular in EP applications. Moldings are used in bobbins, terminal boards, housings, and connectors. The film version is used as wire insulation, coil insulation, and protective layers. Pultrusions are used as bus supports and spaces. The filament wound and compressed types are suitable to fabricate radomes.

Table 19.5 Electrical Properties of Phenolic Plastics

Type	Dielectric Constant (ε_r)	Dissipation Factor ($\tan\delta$)	Dielectric Strength (volt/mil)	Bulk Resistivity (ohm-cm)	Arc Resistance (second)
Intrinsic	12.0 at 60 Hz 6.0 at 10^6 Hz	0.30 0.70	400	10^{13}	50
Glass-filled	50.0 at 60 Hz 10.0 at10^6 Hz	0.30 0.80	350	10^{13}	70
Mineral-filled	6.0 at 60 Hz 6.0 at10^6 Hz	0.07 0.10	400	10^{14}	180

Table 19.6 Electrical Properties of Polyester

Type	Dielectric Constant (ε_r)	Dissipation Factor ($\tan\delta$)	Dielectric Strength (volt/mil)	Bulk Resistivity (ohm-cm)	Arc Resistance (sec)
Polyester (punching grade)	4.5 at 60 Hz	0.05	300	_	>180

Polyimides: These are heat-resistant high grade polymers with minimal outgassing properties compatible for cryogenic and high temperature environments, and space-ambient applications and also suitable for wire insulations, coatings, sleevings, and tapes. Multilayer circuit board coating, chip carriers, laminates, flexible cables, tape-wire wrap wire enamels, etc. use polyimides widely.

Polyimides have a dielectric constant of 3.4 to 3.5 over 10^6 Hz to 60 Hz with a low dissipation factor (0.0025) at 60 Hz increasing to 0.01 at 10^6 Hz. The dielectric strength of polyimides is about 500 volt/mil, volume resistivity is 10^{14} ohm-cm and the arc resistance is 230 second.

Polyurethanes: These are almost elastomeric-like materials formed by reacting a diisocyanate with a glycol. In EP applications, polyurethanes are used as embedding compounds or conformal coatings. Also they can be molded as automative parts or foamed. They are sensitive to solvents like ketones, acids, or bases and have restricted applications below 250°F. Typical electrical parameters of polyurethanes are:

Dielectric constant (ε_r):	6 at 60 Hz 3 at 10^6 Hz
Dissipation factor:	0.10 at 60 Hz 0.04 at 10^6 Hz
Dielectric strength:	500 volt/mil
Volume resistivity:	10^{14} ohm-cm
Arc resistance:	120 second

Silicones: These are high temperature tolerant plastics with a usable range of -86 to 480°F. Also they are characterized by excellent arc resistance. Silicones unlike carbon-hydrogen-based organic compounds, have silicone-oxygen-based molecular chains. Silicones are available as liquids, resins, and elastomers making them useful in EP applications as shaped parts, wire enamels, tapes, sleevings, tubings, adhesives (with elastomers), and varnishes (with solvents). Molding compounds are heterogeneous with the inclusion of resins, mineral fillers and glass fibers.

Silicones are convertible as elastomers *via vulcanization* with peroxide catalysts. Electrical parameters of typical mineral-filled silicones are:

Dielectric constant (ε_r):	3.6 at 60 Hz
	3.7 at 10^6 Hz
Dissipation factor:	0.005 at 60 Hz
	0.003 at 10^6 Hz
Dielectric strength:	425 volt/mil
Volume resistivity:	10^{15} ohm-cm
Arc resistance:	240 second

19.5.2 *Thermoplastic plastics*

Acrylics, fluoropolymers, liquid crystal polymers, nylons, polycarbonates, polyesters and polyacrylates, polyether-imides, polyethylene and polypropylene, polyimide and polyamide-imide, polyether-ether-ketones, polyphenylene oxide, polyphenylene sulfide, polystyrene, polysulfones, and polyvinyl chloride are typically thermoplastic. The electrical characteristics of these materials are presented in Chapter 3. Their specific uses in EP applications are listed in Table 19.7.

Table 19.7 EP Applications of Thermoplastics

Material	General Properties *vis-a-vis* EP Applications
Acrylics	Optically clear and weather resistant. Useful as conformal coating.
Fluoropolymers	Heat and chemical resistant, low moisture absorption, and low dielectric constant. Used extensively in EP applications such as gaskets, high frequency supports, coaxial line dielectric interspace, microwave plumbing parts/spacers (e.g. Teflon)™, heat-resistant coating in wires, etc.
Liquid crystal polymers	High temperature resistant and high physical strength. Used in surface-mount technology and as chip carriers, sockets, connectors, bobbins and relay cases.
Nylon	Used as coil bobbins, wire ties, connectors, and as wire jackets.

(continued...)

Material	General Properties *vis-a-vis* EP Applications
Polycarbonates	Flame retardant, temperature resistant, and highly impact resistant. Dielectric constant is low. Used in multilayered printed circuit boards, connectors, terminal blocks, and fuse holders.
Polyesters and polyacrylates	Possess stable electrical properties over a range of temperature and frequency. Characterized by low dielectric constant and high arc resistant. Used in 3D circuit boards, as insulating films, as terminal blocks and as fuse holders.
Polyether-imides	Low dielectric loss materials over a range of temperature and have high mechanical strength. Easily moldable into complex shapes. Resistant to ionizing radiations. Pure and/or fiberglass reinforced materials are used as circuit-breaker housings, chip carriers, pin connectors, bobbins, automative engine parts.
Polyethylene and polypropylene	Lightest polymers with dielectric constant equal to 2.3 and dissipation factor of 0.0002 at 60 Hz. Useful as primary insulators on wires and cables, capacitor dielectrics, fuse blocks, and as battery cases. These are susceptible to burning.
Polyimide and polyamide–imide	High temperature resistant materials with low dielectric constant. High tolerance to ionizing radiations (10^9 rad). Resistivity is also very high. Chemical resistant but dissolve in polar solvents and/or attack by hot caustic. Used in connectors, radomes, and circuit boards.
Polyetheretherketones	Exhibit good electrical properties even at elevated temperatures. Resist burning with minimal smoke generation. Used in wire/cable insulation and connectors.
Polyphenylene oxide	These resins have high heat resistance and low water absorption characteristics. Usable temperature range: 180 to 317°F. Used in computers, connectors, fuse blocks, relays, and bus bar insulation.

(continued...)

Material	General Properties *vis-a-vis* EP Applications
Polyphenylene sulfide	These are flame retardant plastics with fair resistance to chemicals and high tolerance to heat. They permit wettable blending with glass fibers for reinforcement. There are no solvents for these plastics. Low dielectric constant. Applications include computers, IC encapsulation, and microwave components.
Polystyrene	These offer very low loss tangent (dissipation factor $\cong$0.0001) and the dielectric constant is about 2.45. These properties permit their use in high frequency units and strip lines. Polystyrenes, however, are affected by chemicals, solvents, and high temperatures.
Polysulfone family (polysulfone, polyarylsulfone, polyethersulfone, polyphenylenesulfone)	These have excellent electrical properties. Usable temperature range 30 to 350°F. Stable plastics and least affected by radiation. Easily moldable through extrusion and injection methods. Used in PCBs, television components, and multipin connectors.
Polyvinyl chloride	PVC is a flexible polymer with chemical resistance properties. Usable at modest temperatures only. Mainly used in nominal temperature insulations, cable jackets, tubings, and sleevings.

19.5.3 *Elastomers*

These are macromolecular materials which return rapidly to the approximate initial dimensions and shape after substantial deformation by a weak stress and release of the stress (ASTM D-1566).

The general characteristics of the elastomers are:

- They are plastics with elastic behavior (unlike rigid plastics).
- They are susceptible to environmental influences (temperature, oxidation, ionizing radiation and ultraviolet radiation) with the consequence of rapid aging.
- Elastomeric plastics creep, that is, the strain in them would change even when the applied stress is held constant.
- On a scale of 0 being soft and 100 being hard, elastomers show hardness or resistance to deformation (upon pressure) over a wide range (20 to 70).
- Hysteresis (energy loss per loading cycle) in mechanical and electrical properties exist.
- At low temperatures, elastomers become stiffer and harder with a brittle point at the glass transition temperature.
- Elastomers vary widely in their ability to withstand tearing. That is, their ability to resist tearing (expressed in terms of stress needed to continue rupturing the material in sheet form) has a wide latitude.
- Modulus of elasticity in general is decided by linear stress-strain relation.

The different types of elastomers and their EP applications are described in Table 19.8.

Table 19.8 Description and Characteristics of Elastomers used in EP Applications

Elastomers with ASTM D-1418 Nomenclature	Characteristics/Applications
NR: Natural rubber/*hevea* latex (95% *cis-1, 4-polyisoprene*)	Usable as general purpose insulation materials. Poor aging characteristics. E_{Max} = 800 V/mil; $\tan\delta$ = 0.0025; ε_r = 3; $\rho = 10^{16}$ ohm-cm.
IR: Polyisoprene rubber	Synthetic counterpart of natural rubber. Substitutes the insulation uses of natural rubber. Electric properties similar to those of natural rubber.
ABR: Acrylic butadiene elastomer	These are highly ozone and heat resistant. They age well but are sensitive to water. Used in situations where oil is encountered. Useful as mechanical elastomer.
BR: Polybutadiene elastomer	Useful in tirestocks for copolymerization with SBR and NR.
CO (or ECO): Epichlorohydrin elastomer	The chlorine content makes them heat-retardant. They age well and are fairly chemical resistant. They have high dielectric loss.
COX: Carboxylic elastomer (butadiene-acrylonitrile)	Used with NBR to improve low temperature performance. They have excellent weather performance and wear resistance. E_{Max} = 500 V/mil; $\tan\delta$ = 0.05; ε_r = 10; $\rho = 10^{15}$ ohm-cm.
CR: Chloroprene, neoprene	Weather, flame and chemical resistant. Tolerant to ozone and radiation. Polar dielectric. E_{Max} = 700 V/mil; $\tan\delta$ = 0.03; ε_r = 8; $\rho = 10^{11}$ ohm-cm.
CSM: Chlorosulfonated polyethylene	Similar to CR with improved electrical properties and better heat resistant. Used in high voltage applications. Available in different colors. E_{Max} = 700 V/mil; $\tan\delta$ = 0.07; ε_r = 8; $\rho = 10^{14}$ ohm-cm.

(continued...)

Elastomers with ASTM D-1418 Nomenclature	Characteristics/Applications
EPDM: Ethylene-propylene terpolymer	Similar to EPM. Synthesized from ethylene, propylene, and a diene monomer. The monomer permits sulfur polymerization. Extremely heat and radiation resistant. Glass transition temperature -60°C. Good electrical properties. E_{Max} = 800 V/mil; $\tan\delta$ = 0.007; ε_r = 3.5; $\rho = 10^{16}$ ohm-cm.
EPM: Ethylene-propylene copolymer	Used for wire insulation. Inferior to EPDM in processing qualities.
FPM: Fluorinated copolymer	Outstanding heat and chemical resistance properties. Available as fluorocarbon, fluorosilicone, and fluoroalkoxy phosphazene polymers. Usable up to 600°F. Good thermal stability. Silicone type ages better. Excellent electrical and physical properties. Expensive material. E_{Max} = 700 V/mil; $\tan\delta$ = 0.04; ε_r = 18; $\rho = 10^{13}$ ohm-cm.
IIR: Isobutylene-isoprene-butyl	Butyl rubber is highly impermeable to water vapor. Non-polar dielectrics. With aluminum oxide trihydrate exceptional arc and track resistances are obtained. Outstanding weather resistant but low physical properties. Good aging characteristics and good flexibility at low temperatures. E_{Max} = 600 V/mil; $\tan\delta$ = 0.003; ε_r = 2.4; $\rho = 10^{17}$ ohm-cm.
NBR: Nitrilerubber (Butadiene-acrylonitrite, nitrile, Buna-N)	General purpose elastomer. Poor electrical properties due to polarity. Resistant to most chemicals. Mainly used for mechanical applications.
PVC/NBR: Polyvinyl chloride and NBR	Can be colored and used for wire insulations and jackets. Weather, chemical, and ozone resistant.
SBR: Styrene butadiene, GRS, Buna S	GRS – Government rubber, styrene. Synthesized as an alternative mostly in mechanical applications. E_{Max} = 800 V/mil; $\tan\delta$ = 0.003; ε_r = 3.5; $\rho = 10^{15}$ ohm-cm.

(continued...)

Elastomers with ASTM D-1418 Nomenclature	Characteristics/Applications
SI: Silicone elastomers (FII, PSI, VSI, PVSI)	Useful both at high and low temperatures (-100 to 600°F). Excellent arc and track resistant. Good weather and ozone resistant. Poor physical properties. Excellent electrical properties. E_{Max} = 700 V/mil; $\tan\delta$ = 0.001; ε_r = 3.6; $\rho = 10^{15}$ ohm-cm.
T: Polysulfide	These weather better than other elastomers. Highly chemical resistant. Very low dielectric loss. Physical properties are modest. E_{Max} = 700 V/mil; $\tan\delta$ = 0.005; ε_r = 9.5; $\rho = 10^{12}$ ohm-cm.
U: Polyurethane	These are either esters or ester-based. Ester-based are poorly water resistant. Excellent electrical properties and outstanding physical properties. Highly abrasion resistant. They get stiffened at low temperatures. They can be cast or injection molded. E_{Max} = 500 V/mil; $\tan\delta$ = 0.03; ε_r = 5; $\rho = 10^{12}$ ohm-cm.

Note: E_{Max}: Dielectric strength; $\tan\delta$: loss tangent; ε_r: dielectric constant; ρ: volume resistivity.

19.6 Processing of Plastics

In the manufacturing methods, the following are the conventional processes adopted in facilitating plastics as useful EP end products.

- Compression and transfer molding
- Injection molding
- Extrusion process
- Thermoforming
- Laminating
- Pultrusion
- Reaction-injection molding

Descriptions of these processes are available in [1].

19.7 Specific Processes of Plastics *vis-a-vis* EP Products

The plastic parts constituting EP products can be fabricated by a number of processes. They are as follows:

Casting: This refers to a plastic electric part being made by pouring a liquid resin system into a mold, curing the part, and removing it for use

Embedding: Superficially surrounding an electronic part with a liquid resin system so that the electronic part forms an embedment

Encapsulation: Totally enclosing a component or a device with a viscous resin system by dipping, spraying, or embedding with or without a mold

Impregnation: Filling the interstices in an electrical part (such as a coil or transformer winding, fabric screen etc.) with a low-viscous resin

Potting: This is similar to molding except that the mold remains as an integral part of the unit
Transfer molding: A special type of embedding

The various purposes of using the above processes on electrical parts are:

- Environmental protection
- Maintenance of electrical integrity
- Preservation of mechanical integrity

The major characteristics of casting materials can be enumerated as follows.

- They have a viscous resistance to flow and commonly the casting materials have a viscosity on the order of 30,000 cp. Fillers increases the viscosity. Also temperature would largely influence the viscous flow. Lower viscosity permits the flow of impregnating materials into the interstices of coils, etc.
- While curing or polymerizing, the casting materials get packed at molecular level causing an overall shrinkage (which could be on the order of 0.1 to 7% of the volume). Shrinkage could induce stresses leading to cracks in the casting.
- The curing process or polymerization can be an exothermic chemical reaction with the emission of heat. This could lead to uneven temperature gradients on the casting facilitating cracks to occur.
- The stay-in time of the cast in the molds refers to the cure time. Sufficient curing time is required to attain a good casting integrity, though a long-term curing is not cost-effective.
- Voids in casts could permit stagnation and trapping of moisture. When the molded product is then used under high voltage environments, partial (corona) discharges may occur reducing the lifetime of the electrical part.
- Cyclic exposure to varying temperature ambient could cause thermal shocks in casting-based electrical parts leading to eventual failures. Specifications such as MIL-1-169 23 call for the thermal cycling of casting from –55 to 126°C ten times without failure.
- Resistance to heat flow in castings is decided by the thermal conductivity of the cost materials. Poor thermal conductivity of plastics can be improved with fillers (such as metal powder).

Typical resins compatible for casting applications are as follows:

1. Epoxies (most popular for casting and embedments)
2. Polyesters (being of low viscosity, useful for impregnation and embedment purposes on low-cost budgets)
3. Silicones (highly flexible embedment materials)
4. Urethanes (useful as tough and tear-resistant embedments)
5. Foams (for *cast-in-situ* structures with low electric losses such as in supports for microwave plumbings, etc.)

19.8 Fillers Used in EP Plastics

Fillers are added to plastics adopted for EP applications for the purpose of realizing specific casting and embedment characteristics. Such fillers, in general, have the following properties:

- Fillers augment the resistance to the flow of casting materials
- They reduce the exothermic reactions in the casting materials
- They can control shrinkage of the casts

- Fillers (such as metal particles) enhance the thermal conductivity of the casting materials
- Addition of fillers improve the physical reinforcements of the cast structures
- Fillers can stabilize the embedment materials, increasing their shelf-life
- Hardness of casting (host) materials can be improved with fillers
- Fillers, in general, are less expensive than the casting resins
- Addition of fillers may improve or deteriorate the electrical properties of the castings. Specifically, the complex permittivity of the end product is dependent on the volume fraction of the filler added

The various types of filler used in practice are:

Mineral fillers

- Aluminum oxide trihydrate: –added to eliminate moisture so that the part becomes electric arc resistant
- Beryllium oxide: –high resistivity material added to improve the thermal conductivity
- Calcium silicate: –a fibrous material added for reinforcement purposes
- Aluminum silicate: –clay-like material which can improve shrunkage of the end product
- Muscovite: –a flaky mica material which offers improved impact strength and dielectric properties to the cast products
- Silica: –used for viscosity control
- Tale: –used to control cracks and machinability

Metallic Fillers

- Aluminum
- Copper
- Bronze
- Gold
- Silver
- Platinum

These are used to regulate the electrical and thermal conductivity characteristics of castings. Also they influence the complex permittivity of the end product.

Miscellaneous Fillers

- Antimony oxide: –added to improve flame-retardant properties
- Graphite: –useful as reinforcement fillers for elastomers
- Glass spheres: –use to form syntactic, lightweight foams

Fibers

- Glass fibers are used as reinforcement agents in insulating materials.

Organic fillers

- Wood/cellulose
- Flour/starch
- Nut shells

- Fiber-like jute: –A useful as reinforcement materials as well as EM energy absorbing media (see Chapter 22)

19.9 EP Materials in Nonbulk Forms

Under many circumstances, the practical EP materials are in nonbulk forms such as films, tapes, coatings, and sheets.

While a coating is obtained from a liquid material processed on a surface, the other three forms of nonbulk EM materials are solid materials sized to be two-dimensional. Normally, their thickness is taken for standard considerations:

(polymer)	Film:	Thickness $<$ 15 mil (ASTM D - 2305)
	Sheet:	Thickness $>$ 15 mil
	Tape:	Slit film or fabric coated with an adhesive

19.10 Film Materials and Fabricational Aspects

All thermoplastic resins can be made into films using:

- Extrusion process: –where a film is blown out of a suitable die attached to an extruder
- Casting process: –where the film material dissolved in a solvent is spread out as a wet film on a metal belt running between rollers to control the film thickness
- Calendering process: –where the film is formed by passing a polymer sheet between hot rollers
- Skiving process: –where the film is mechanically cut out from round billets of plastic mounted on a lathe

Electrical properties of film and/or sheet materials are specified *via* sheet resistance (ohm/square). Electrical designs use polyester, polyimide, fluorocarbon, polyethylene, polycarbonate and polysulfone commonly as compatible film materials.

19.11 Adhesive Materials in EP Applications

Adhesives are used in electronic applications mostly in the manufacturing stage as bonding materials. There are essentially five types of such bonding materials, namely:

Thermosetting adhesives

- Epoxies: –available as hardeners plus resin with or without metallic inclusions (fillers) such as silver particles curable over a wide temperature range (80-150°C).
- Polyimides: –low viscous adhesives curable at 180°C with low shrinkage characteristics. These can also be used with metallic fillers.
- Silicones: –available from water-like to thixotropic viscous properties. Pure versions are useful in bonding electrooptic devices. They do not off-gas to an appreciable extent but RTV silicones emit acetic acid while curing at room temperatures.

Thermoplastic adhesives

- Hot-melt adhesives: –available in stick form with a hot-gun facilitation for extrusion while applying *in situ*. These materials are acrylic, nylon, phenoxy, or olefins. Hot melts have the merits of being cheap, have a fast-set time, and need no special fixtures. They can also bond materials like polyethylene and polypropylene.
- Sheet adhesives: –these are ethylene acrylic acid copolymers which can be hot-melted (at 150°C) and cooled to freeze.
- Pressure-sensitive adhesives –these are electronic, silicone, or acrylic tapes.

Elastomeric adhesives

- These are solvent-based adhesives which can be applied *via* spraying or brushing.

Ceramic adhesives

- These are thixotropic pastes with silver filling and an organic base formulated as low-melt glasses. Sublimation of the polymer and melting of glass facilitate substrate wetting at controlled high temperatures (≅390°C).

Cyanoacrylate adhesives

- These are heat curable or they cure *via* solvent removal with rapid set times and are usable as adhesives to bond a wide variety of substrates.

19.12 Uses of Adhesives in EP Technology

- Surface-mounts/DIP packages: These are useful as thermally conductive bonding media.
- Enclosures: In the fabrication of enclosures and the attachment of components on the enclosures, seams of the enclosures can be filled with conductive adhesives to prevent EMI leakage.
- Heat sinks: Bonding of heat sinks on PCBs.
- Repairing PCBs: Board-level cracks can be repaired with adhesives.
- Die-attach: Dies are bonded to substrates with suitable adhesives.
- Flat cable: Flat-cable films can be bonded to substrates with adhesives.
- Conductor-jacket bonding: It is done at cable assembly levels with appropriate bonding materials.
- Cathode-ray tubes: Components are attached in the CRT with adhesives.
- Board-to-board connector bonds.
- Fixing large-sized components on boards for mechanical rigidity.
- Wire-hold-down: Leads and wires are unmobilized with proper adhesion to board or enclosures.
- Attachment of hybrid components to substrates.
- Microwave plumbing and waveguides are bonded in place with adhesives.

19.13 Polymeric Thick Films

Polymeric pastes with semiconductive fillers and a solvent are suitable to form thick-film links across components. With appropriate (and proportionate) use of fillers (such as metallic inclusions), the conductivity of the thick-film conducts can be controlled. These films can be formed at low temperatures unlike the ceramic-type thick films which warrant firing at 850°C. Polymeric thick films (PTF) can also be adopted for multilayer applications. With suitable plating, soldering can also be facilitated on PTFs.

19.14 Uses of Metals in EP Applications

Metals and alloys are widely used in EP applications. Both ferrous and nonferrous metals find vital role, to play in constituting electric/electronic parts and the associated packaging strategies.

Among ferrous metals, steels of different compositions are commonly used. The nonferrous metals, a wide gamut of candidates, prevail as listed in Table 19.8. Such elements (and/or their alloys) are useful in a variety of electronic parts/components/systems on need-based criteria. Salient characteristics and applications of ferrous and nonferrous metals/alloys *vis-a-vis* EP applications are furnished in Table 19.9.

Table 19.9 Ferrous Metals/Alloys Used in EP Applications

Ferrous Metals/Alloys	Characteristics and EP Application Potentials
Carbon steel	Available as hardenable, carburizable, and nitridable varieties. Used in small machine parts, screw fastenings, specific structural shapes, electromagnetic cores/stampings.
High-strength, high-speed steel with Mo or Wo contents	Available as highly hardenable, tough, and machinable materials. Used in heavy machine parts, cutting surfaces, forging dies, molds, etc.
Metallurgical powders, ferrous Fseries, stainless steel SS series	Magnetic material corrosion resistant. Useful as coil cores, armatures, magnetic linkages.

Typical nonferrous metals/alloys which find applications in EP technology are listed below. Their characteristics are presented in Chapter 9.

- Aluminum and aluminum alloys
- Beryllium, beryllium-copper, beryllium-nickel
- Cobalt and cobalt alloys
- Copper and copper alloys
- Lead and lead alloys
- Magnesium and magnesium alloys
- Nickel and nickel alloys
- Zinc and zinc alloys
- Precious metals (platinum, gold, palladium, iridium, rhodium, osmium, ruthenium and silver)
- Tungsten
- Thorium
- Molybdenum
- Tantalum
- Columbium
- Titanium
- Zirconium

19.15 Ceramics as EP Materials

Ceramics and glasses are dielectrics which find wide applications in electronic packaging mainly as insulators. Their electrical applications are based on their compatible mechanical, thermal, physical and/or chemical properties, availability, and assembly characteristics *vis-a-vis* the application.

Among the various ceramics, the following are the most widely considered candidates in EP technology:

19.15.1 Alumina (Aluminum oxide)

- Electrical characteristics

  Dielectric constant (ε_r): 9.7

  Loss tangent $(\tan\delta)$: 0.06

- Mechanical/physical characteristics
  - Density *(d)*: 4.0 kg/m^3
  - Flux strain *(ρ_F)*: 385 MPa
- Thermal characteristics
  - Thermal conductivity *(σ_T)*: 40 watt/m K
  - Thermal expansion coefficient *(α_T)*: 7.2 ppm/^{o}C

Aluminum oxide-based ceramics are generally the porcelain clay material and are formulated as several grades for electrical applications. Though these ceramics have high thermal conductivity, their main drawback is the high thermal expansion behavior.

Aluminas permit the products to be made *via* pressing or by sparry-based casting; surface finishing is done with high temperature glazing.

Bonding of alumina products with metal surfaces (metallization) is done either by high temperature firing or by low temperature thick-film processing.

The quality of alumina ceramics is decided by the extent of the glassy phase identified by the percent of aluminum oxide contained. The weight percentage of chemically analyzable Al_2O_3 in the ceramic affects the physical, mechanical, thermal, and electrical properties of the ceramic. Table 19.10 gives typical electrical parameters *versus* weight percentage of Al_2O_3 content.

Table 19.10: Electrical Properties of High Alumina Ceramics

Electrical Properties	Aluminum Oxide Content,% (by Weight)							
	85	90	90*	94	96	99.5	99.9	99.9
Dielectric strength (V/mil)								
0.25 in.	240	235	135	220	210	220	240	230
0.050 in.	440	450	415	425	370	430	460	510
0.01 in.	720	760	720	720	580	840	800	_
Dielectric constant								
1 kHz	8.2	8.8	22.0	8.9	9.0	9.8	9.9	10.1
1 MHz	8.2	8.8	9.8	8.9	9.0	9.7	9.8	10.1
100 MHz	8.2	8.8	_	8.9	9.0	_	_	10.1
Dissipation factor								
1 kHz	0.0014	0.0006	0.3000	0.0002	0.0011	0.0002	0.0020	0.00050
1 MHz	0.0009	0.0004	0.0200	0.0001	0.0001	0.0003	0.0002	0.00004
100 MHz	0.0009	0.0004	_	0.0005	0.0002	_	_	0.00006

(continued...)

Electrical Properties	Aluminum Oxide Content,% (by Weight)							
Loss index								
1 kHz	0.011	0.005	6.600	0.002	0.010	0.002	0.020	0.0050
1 MHz	0.007	0.004	0.200	0.001	0.001	0.003	0.002	0.0004
100 MHz	0.007	0.004	–	0.004	0.002	–	–	0.0006
Volume-resistivity (ohm-cm)								
25°C	$>10^{14}$	$>10^{14}$	$>10^{14}$	$>10^{14}$	$>10^{14}$	$>10^{14}$	$>10^{15}$	–
1000°C	–	8.6×10^{5}	4.0×10^{4}	5.0×10^{5}	1.0×10^{6}	–	1.1×10^{7}	–
T_c value, °C	850	960	–	950	1000	–	1170	–
* opaque								

Adapted from [1].

Alumina ceramics are useful as substrates in high density packages, and in chips with multilayer configurations. However, the high dielectric constants of alumina restrict the signal speed in the chips. Also the lead traces pose high resistivity. Further, thermal management is difficult with alumina substrates due to thermal expansion mismatch problems and poor thermal conductivity. Therefore, alternative ceramics have emerged as replacement materials.

19.15.2 Beryllium (Beryllium oxide)

- Electrical characteristics
  $\varepsilon_r = 6.8$; $tan\delta = 0.100$
- Mechanical/physical characteristics
  $d = 2.9$ kg/m^3; $\rho_F = 170$ MPa
- Thermal characteristics
  $\sigma_T = 300$ watt/meter °K; $\alpha_T = 7$ ppm/°C

Beryllium oxide ceramics are superior to alumina in many respects but are highly hazardous materials to handle due to their toxicity. The product manufacturing strategies are the same as those of alumina. Relatively low dielectric constant makes BeO substrates usable for high speed circuit substrates. Also high conductivity of these ceramics permits better thermal managment than alumina oxide. Cost of this material is, however, considerably higher.

19.15.3 Aluminum nitride ($A\ell N$)

This is a low toxic and a better alternative material than BeO. The properties of $A\ell N$ are:

- Electrical characteristics
  $\varepsilon_r = 10$; $tan\delta = 0.100$
- Mechanical/physical characteristics
  $d = 3.2$ kg/m^3; $\rho_F = 300$ MPa

- Thermal characteristics
  σ_T = 150 watt/meter °K; α_T = 2.7 ppm/°C

Though the dielectric constant is high (posing a liability on signal speed characteristics), AℓN has low thermal expansion behavior compatible with silicon substrate attachment used in multichip designs.

AℓN products are made *via* high temperature sintering. This process, however, augments the product cost.

AℓN is still an emerging technology. Its applications as a *gate-insulating material* in unipolar devices *in lieu* of conventional SiO_2 has been studied.

19.15.4 Silicon Carbide (SiC)

- Electrical characteristics
  $\varepsilon_r = 40$
- Mechnical/physical characteristics
  d = 3.2 kg/m^3; ρ_F = 450 MPa
- Thermal characteristics
  σ_T = 270 watt/m K; α_T = 4.3 ppm/°C

SiC has unique application potential due to its high thermal conductivity. Under compaction SiC is a conductor. With porous packing with polycrystalline sintered material, it, however, behaves as an insulator due to discontinuities in contiguous packing. This material has a high dielectric constant (40) which restricts its usage in high speed circuit substrates. It is also more expensive than alumina and costwise comparable with BeO or AℓN.

19.15.5 Boron nitride (BN)

- Electrical characteristics
  $\varepsilon_r = 4.1$; $\tan\delta = 0.100$
- Mechanical characteristics
  d = 2.2 kg/m^3; ρ_F = 110 MPa
- Thermal characteristics
  σ_T = 60 watt/m K; α_T = 3.8 ppm/°C

19.15.6 Other Ceramics

Steatites: Used as insulators, these are low firing ceramics and costwise are better than alumina.

Forsterite: It is a low-loss insulating material with moderately high thermal expansion coefficient and hence is compatible for bonding with metals.

Cordierites: These are magnesium silicates with low thermal expansion properties. They are conducive to applications where thermal shocks and uneven thermal stresses are expected.

Lava: These are naturally occurring ceramics with alumino– or magnesium silicate constitution. They are soft, moldable, machinable, and cheap, and can be fired at moderate temperatures. They are useful as *in situ* moldable ceramics for laboratory and/or as prototype development applications.

19.16 Glasses as EP Materials

Glasses are important EP materials. Their relevant application potentials are more comprehensive than ceramics. Essentially, glasses serve as insulators and optical materials. Their controllable thermal deformability and their ease in providing a bonding phase plus the associated electrical characteristics permit their applications in substrates, delay lines, passivation layers, capacitors, resistors, conductor bonding phases, package sealants, and insulation bushings. Their applications in vaccum-based devices (electronic valves) are well known. In modern applications at device level, glasses are used to provide appropriate optical transmission in packages of electrooptic devices. Glass fibers and optical domes (such as irdomes) in optical communication technology mostly rely on quality glass materials.

There are a variety of glass materials with different chemical compositions. Some typical glasses and their electrical, mechanical/physical, and thermal properties are listed in Tables 19.11 and 19.12.

Table 19.11 Electrical Properties of Glasses

		$Log_{10}(\rho_v)$			Dielectric properties at 1 MHz and 20°C		
Glass* Code	Type	25°C	250°C	350°C	Power Factor, %	Dielectric Constant	Loss Factor, %
0010	Potash soda lead	17.4	8.9	7.0	0.16	6.7	1.0
0080	Soda lime	12.4	6.4	5.1	0.9	7.2	6.5
7052	Borosilicate	17.0	9.2	7.4	0.26	4.9	1.3
7570	High lead	_	10.6	8.7	0.22	15.0	3.3
7740	Borosilicate	15.0	8.1	6.6	0.50	4.6	2.6
7900	96% silica	17.0	9.7	8.1	0.05	3.8	0.19
7940	Fused silica	_	11.8	10.2	0.001	3.8	0.0038
9010	Potash soda barium	_	8.9	7.0	0.17	6.3	1.1

* Corning Glass Works; ρ_v = Volume resistivity.

Adapted from C.A. Harper (Ed.) Electronic Packaging and Interconnection Handbook. McGraw-Hill, Inc.: 1991. With permission.

Glasses are super-cooled liquids with noncrystalline structures. In molten state, they have very high viscosity which renders them amorphous (instead of crystalline) when frozen.

The highly viscous molten glass can be used as cavity fillings and feed-throughs in appropriate EP applications such as in making capacitors, thick-film structures, etc.

Glasses are chemically based on four glass-forming oxides (or their mixtures), namely:

- SiO_2

- B_2O_3
- P_2O_5
- GeO_2

Table 19.12 Physical/mechanical and Thermal Properties of Glasses

Glass* Code	Type	Thermal Expansion 0-300°C ppm/°C	Viscosity data, °C			
			Strain Point	Annealing Point	Softening Point	Working Point
0010	Potash soda lead	93	395	435	625	985
0080	Soda lime	92	470	510	695	1005
7052	Borosilicate	46	435	480	710	1115
7570	High lead	84	340	365	440	560
7740	Borosilicate	33	515	565	820	1245
7900	96% silica	8	820	910	1500	_
7940	Fused silica	5.5	990	1050	1580	_
9010	Potash soda barium	89	405	445	650	1010

(continued...)

Glass* Code	Type	Thermal Expansion 0-300°C ppm/°C	Density g/cm^3	Young's Modulus, 10^6 lb/in.^{2}	Poisson's Ratio
0010	Potash soda lead	93	2.86	8	0.21
0080	Soda lime	92	2.47	10.0	0.24
7052	Borosilicate	46	2.28	8.2	0.22
7570	High lead	84	5.42	8.0	0.28
7740	Borosilicate	33	2.23	9.1	0.20

(continued...)

Glass* Code	Type	Thermal Expansion 0-300°C ppm/°C	Density g/cm^3	Young's Modulus, 10^6 lb/in.^{2}	Poisson's Ratio
7900	96% silica	8	2.18	10.0	0.19
7940	Fused silica	5.5	2.20	10.5	0.16
9010	Potash soda barium	89	2.64	9.8	0.21

* Corning Glass Works.
Adapted from C.A. Harper (Ed.) Electronic Packaging and Interconnection Handbook. McGraw-Hill, Inc.: 1991. With permission.

In addition to these oxides which form the base materials for the glasses, "fluxes" (softeners), extenders, and colorants are added to realize a desired end product. Commonly used additives are:

- Softeners: Oxides of Li, Na, K, Rb, Cs, Pb
- Extenders: Oxides of Mg, Ca, Sr, Ba
- Colorants: Oxides of Co, Mn, Fe, Cr

19.17 Characteristics and EP Applications of Typical Glass Materials

Selection of a glass material for EP applications is mainly based on:

- Thermal expansion characteristics: –Glasses have thermal expansion coefficients in the range ≅ 0 to 12 ppm/°C depending on their composition.
- Transformation point (T_g): –Glasses exhibit abrupt increases in thermal coefficient around 20 to 50°C of their softening point. Glass-metal bonding takes place at temperatures higher than T_g. Therefore care is exercised in choosing the glass material *vis-a-vis* T_g for such bonding applications.
- Electrical conductivity: –The bulk resistivity of glasses is decided by the extent of alkaline level in the composition.

Details on typical glass materials follows.

19.17.1 *Pure SiO_2 glass*

This is also known as *vitreous silica* or *fused silica.* Being essentially a SiO_2, sometimes it is also referred to as *fused quartz.* Its applications include coated and uncoated windows in electrooptical devices. Low infrared absorption characteristics permit its use as irdomes (infrared domes) used in heat-seeking missiles. An almost zero thermal coefficient enables its application where thermal shock is the primary consideration. It is a very expensive material with an exceptional melting point (>1700°C). It has the lowest value of dielectric constant (3.8) of any other glass material.

19.17.2 *Fractional silica glass*

This refers typically to a 96% SiO_2 which has almost the same properties as pure SiO_2 but is less expensive. Also, relatively this material can be more easily shaped than pure SiO_2. It is formed from a low temperature glass which exhibits phase separation (of two intertwined phases) at an elevated temperature. By chemically digesting one phase (with an acid), the other phase constitutes a 96% SiO_2 selection which can be sintered to a denser state below 900°C.

19.17.3 *Soda-lime glass*

This an alkaline-earth oxide-based silica glass. Typically SiO_2, Na_2O, and CaO are the constituents with lesser amounts of K_2O, MgO, or Al_2O. Soda-lime glasses are the most widely used glass materials. They do not withstand thermal shocks due to high temperature coefficients. Alkaline content (12 to 15% by weight) controls their bulk electrical resistivity. Typically, dielectric constant varies from 6 to 9. Addition of PbO alters the resistivity (or lossy dielectric behavior) of these materials.

Typical versions used for electrical applications have the following percentage compositions.

- SiO_2 (63%), Na_2O (8%), K_2O (6%), CaO (0.3%), PbO (22%), B_2O_3 (0.2%), and Al_2O_5 (0.5%)
- SiO_2 (73%), Na_2O (2%), K_2O (2%), CaO (2%), PbO (6%), and B_2O_3 (17%)

19.17.4 *Lead alkali borosilicate glass*

These are *soft glasses* with low melting and softening temperatures and are composed of SiO_2 plus PbO, Al_2O_3, B_2O_3, BiO, and ZnO.

They are useful in sealing and/or adhesive applications. They can be used in powder form as an additive material to thick film pastes. The low melting point glasses are used as "hermetic sealing glues". If formulated to crystallize (or devitrify), these glasses have the added advantage of controllable thermal expansion coefficient.

19.17.5 *Glass ceramics*

Glasses can be formulated to crystallize after forming at temperatures below the glass melting and deformation points. Such glasses are known as *glass ceramics*. They are predominantly crystalline. (Corningware™ glass is a typical example of this category.) Glass ceramics are almost twice as strong as ordinary glasses and find special applications in electronics. In some cases, they are also machinable.

19.18 Synthetic Diamonds as EP Materials

These are graphite materials synthesized at high temperatures and pressures. Synthetic diamond films can be formed *via* the chemical vapor deposition (CVD) process using 95% hydrogen plus 5% hydrocarbon (such as methane). The process reaction leads to a plasma state which allows the decomposed carbon to deposit on a substrate as a film. Such films are of use in electronic packaging as heat sinks due to their thermal conductivity being 20 watt/cm °K which is five times that of copper and silver. Solid versions of synthetic diamonds are useful as heat sinks for larger diodes. Diamond, in essence, is one of the very few materials which is a very good dielectric (electric insulator) and an excellent thermal conductor.

Synthetic diamonds are emerging materials useful as semiconductors. Doping with boron and phosphorus atoms, such diamonds can form a typical pn junction. The mobility of electrons in diamonds is more than that in gallium arsenide. This enables high frequency application of diamond semiconductors. Diamonds have a very large dielectric breakdown strength and can handle high power levels. These properties cohesively indicate the applications of diamonds for high frequency, high power semiconductor devices which have the potentials of upcoming technology. Further, diamonds withstand high levels of ionizing radiation dosage. Therefore, they are highly suitable for space applications.

19.19 Conclusions

The materials of EP technology as evinced above are of different varieties, and therefore their characteristics and applications involve multidisciplinary aspects of material science and technology. These materials play a vital role in modern electrical and electronic technology, not only do they decide the functional aspects of the end products but also they control the pragmatic life span of the packaged assembly. Further, EP materials as a whole constitute the largest piece of pie in the gamut of electromagnetic materials.

References

[1] R. N. Sampson and D. M. Mattox: Materials for electronic packaging in *Electronic Packaging and Interconnection Handbook.* (Editor-in-Chief: C. A. Harper), (McGraw Hill, Inc., New York: 1991), Chapter 1, pp. 1.1-1.72.

[2] W. C. Bosshart: *Printed Circuit Boards-Design and Technology.* (Tata McGraw-Hill Publishing Co. Ltd., New Delhi: 1983).

[3] G. L. Gimsberg: *Electronic Equipment Packaging Technology.* (Van Nostrand Reinhold, New York:1992).

Defining Terms

Adhesives: Bonding materials, normally insulative in characteristics, unless specified as conductive adhesives.

Electronic packaging conducting materials: Subset of conductive media of EP materials.

Electronic packaging insulating materials: Dielectric subsets of EP materials.

Electronic packaging materials: Family of materials intended for combining engineering considerations and technological aspects to realize a manufacturable end product constituted by electronic circuits and/or subassemblies.

Electronic packaging semiconducting materials: Semiconductor subsets of EP materials.

Fillers for plastics: Powder materials deliberately added to plastics to realize certain specific casting and embedment characteristics.

CHAPTER 20

Static Control Materials

20.1 Introduction

Static control materials are those which are useful to prevent or minimize the buildup of *electrostatic charges* on conducting and/or nonconducting bodies. These materials are essentially of three types: (1) *Antistatic* materials; (2) *static dissipative* materials and (3) *static conductive* materials.

Antistatic materials (or *antistats*) act as the static control media in three different ways: First, they reduce the *coefficient of friction* to help minimize the frictional (*triboelectric**) static charge generation. Second, they increase surface resistivity to dissipate (*or bleed off*) the charges. Third, they interact with the environmental factors to neutralize the generated charges. Any grounded conducting material is antistatic as it instantaneously bleeds off the charge accumulation on it. In general, antistatic materials should resist triboelectric charging and produce minimal static charges when separated from themselves or from other materials.

Typically, an antistatic material could be a medium impregnated with *migratory antistats* or it may correspond to those treated by spraying, dipping, painting or wiping with a *topical antistat agent* to render them surface conductive thereby facilitating an "easy bleed-off" of the charges. Antistatic surfaces so treated have a thin lubricious layer caused by the antistat agent which reduces the frictional effects during rubbing and/or separation of the bodies.

Static dissipative materials refer to those which are seldom *static propensive* and do not accumulate charges on the surface by letting the charges if acquired to bleed off instantaneously to the ground. Normally, metallic media or those impregnated with metals or conductors like carbon would seldom be charged when separated from one another. (However, they can transfer charge to a nonconductive surface when separated from it.)

Static conductive materials: These are functionally same as static dissipative materials but with higher *surface conductivity*. They represent bulk forms of metals and conductor-included plastics, etc. of larger sizes (by volume) in comparison with the static dissipative materials which are invariably thinner in dimension.

20.2 Need for Static Control Materials

Static control materials have gained importance as a result of observed failures and/or performance degradation of electrical and electronic parts due to the discharge of electrical charges through them. Known as *electrostatic discharge* (ESD), this phenomenon has been detrimental to electronic components such as *metal-oxide semiconductors* (MOS). Related to ESD are two other phenomena namely, the static-discharge (spark)-induced *electromagnetic interference* (EMI) and *charge-induced electrical overstressing* (EOS). These indirect effects due to static charges have also been observed to cause failures at component and/or subsystem levels in electronic systems. Therefore, static control materials have emerged essentially as static protective materials in static control strategies adopted in handling semiconductor products.

20.3 Static Propensity

Some materials are inherently *static propensive*. That is, upon triboelectrification they acquire a significant amount of charges which remain as "puddles" on the material surface. These charge puddles cause intense ESD by coming in contact with a grounded conducting medium. The static propensive materials are classified as two types of *triboelectric series* as listed in Table 20.1.

---

* *Triboelectricity* : This refers to electric charge transfer in bringing two different materials into intimate contact and separating them (tribo $\Rightarrow$ action of rubbing).

Table 20.1 Triboelectric Series

Air	+	Sealing wax	
Human body		Hand rubber	
Asbestos		Nickel, Copper	
Fur		Brass, Silver	
Glass		Gold, Platinum	
Mica		Sulfur	
Hair		Acetate, Rayon	
Nylon		Polyester	
Wool		Celluloid	
Lead		Orlon	
Silk		Saran	
Aluminum		Polyurethane	
Paper		Polyethylene	
Cotton		PVC (vinyl)	
Steel		KEL-F (CTFE)	
Wood		Silicon	
Amber		Teflon	−

The materials labeled "positive" + in Table 20.1 will take on a positive charge every time they come in contact with a material lower on the scale. The static propensity of a material can be assayed by a figure-of-merit parameter, namely, the *surface electrical resistance* per unit area (ρ_s). The materials with $\rho_s > 10^9$ ohm/square are likely to develop electrostatic charges which will not bleed off by themselves due to the high electrical insulating property of the surface. The relative propensity scale of materials is approximately depicted in Table 20.2.

Table 20.2 Relative Static Propensity of Materials

Static Propensity Figure-of-Merit Scale (ρ_S ohms/square)	Classification of the Materials
10^9–∞	Static-prone materials
10^6–10^9	Static-free materials
0–10^6	Conductors

20.4 Static Propensity *versus* Dielectric Property of Materials

Dielectric property is an another way of depicting (or quantifying) the *static proneness* of materials. The charge density (q_s) that accumulates triboelectrically on a surface between two dielectric (insulating) materials is given by:

$$q_s = 15 \times 10^{-6} (\varepsilon_{r1} - \varepsilon_{r2}) \qquad \text{coulomb/meter}^2 \tag{20.1}$$

where ε_{r1} and ε_{r2} are the relative permittivities of the two dielectrics in contact.

20.5 External Influences Affecting the Static Propensity

Humidity of the surrounding air would tend to lower the surface resistance of a material due to moisture deposition on the surface. Hence, the material has less static propensity. Surface film of moisture could lower the surface resistivity as low a value as 10^{-9} ohms/square of materials like tiles, carpets, table mats, etc. Thus, relative humidity has a significant effect on the static accumulation on material surfaces. On the contrary, dry air or wind passing over a material would cause a triboelectric generation of charges on the exposed surface of a material.

20.6 Static Voltage Induction on Insulating Materials

Suppose an insulating material constitutes a capacitance *(C)* with respect to the ground; the corresponding electrical potential *(V)* developed on the material as a result of surface static charge of density (q_s) is given by

$$V = q_s A/C$$

$$= 15 \times 10^{-6} (\varepsilon_{r1} - \varepsilon_{r2}) A/C \quad \text{volt} \qquad (20.2)$$

where A is the surface area of the material.

In the presence of moisture, however, the static voltage of the material drops as a function of the relative humidity (R_H). It should be noted that relative humidity is also dependent on the ambient temperature. As such, the static propensity also varies with the local environmental temperature.

20.7 Characteristics and Types of Static-Control Materials

Typically, the materials useful for static-control applications refer to those which prevent the invasion of static bleed-off from one material through another material. For example, electronic components sensitive to electrostatic zaps should be protected from the effect of static charges on other materials (such as a human body handling the components). As indicated earlier, there are three modes of static-induced (ill) effects on sensitive semiconductor products as illustrated in Figure 20.1.

In each of the three cases as shown in Figure 20.1, a static-control material is needed for interpositioning between the static-prone body and the sensitive component so that static bleed-off or EOS or EMI effects could be minimized.

For this purpose, electrostatic-sensitive electronic components are packaged in static-control materials. Examples are:

- Static protection flexible bags and pouches
- Static protection packaging boxes, IC carriers and vials
- Antistatic cushioned packaging materials
- Conducting foams
- Conducting printed circuit board (PCB) shunts
- ESD protective tote boxes, bins and trays
- Static preventive conductive floors
- Static nonpropensive floor mats/carpets
- Static conducting footwear
- Antistatic garments/clothing
- Topical antistats, sprays and paints

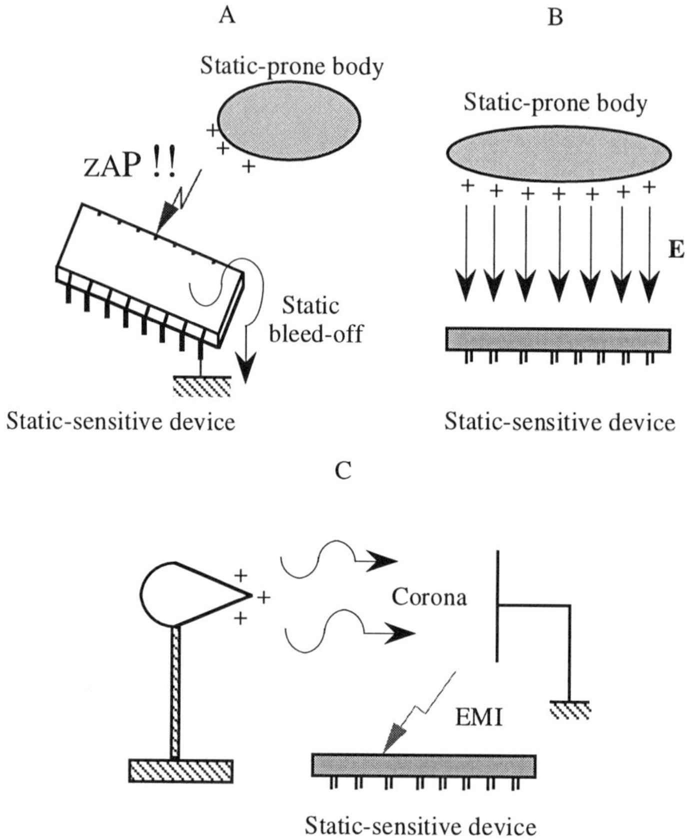

Figure 20.1 Modes of ESD.
A: Static bleed-off through a static sensitive component by direct contact.
B: Static-induced overstressing electric field (**E**).
C: EMI due to static-induced corona discharge.

20.8 Basic Functions of ESD-Protective Materials

The ESD-protective materials functionally have one or more of the following attributes:

- Low triboelectric charge generation and low static propensity
- Controlled static bleed-off characteristics
- Shielding off electrically overstressing, charge-induced electrical fields
- Fast dispersion and dissipation of charges

The descriptions of the above attributions pertinent to the static-control materials enumerated before are presented in Table 20.3.

20.9 Measurable Parameters of ESD-Protective Materials

The performance of materials used as ESD-protective media is decided by:

- Volume resistivity of the bulk material (ρ_v ohm-cm)
- Surface resistivity (ρ_s ohm/square)
- Charge decay time τ in seconds

Table 20.3 Functional Characteristics of Static-Control Materials

Characteristics	Description	Specification, Standards and Remarks
Low triboelectric charge generation	This refers to the charge generation and accumulation on the surface of two materials. The polarity of the charges acquired by a material depends on its inherent property of being in the triboelectric series listed in Table 20.1. Low triboelectric materials are bulk conductive and antistatic plastics, wood and wood-based composites, paper, cardboard, untreated cotton, and melamine laminates, etc.	
Controlled-static charge bleed-off	Depending on surface resistivity, flow of charges on the material surface is dictated by an RC time constant factor. Large value of this time constant prevents the flow of charges to a locale being exposed to electrification by a charged body. Antistatic materials ($\rho_s = 10^9 - 10^{14}$/square) match this requirement.	MIL-B-81705B, Type II: "Barrier Materials, Flexible, Electrostatic-Free Heat Sealable" August 15, 1974.

(continued...)

Characteristics	Description	Specification, Standards and Remarks
Shielding of ESD-induced overstressing electric fields	Electrostatic shielding is governed by the surface resistivity of the material. Materials with surface resistivity $\cong 10^9$/square provide adequate shielding. Shielding is based on Faraday-cage principle. For shielding, mesh-like structures, interwoven materials can be used *in lieu* of continuous materials. EMI due to ESD or corona can be shielded by highly conductive materials ($\rho_s \cong$ 10/square). Perforated structures can also be used for this purpose.	DOD-HDBK-263 : "Electrostatic Discharge Control Handbook for Protection of Electrical and Electronic Parts, Assemblies and Equipment (Excluding Electrically Initiated Explosive Devices)". May 2, 1980.
Fast dissipation of charge accumulation	This characteristic again depends on the surface resistivity of the material. Conductive ESD protective materials with $\rho_s \cong 10^5$ ohm/square or less are useful for this purpose. Metals and bulk conductive plastics (plastic composite with copper or nickel or carbon impregnation) are base materials for fast static dispersion and dissipation.	MIL-P-82646 : "Plastic Film, Conductive, Heat-Sealable, Flexible." March 26, 1976.

Volume resistivity is related to the bulk resistance *(R)* of a material of cross-sectional area A cm^2 and length ℓ cm as given by :

$$R = \rho_v \ell / A \qquad \text{ohm} \qquad (20.3)$$

Surface resistivity (ρ_s), as defined earlier, is the electrical resistance offered by a square section of a material. It is independent of the size of the square and related to the bulk resistivity as follows:

$$\rho_s = \rho_v / t_m \qquad \text{ohm/square} \qquad (20.4)$$

where t_m is the thickness of the material.

Decay time (τ) refers to charge dissipation time (say, to 10% of the original value). FED-STD-101B Method 4046 describes relevant test methods to evaluate τ of ESD-protective materials.

20.10 Physical Forms of Commonly Available ESD-Protective Materials

The three classes of ESD-protective materials, namely, static conductive, static dissipative, and antistatic solid materials are normally constituted by monolithic or laminated materials with different forms/shapes conducive for practical products. Typically available shapes are: Sheets/plates (tile-like flooring structures, mats, tabletop laminations etc.); vacuum-formed shapes such as trays, vials, boxes etc.; rigid bars and clips; foams; and bubble-pack or open cell plastics, flexible materials, and straps.

Antistatic agents are available as sprays, paints and topical coating materials.

20.11 ESD-Protective Products and Their Material Characteristics

Protective Bags: There are three generic types of ESD-protective bags, namely, (i) bulk conductive plastic bags; (ii) antistatic impregnated plastic bags, and (iii) metallic film blended material bags. The relative properties of these bag materials are presented in Table 20.4; and the general engineering needs of the materials are specified in Table 20.5.

20.12 Cushioned Antistatic Packaging Materials

These are made of antistat-impregnated plastics either in bubble-pack or in expanded form with polyethylene or polypropylene. Cushioned packaging materials provide both antistatic protection as well prevent shock and damages during handling, shipping, and storage of electronic devices contained in them. The antistatic effect includes protection against both triboelectric charges and external electrostatic fields (specifications: MIL-P-81997A and PPP-C-1842A).

The bubble and foam construction forms of cushioned materials show a dependency of antistatic effectiveness against relative humidity (R_H). The surface resistivity of these materials, in general, increases with the decrease in R_H. Consequently at low R_H values, some of these materials exhibit poor static decay time and may not meet the requirements of standards such as MIL-B-81705B. Typical variations of resistivity and static decay *versus* R_H are indicated in Figure 20.2A and 20.2B.

It has been observed that antistatic-treated packaging materials may outgas substances that may affect the transparency of precision optic surfaces packages. Also, detergent action of antistatic chemicals may have adverse effects on some packaged components.

Table 20.4 Static-Protective Packaging (Bag) Materials

Type and Description	Name/Model and Specification Standards	Inside Surface Resistivity ρ_s:ohm/square	Triboelectric Protection	Faraday Caging (Electrostatic Field Protection)	References and Remarks
• Opaque $A\ell$ foil film laminate	Type: MIL-B-81705 I	1×10^{16}	No	Excellent	MIL-B-81705: Barrier materials, flexible, electrostatic free, heat sealable
• Antistat-treated polyethylene ("sweat layer" of moisture derived from the atmosphere controls the surface conductivity)	Pink or blue poly bags Type II: MIL-B-81705 II	$10^9 - 10^{12}$	Yes	None	MIL-B-117: Bags, sleeves and tubing, interior packaging.
	Cushioned, reusable zip-lock bags: MIL-B-81997			None	
• Conductive-carbon impregnated polyolefins (bulk-conductive material)	Padded conductive bags	$1 \times 10^4 \sim 3 \times 10^4$	Yes	Fair	
• Conductive nylon		3×10^4	Yes	Fair	
• Conductive multilayers with outside antistatic interlayer of metallized polyester/antistatic polyethylene composite. (sandwich construction)	Static-barrier bags	$10^{11} - 10^{13}$	Yes	Good	
• Antistatic treated $A\ell$ foil		10^{12}	Yes	Excellent	

Table 20.5 General Engineering Requirements of ESD-Protective Bags

Material Type	Optical Transparency	Water Vapor Permeability gm/100 in.^{2}/ 24 Hr	Abrasion Resistance	Heat-Seal Strength lb/in. of Width	Puncture Resistance Failure (lb)
	Requirements & Compatibility Ratings				
Type II pink poly bags	Excellent (E)	0.6 (Fair)		10.3 (E)	1.5 (F)
Type II blue poly bags	Good (G)	1.5 (Fair)		7.3 (E)	2.0 (F)
Ni-coated conductive bags	Poor (P)	1.1 (Fair)	10-50 (Fair)	0.5-9.0 (P to E)	2.5 (G)
Al-coated conductive bags	Fair (F)	2.3 (Fair)	~100 (Fair)	15.0 (E)	3.0 (G)
Carbon-loaded plastic bags	None	3.0-5.0 (Poor)		8.0-9.0 (E)	~1.6 (F)
Type I: Aℓ-foil film bags	None	< 0.003 Excellent	~1000 Excellent	6.0 (G)	5.0 (E)
Antistatic treated Aℓ-foil bags	None	< 0.002 Excellent	~1000 Excellent	4.5 (G)	4.0 (E)

Definitions:

- Optical transparency: See-through compatibility of the bags to view the bag contents forbidding zapping the contents *via* touching.
- Moisture permeability: This refers to the permeance of water vapor through the bag material as controlled by atmospheric conditions.
  (MIL-STD-202E/METHOD 106). Requirement is specified by MIL-B-81075
  (for Type I materials): 0.02 gm/100 sq in./24 hr at 100°F and $R_H = 90\%$.
- Abrasion resistance: Cyclic repetition of abrasive force on the bag to check its wear and tear withstandability.
- Heat-sealability: Compatibility for being sealed *via* hot tools.
- Puncture resistance: Maximum withstandability towards the application of puncturing force.

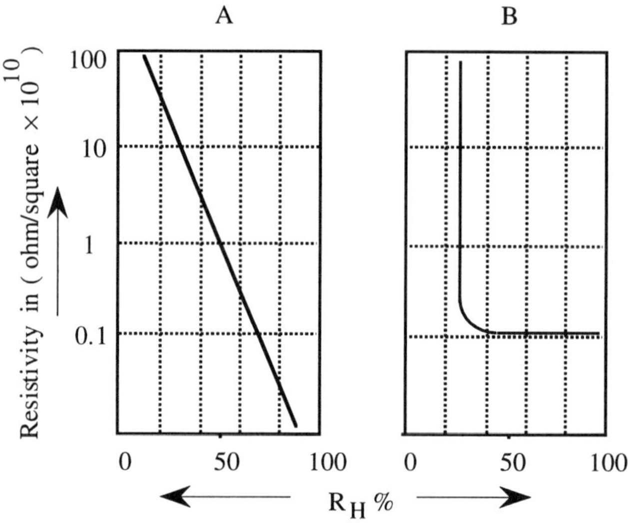

Figure 20.2 Static decay *versus* relative humidity (R_H).
A: In bubble-pack construction of plastic materials.
B: In expanded-foam construction of plastic materials.

Another form of antistatic cushioning material is the conductive foam which acts as a short-circuit (shunting) medium when the integrated-circuit (IC) pins are inserted in it. Its main requirements are:

- High conductivity for effective shunting
- Low triboelectric generation upon insertion or removal of IC pins
- Noncorrosive and no effects on pin surface *vis-a-vis* solderability
- No shedding of particles, (lest short circuit may be caused between electrical connections)
- Lower resistance than the resistance across the pins.

There are two types of conductive foams. They are carbon-filled low density type to provide optimum cushioning against physical damages and high density type to provide equipotential bonding while holding IC and other discrete device-leads securely without bending. High density foams with fire retardants may source corrosive sulfide and chloride ions under moisture conditions.

A comparision of cushioning materials is presented in Table 20.6.

20.13 Conductive Shunts

These are used on printed circuit boards (PCB) to short-circuit the PCB terminal pins by shunting them together during handling, storage, and assembly so as to keep all the pins at equipotential levels and preventing static flow from pin-to-pin. General characteristics of these conductive shunts are :

- Resistance of the shunting material should be an order of magnitude less than the minimum impedance between any two pins of the ESD-sensitive item/PCB
- Low corrosiveness and low triboelectric generation

Table 20.6 Relative Characteristics of Antistatic Cushioning Materials

Type of Material	Resistivity (ohm/square)	Corrosive Effects
Antistatic bubble wrap	–	–
Low-density, carbon-filled conductive foam	10^4–10^5	Present
Antistat-impregnated foam	10^9–10^{11}	
High-density carbon-filled conductive foam	10^4	Present

- The shunt materials are therefore static conductive or static dissipative rather than antistatic materials
- PCB shunts should be semi-rigid and rubber-like clips for easy installation and removal without damaging the pin outlets or their solderability

Typically conductive shunts are made of carbon-filled conductive plastics or rubber and are popularly known as *PCB shunt bars* or *board shorts*. Resistivity of these materials is on the order of 0.5 milliohm/36 in.

20.14 ESD-Protective Materials of Tailored or Formed Shapes

For specific needs, ESD-protective materials are made in formed shapes such as tote boxes, trays, vials, shipping tubes (sticks), carriers, boxes, bottles, part bins, loaders, and hand tools (like solder suckers, IC inserters/extractors).

To meet the above applications, metals such as aluminum without anodic coating provide the best protection including EMI and electrostatic shielding. Alternatively, conductive and static dissipative bulk conductive plastics can be used for these applications with protection against triboelectricity and static fields. Antistatic materials on the other hand if used, would provide adequate triboelectric static protection but do not yield satisfactory static field protection. Typically characteristics of ESD-protection materials for formed shape applications are listed in Table 20.7.

20.15 Conductive Floors, Floor Mats, and Footwear

These are cohesively used in ESD control programs to provide a total ESD protection in work environments where ESD-sensitive materials are used, handled and/or stored.

20.15.1 *Flooring*

Common vinyl flooring, sealed concrete, and finished or waxed wood floors are prime sources of static charges. Static-protective flooring should:

- Promote low triboelectric generation
- Have high conductivity
- Possess high static dissipative properties

Available types:

- Carbon-loaded polyolefin, polyethylene, vinyl tiles
- Poured terrazo

Table 20.7
Characteristics of ESD-Protection Materials For Formed-Shape Applications

Application	Material	Problem	Resistivity (ohm/sq)	Transpar--ency/ Viewing Slots	Faraday Caging	Triboelectric Charge Generation
Dual-in-line IC-packaging tubes and IC carriers	Formed aluminum (without anodic coating)	Heavy formation of oxide coating on pins			Excellent	Minimum
	Carbon-filled		200-300 ohm-cm	Yes	Fair	
	Conductive plastics					
	Antistatic plastics		1.8×10^9	Yes	None	Low
	Antistatic-treated plastics		1.5×10^9	Yes	None	High to low
Flat-pack IC carrier	Aluminum foil/cardboard			Yes	Good	Not applicable
Antistatic tote boxes, bins, and trays	Antistat-impregnated		10^{10} - 10^{12}			
	Carbon-filled		200 ohm-cm			
	Conductive-nylon tote box liners		30×10^3			
Carriers	Cardboard with conductive lining, coating and/or foam insertion					

- Static-dissipative carpeting
- Floors treated with topical antistats

Characteristics of these flooring materials are presented in Table 20.8.

20.15.2 *ESD–protective footwears*

Conductive footwears go together with conductive flooring. They deplete the electrostatic charge on persons by discharging them to the conductive flooring. They are two types of conductive footwears: (i) *Conductive shoes*, and (ii) *heel grounders*. The generic types commercially available are heel straps, heel protectors, neurostat boots, heal grounders and shoe-grounding conductive straps. All these are typically made of a flexible band of carbon-impregnated polyolefin or polycarbon-ethylene which covers the bottom of the heel of the shoe and tucks into the shoe to make contact with the person's sweat layers.

The ANSI Z41-1983 gives the standard specifications on conductive footwears intended for static protection as Type I designed to dissipate static electricity. Conductance of Type I footwear is specified in the range 0 to 500 K ohm.

20.16 Static-Control Garments and Clothing

Laboratory coats and smocks should meet ESD-protection considerations. Normally five generic types of clothing materials are considered for this purpose, namely: Cotton, polyester, cotton plus polyester blend, spun-bonded olefin, and cotton plus polyester blend interwoven with stainless steel threads.

Virgin cotton is in the middle of triboelectric series and has conductivity sufficient to prevent electrostatic accumulation, but textiled garments made for work-and-wear and permanent-press purposes may have enhanced or degraded conductivity properties.

Polyester or its blends with cotton are more static propensive. Washing and drying these materials causes an enhancement of static buildup. Therefore, polyester/cotton and polyester/Dacron blends are treated with antistats for ESD protection, such antistat treatment to be done after every time the garment is laundered.

Blending cotton or polyester/cotton materials with 1 to 0.5 percent of stainless steel effectively prevents charge buildup.

Typical garment materials for ESD protection and their properties are listed in Table 20.9.

20.17 Topical Antistats

These materials (available as sprays, paints or coatings) reduce the coefficient of friction between materials (by increasing surface lubricity) and increase surface conductivity facilitating a fast static bleed-off. Some antistats function only along with ambient moisture (hygroscopic) forming a conductive vapor layer on the surface of the materials. In liquid form topical antistats consist of:

- Carrier or a vehicle that enables pervading of antistaticity on the surface (water, alcohol, or other solvents used for this purpose)
- Primary antistat material which upon deposition on a material surface performs static control and anticharge propensive functions

Topical antistats are used for:

- Surface cleaning plus charge elimination
- Static control on surfaces not amenable for ESD control techniques otherwise (examples: CRT displays, computer terminals, etc.)

Physical characterizations of topical antistatic materials refer to:

Table 20.8
ESD-Protective Conductive Floorings

Type/Description	Surface Resistance (ohm/square)	Static Decay Rate: Static Voltage (volt)	Static Decay Rate: Time in (sec.)	Safety (Whether Resistor Required*)
Floor Mats:				
Carbon-filled conductive plastic/rubber	10^3-10^4			Yes
Conductive (static-dissipative) vinyl laminate	10^6-10^9	5000	0.06-0.45	Yes
Tiles:				
Conductive vinyl flooring	25 K to 1 M ohm	5000	0.02	
Vinyl bonded to aluminum floor plate	< 1 M ohm			
Carpeting:				
Nylon with conductive monofilament				
Nylon with conductive copper threads				
Nylon interwoven with conductive filaments				
Paint:				
Conductive floor coatings	5×10^3-5×10^5			

* For operator safety, the protective work surface is grounded through a current-limiting resistor of 250×10^3 to 1×10^6 ohm. The bonding and grounding should meet ANSI/NFPA 77 Specification on Static Electricity, 1983.

Table 20.9
ESD-Protective Garments and Clothing

Material	Typical Surface Resistance (ohm/square)	Static Voltage (kilovolt)	Decay Time (sec)	Temp. °F/ %Humidity	Max. Static Typical Voltage Buildup
Virgin cotton	3.0×10^8–8.0×10^{11}	5	0.25-5.35	70/50	~10 KV
Cotton (fire-retardant treated)					<100 V
Dacron polyester (antistat treated)	~2.0×10^{10}	5	0.20	70/50	
Polyester/conductive nylon	1.6×10^6		0.1	–/50	
65% Polyester + 34% cotton + 19% stainless steel	2.3×10^{10}			–	
Polyester/cotton plus 0.5% stainless steel	2.0×10^{10}			70/50	~1 KV

- Compatibility of their use with material surfaces being treated for ESD control
- Inherent contamination and corrosive behavior
- Longevity and wear characteristics *vis-a-vis* repetition requirements of their application
- Static decay performance and controlling static propensiveness
- Ease of application

Antistats may cause low resistance bridge action. Therefore, their application at high impedence circuit nodes may cause circuit malfunctioning. Listing and properties of typical topical antistats are presented in Table 20.10.

Table 20.9 Topical Antistats

Material	Typical Surface Resistivity (ohm/square)
Antistatic spray	$(25–1000) \times 10^6$ ohm
Antistatic freeze spray	10^{10}
CRT screen and terminal cleaning spray	
Work-surface cleaner	

Topical antistat agents can be sprayed, brushed, rolled, dipped, mopped, wiped, or otherwise applied to floors, carpets, workbench tops, parts bins, chairs, walls, ceiling, tools, paper, plastics, and clothing to render them antistatic. The longevity of applied antistats depends on wear and tear situations and the substrate on which they are applied.

Another class of antistat agents is referred to as *internal antistats*. They are additive types compounded into plastics during processing. These are of two types as per the following functions:

- Those internally enhancing conductive paths in the plastics for static migration
- Those *blooming* to the surface and actively participating in static dissipation with the assistance of ambient moisture

Typically internal antistats form 5 to 30% as additives. High loading may affect the physical properties of the plastics. Surface blooming types are also added only in very low fractions – 0.1 to 4.0%.

20.18 General Considerations of Designing Static Control Materials

As discussed in the previous sections, a variety of synthetic and organic composites are used in the industrial environment as bench-top materials, floor finishes, containers, carpets/floor mats, workroom apparels/garments, etc.; and these materials, in general, are highly electrostatic propensive unless properly treated for static control. Whenever there are two materials moving in opposite directions abrading against each other, a high triboelectric potential would build up between the abrading surfaces on separation; as a result, electric charges of opposite polarity accumulate upon these surfaces and they do not bleed off easily if the abrading media are insulators with high resistivity. The static charge can stay put upon the surfaces as puddles over a long duration of time until a conducting medium comes in contact with the surfaces.

In electronic industries triboelectricity is regarded as a menace because any accidental static voltage transfer/buildup occurring in a semiconductor device may cause catastrophic or latent device failure. Especially microelectronic devices pose high reliability problems arising from sneaky failures due to electrostatic discharge (ESD) which is considered as a new contaminant of the age of chips. One of the preventive measures adopted to control static electrification in microelectronic industries is to use a distinct class of synthetic composites which are less prone to triboelectric effects. Such static propensity-controlled composites are of two types, namely, (i) antistatic or static-repulsive materials and (ii) static-dissipative or static-conductive materials.

Static-conductive composite materials help to solve the problems of electrostatic discharge by controlling the generation, accumulation, and dissipation of static charges. They offer proven static protection in electronic manufacturing, assembly, and test areas; and in hospitals and in computer facilities where sensitive electronic equipment is installed and handled.

Static dissipative composites are, in general, composed of conductive materials (such as carbon, metallic particles, etc.) which are diffused into an insulating medium like ceramic, rubber or plastic, etc. The conductive elements are randomly distributed throughout the surface as well as in the bulk portion of the material so that a required amount of volume and surface electrical resistivity are realized, and this resistivity generally determines the ability of the material to dissipate the static charge. Though it can be expected that elecrtostatic decay performance would bear a linear relation with the conductivity, this hypothesis may not be wholly correct in respect to a composite material. This is because of the capacitance effects associated with the material which would "slow down" the charge dissipation rate.

The following section provides a stochastic model to predict the electrostatic propensity and bleed-off properties of a static dissipative composite in terms of quantifiable terms suitable for design calculations pertaining to the fabrication of composites having desired static-dissipative characteristics.

20.19 Stochastic Characteristics of Static-Dissipative Composites

In order to design a composite medium which has a high electrostatic dissipative property, it is necessary to consider the electromagnetic response of the material in terms of both electrical conductivity and permittivity of the medium. For this purpose, the test composite is regarded presently as a two-phase stochastic mixture in which the insulating medium forms the dispersing continuum and the conducting phase constitutes the random inclusions.

The electrostatic propensity of this composite/mixture can be quantified in terms of electrical polarizability of the medium which depicts the surface density of bound charges therein. Also the polarizibility can be assessed in terms of dielectric susceptibility or permittivity characteristics of the statistical mixture. To quantify the static-bleed-off abilities of the test medium, one has to consider the resistivity of the medium which is primarily determined by the conducting inclusions.

To evaluate the effective permittivity and/or conductivity of the test material, the relevant parameters to be considered are therefore, (i) the permittivity (ε_1) and the conductivity (σ_1) of the dispersing inclusions, (ii) the volume fraction of the inclusions (φ), (iii) the permittivity (ε_2) and conductivity (σ_2) of the dispersing insulator, and (iv) a shape factor (g) depicting the geometry of the inclusions.

There are a host of formulas available in the literature to calculate the effective permittivity and/or conductivity of multiphase systems as listed in Chapters 4 and 5.

20.20 Electrostatic Propensity and Bleed-Off Characteristics

To study the electrostatic propensity and bleed-off characteristics of a static-dissipative composite, the two electrical properties which are of interest are the permittivity (ε) and the conductivity (σ) of the mixture state which can be quantified by the formula in [23 of Chapter 5] by considering the equation to represent ε or σ as appropriate. When a composite medium with an effective permittivity ε is subjected to triboelectrification, the corresponding surface-charge (q_s) induction can be related to the electrical polarization $\boldsymbol{P}$ as follows:

$$q_S = |\boldsymbol{P}| = (\varepsilon - 1)\,\varepsilon_o\,|\boldsymbol{E}| \tag{20.5}$$

where ε_o is the free-space permittivity and $\boldsymbol{E}$ denotes the electrical field intensity associated with the triboelectric potential of the medium.

Hence, the electrostatic propensity arising from triboelectric polarization is directly proportional to the dielectric constant or the effective permittivity of the composite (assuming that the two materials involved in the abrading process are identical); however, if the materials involved are dissimilar, the triboelectrification would depend on the ratio of the dielectric constants of the materials concerned. That is, the relative triboelectrification in the materials A and B can be specified as:

$$|\boldsymbol{P}_A| / |\boldsymbol{P}_B| = (\varepsilon_A - 1)/(\varepsilon_B - 1) \tag{20.6}$$

In view of the above considerations, it follows that materials with low effective permittivity are less susceptible to triboelectric propensity. This property should be duly considered in the design of static-conductive materials as described below.

Static-conductive materials can be characterized by their high static-dissipative abilities. Quantitatively, the time constant (τ) of static bleed-off can be regarded as an indicator of the static-dispensing nature of the test medium. This time constant τ can be expressed in terms of the effective values of permittivity (ε) and the electrical conductivity (σ) of the composite as follows:

$$\tau = \varepsilon_o \varepsilon / \sigma \tag{20.7}$$

where the values of ε and σ are effective values of the composite and ε_o is the free-space permittivity.

In order to achieve a fast bleed-off, the time constant τ must be minimal. However, minimization of τ is subjected to certain practical constraints. The constraints are: (i) The test composite is a stochastic mixture and therefore the effective values of ε and σ should be specified by the expression such as that given in [23 of Chapter 5]. (ii) The maximum value of the volume fraction (φ) of the conducting inclusions is equal to 1. (iii) The minimum value of the volume fraction (φ) (threshold value) is determined and limited by the amount of conducting inclusions required for the establishment of the electrical percolative current paths in the mixture matrix. (iv) Considering a test material of cross-sectional area 'a' and length 'ℓ', the resistance per unit length, namely, $R/\ell = (1/\sigma a)$, should be greater than a minimum value specified by certain mandatory regulations/codes *vis-a-vis* fire-hazard/short-circuit protection specifications stipulated for industrial applications of these materials. With the aforesaid constraints, an optimum value for φ can be obtained by minimizing the bleed-off time constant (τ) as detailed below:

The electrical capacitance (C) and the resistance (R) of the test material of cross-sectional area 'a' and length 'ℓ' are given by:

$$C = \varepsilon_o \varepsilon\, a/\ell \tag{20.8}$$

and

$$R = \ell/\sigma a \tag{20.9}$$

Considering the fire-hazard/short-circuit protection limitations on the resistivity of a test material, the relevant constraint can be explicitly written as:

$$\rho_{min} \leq 1/\sigma \tag{20.10}$$

where ρ_{min} is the minimum value of bulk resistivity of the composite material prescribed by fire proof regulations.

Using effective value formulation for ε and σ and with relevant simplifications, the constraint specified by Equation 20.10 can be rewritten as:

$$\varphi_{min} < \varphi < -log_e(\rho_{min}\sigma_2)/log_e(\sigma_1/\sigma_2) \tag{20.11}$$

where φ_{min} specifies the threshold value required for the current percolation.

Considering the time constraint τ (equal to RC), its approximate value determined by Equation 20.11 and can be expressed as follows:

$$\tau = \varepsilon_o\, [\varepsilon_2/\sigma_2]^{1-\varphi}/\sigma_2^{\varphi} \tag{20.12}$$

Hence, to obtain a minimum value for τ, ε_2 should be close to 1; and, since $\sigma_2 >> \sigma_1$, it is necessary to take the largest possible value of φ. Therefore, the design value of φ (as given by Equation 20.11 should be:

$$\varphi = -log_e(\rho_{min}\sigma_2)/log_e(\sigma_1/\sigma_2) \tag{20.13}$$

20.21 Design Example

Consider a composite material formed by blending a dielectric and recycled aluminum powder. Let the dielectric material (insulator) have the following values for the electrical constants: $\varepsilon_2 = 4.5$ and $1/\sigma_2 = \rho_2 = 2 \times 10^{14}$ ohm-meter; and for aluminum, $\varepsilon_1 = 1$ and $\sigma_1 = 3.53 \times 10^7$ siemen/meter.

Suppose this composite material is used as a static-conductive floor covering. Then, it has to meet the electrical resistance requirements of the National Fire Protection Association Bulletin 56A, "Standard for the Use of Inhalation Anesthetics". This standard specifies that the average electrical resistance of an installed floor shall be between 25000 ohms and 10^6 ohms as measured between two electrodes placed 3 feet apart. The average resistance to ground shall be more than 25000 ohms as measured between a ground connection and an electrode placed at any point on the floor. The resistances represent the average of five or more reading per room or installation and are measured according to the procedures outlined in NFPA 56A which are essentially the same as that of ASTM F 150-72, "Standard Test Method for Electrical Resistance of Conductive Resilient Flooring".

Hence, taking the specified minimum value of 25000 ohms, the corresponding value of resistivity (ρ_{min}) of the test composite can be calculated by assuming the thickness of the floor covering as 1/8 inch. Then the calculated value of ρ_{min} is equal to 8×10^6 ohm-meter.

Using Equation 20.13, the optimum value of φ can be determined. In the present example, it is equal to 0.3386. The corresponding value of decay time constant τ is 191.53 microsecond. This value is acceptable as per MIL-B-81705B specification which stipulates a decay rate of 2.0 seconds as maximum permissible.

The static decay time constant (τ) of materials is normally measured by the procedures outlined in Method 4046 of Federal Test Method 101B, dated 8/15/74. The static accumulation or propensity can be determined similar to the test procedure of AATCC-134, "Electrostatic Propensity of Carpets"(or ANSI-ASTM D 2679-73).

The relative electrostatic propensity of two materials (A and B) can be specified in terms of their dielectric constants (ε_A and ε_B) as illustrated in Figure 20.3. For identical materials ($N = \varepsilon_A/\varepsilon_B = 1$), the charge propensity is the same in either of them as expected; and for large values of the dielectric constants of any one of the materials (say, ε_A), the relative propensity approaches asymptotically the ratio of the dielectric constant, namely, *N*, irrespective of the magnitude of *N*. However, for low values of ε_A, the relative propensity tends to infinity for any given value of *N*. Therefore, it follows that, when a material of low permittivity abrades with a material of higher permittivity, the triboelectrification would be intense. This is true for composite materials also.

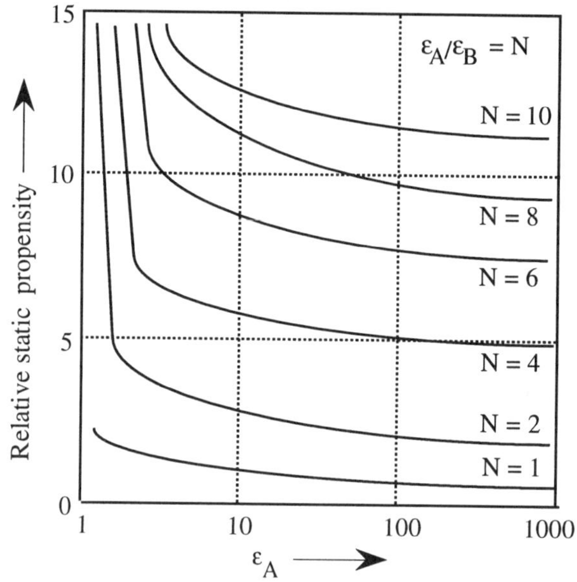

Figure 20.3 Relative electrostatic propensity of two abrading (nonconducting) materials A and B with dielectric constants ε_A and ε_B, respectively.

Considering the design of a composite with controlled static propensity, the choice of optimum value of volume fraction *(φ)* (with the constraint on resistivity specified by fire-hazard limitations) depends on both the conductivity of the inclusions as well as on the ratio of the conductivities of the dispersing insulator and the dispersed inclusions (Equation 20.13). Figure 20.3 illustrates the typical ranges of the practical values of the material constraints and the corresponding design values of the volume fractions.

If low volume fraction of inclusions is preferred (so as to obtain, for example, certain desired mechanical/elastic properties), then as could be inferred from Figure 20.3, it is necessary to choose the dispersing material with higher conductivity. Thus the present formulation has a design flexibility to suit the practical situations.

Bleed-off time of a composite material as a function of conductivity of dispersed inclusions is presented in Figure 20.4 for two different volume fractions of the inclusions. The delay or capacitive effects of the dispersing insulating medium is determined by the dielectric constant (ε_2) and Figure 20.4 corresponds to a (practical) parametric value of ε_2/σ_2 equal to 10^{15} ohm-meter. It can be observed from Figure 20.4 that both σ_1 and σ_2 control the bleed-off time to a significant extent and that the role of ε_2 is implicit. However, compatible design can be achieved as illustrated by an example given before regarding a mixture composite of a dielectric and aluminum powder.

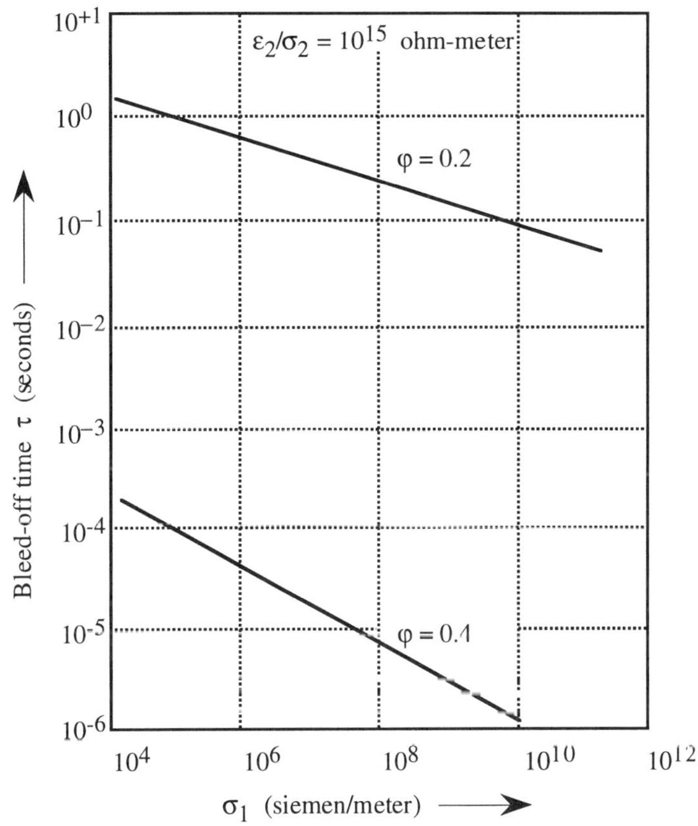

Figure 20.4 Bleed-off time *versus* the conductivity of metallic loading

20.22 Conclusions

Electrostatic control materials largely form a distinct class of electromagnetic materials vitally important in modern electronic engineering practice. Though extensive studies have been done in the development and applications of such materials, there remains a great potential to meet the demands posed by ULSI and other upcoming semiconductor devices which face greater ESD/EOS threat due to their ultraminiature scaled-down geometries.

References (General Reading)

[1] B. S. Matisoff: *Handbook of Electrostatic Discharge Controls: Facilities, Design and Manufacturing Procedures.* (Van Nostrand Reinhold Co., New York: 1986).

[2] M. Mardiguian: *Electrostatic Discharge: Understand, Simulate and Fix ESD Problems.* (Interference Control Technologies, Inc. Gainesville, VA: 1986).

[3] T. N. Bhar and E. J. McMahon: *Electrostatic Discharge Control.* (Hayden Book Co., Inc., Rochelle Park, NJ: 1983).

[4] Electrostatic Discharge (ESD): Protection Test Handbook by the Technical Staff, KeyTek Instrument Corp., Burlington, MA 01803, 1983.

[5] Triboelectric Testing, Final Report: RAC–TR–83–03–E01, (Reliability Analysis Center, Rome Air Development Center, Grittiss Air Force Base, NY) Dated August 15, 1983.

[6] J. E. Lingousky and V. E. Holt: Analysis of Electrostatic Charge Propensity of Floor-Finishes, EOS/ESD Symp. Proc., vol. EOS-5, pp. 17-20 (1983).

[7] F. S. Felt: Coplanar Triboelectrification of Selected Materials, EOS/ESD Symp. Proc., vol. EOS-5, pp. 95-101 (1983).

[8] C. Briggs Jr: Electrostatic Conductivity Characterisation of Workbench-top Surface Materials, EOS/ESD Symp. Proc., vol. EOS-1, pp. 7-12 (1979).

[9] A. Halperin: Static Control Using Topical Antistats, EOS/ESD Symp. Proc., vol. EOS-1, pp. 13-21 (1971).

[10] J. M. Kolyer and W. E. Anderson: Selection of Packaging Materials for Electrostatic Discharge-Sensitive Items, EOS/ESD Symp. Proc., vol. EOS-3, pp. 75-84 (1981).

[11] D. E. Frank: ESD Considerations for Electronic Manufacturing, presented at American Society of Manufacturing Engineers: Westec Conference, Los Angeles, CA, March 21-24, (1983).

[12] P. S. Neelakanta: What shapes the properties of ESD–protective products? Electronic Packaging and Production, May 1969: 64-66.

[13] P. S. Neelakantaswamy and R. I. Turkman: Suceptibility of PCB–mounted microelectronic devices to failures caused by electrostatic dicharges. Electronic Packaging and Production, February 1987: 132-134.

[14] P. S. Neelakantaswamy and R. I. Turkman: Coping with static electricity – Part LXXX II; ESD/EOS susceptibility of stripline-opposed-emitter transistors, Evaluation Engineering, vol. 26(7), 1987: 60-68.

References on Specifications (Specification number and general topics covered)

• **ASTM Test Methods**
ASTM–D–257: Surface resistivity measurement.
ASTM–D–991: Volume resistivity measurement.
ASTM–D–2679: Standard test method for electrostatic change.
ASTM–D–3509: Standard test method for electric field strength.

• **DOD Specifications**
DOD–HDBK–263: Handbook on the properties of ESD–protective materials, etc.
DOD–STD–1686: Electrostatic potential levels in protected areas.

• **Electronics Industry Association (EIA) Standards**
EIA Interim Standard P. N. 1525: Static bleed-off rate measurement.
EIA Standard RS–471: Symbol for static sensitive products.
EIA Standard IS–5: Faraday cup test.

• **National Fire Association Standards**
NFPA Standard 56A: Ohmic continuity testing.

- **Federal Standard Specifications**

Std. 101, Test Method 4046: Static decay time/rate measurements on conductive coatings/pairs.
Std. 595: Colors of plastics.
Std. 406: Physical strength of plastics.

- **Federal Specifications**

SS–T–312B: Dimensional stabilities, appearance, etc. of flooring material.

- **Military Standards**

MIL–STD–454: Effects of electrical current on human body.
MIL–P–82646: Conductive plastic films to make bags, etc.
MIL–P–82647: Heat-sealable conductive bags.
MIL–B–117: Bags, sleeves, tubings, and interior packaging.
MIL–B–81705: Barrier materials, flexible, electrostatic fire, heat sealable.
MIL–M–38510: Surface resistivity of antistatic package.

- **Underwriters Laboratories Specifications**

UL 779: Standard for electrically conductive floorings.

Defining Terms

Bag (static control): A preformed container made of flexible material generally enclosed on all sides except one which forms an opening that may not be sealed after loading. It is normally constructed from one piece of material that has been folded over and sealed on two edges. It is intended to protect ESD sensitive electronic parts contained in it.

Carrier (static protected): Holder for electronic parts and devices which facilitates handling during processing, production, imprinting, or testing operations and protects such parts under transport.

Conductor (conductive): A substance or body that allows a current of electrons to pass continuously along it or through it when a voltage is applied across any two points. Such materials exhibit relatively low resistance.

Dielectric breakdown: A threshold effect in a dielectric medium where, at some electric field strength across the medium, bound electrons become unbound and travel through the medium as a current. In solid media, the region of the current is permanently damaged. The units of measurement are usually volts per unit of thickness.

Electrically continuous surface: A surface that is electrically conductive, in that current can be passed as the result of an applied voltage between any two points on its physical surface, and discontinuities, slots, or holes do not occupy more than 10% of the material's surface.

Electrostatic field: The region surrounding an electrically charged object in which another electrical charge can be induced and will experience a force of interaction. Quantitatively it is the voltage gradient between two points at different potentials.

Electrostatic shield: A barrier or enclosure that prevents the penetration of an electrostatic field. An electrostatic shield, however, may not offer much protection against the effects of electromagnetic interference (EMI). EMI shields, however, are good electrostatic shields.

EMI electromagnetic interference: Sources are static sparks, lightning, radar, radio and TV transmission, brush motors, line transients, etc. By line conduction or air propagation, EMI can induce undesirable voltage signals in electronic equipment causing malfunction and occasionally component damage. Protection against EMI usually requires the use of shields, filters, and special circuit design.

Faraday cage: An electrically continuous, conductive enclosure which provides electrostatic shielding. The cage or shield is usually grounded, although it need not be.

Ground: A metallic connection with the earth to establish zero potential or voltage with respect to ground or earth. It is the voltage reference point in a circuit. There may or may not be an actual connection to earth, but it is understood that a point in the circuit said to be at ground potential could be connected to earth without disturbing the operation of the circuit in any way. Grounds that can be used for static control work stations include water pipes, any power ground, or any large metal structure member of a building.

Grounding: Connecting to ground or to a conductor that is grounded. A means of referencing all conductive objects to a zero voltage equipotential surface. This is the surest method of eliminating ESD, since everything is maintained at the same potential.

Induction static eliminator: Generally, a series of conductive grounded points or brushes. When a single sharp grounded needle point is brought into the proximity of any highly charged surface, it has induced in it a charge opposite to that of the surface. When a high enough charge concentration has been developed, the surrounding air will break down. A vast number of charge-balancing ions are formed. The simple "tinsel" static eliminator is an example of an induction static eliminator.

Insulator: A material generally that does not conduct electricity. A nonconductor.

Ionization: The process by which a neutral atom or molecule, such as air, acquires a positive or negative charge.

Nuclear static eliminator: Nuclear static eliminators create ions by the irradiation of the air molecules. Some models use a safe alpha-emitting isotope to create sufficient ion pairs to neutralize a charged surface. The high-speed particle interacts with air molecules with sufficient energy to actually strip off one of its outer electrons (see Ionization).

Package: A package is the enclosure of products, devices, or other packages in a wrap, pouch, bag, slide, magazine, or other container form so as to perform one or more of the following functions:

- Containment for handling, transportation, and use.
- Preservation and protection of the contents for the life of the item.
- Identification of contents, including quantity and manufacturer.
- Facilitate the dispensing and use of the contents.

Packaging: In the electronics industry, packaging refers to the process of locating, connecting, and protecting various devices, components, etc. For example, an entire circuit can be printed onto thin film wafers which are electrically interconnected during fabrication. These miniature circuits are then fused into an easy-to-handle element called a package. This package both protects the semiconductor circuitry and permits convenient external connections to be made to it.

Potential: Measured in millivolts, volts, or kilovolts. Potential or voltage is measured from a base point. This point can be any voltage but is usually grounded, which is theoretically zero voltage.

Pouch: A small or moderately sized baglike container constructed by the sealing on three edges of two flat sheets of flexible material, or by sealing one end of a tube of flexible material.

Surface resistance: The ratio of DC voltage to the current that passes across the surface of the system. In this case, the surface consists of the geometrical surface and the material immediately in contact with it. In effect, the surface resistivity is the resistance between two opposite sides of a square and is independent of the size of the square and its dimensional units. Surface resistivity is relevant only for materials where current conduction is virtually on the surface. Surface resistivity is not meaningful for volume-conductive materials. The unit of surface resistivity is *ohms per square*. The unit *ohms* is occasionally used, since the value is the resistance across a square. The use of ohms per square is recommended to distinguish surface resistivity from arbitrary resistance.

Triboelectric: Pertaining to an electrical charge generated by frictional rubbing or separation of two surfaces.

Triboelectric series: A list of substances arranged so that any of them can become positively charged when rubbed with one farther down the list, or negatively charged when rubbed with one farther up the list. Generally, the farther apart such materials are in the triboelectric series, the greater their tendency to charge one another. This series is derived from specially prepared and cleaned materials tested in very controlled conditions. In everyday circumstances, materials reasonably close to one another in the series can produce charge polarities opposite to that expected. This series is only a guide. See Table 20.1.

Voltage suppression: A phenomenon where the voltage from a charged object is reduced by increasing the capacitance of the object rather than decreasing the charge on the object. The relation Q = CV describes the phenomenon. It occurs most frequently when a charged object is close to a ground plane, but not in resistive contact with the ground plane.

Volume resistivity: The ratio of the d.c. voltage per unit of thickness, applied across two electrodes in contact with or embedded in a specimen, to the amount of current per unit area passing through the system. Volume resistivity is generally given in *ohm-centimeters*.

CHAPTER 21

Electromagnetic Shielding Materials

21.1 Introduction

Electromagnetic shielding materials are the structural constituents of the so-called *electromagnetic shields* used for the purpose of confining electromagnetic energy within the bounds of a specific region and/or to prevent the proliferation of such energy into a designated locale. Electromagnetic energy, in general, manifests as:

- Energy associated with static and/or time-varying electric force field
- Energy associated with static and/or time-varying magnetic force field
- Radiated electromagnetic energy

Depending on the shielding requirements *vis-a-vis* the interference due to any of the above forms of electromagnetic (EM) energy, a variety of materials have been developed and deployed in the fabrication of shielding partitions or enclosures which confine the EM energy within a specified region, thereby preventing its invasion elsewhere. Apart from the partition or enclosure configurations, shielding is also facilitated for cables and connectors to avoid electromagnetic interference through mutual and/or external coupling. Tailoring shield designs for use in cables and connectors dictates specific material needs to match the geometrical and mechanical constraints.

The basic requirement of an EM shielding material is that a shield fabricated with this material should meet the *electromagnetic compliance* (EMC) aspects of achieving a specified extent of shielding under a given *electromagnetic interference* (EMI) ambient. The choice of EM shielding material is in general decided by:

- Electromagnetic properties of the material to provide a given *shielding effectiveness* (SE)
- Its compatibility for specific shielding applications *vis a vis* interference due to electric, magnetic, or electromagnetic (radiated) fields
- Geometrical considerations, namely, shape and size
- Mechanical considerations such as rigidity, flexibility, weight, structural mating (fastening and joints), and withstandability against shocks and vibration
- Performance under hostile thermal environments
- *Bandwidth* of operation; that is, the effective frequency range with acceptable shielding performance
- Ease of fabrication of the shields
- Cost-effectiveness

21.2 Mechanisms of EM Shielding

21.2.1 Shielding the time-varying electromagnetic waves [1,2]

As indicated earlier, the electromagnetic energy could be associated with static and/or time-varying electric and magnetic force fields (known as *induction fields*) in the vicinity of the source; or it could be a radiated field detatched from the source. Illustrations of these force fields are presented in Figure 21.1.

Pertinent to the various modes of EM field forces as depicted in Figure 21.1 the choice of a material in each case to shield the relevant field is distinct, inasmuch as the shielding mechanism involved is different for each of the EM force fields illustrated.

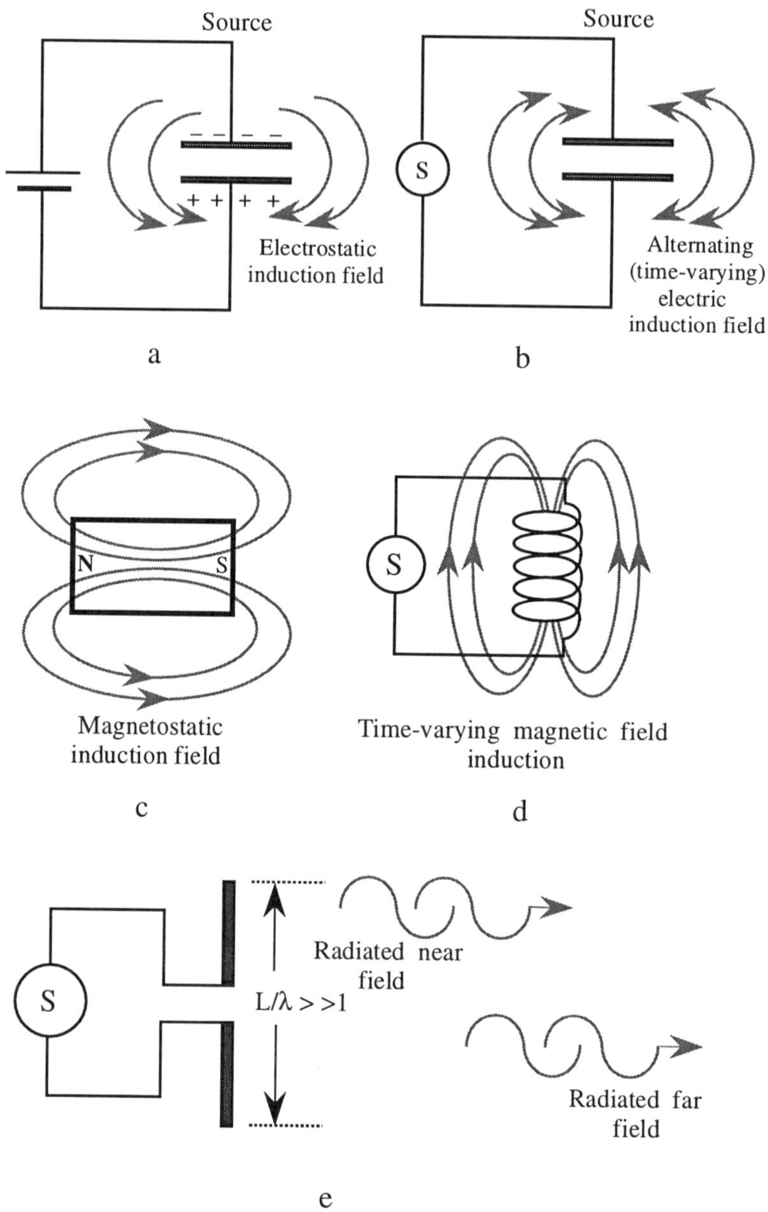

Figure 21.1 Electromagnetic interference fields.
(a) Electrostatic (d.c.) induction field. (b) Time-varying electric induction field.
(c) Magnetostatic induction field. (d) Time-varying magnetic induction field. (e) Radiated near and far electromagnetic fields.

To accommodate the electromagnetic field conditions as depicted in Figure (21.1), the electromagnetic field can be distinguished as two regimes:

- Quasistatic energy (with very low frequency, f or very large wavelength λ) specified as the induction or near-field components
- Radiated far-field components pertinent to high frequency or small wavelength electromagnetic energy

In reference to time-varying fields, a *wave impedance (intrinsic impedance) parameter* (Z_w or Z_o) can be defined for the medium in terms of the electric ***(E)*** and magnetic ***(H)*** field intensities. That is, Z_w or Z_o = $|E/H|$ ohms. When the source is a capacitive (potential-dependent)-type dipole (as shown in Figure 21.2), the associated E field is large leading to a high wave impedance ambient. On the other hand, a current-dependant loop source offers an extensive H field with the result the corresponding wave impendance of the medium tends to be small.

The mechanism of shielding is that the shielding medium (of appropriate material and geometry) when placed in the region of electromagnetic field should offer a barrier impedance Z_m sufficiently large in comparision with the wave impedance Z_w in the case of an electric field dominant ambient so that the wave is impeded sufficiently and shielded off from entering the region beyond the shielding barrier as illustrated in Figure 21.2.

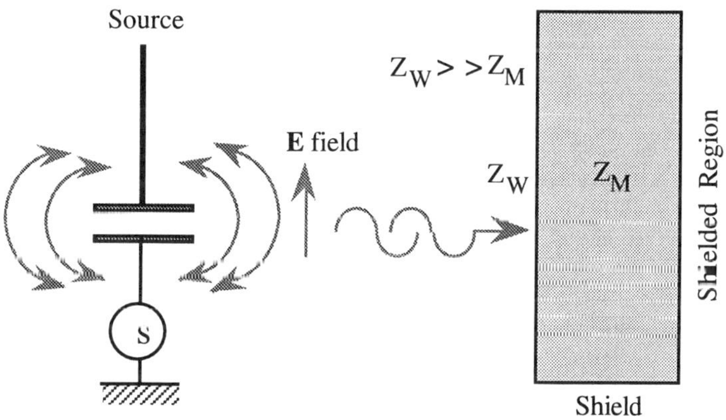

Figure 21.2 Electric field dominant ambient.

Likewise, in the case of magnetic field caused by current-dependent loop source, the effective shielding can be achieved by a shield (of appropriate material and geometry) with a barrier admittance Y_m well in excess of Y_w (Figure 21.3) where $Y_w = 1/Z_w$.

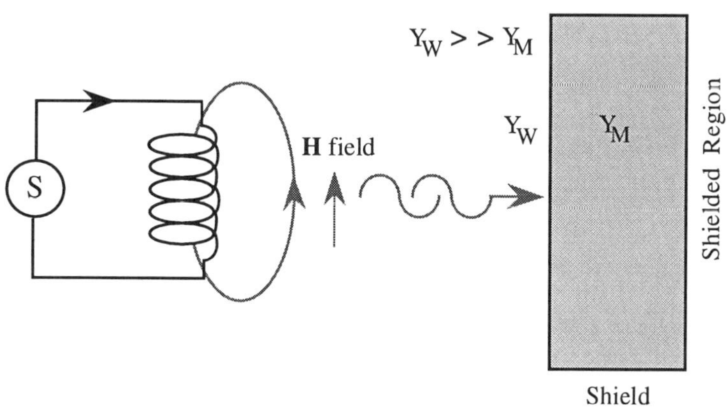

Figure 21.3 Magnetic field dominant ambient.

The primary considerations of electromagnetic shielding for time-varying fields are the proper choice of material and the geometry of the shield to provide the necessary barrier-immittance parameter. The shield geometry implicitly includes the thickness of the shield and the disposition of the shield (in terms of distance/wavelength ratio) from the source. The material and the geometry cohesively accomplish the shielding by the following mechanisms:

- Absorption of electromagnetic energy in the shielding medium (absorption loss)
- Reflection of the electromagnetic energy back into the source side with minimal transmission into the region being shielded (reflection loss)

21.2.2 Static fields

There are circumstances which warrant the shielding of static electric and magnetic fields. In such conditions, the relevant field is "blocked" from entering the region to be shielded by techniques distinctly different from the absorption loss and/or reflection loss modes. This is because, with static fields, there is no field energy being dissipated (and absorbed) and no reradiation (reflection) is feasible. Therefore, for electric and/or magnetic static fields the following shielding principles are advocated: Considering a static electric field, by placing a metal (or a good conductor) medium in the field area, the electric flux lines would terminate on the induced surface charges (normal to the plane of the surface) as illustrated in Figure 21.4.

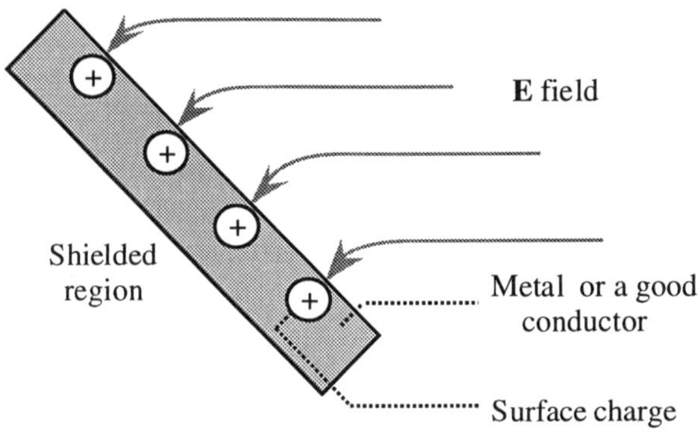

Figure 21.4 Electrostatic shielding.

The metal (or the good conductor) is a *equipotential medium* with zero potential gradient (or E field) in it. Therefore, the electric field terminates on the surface of the material and does not penetrate inside. Thus, a simple, high-conductivity medium can effectively provide *electrostatic shielding*.

Pertinent to static magnetic field, the shielding medium made of high permeability material accommodates the flux lines within itself by providing a low *magnetic reluctance* path. Thus, the field leakage into the region being shielded is negligible if relative permeability and the thickness of the shield are chosen to confine the flux lines as depicted in Figure 21.5.

Thus, to meet specific conditions (static or time-varying fields), shielding materials should be chosen appropriately as per the details furnished in the following sections.

21.3 Characteristics of Shielding Materials

As discussed earlier, the properties of shielding materials *vis-a-vis* time-varying electromagnetic fields are specified by the barrier immittance parameters, namely, Z_m and Y_m. Required values of these parameters are decided by the design value of the shielding effectiveness (SE) defined as the ratio of incident EM power density on the shield and the transmitted power density, normally expressed in *decibels* (dB). That is:

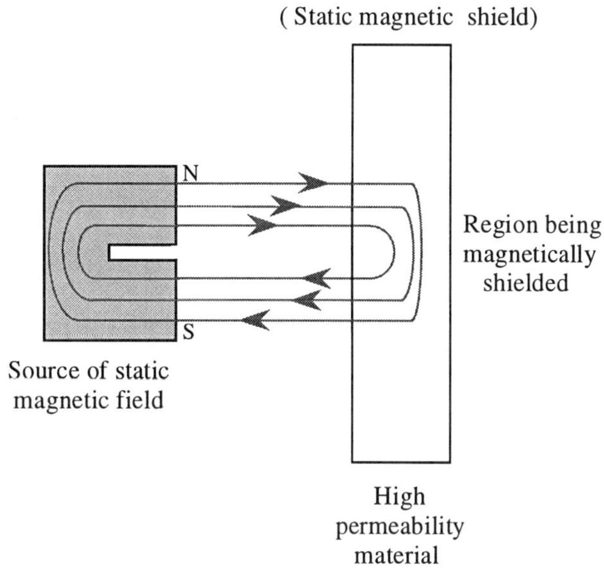

Figure 21.5 Magnetostatic shielding.

$$(SE)_{dB} = 10 \log_{10} \text{(Incident EM power density/Transmitted EM power density)} \quad (21.1)$$

where the incident power density refers to the power density at a measuring locale before the shield is in place and the transmitted power density corresponding to the measured power density at the same locale after the shield is in place. The power density ratio can also be expressed in terms of the ratio of field strengths as follows:

$$\begin{aligned}(SE)_{dB} &= 20 \log_{10} (E_1/E_2) \\ &= 20 \log_{10} (H_1/H_2) \quad (21.2)\end{aligned}$$

where (E_1,H_1) are electric and magnetic field strengths at a point, respectively, prior to installing the shield and (E_2,H_2) refer to the corresponding fields at the same point after the shield is installed.

Depicting the incident wave immittance parameters as (Z_w, Y_w) and the barrier immittance parameters of the shield as (Z_m, Y_m), the complex reflection coefficient at the shield boundary (Figure 21.6) is given by:

$$\begin{aligned}\Gamma &= (Z_m - Z_w)/(Z_m + Z_w) \\ &= (Y_m - Y_w)/(Y_m + Y_w) \quad (21.3)\end{aligned}$$

assuming no multiple reflections within the shield at its boundary surfaces. More rigorously, by considering multiple reflections as well as absorption due to EM energy dissipation with in the shield, Γ can be rewritten as:

$$\Gamma = [\exp(-\alpha d)]\,[4K/(1 + K)^2]\,[1 - \{(K - 1)/(K + 1)\}^2 \exp(-2\gamma d)]^{-1} \quad (21.4)$$

where $K = Z_w/Z_m$ (*or* Y_w/Y_m), $\gamma = (\alpha + j\beta)$ with α, β being the *attenuation and phase constant,* respectively, and d is the shield thickness. In Equation 21.4, $[\exp(-\alpha d)]$ depicts the EM energy absorption in the shield, $4K/(1 + K)^2$ refers to the reflection back to the source side and the third term is a coefficient arising from multiple reflections within the shield.

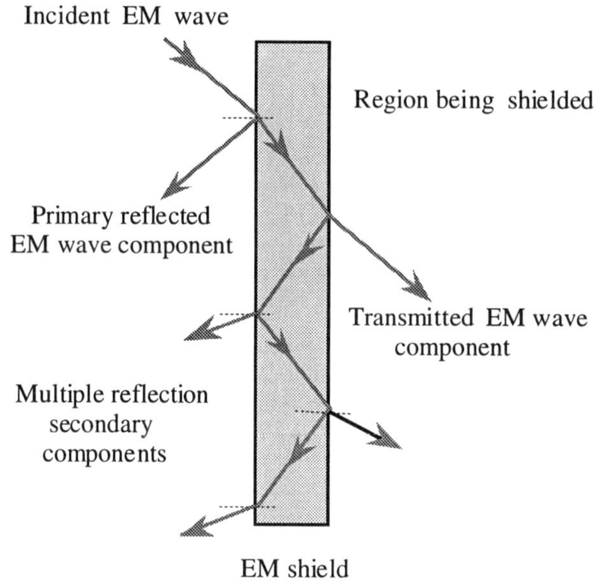

Figure 21.6 EM wave reflection at a shield boundary.

For effective shielding therefore the three major design parameters are:

- Z_w/Z_m *(or* Y_w/Y_m*)* ratio pertinent to the shielding materials
- Attenuation constant of the shield
- Thickness of the shield

Further, the choice of Z_w/Z_m or Y_w/Y_m is dictated by whether the shielding required corresponds to high frequency or low frequency (quasistatic) conditions as discussed earlier.

Shielding material for time-varying EM shielding can be a monolithic medium or a composite medium which offers optimum values of α (field attenuation constant) and Z_w/Z_m (or Y_w/Y_m) to realize the design specifications on shielding effectiveness. The attenuation constant (α) is decided by the effective conductivity (σ) and permeability μ of the shielding medium and the frequency (f) of operation. It is given by:

$$\alpha = (\pi f \mu \sigma)^{1/2} \tag{21.5}$$

if the material has an effective conductivity $\sigma >> \omega\varepsilon$, where ε is the effective permittivity of the medium and $\omega = 2\pi f$. The parameter m in Equation (21.5) is equal to $\mu_o\mu_r$ where μ_o is the absolute permeability of free space and μ_r is the relative permeability of the medium.

The absorption loss (A) part of the shielding effectiveness, $(SE)_{dB} = 20 \log_{10} (1/\Gamma)$ is given by:

$$\begin{aligned}(A)_{dB} &= 8.68\,\alpha d \\ &= 3.338(d)_{mils}\,(f_{MHz}\,\mu_r\,\sigma_r)^{1/2}\ dB \\ &= 1314.3\,(d)_{cm}\,(f_{MHz}\,\mu_r\,\sigma_r)^{1/2}\ dB\end{aligned} \tag{21.6}$$

where $\mu_r = \mu/\mu_o$, with $\mu_o = 4\pi \times 10^{-7}$ henry/meter (absolute permeability of free space) and $\sigma_r = \sigma/\sigma_c$ with σ_c the conductivity of copper at room temperature equal to 5.80×10^{-7} siemen/meter.

The barrier immittance parameters (Z_m, Y_m) depend on the *skin depth* (δ) of EM energy penetration into the shielding medium, frequency of operation (f), and the conductivity (σ) and permeability (μ) of the shielding material. The barrier immittance parameters are given by:

$$Z_m = 1/Y_m = 369\,(\mu_r f_{MHz}/\sigma_r)^{1/2}[1 - exp(-\tau/\delta)]^{-1} \text{ microohm/square} \qquad (21.7)$$

where δ is the skin depth defined as the surface thickness of a material at a given frequency at which the EM energy penetrating into the medium attenuates to an extent of $(1 - 1/e) \Rightarrow$ 63.2%. It is given by :

$$\begin{aligned} \delta &= 0.0066/(f_{MHz}\,\mu_r\,\sigma_r)^{1/2}\ cm \\ &= 2.6/(f_{MHz}\,\mu_r\,\sigma_r)^{1/2}\ mils \end{aligned} \qquad (21.8)$$

Pertinent to the quantitative aspects of shielding as described above, there are three phenomena which can be regarded as responsible for effective shielding. They are:

- Conductive reflection: The time-varying magnetic field component of the incident EM energy induces electric current in the shielding material and these currents in turn provide opposing magnetic field (as per the Faraday-Lenz's law) minimizing the total field beyond the shield.
- Magnetic reflection: In the event that the shielding material offers high magnetic permeability, the magnetic flux lines (time-varying or static) are confined as conductive (low reluctant) paths through the shield and do not link to the region being shielded.
- Conductive energy absorption: This refers to the energy dissipated in the conductive shielding medium manifesting as the field attenuation.

Choice of material for shielding time varying EM field depends on the aforesaid phenomenological considerations and a chosen material could accomplish effective shielding on either one or combinations of these mechanisms.

To comply with the loss mechanisms stated above, the materials adopted commonly for effective shielding fall under the following wide categories:

- Metallic or alloy-based conductors
- Non-metallic conductors
- Conducting polymers and metal-infused paints
- Polymeric composites with metallic and nonmetallic (conducting) inclusions
- Concrete/polymer concrete hybrid composites
- Metal-included ceramics
- Multilayered laminates of boron/boron tungstates or graphite fibers in an epoxy matrix
- Intercalated graphite fiber composites
- Ferromagnetic materials
- Metal-included fabrics

21.4 Metallic and Alloy-Based Shielding Materials

These include both ferromagnetic and diamagnetic materials with significant electrical conductivity (σ). Typical shielding metals and alloys are listed in Table 21.1.

Table 21.1 Metals Used in EM Shields

Metal	Conductivity Relative to Copper $\sigma_r = \sigma/\sigma_{cu}$ ($\sigma_{cu} = 5.26 \times 10^6$ siemen/meter)	Remarks
Silver	1.05	
Copper	1.00	
Gold	0.70	
Magnesium	0.36	
Zinc	0.29	Diamagnetic
Brass	0.26	
Bronze	0.18	
Tin	0.15	
Lead	0.08	
Aluminum	0.61	Paramagnetic
Nickel	0.20	
Iron	0.17	
Steel (SAE1045)	0.10	Ferromagnetic
Stainless steel	0.02	

Metallic shielding could be done in various forms as indicated below:

1. Metal sheet or foils
2. Zinc arc-sprayed coating
3. Vacuum metallization coating
4. Cathode-sputtered coating
5. Electroless plating
6. Apertured sheets
7. Metallic wire screens
8. Metallized fabrics

21.5 Description of Metal-Based Shields

Homogeneous metal sheets or foils for enclosure-type or overlay applications are used. Popularly, shielding enclosures have been constructed with welded steels with the attempts to realize a standard shielding effectiveness of 60 to 80 dB (as per NSA 73-2A specification) possibly over a frequency range of 10 KHz to 1 GHz. Copper and aluminum sheets or foils enjoy prominence in RF shielding techniques [3].

The nonmagnetic metals and alloys provide relatively low attenuation to transmission of EM wave as decided by the $(\sigma_r \mu_r)^{1/2}$ product (where σ_r is the relative conductivity of the material with respect to copper and μ_r is its relative permeability). The ferromagnetic metals/alloys (with large μ_r) therefore offer higher attenuation, especially at low frequencies. At higher frequencies, (>100 KHz) μ_r degrades for most of the ferromagnetic metals/alloys with the consequence of absorption attenuation being comparable or less than those of nonmagnetic metals/alloys.

Metals and alloys available in sheet stock form have a range of thickness from 1/64 in. (0.4 mm) or less to about 1/8 in. (3.2 mm) or more. In foil form (with thickness from 1 (25.4 μm) to 10 mil (254 μm)) shielding metals are available as sheets and tapes and as adhesive-backed foil rolls. Nonmagnetic foils are widely used in RF shielding. At low frequency magnetic fields such foils, however, pose little attenuation.

Metal-based shielding of large spaces (such as rooms) is done normally with metal-foil wallpaper (MFWP) in conjunction with pressure-sensitive metal-foil tapes and conductive adhesives/epoxies/caulking compounds. Typical MFWP shieldings are characterized in Table 21.2.

Table 21.2 MFWP Shielding Characteristics

MFWP Material	Thickness (mils/microns)	Electrical Characteristics	Shielding Effectiveness (dB) MIL-STD-285
Aluminum		Paramagnetic Good conductor	25-40 dB for magnetic field at 200 KHz
Copper	2-3/51-76	Diamagnetic Very good conductor	80-100 dB for electric field over 200 KHz to 10 MHz
Stainless steel		Ferromagnetic Poor conductivity	60-80 dB for plane wave above 400 MHz

21.6 Process-Based, Inhomogeneous Metal Shields

The different types of metallic coating on surfaces to realize EM shielding have invariably the process-induced inhomogeneity in them. The extent of such inhomogeneity determines the effectiveness of shielding offered by these materials. The corresponding shielding characteristics are controlled by the surface resistance of the coating, the quality of which is decided by the process adopted, surface characteristics of the substrates, type of metal being coated, and the thickness of the coating advocated.

The various coating processes of metallic shielding materials are characterized as follows [4-6]:

1. *Zinc-arc spraying*: This process involves electrically isolated wires which are continuously fed into a gun so that only the tips of the wire come into contact. Upon reaching critical distance from each other an intense arc across the tips melt them so that an air jet from the gun carries the metal particles and deposits them on the surface being coated.

Zinc-arc sprays give good conductivity and high-dB attenuation to EM fields and accommodate dense coatings. However, it is expensive and arcing products are highly toxic warranting special precautions. Also, zinc sprays are poorly adhesive with cracking and pitting problems.

2. *Vacuum metallizing*: This process allows the deposition of a pure metallic films (such as aluminum) in a vacuum chamber. This method also offers surfaces with good conductivity and continuously homogeneous coatings. Lack of abrasion and poor corrosion resistance specify the demerits of this technique.

3. *Cathode sputtering*: This is similar to vacuum evaporation. An inert gas is set within the coating chamber. When the chamber pressure is reduced, an arc strikes at the coating

material to vaporize it. Upon condensation, the vaporized metal forms a film on the substrate. This method again yields surfaces of good conductivity and offers a good adhesivity of coating. Expensive equipment and cracking of the coatings at higher temperatures are the limitations of this process.

4. *Electrodeless plating*: This refers to an immersion technique of the substrate in a suitable aqueous solution wherein a controlled autocatalytic chemical reduction permits the deposition of metal films (of copper and/or nickel) onto a substrate. The resulting coating offers excellent shielding effectiveness over a wide range of frequencies (65 to 120 dB). The coating is also uniform in thickness with adequate recess and side wall coverage. It provides an excellent contact for grounding and also offers a good corrosion resistance. It is applicable to substrates with complex surfaces.

21.7 Apertured Metal Panel Shields

Perforated metal sheets with apertures of small sizes (relative to wavelength) are useful as lightweight shields. The performance of these shields is dictated by the aperture-to-metal area ratio and the thickness of the metal used. Metals with punched-hole perforations or of honeycomb structures and as interwoven lattices constitute feasible shielding structures of this category. For the perforated panels, the shielding effectiveness is given by:

$$(SE)_{dB} = 32t/g + 4 + 20\log_{10}[(D/g)^3/N] \tag{21.9}$$

where t is the thickness of the shield, g is the size of the perforations, D is the lateral size of the square panel and N is the total number of perforations. The above formulation refers to low impedance magnetic fields independent of frequency.

21.8 Wire-Mesh Screens as EM Shields

These are even more lightweight in comparision with apertured metal shields. Flexible wire-mesh screens (such as chicken wire-mesh screens) with metal-to-air area ratio on the order of 0.05 to 0.5 are popular as shields constituting the so-called *Faraday cages.* The mesh size is decided by the wavelength (λ) of operation such that, at the operating frequency, the screen offers a cutoff window at each air-gap region to the EM wave incident on it. The shielding effectiveness is given by:

$$\begin{aligned}(SE)_{dB} &= 20\log_{10}(0.5\lambda/g)\ dB\ for\ g \le \lambda/2 \\ &= 0\ for\ g \ge l/2\end{aligned} \tag{21.10}$$

where g is the size of the airgap in the mesh.

When the frequency decreases (or wavelength increases), the shielding effectiveness does not increase indefinitely but levels off at about 110 dB for copper and aluminum screens and at about 150 dB for galvanized steel. These limiting values correspond to the situation that the screen can be regarded as a homogeneous material at wavelength $\lambda >> g$.

21.9 Metallized Fabrics/Textiles as EM Shields

The shielding effectiveness of metallized fabrics and textiles (both woven and nonwoven) depends on the geometry of the fabric (for example, pore size and thickness), and the amount of metal present in the fabric. These materials are useful as personnel wear to protect people from EM irradiation (also known as *nonionizing radiation*). Also, they are used as sheet covers for equipment or a space to be protected from electromagnetic fields. The uses of metallized fabric EMI shields are [7-9]:

- (Partial) shields for equipment cases
- Shielding and grounding curtains

- Torso contacts for electromedical sensors, and probes
- Electrostatic discharge wipers
- Protective clothing for personnel working under high-voltage magnetic fields and/or in RF/microwave environment
- Flexible shielded shrouds, smocks, stockings, and boots
- Accordian-type (collapsible) EMI-protected walkways

21.10 Generic EMI Shielding Fabrics

The major divisions of EMI shielding fabrics are: (i) Metal-coated fabrics and (ii) metal interwoven fabrics. Silver and copper are the candidate materials normally blended with the textiles. The basic requirement of the shielding fabrics is that they are electrically conductive. Examples of EMI shielding fabrics are:

- Silver-metalized woven nylon fabrics
- Blend of 75% wool plus 25% of conductive (metallic) alloy fibers
- Blend of 85% polyster plus 15% of conductive (metallize fibers)
- Blend of 85-90% of syntheic textile plus 15-10% of stainless steel fibers

Typical surface resistances of different shielding textile materials are listed in Table 21.3.

Magnetic field shielding: Woven type shielding fabrics (of the types described in Table 21.3) provide almost nil shielding effectiveness to magnetic fields below 10 MHz.

Electric field shielding: Metal-coated fabrics are akin to metallic foils in offering electric field shielding

Characteristic requirements of shielding fabrics should meet the following general requirements:

- The clothing should offer effective and adequate EM shielding.
- The clothing itself should not pose an hazard.
- The shielding limit should be flexible, comfortable, light in weight, and unrestricting to the wearers.
- The fabric must be conducive for the making of suits such as an overalls with integral hood, gloves, and oversocks with no leakage of EM energy through zippers etc. Caution should be exercised in compliance with: ANSI C95 1973 standard which specifies "In very intense fields, arcs will occur between folds in the fabric and between the arm and body, etc. These arcs will burn holes in the suit, exposing the wearer to harmful radiation".

Corresponding to the four types of shielding fabrics blends as detailed in Table 21.3, the measured values of shielding effectiveness at microwave frequencies are presented in Table 21.4.

Allowable limits of exposure of suits to radiation to avoid arcing are: 200 mw/cm^2 (U.S Navy Limit) and 1 mw/cm^2 averaged over any 60 second period for frequencies above 30 MHz (Australian RF Exposure Standard AS 2772-1985 is specified as the corresponding personnel exposure). The above stipulations warrant the suit to provide at least 23 dB of shielding effectiveness with all its opening, if any.

Table 21.3 Surface Resistance of Shielding Fabrics

Type of Fabric Blend	Mean Surface Resistance (ohm/sq)	Fiber Size (μm)	Wave Orientation
1. 75% Wool + 25% conducting fiber	~1.0	~10.0	
2. 85% Polyster + 15% conducting fiber	~8.0	~8.0	Warp/weft/diagonal
3. 85% Synthetic fabric + 15% stainless steel fiber	~65.0	~7.5	
4. 90% Synthetic fabric + 10% Stainless steel	~55.0	~7.5	

Table 21.4 Measured Shielding Effectiveness at Microwave Frequencies

Fabric Blends (of Types in Table 21.3)	SE (dB) at 4 GHz	SE (dB) at 10 GHz
Type 1	48	43
Type 2	43	42
Type 3	41	38
Type 4	33	37

Normally at high frequencies, the conductive suits offer 20 dB or more shielding effectiveness. At low impedance (low frequency) conditions, the fabrics should be more conductive to provide a shielding effectiveness of 20 dB or more. Further requirements are:

- The conductive suit should be nonflammable.
- The suits should assure normal ventilation.
- The personnel-wear fabrics should offer no threat of corona or other breakdowns under high voltage operations.

21.11 Modeling Metallized Fabric Shields

Due to the extensive geometry dependence of the shielding characteristics *vis-a-vis* the conductive fabric, exact formulations to deduce the shielding effectiveness are rather sparse. Expressions have been developed on the basis of plane wave shielding theory of meshed or perforated conducting panels with appropriate empirical changes to accommodate the surface resistivity, pore size, and thickness of the fabric material. A functional form of relation derived to depict the shielding effectiveness of conductive fabrics both at high and low frequencies is given by:

$$(SE)_{dB}\ (Fabric\ Material) = exp\ (-0.129\ L\sqrt{f})\ (SE)_{Foil} +$$

$$[1 - exp\ (-0.129\ L\sqrt{f})\ (SE)_{Aperture} \quad (21.11)$$

where L is the aperture dimension of the fabric and f is the frequency of operation. Further $(SE)_{Foil}$ and $(SE)_{Aperture}$ depict the shielding effectiveness of a metallic foil (of the same thickness (t) as the fabric) and that of an aperture (of size $L \times D$) subjected to a plane wave excitation. They are given by:

$$(SE)_{Foil} = 20\ log_{10}\ [(1 + K)^2/4K]\ [1 - (K - 1)^2/(K + 1)^2]\ exp(-2t/\delta) \quad dB \quad (21.12)$$

where $K = Z_w/Z_s$ (ratio of wave impedance to shield impedance) and δ is the skin depth.

$$(SE)_{Aperture} = 100 - 20\ log_{10}(L \times f) + 20\ log_{10}\ [1 + ln(L/s)] + 30\ D/L \quad dB \quad (21.13)$$

where L is the maximum pore size, s is the minimum pore size and D is the depth of the aperture.

Alternative form of expression derived for the shielding effectiveness of a conductive fabric is :

$$(SE)_{Fabric} = (A_a + R_a + B_a + K_1 + K_2 + K_3) \quad dB \quad (21.14)$$

where
A_a = attenuation introduced by a discontinuity in dB
R_a = aperture (single reflection) loss in dB
B_a = multiple reflection correction term in dB
K_1 = correction term to account for the number of like discontinuities in dB
K_2 = low frequency correction term to account for skin depth (in dB), and
K_3 = correction term to account for coupling between adjacent holes in dB

Explicit expressions for the terms in Equation 21.14 are:

$A_a = 27.3\ (D/L)$ for an incident wave below cutoff for a rectangular opening (D and L in meters)
$R_a = 20\ log_{10}\ |(1 + 4K^2)/4K|$
($K = j\ 6.69 \times 10^{-3}\ f \times L$, f in MHz, L in meter)
$K_1 = -10 log_{10}\ (a \times n)$
(a = Mesh area in square meters and n is the density of the meshes per square meter)
$K_2 = -20 log_{10}\ (1 + 35\ p^{-2.3})$
(p = fiber diameter/skin depth)
$K_3 = 20\ log_{10}\ [coth\ (A \times a/8.686)]$

Typical variations of shielding effectiveness (SE) *versus* the metal content, and geometrical parameters over the frequency of fabric-type shields are presented in Figure 21.7.

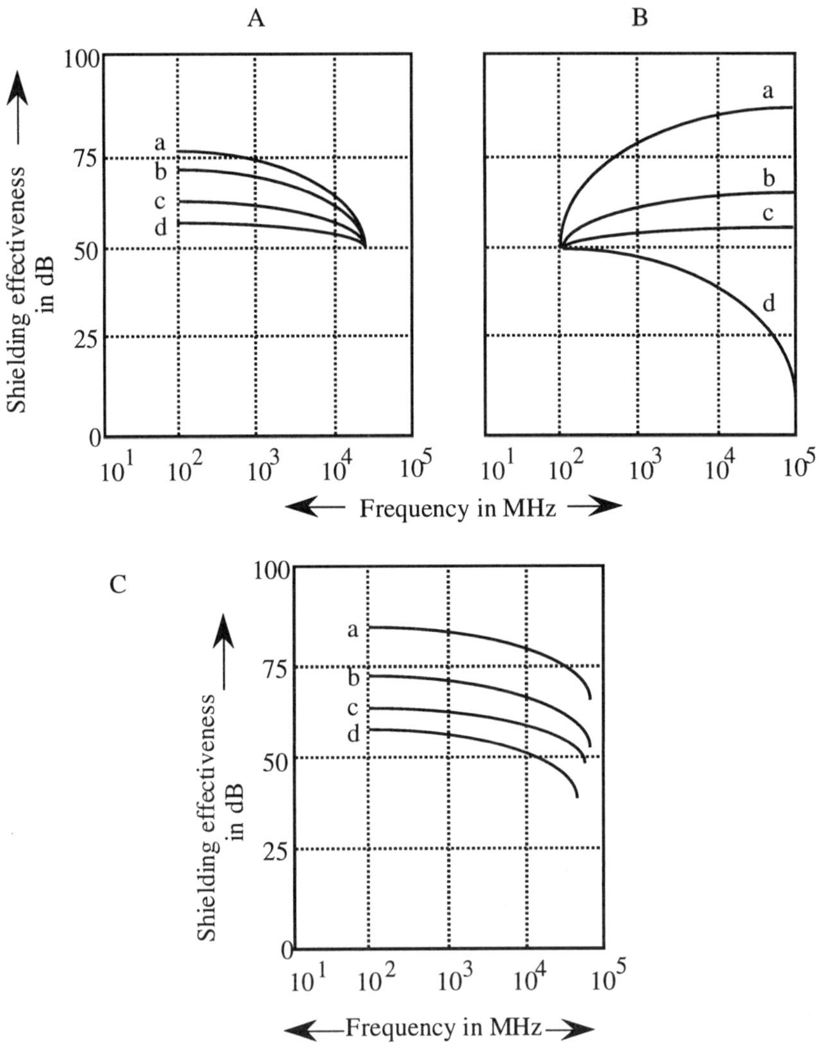

Figure 21.7 Shielding effectiveness *versus* frequency of a fabric-type EM shield.
A: Effect of metal content – Examples with L: Maximum pore size ≈ 160 μm; s: Minimum pore size ≈ 106 μm; D: Depth of the aperture ≈ 90 μ/m.
(a) Copper content /meter^2 = C_{cu} = 1.0; (b) C_{cu} = 0.8; (c) C_{cu} = 0.6, and (d) C_{cu} = 0.4.
B: Effect of maximum pore size – Examples with C_{cu} = 1; D ≈ 90 mm and L/s ≈ 1.5.
(a) L = 50 μm; (b) L = 100 μm; (c) L = 200 μm and (d) L = 400 μm.
C: Effect of aperture depth – Examples with L ≈ 250 μm; L/s = 1.5 and C_{cu} ≈ 1.
(a) D = 400 μm; (b) D = 200 μm; (c) D = 100 μm, and (d) D = 50 μm.

21.12 Conductive Paints for Shielding

Conductive paints are used as conductive surface coatings on carefully prepared surfaces such as plastics, woods, ceramics, and other base materials so that the coated material offers EM shielding. Normally these paints are prepared with the suspension of conductivity inclusions such as graphite and silver particles in a medium like lacquer, elastomer, silicon resin, vinyl base, acrylic fluid, or latex. Conductive coatings are done on prepared surfaces *via* dipping, spraying, silk-screening, roller-coating, brushing, or aerosol spraying. Typical conducting paints and their characteristics are presented in Table 21.5.

The need for conductive coating arises mainly due to the current practice of using plastics for electrical/electronic equipment enclosures (*in lieu* of metallic boxes). Such plastic enclosures as such do not offer EMI shielding and therefore warrant an application of conductive coating on their surfaces. Further, plastics are prone to *electrostatic propensiveness* (see Chapter 20). Conductive coating facilitates the bleed-off of static charges accumulated on the plastic surfaces.

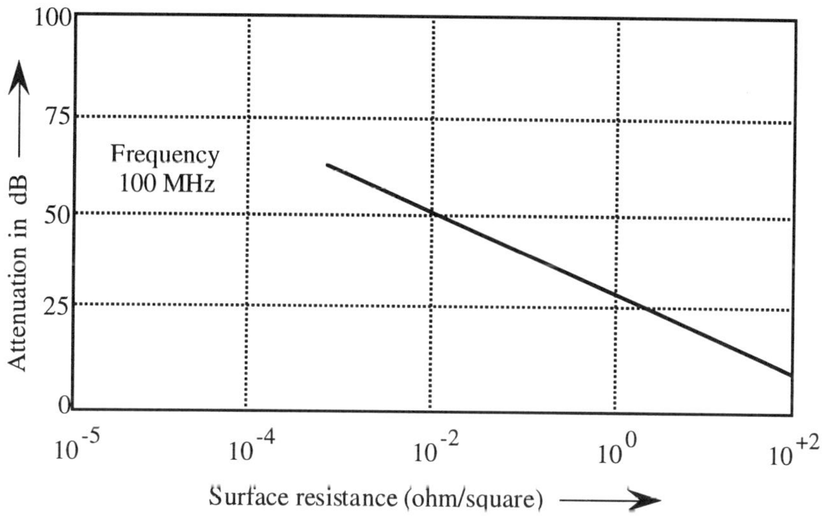

Figure 21.8 Attenuation of EM wave *versus* surface resistance of conductive coating at 100 MHz.

Table 21.6 Coating Thickness Requirements

Type of Coating	Coating Thickness (micrometer)	Function	Surface Resistance (ohm/square)
Silver paint	40-80	EMI/RFI Shielding	1
	50-80	Grounding	10
	20-100	Electrostatic Bleed-off	50-150

The general requirements of a conductive coating are:

- Adequate surface conductivity
- Operation over a wide range of temperature
- Stability in most environments
- Stability against mechanical shocks and abrasions
- Facilitation of easy grounding
- Presenting an aesthetic appearance
- Uniformity of coating thickness
- Ease of application

The coating thickness required depends on the type of shielding required. For example, requirements of silver-coating thickness are specified in Table 21.6.

21.13 Surface Resistance of Conductive Paints

Surface resistance (R_s) of conductive paint coating is expressed in ohms/square. In terms of the specimen's length (ℓ) and width (w) and the bulk resistance R, the surface resistance (R_s) is expressed as

$$R_S\ (ohms) = R\ (ohms)\ (\ell/w) \tag{21.15}$$

(ℓ and w should have the same units, cm, meter, or inches).

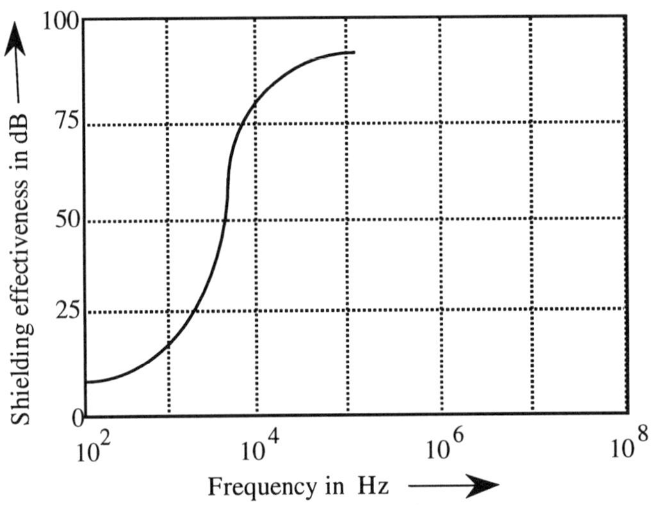

Figure 21.9 Magnetic field shielding effectiveness of a conductive paint coated enclosure with full integrity and no openings *versus* frequency.

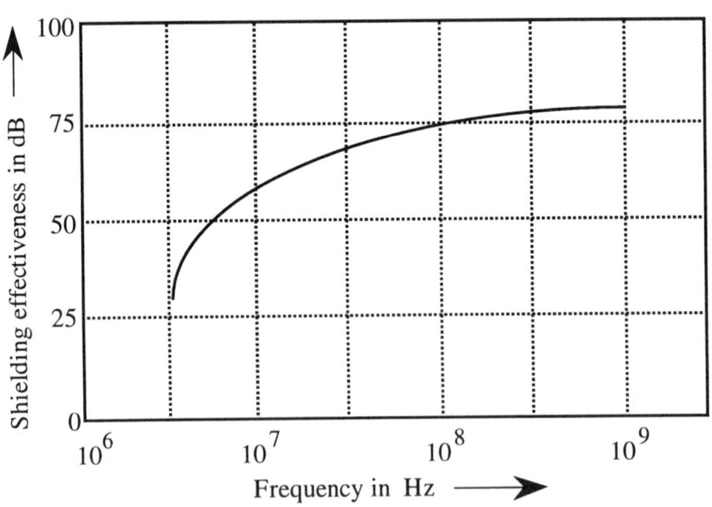

Figure 21.10 Electric field shielding effectiveness *versus* frequency of a conductive paint coated enclosure with full integrity.

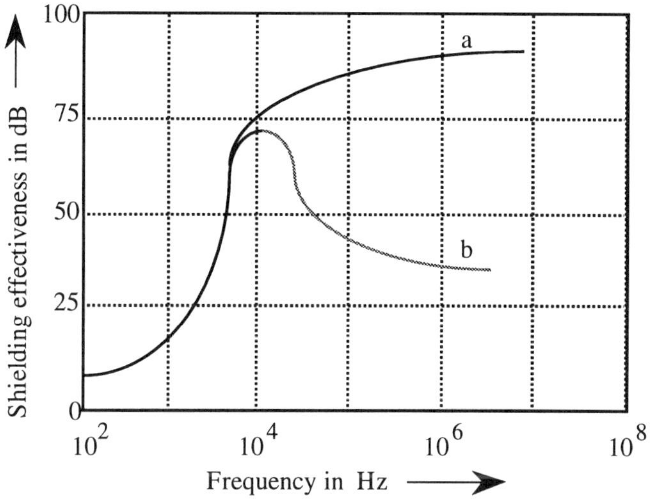

Figure 21.11 Plane wave attenuation *versus* frequency offered by a conductive paint coated enclosure. (a) Enclosure with full integrity; (b) Enclosure with partial openings.

Surface resistance is the quality index of conductive coatings specifying the uniformity of coating and the extent of shielding effectiveness of the specimen. The dependence of EM field attenuation *versus* surface resistance is a function of frequency and at 100 MHz, typical attenuation as a function of sheet resistance is shown in Figure 21.8.

Shielding effectiveness *(SE)* at low impedance (low frequency) conditions offered by conductive paint coating may degrade with openings in the shielding enclosures. A typical low frequency magnetic field shielding effectiveness *versus* frequency of conductive paint (such as acrylic-based silver paints) is depicted in Figure 21.9.

The electric field attenuation (expressed in terms of shielding effectiveness) due to conductive paint (acrylic-based silver paint) coated structures as a function of frequency is presented in Figure 21.10.

Plane wave attenuation due to conductive coating is also affected by the loss of integrity due to openings in the shielding enclosures. For a typical conductive paint coated enclosure, variation of shielding effectiveness *versus* frequency is shown in Figure 21.11.

21.14 Properties of Conductive Pigments/Fillers

Paints, like plastics on which they are coated, are inherently nonconducting and therefore provide no shielding effect unless suspended with conductive fillers/pigments. As indicated by Hart [10] that no single property of the conductive pigment assumes overriding importance and the materials used are normally chosen because they exhibit a combination of properties which fit the requirements of the specific application. The cost of the pigment is also very important depending on the mission involved. The pigments chosen are of materials with high electrical conductivity. Table 21.7 gives the electrical conductivities of typical pigment materials relative to that of copper.

When the conducting particles are suspended in a paint, the coating realized is a thin organic film in which the particles are dispersed as random chains in the organic vehicle. Therefore, the effective surface resistance offered by the coating depends on the mode dispersion of the conductivity particles as decided by their volume fraction, shape, size, and surface conditions of the substrate.

Table 21.7 Relative Electrical Conductivities of Pigment Materials

Pigment Materials	Relative Conductivity (σ/σ_{cu})
Silver	1.05
Copper	1.00
Gold	0.70
Aluminum	0.61
Nickel	0.21

Adjunct considerations that determine the shielding properties of the coating the chemical properties of the pigment materials. Such properties determine the surface condition of the particles. Materials which are resistant to corrosion and oxidation are capable of presenting clean metallic particulate surfaces with minimal contact resistance between the dispersed particles in the vehicle medium, whereas materials like aluminum particles due to surface oxidation contribute high particle-to-particle contact resistance, thereby offering a low surface conductance. Silver on other hand presents good chemical properties and mostly nonreactive with the vehicle medium. Therefore silver-based paints show better storage stability and facilitate stable conductive coatings.

Though copper particles offer excellent conductivity in the suspended phase, their chemical stability is rather poor leading to deterioration of initial high conductivity properties on storage and in aggressive environments due to oxidation and corrosion. Surface-treated copper pigments have been developed to combat against the aforesaid chemical activity *versus* conductivity characteristics.

Pigment materials should also offer stable mechanical properties to the coating against wrinkling, blistering, pitting, corrosion, cracking, and peeling.

Though gold pigments are excellent conductors, their usage in conductive paints are limited due to prohibitively high cost considerations.

Aluminum has inherently high electrical conductivity and presents a stable coating. However, aluminum particles form insulating oxide films degrading the effective conductivity of the films.

Despite nickel having lower conductivity, it has become a popular pigment material due to its excellent chemical properties. Its chemical integrity offers low contact resistance between the particles suspended in an organic vehicle. Nickel withstands extreme and aggressive environments and remains stable over a long storage period.

Though graphite has good chemical properties, its applications are limited due to its poor conductivity and its application is not aesthetically pleasing on surfaces intended for consumer products. Further, graphite has a tendency to "loosen out" and gets shed from the coated surface.

21.15 Composite Shielding Materials

Composite shielding materials can be classified into two major groups, namely, the host-inclusion system and multilayered "stack-up" system. The first category refers to a host material in which another material (in the form of particles, fibers, or flakes) is dispersed so that the volume fraction of the constituent materials, their electrical characteristics, and the shape and size of the inclusions determine cohesively the shielding effectiveness of the composite. In the second type, selective materials are stacked up as layers to yield a desired shielding property [11-14].

Particulate-blended shielding composites are constituted by metallic inclusions like silver, aluminum, nickel, or copper particles heterogeneously mixed in a host medium such as polymer/plastic, epoxy resin, or concrete. Nonmetallic inclusions like graphite, pyrolized organic fiber (polyacrylonitrile), boron/boron tungstate, or conductive polymers have also been advocated for similar applications. For high temperature applications suitable candidate materials for the host matrix are:

1. MIL-C-28840 Series (thermoset and thermoplastic resins)

Polyether-ether-ketone	282°C
Liquid crystal polymer	240°C
Polyphenylene sulfide	232°C
Polyamide-imide	220°C
Polyimide	204°C

2. MIL-C-38999 Series IV (thermoset and thermoplastic resins)

Polyther-sulfone	180°C
Polyaryl-sulfone	180°C
Polyther-imide	180°C

3. Portland Cement

4. Epoxy Resin, PVC, nylon

5. Ceramics
   Barium titanate
   Titanium dioxide
   Ferrite materials

As inclusions a variety of metals, metal-coated nonmetals and alloys in different particulate shapes have been studied in synthesizing composite shielding materials. A list of such materials are as follows:

Doped conductive polymers
(poly-p-phenylene-benzobis-thiozole, PBT)
Carbon/graphite fiber, carbon powder
Pyrolyzed organic fiber
(Polyacrylonitrile)
Nickel-coated graphite/polycarbonate fibers
Iron oxide powder
Indium/tin oxide (ITO) powder
Nickel flakes/powder
Aluminum powder/flakes/fibers
Stainless steel fibers (300 and 400 series)
Alloy fibers (Nichrome, Inconel, Hastelloy X, Carpenter 20CB3, 80/20 nickel chromium)
Titatium Tantulum fibers
Boron and boron tungstate fibers
Chopped copper wires

The shielding characteristics of different combinations of host-inclusion systems as measured and reported are presented in Table 21.8.

The choice of host material and/or inclusions depends on the type of the shield, its application, and the shielding effectiveness warranted. Specific to the inclusions, the main requirements of their suitability in a shielding composite can be listed as follows:

- Significant electrical conductivity

- Availability in different forms (powder, fibers, or flakes)
- Minimum chemical interaction with the base matrix
- Minimal alteration of base material properties
- Shrinkage compatible with the host medium
- Excellent abrasion strength
- Corrosion resistance
- Shelf-life durability with no "shedding off" from the composite
- Cost-effectiveness

The host medium is expected to have:

- High dielectric and/or magnetic loss characteristics
- Good bonding with the inclusions
- High strength and impact resistance
- Moldability
- Chemical stability
- Noninteraction (chemically) with the inclusions
- Heat resistance (against warping and cracking)
- Coloration with pigments for aesthetic appearance

The physical forms of inclusions significantly influence the effective conductivity of the composite material. Therefore, the design of shielding composites can be controlled (to achieve a specified shielding effectiveness) with proper choice of particulate shapes of the inclusions (see Chapter 6). Typically, the following particle geometries are considered in practice:

1. Spherical or near-spherical fine or coarsed particles
2. Spheroidal or elliposoidal particles
3. Flakes or disks
4. Fibers or needles
   - *Tows* –fibers of continuous strands of multiple-end filaments
   - Sized and chopped fibers
   - Air-laid *web* –a nonwoven, randomly arrayed fiber web
   - Continuous filament yarn

21.16 Shielding Effectiveness of Particulate-Blended Composites

The shielding performance of these composites is decided by the effective conductivity (σ_{eff}) of the composite material. As discussed in Chapters 6, the value of σ_{eff} is controlled by the volume fraction, conductivity, and shape of the inclusions, and by the dielectric loss of the host medium. Further, the dispositions of the dispersed particles (either in totally anisotropic or isotropic random fashion or in a textured selective orientation) also play a significant role in attenuating the electromagnetic energy passing through the composite. They also decide the polarization dependency of the attenuation realized.

In the design considerations, proper choice of constituent materials and their volume fractions, shapes, and dispositions are crucial factors to realize a specified value of shielding effectiveness. The algorithms pertinent to composite multiphase dielectrics and/or conductor-included dielectric mixtures described in Chapter 4-6 are useful in synthesizing a test composite of a specified effective conductivity (and hence the shielding effectiveness).

The choice of host medium could be based on mechanical properties of the composite being synthesized as well as on the thermal withstandability, chemical stability, and corrosion-resistant characteristics.

Following are some typical recipes of particulate-blended shielding composites and the test results on the shielding performance:

Shielding composites with doped polymers as conducting inclusions: A number of lightweight polymers which are intrinsically nonconductive become conductive upon doping are useful as EMI shielding materials. For example, when the polymer poly-p-phenylene-benzobis-thiazole (PBT) available as Pristine PBT 2002-2™, is doped with iodine *via* ion-implantation, the resulting composite offers a sheet resistivity on the order of 350 ohm/square at microwave frequencies. Typical shielding effectiveness *versus* frequency of a PBT film is depicted in Figure 21.12.

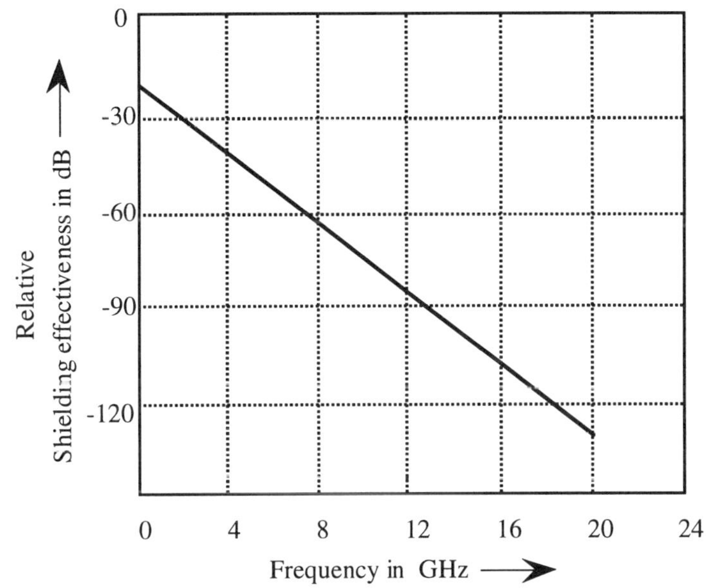

Figure 21.12 Shielding effectiveness *versus* frequency of PBT composite shield.

In addition to PBT, a variety of other conductive polymers (such as iodine-doped polyacetylene) have been reported in literature [17] (and described in Chapter 8) which can be useful for the purpose of synthesizing EMI shielding composites.

Apart from using conducting polymers as such for shielding applications, they can also be blended with thermoplastics to achieve a very high level of shielding performance. Typical blends are constituted by a matrix polymer such as polyvinyl chloride (PVC) or nylon compounded with an inherently conducting polymer (ICP) with a conductivity on the order of 1 to 10^5 siemen/cm. The resulting thermoplastic blends have conductivities as high as 20 siemen/cm which are an order of magnitude higher than those that can be obtained with carbon black-included polymers.

Volume fraction of the conducting polymer inclusions decides the effective bulk conductivity of the composite blend and hence the shielding effectiveness of the material. Shown in Figure 21.13 is the typical variation of shielding effectiveness due to conducting-polymer included blends (of different bulk conductivities) *versus* frequency. For comparison, relevant data on a blend using 6% stainless steel fiber is also presented in Figure 21.13.

The results indicate that this type of composites is suitable for low frequency electric field shielding. However, coating the shield with nickel plating provides both plane wave and electric field shielding effectiveness well in excess of 100 dB.

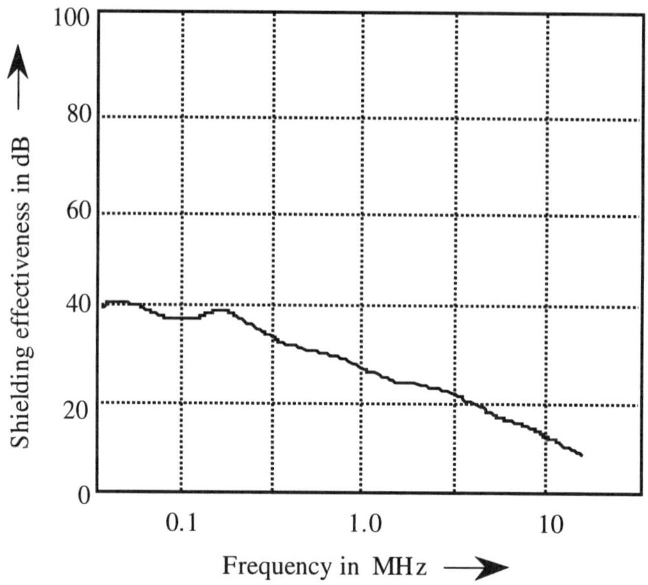

Figure 21.13 Measured data on electric field *versus* frequency of 70 mil thick shield of epoxy resin filled with 40% by weight of carbon fibers.

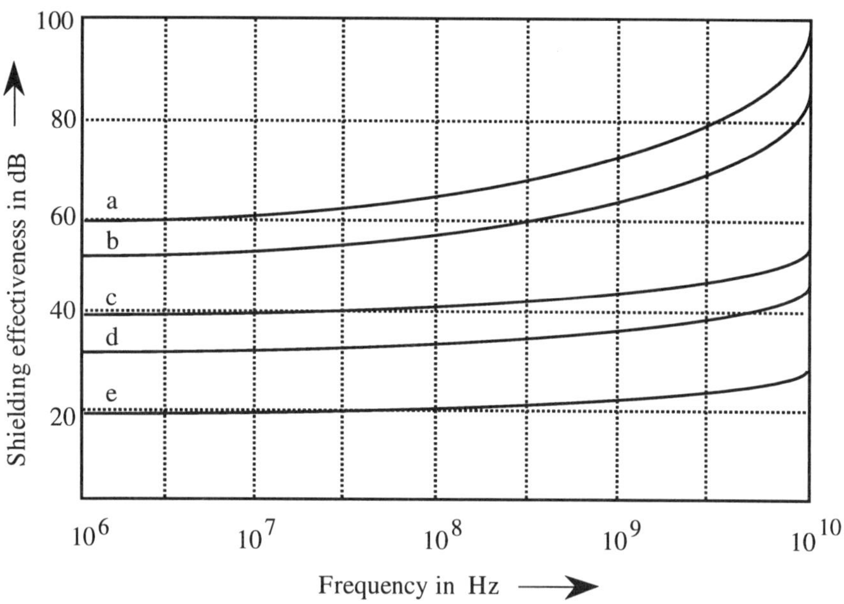

Figure 21.14 Measured far-field shielding effectiveness of conducting plastic included blends *versus* frequency.

(a) Conductivity (σ) = 7.50 siemen/cm.

(b) σ = 3.75 siemen/cm.

(c) σ = 1.00 siemen/cm.

(d) σ = 0.20 siemen/cm.

Table 21.8 Measured SE of Shielding Composites (Frequency up to 100 MHz)

Material	Shielding Effectiveness in dB	Remarks
1. (Polyether-ether-ketone, 75%) + (indium-tin oxide, 15%) + (Ni flakes, 10%)	43-45	Cu plated
2. (Polyther-ther-ketone, 75%) + (indium-tin oxide, 15%) + (Ni flakes, 10%)	36	Ag painted
3. (Polyether-ether-ketone, 75%) + (graphite, 15%) + (polycarbonate, 10%)	28	Cu plated
4. (Polyether-ether-ketone, 75%) + (graphite, 15%) + (Ni flake, 10%)	25	Cu plated

Other shielding composites of filler-added resins: Polymeric materials with a variety of conducting fillers such as indium-tin oxide (ITO), nickel-plated graphite, aluminum flakes, iron oxide powder, graphite fibers, etc. have been studied as shielding materials and Table 21.8 illustrates some of the pertinent results on the shielding effectiveness.

Carbon-filled epoxy resin shielding composites: Epoxies blended with carbon particulates (either in powder or in fiber form) constitute simple and cost-effective shielding composites. These composite materials can also be coated with silver or nickel paints to improve the shielding effectiveness. Measured data on a typical composite (as per MIL-STD-285 are shown in Figure 21.14).

21.17 Intercalated Graphite Fiber Composites

Carbon/graphite fiber composites have been successful structural materials in aircraft and spacecraft due to their low-density and high strength considerations. However, conventional carbon fiber/epoxy composites do not have sufficient electrical conductivity to offer adequate shielding performance. Therefore, such composites are made with *intercalated* graphite fibers.

Intercalation is the process of introducing guest atoms or molecules between the graphene layers of graphite. The guest species can contribute carriers (either electrons or holes) to the graphite lattice and thus increase its conductivity significantly without seriously degrading its physical/mecahnical properties. Although most intercalation compounds are usable at the temperatures needed to cure epoxies, the residual intercalation compounds which use bromine as the guest molecule have been shown to be quite stable. Lamina composites fabricated using bromine intercalated fiber show enhanced conductivity and improved shielding effectiveness. The far-field and near-field shielding effectiveness of several grades of graphite fiber/epoxy composites compared to those of metals are depicted in Figures (21.15 and 21.16) as functions of frequency [12].

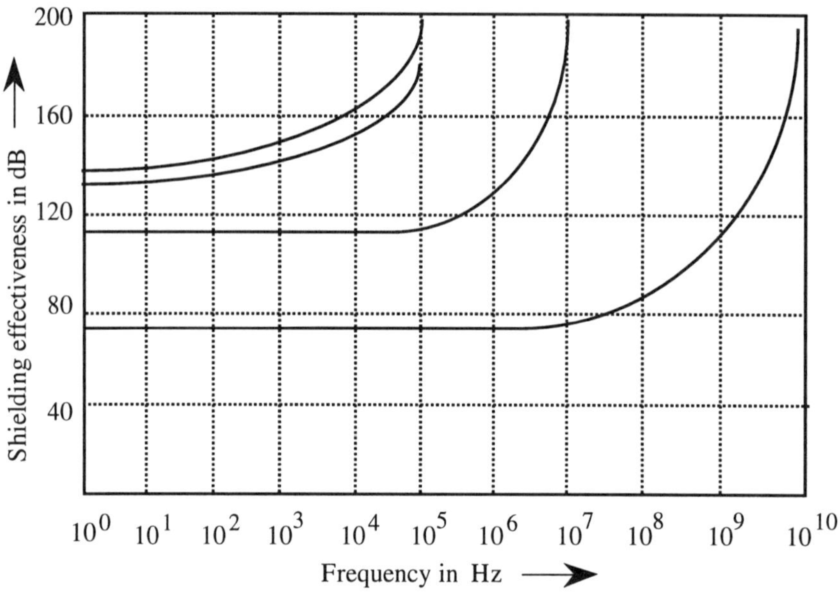

Figure 21.15 Far-field shielding effectiveness of intercalated graphite fiber composite shields *versus* frequency.
(a) PAN-based carbon fiber/epoxy; (b) P-100+Br/epoxy and intercalated pitch-based graphite fiber/epoxy; (c) Aluminum shield; (d) Copper shield.

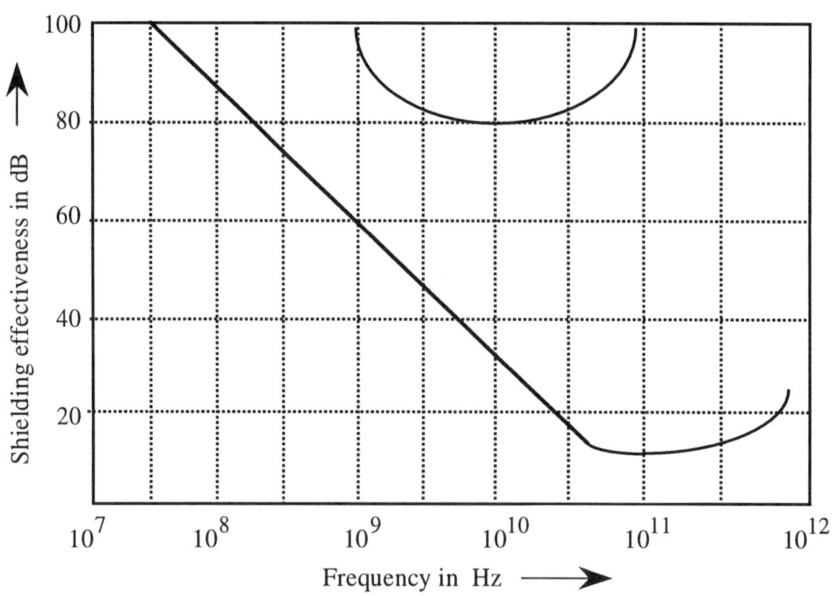

Figure 21.16 Near-field shielding effectiveness of intercalated graphite fiber composite shields *versus* frequency.
(a) p-100-Br_2; (b) PAN.

Studies indicate that far-field attenuation of EM fields of at least 70 dB in 1–12 GHz range is offered by p-100 + Br/epoxy composites. Adoption of this technology is, however, likely to be cost-critical, but with lower grade pitch fibers such as p-55 may permit synthesizing cheaper composites. Use of intercalated p-55 in a test composite has shown to yield far-field attenuation of about 55 dB over the frequency range 30-1000 MHz.

21.18 Shielding Composites with Conducting Flakes

As discussed in Chapter 6, addition of flaky or disk-like conductors to a nonconducting host medium substantially alters the effective conductivity of the blend even at low volume fractions of the inclusions. Flakes in general are superior conductivity modifiers in comparison with conductive carbon black or graphite powder or doping agents being added to plastic materials.

Typical conducting flakes are of aluminum which in a injection resin molding system can offer a resistivity variation with its percentage weight in the blend as presented in Table 21.9.

The shielding effectiveness *(SE) versus* bulk resistivity (ρ_B) of the conductive plastic shield made by a combination of aluminum flakes (such as Transmet™) plus a thermoset plastic is illustrated in Figure 21.17.

Conducting fiber-added plastics as shielding composites: Another effective method of realizing conducting plastics is to blend metal fiber with polymeric hosts. Either pure metallic or metallized glass fiber can be used for this purpose, in the form of chopped fibers.

Typical electrical resistivites of the metal fiber-filled composites fall in the range 10^{-2} to 1 ohm-cm with a shielding attenuation well over 40 dB up to 100 MHz and about 25 dB at higher frequencies.

Table 21.9 Resistivity of the Aluminum Flake-Added Plastics

Weight % of Aluminum Flakes	Resistivity (ohm-cm)
5	10^{15}
10	10^{13}
15	10^{6}
20	10^{2}
25	10^{0}
30	10^{-1}
35	10^{-2}
45	10^{-3}

Major variables crucial to achieve success in molded metal fiber-filled composites and to realize a good electrical conductivity, effective shielding performance, acceptable physical performance, and aesthetic appearance are:

- Fiber concentration
- Fiber aspect ratio and orientation
- Uniformity of dispersion

The concentration of fiber inclusions decides the probability of the fibers making effective contact with each other in establishing electrical connectivity. The aspect ratio decides the effective conductivity and permittivity properties as discussed in Chapter 6. Fiber orientation has a significant impact on shielding effectiveness. The response of the shield to EM wave of certain polarization depends on the anisotropic orientation of the fiber relative to the polarization direction of the EM energy.

Uniformity of fiber dispersion is also important to maintain a uniform shielding effectiveness over the entire area of surface exposed to EM radiation. Clumping is undesirable and fiber dispersion is controlled by the molding process involved.

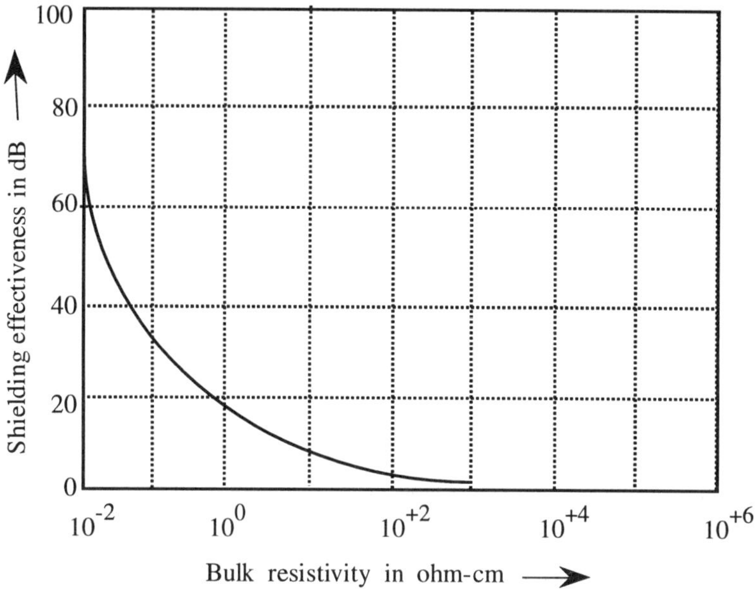

Figure 21.17 Shielding effectiveness *versus* ρ_B (frequency 0.1 to 1000 MHz) of aluminum flake-inclused thermoset plastic shields.

Conducting-filler added ceramic shielding materials: In this category, the host material chosen is one of the following ceramic materials instead of the polymeric matrix base:

- Heat resistant ceramics such as barium titanate and titanium dioxide [18]
- Portland cement concrete
- Ferrites

Heat-resistant ceramic based shielding composites: A class of composite materials constituted by conducting inclusions dispersed in a host ceramic medium (such as TiO_2) is a potential EMI shielding medium. In the existing art of electromagnetic shielding with composite materials as mentioned earlier the shielding material is constituted invariably by a polymeric base, dispersed with conducting inclusions [10] or with laminates of fibers of conducting materials (such as boron tungstate, graphite, etc.) stacked as multilayer layups in an epoxy matrix [10]. As discussed in the previous sections, the performance of such materials in shielding the radio frequency interference effectively has been adequately elucidated over a limited range of frequencies and/or at room temperature conditions. However, in view of the state-of-the-art requirements in military and space applications involving high-temperature ambients as well as interference arising from signals covering a broad spectrum of frequencies, there has been a quest for newer materials with higher temperature withstandability and better broadband EMI shielding capabilities.

Essentially, such EMI shields can be constituted by a ceramic base (of titanium-dioxide) with two categories of conducting inclusions, namely: (i) Spiky copper fibers and (ii) flaky (disk-like) aluminum foils. The TiO_2 base provides a ceramic receptacle for the composite with a high-temperature withstandibility and the conducting particles are chosen to offer different geometrical aspects and hence a controllable effective conductivity of the composite. The spiky, fibrous inclusions have a geometrical aspect ratio >>1; and the flaking inclusions have an aspect ratio <<1. The geometrical aspect together with the volume fraction and the conductivity of the inclusions (as well as the permittivity of the host medium) decide the attenuation (shielding effectiveness) offered by the composite material as a function of the frequency of the electromagnetic wave.

The host medium (TiO_2) is a dielectric (of relative permittivity ε_r) dispersed with conducting inclusions of (conductivity σ_m in siemen/meter). These inclusions are shaped either as spiky rods with an aspect ratio $a/b >> 1$ or flaky disks with $a/b << 1$. For these types of two-phase mixture system(s), the effective permittivity and conductivity can be deduced by the considerations described in Chapter 6.

Effective (relative) permittivity of the mixture:

$$\varepsilon_{eff} = (\varepsilon' - j\varepsilon'') \tag{21.16}$$

where

$$\varepsilon' = \{\varepsilon_r/[1 + (\varepsilon_r - 1)^{2u}]\}\ \{[(\sigma_m/\omega\varepsilon_o)^{\theta}(\varepsilon_r - 1)^{1-\theta}\cos(n\theta/2)]^{2u} + 1\}$$

with $\varepsilon_o = (1/36\pi) \times 10^{-9}$ F/m being the permittivity of free space. Further, θ refers to the volume fraction of the inclusions.

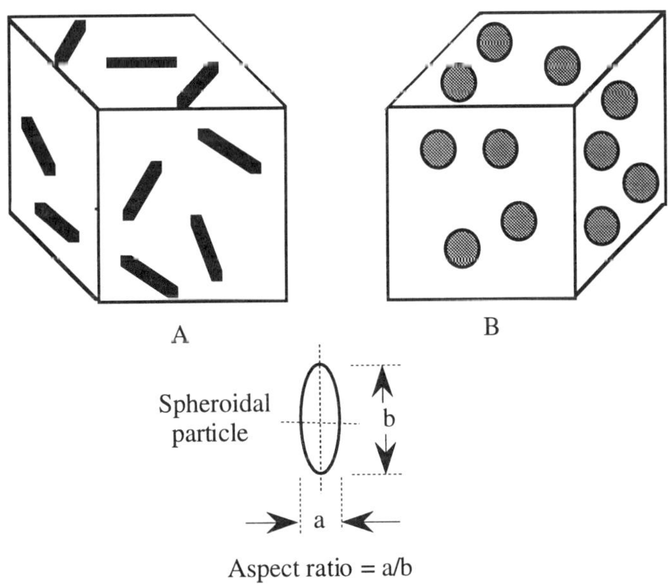

Figure 21.18 Random nonspherical conducting particulates included dielectrics. (A) Needle-like conducting inclusions. (B) Disk-like conducting inclusions.

Effective conductivity of the mixture:

$$\sigma_{eff} = \sigma_m \{[\omega\varepsilon_o(\varepsilon_2 - 1)/\sigma_m]^{(1-\theta)} \sin(n\theta/2)\}^{2u} \tag{21.17}$$

and

$$\varepsilon'' = \sigma_{eff}/\omega\varepsilon_o\varepsilon_{eff}$$

In the above equations (Equations 21.16 and 21.17), u refers to an order parameter specifying the anisotropic state of randomness of the particulate dispersion [18]. It is equal to 1/3; and $\omega = 2\pi \times$ frequency of operation.

In terms of these effective mixture parameters, the transmission loss (attenuation) depicting the shielding effectiveness of the test medium of thickness τ (in meters) is given by:

$$\alpha\,(in\ dB) = 20log[(\pi/\mu_o\sigma_{eff})^{1/2}] + 20log[\tau(\varepsilon_{eff} + 1)] - 20log[2\varepsilon_{eff}] \qquad (21.18)$$

where $4\pi \times 10^{-7}$ henry/meter, being the absolute permeability of free space.

For the test samples using the material constituents of Figure 21.18, the shielding effectiveness measured is shown in Figure 21.19. Depending on the conducting particulate loading, the shielding effectiveness of the composite(s) falls in the range 5-20 dB over the frequency spectrum 500 MHz to 1 GHz. The mixture formulas (Equation 21.16 and 21.17) enable the elucidation of the effective conductivity and permittivity of the composite(s); and hence the theoretical evaluation of the shielding effectiveness based on Schelkunoff's theory is made feasible *via* Equation 21.18. Sample theoretical results are also furnished in Figure 21.19.

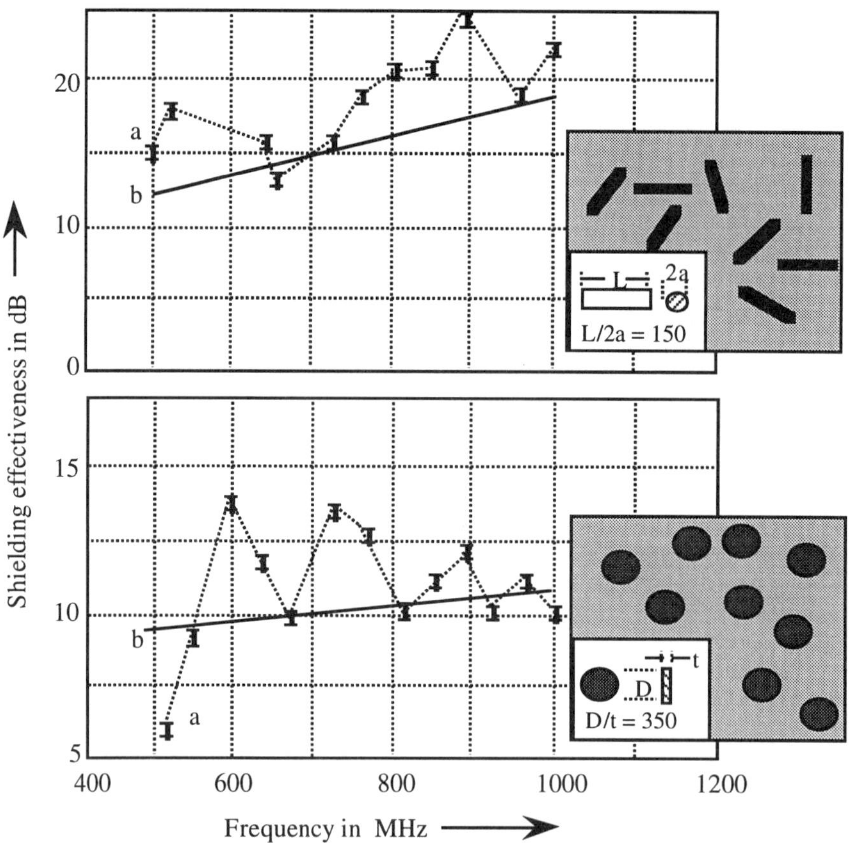

Figure 21.19 Shielding effectiveness *versus* frequency of needle-like or disk-like conducting particulates included dielectric composite shields.

Portland-cement concrete with metallic inclusions as an EMI shielding material: Concrete/polymer concrete-based hybrid composites as EMI shielding materials have been studied [15]. Being a low-cost construction material Portland cement concrete/polymer concrete mixed with chopped electrically conductive fibers has proved to be economical and viable EMI shielding composites with easily and rapidly castable characteristics endowed with unique properties to survive harsh environments over a long period of time.

Electrical conductivity of hydraulic cements in general depend on moisture content and upon over-drying. Impregnating with subsequent polymerization, the cement-based concretes can be regarded as good insulators with a dielectric constant falling in the range of 4 to 5. Apart from hydraulic cements, the so-called polymer concretes constituted of a carefully graded mixture of coarse aggregates and fine fillers bonded together by means of an ambient temperature cured, low-viscosity organic resin system (of 6 to 15% by weight) are more popular candidates for blending with metallic inclusions and culminating as a shielding composite. These polymer concretes have better mechanical properties than hydraulic cement concretes; and they have inherently high corrosion-resistant, low or *nil* capillary porosity with low moisture permeability, rapid processing and strength gain (facilitating fast demolding), and attractive cost-effectiveness. These concretes can be blended with chopped metallic fibers such as stainless steel, nickel-coated carbon, PAN carbon or with aluminum flakes. Conducting particulates such as carbon black can also be used as a filling agent. The mixing can be done with conventional vibratory-type mixers.

Ferrite-based shielding composites: Ferrites being essentially ceramics but with a high magnetic permeability can be blended as fillers with a polymeric host plus other conducting alloys having high magnetic permeability (in disk or lamellar form) to constitute a magnetic shielding medium. Such multiconstitutent shielding materials have not been studied thoroughly though, yet some theoretical considerations pertinent to the evaluation of the effective permeability of multiphase materials are available in the literature [22]. Magnetic ore falling in the category of ferrite classes such as franklinite, chromite, and ilmenite can constitute a highly desirable filler in realizing ferrite based shielding composites.

21.19 Multilayered Shielding Composites

Multilayered structures have proved to be effective EMI shields both for high frequency (RFI) shielding purposes as well as for low frequency magnetic shielding [16].

Advanced composite shielding materials have become increasingly important in recent years because of their significant strength and relatively light weight at the same time providing adequate EMI shielding. Typically, such materials are laminates or multilayer "layups" of individual laminae. A single lamina consists of a planar array of fibers (such as graphite, boron/boron tungstate, etc.) in an epoxy matrix together with appropriate interlayer embedding of conducting (metallic or alloy) screens/wire-mesh or interlaid fiber webs or woven fiber yarns. The arrangement of the layers can be varied to suit the strength requirement as well as to control the polarization dependability of the shielding effectiveness of the laminate in given directions. Usually 0°–90° or 0°–45°–90° layups are recommended.

The effective permittivity, permeability, and conductivity of the multilayered shields decide their shielding performance. The geometry of the multilayer layup, the electrical characteristics and volume/area fractions of individual layers cohesively control the effective electrical parameters of the composite which are also frequency dependent and largely anisotropic in nature. Therefore, modeling multilayered composites has been sparsely perfected and in many situations designing such structures is based on empirical algorithms formulated *via* experimental data. The following are examples of typical multilayered structures studied as EMI shields.

Multilayered metallic sheets (Cu + ferromagnetic alloy + Cu) sandwiched configuration: This combination has been observed to yield improvement in shielding effectiveness in comparison to using the ferromagnetic alloy alone as illustrated in Figure 21.20.

The sandwich structure has the following functional characteristics: (1) The magnetic layers provide some extent of shielding against static and/or quasistatic magnetic fields. (2) Eddy current induced in the copper sheets cancels the effects of high frequency interference. Thus, the combination of magnetic and diamagnetic materials facilitate magnetic reflection at low frequencies and conductive reflection at high frequencies.

A judicial choice of steel or a magnetic alloy and aluminum can also be considered for the sandwich structures in RF shielding systems since aluminum foil can yield RF shielding effectiveness of a medium 50 dB level (as required by NSA-73-2A specifications).

Fiber-embedded epoxy lamina plus wire-mesh layered composites: These structures can be required as alternative layups of lossy dielectrics (constituted by fiber-embedded epoxy layer and conducting screens (wire-mesh structures). EM penetration through this composite can be regarded as due to the low-pass behavior of the lossy dielectric and the high-pass behavior of the wire-mesh screens. Typical fiber-embedded layers used in advanced composites for aircraft structures are made of boron/boron tungstate in epoxy matrix or graphite (in the form of a pyrolyzed organic fiber such as polyacrylonitrile)-embedded epoxy base; and screens with 20 × 20 to 200 × 200 per inch mesh sizes made of stainless steel or aluminum have been used in the layups.

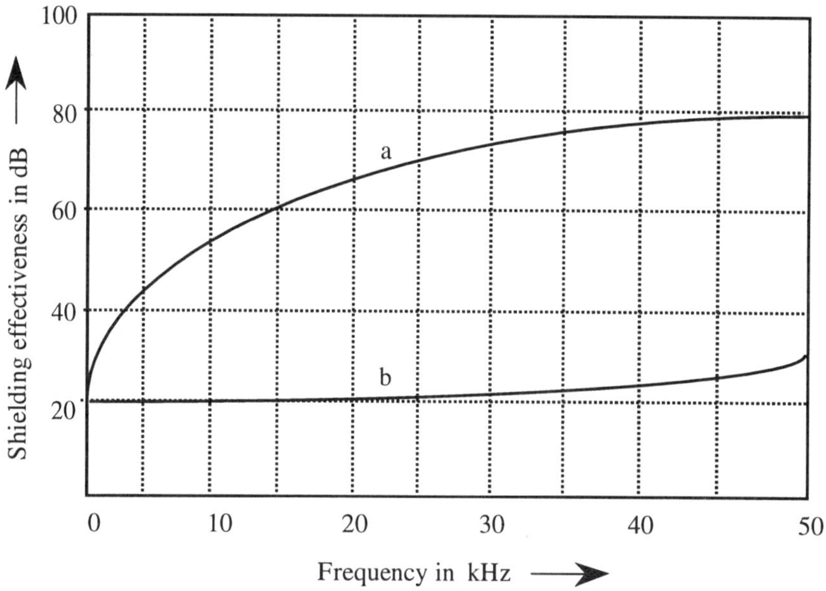

Figure 21.20 Shielding effectiveness *versus* frequency: Magnetic alloy shield sandwiched between copper sheets:
(a) Cu + 42% Ni + Cu (3.5 mil + 5 mil + 3.5 mil); (b) 42% Ni alloy (5 mil).

The shielding effectiveness of these composites has been assessed by modeling the structure as follows: The fiber-included epoxy matrix is regarded as a lossy dielectric panel and the wire-mesh screen is considered as a perforated screen. At low frequencies, an equivalent (conductivity × thickness) parameter is designated as the descriptor of the shielding performance and at higher frequencies, the mesh size becomes a critical parameter. These composites have also been analyzed for their EMP (electromagnetic pulse) shielding performance. The time-domain field transmitted through the screened composite is found to contain a contribution proportional to the derivative of the input pulse waveform as a result of the inductive component of the screen impedance. For standard EMP incident fields, the high-pass or differentiating nature of this shield leads to an early-time response (dominated by the derivative term), while the later-time response is influenced most strongly by the wire conductivity of the mesh.

The shielding performance of graphite-epoxy fiber reinforced composite (as used in modern aircraft systems) without wire-mesh layer is depicted in Figure 21.21. The shielding effectiveness in Figure 21.21 refers to the magnetic field attenuation because magnetic field penetration is much more serious in certain EMI environments.

21.20 EMI Shielding with Chiralic Media

Chiral materials are represented by the cross-coupled constitutive relations $\boldsymbol{D} = \varepsilon \boldsymbol{E} + j\xi_c \boldsymbol{B}$ and $\boldsymbol{H} = \boldsymbol{B}/\mu + j\xi_c \boldsymbol{E}$ where ***D***, ***E***, ***B***, and ***H*** are the time-harmonic electromagnetic vectors representing electric displacement, electric field intensity, magnetic flux density, and magnetic field intensity, respectively, and ξ_c is the chiral cross-coupling (admittance) coefficient. (More details on chiralic materials are presented in Chapter 25.)

Inasmuch as shielding performance depends on the reflection coefficient of the shield, the conventional approach to control the shielding effectiveness is to modify the geometry and the electrical parameters (ε, μ, σ) of the shielding medium. Relevant to this strategy, the chiralic medium offers an additional parameter ξ_c for the synthesis of a layered structure (constituting a *Salisbury shield*) with a desired shielding effectiveness. A typical example of this type of shielding is the Chiroshield™ due to Jaggard et al. as described in Chapter 25.

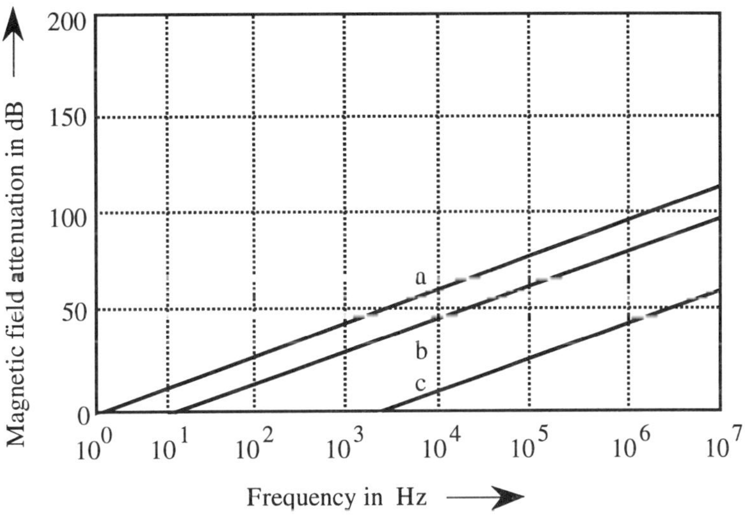

Figure 21.21 Shielding performance of graphite/epoxy composite relative to titanium and aluminum panels with number of layer – N, thickness of the layer = d, and curvature of the Nth layer is R_N.

(a) d = 0.5 mm; N = 4, graphite/epoxy; (b) d = 2 mm; N = 1, Ti; (c) d = 2 mm; N = 1, Al.

21.21 EMI Shielding *via* Active Surfaces: Concept of Smart Shielding

A class of composite materials constitutes the so-called active electromagnetic media whose electromagnetic characteristics can be altered with external stimuli. For example, a fast-ion conducting composite (see Chapter 16) is pyrosensitive so that upon thermal stimulation, it changes to a good conductor from being a dielectric medium. Appropriate incorporation of this material in a multilayered medium could facilitate realizing an active shield (with controllable performance through external stimulation). Suitable adaptive feedback can render such a medium as a smart shield changing intelligently its reflective/absorption of EM energy as and when required, posing thereby smart shielding characteristics. (See Chapter 23.)

21.22 Magnetic Shielding Materials

Magnetic field refers to the force-field setup by a current carrying solenoidal (loop) conductor source. Considering the ratio of the associated total electric field to the total magnetic field at a given point, it refers to the wave impedance which varies as a function of r/λ from the source where r is the physical distance between the source and the point of observation and λ is the wavelength of source excitation. Shown in Figure 21.22 is the wave impedance *versus* r/λ of a magnetic source.

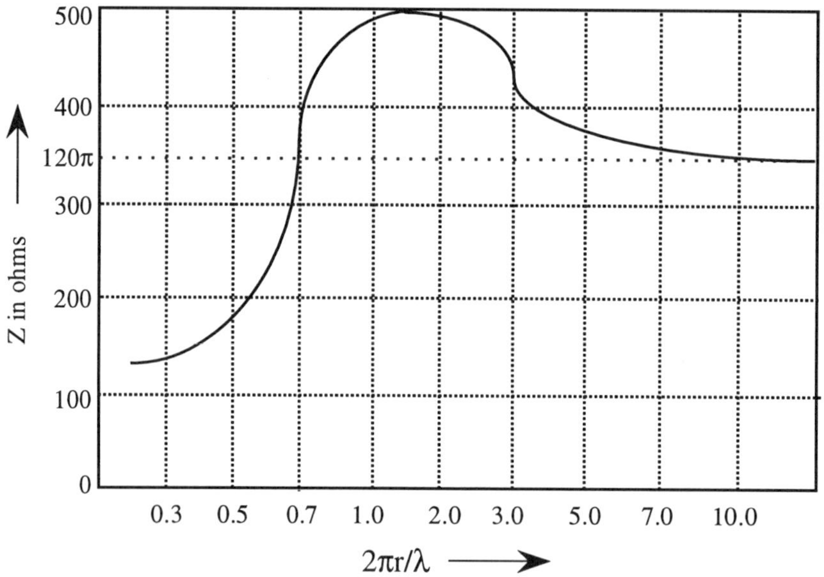

Figure 21.22 Wave impedance *versus* $2\pi r/\lambda$.

Magnetic field source exhibits low wave impedances nearby and approaches the free-space impedance (120π ohm) at large r/λ ratios.

Pertinent to this wave impedance characteristic of magnetic fields, the shielding effectiveness *(SE)* offered by a barrier of thickness d at a distance x_o and having a magnetic relative permeability μ_r is approximately given by:

$$(SE)_{dB} = 20 \log_{10}(1 + \mu_r d/x_o) \tag{21.19}$$

A more rigorous study which accounts for nonlinear and magnetic separation behavior of the barrier (shielding) medium (of thickness d and located at x_o from the source) yields the following expression for the shielding effectiveness:

$$(SE)_{dB} = -20 \log_{10}\{[4\mu_r/(\mu_r + 1)^2] \sum_{n=0}^{\infty} [(\mu_r - 1)/(\mu_r + 1)]^2/ [x + x_o/(x + x_o + 2nd)]\} \tag{21.20}$$

where x is the point of observation.

From the above formulations, it is obvious that the relative permeability and the thickness of the material dictate primarily the magnetic shielding performance. For effective shielding, ferromagnetic materials with $\mu_r >> 1$ are the right choice. However, materials with very high permeabilites tend to saturate at lower flux densities than materials with lower permeabilities.

When a material saturates, its relative permeability approaches unity. Therefore, controlling the magnetic flux level is essential for effective magnetic shielding performance. This can be done by proper choice of the thickness of the shield or by a multilayer design such that the layer nearer the source should have a very high saturation level. Therefore aluminum foil is an appropriate choice as the source-side layer of a magnetic shield design.

Typical ferromagnetic materials and the accompanying shielding effectiveness are presented in Table 21.10 for $x_o = 1$ meter and $x = d$.

Table 21.10 Shielding Effectiveness of Typical Ferromagnetic Materials

Material	Relative Permeability	$(SE)_{dB}$	Remarks/Availability
Transformer core (steel) 0.01 in. lamination	15,000	6.54	–
Mild steel (2% carbon) 0.028 in. shim	2,000	3.59	–
Conetic AA™ 0.01 in. foil	30,000	9.28	Magnetic Shield Corp.
Conetic AA™ 0.006 in. foil	30,000	7.21	Magnetic Shield Corp.

Note: See Figure 21.23; $x = d$; $x_o = 1$ meter.

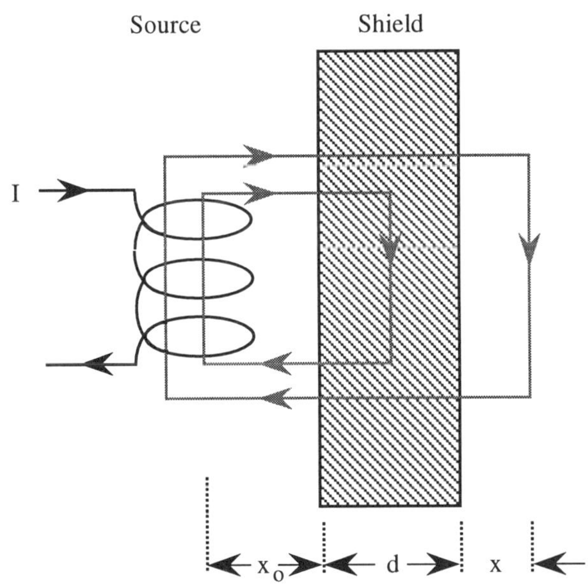

Figure 21.23 Relative dispositions of the magnetic field source with respect to the magnetic shield.

Accurate prediction of magnetic field shielding effectiveness is often hampered by the fact that the permeability of ferromagnetic materials varies nonlinearly by more than an order of magnitude as a function of the induction. For this reason, large errors are encountered in predicting the SE values unless the magnetic state inside the magnetic material is known accurately.

In designing a magnetic shield, the cost of high permeability material is often overwhelming. Thus the amount of ferromagnetic material should be minimized *via* optimum configurations.

In summary, magnetic shield design is decided by :

- High permeability characteristics
- Nonlinear variation of μ_r as a function of magnetic induction
- Saturation behavior of high permeability materials at low magnetic induction
- High cost of the magnetic materials of high permeability
- Geometry of the shield

21.23 Composite Magnetic Shields

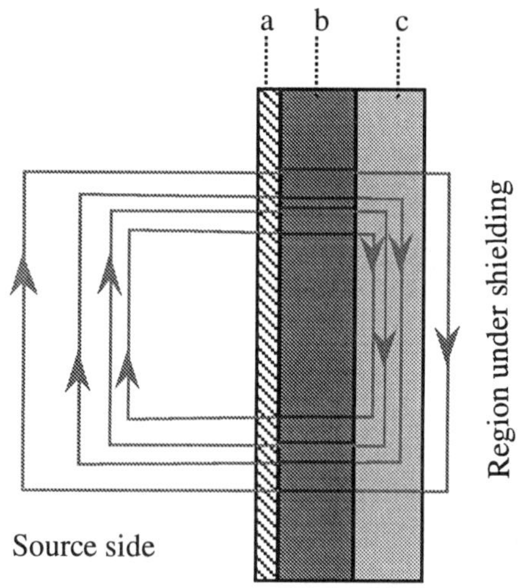

Figure 21.24 Multilayered magnetic shield.
(a) Diamagnetic (copper) or paramagnetic (aluminum) layer. (b) Low permeability steel.
(c) High permeability Ni alloy or ferrite.

The design optimization warranted to meet the above considerations has led to the development of three classes of composite magnetic shielding structures. They are:

- Multilayered metallic/alloy sheets and/or ferrite slabs
- Multiconstituent (in sheet from or otherwise) medium
- Actively compensated medium

Multilayered sheet structure: Instead of a single sheet of ferromagnetic material, if multilayers of metals/alloys/ferrites are used such that the sheet nearest to the source has the lowest permeability, the effects of saturation and nonlinearity can be reduced.

Aluminum and copper having paramagnetic and diamagnetic properties, respectively, can be the surface layer of the shield on the source side. These high conductivity materials induce eddy currents and offer a shielding barrier. Also, they provide a graded flux permeation across the multilayered structure (Figure 21.24) reducing the saturation of the high permeability inner layers.

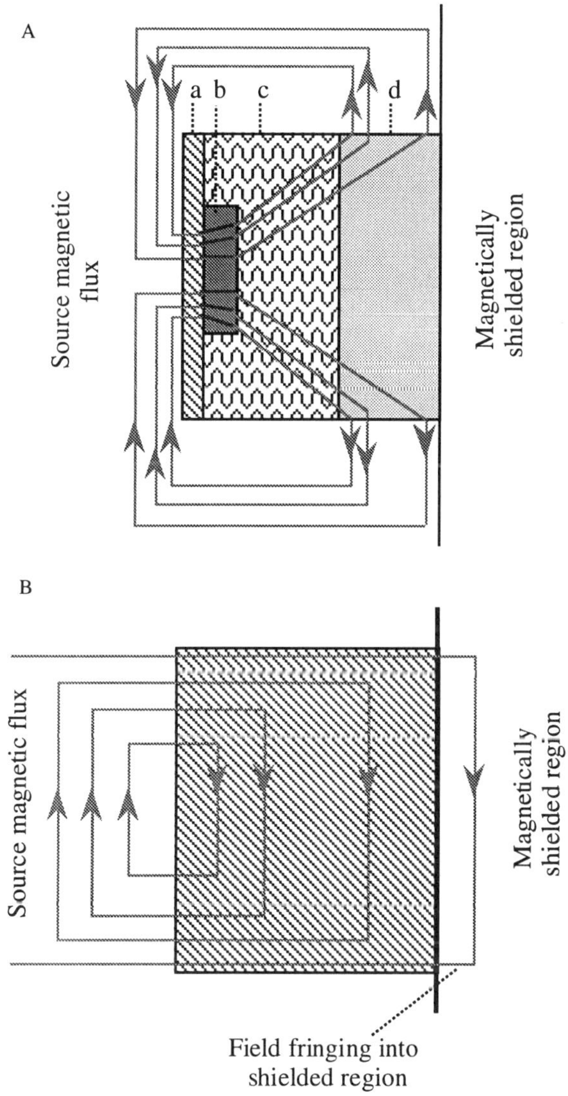

Figure 21.25 Panel-type composite, multilayered magnetic shield.
A: Composite shield; B: Convention monolithic shield.
(a) Aluminum foil; (b) High-permeability material strip; (c) Steel wool plus epoxy and ferrite oxide mixture; (d) Medium permeability steel.

Multiconstituent composite shields: Multilayered, thin-slab (flat), composite structures useful as panel-type magnetic shields are illustrated in Figures 21.25 and 21.26. The basic concept of a multilayered structure is illustrated in Figure 21.25 and 21.26. The functional aspects of each layer are as follows:

(1) **Aluminum foil on the source side**: The underlying principles of magnetic shielding indicate that the mechanism of shielding is twofold, namely, reflection loss and absorption loss, as described earlier. The reflection loss can be enhanced by providing a high conducting surface (regardless of permeability), such as aluminum (which is a high conducting paramagnetic material) as Layer 'a'.

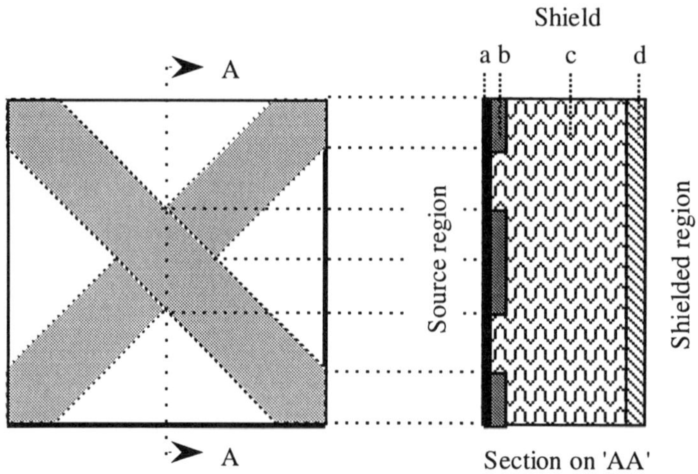

Figure 21.26 Structure of a panel-type composite magnetic shield.
(a) Aluminum foil; (b) High permeability material strips; (c) Steel wool blended with epoxy and ferric oxide; (d) Medium permeability steel.

(b) Layer 'b' is a composite medium in which two diagonal strips of high-permeability material of definite thickness (for example, 4 to 10 mil) and width are kept submerged in a lossy magnetic material composed of steel wool and a conducting polymer or epoxy (Region c).

The high-permeability metal collimates (collects) the incident magnetic field flux on the aluminum foil surface. The collected flux is "diffused" through the composite steel wool wherein magnetic absorption takes place. Different grades of this structure can be obtained by varying the width and the thickness of the diagonal high-permeability strips since the largest percentage of the cost is contributed by this layer.

(c) Layer 'd' is a medium-permeability iron/steel shim/plate (for example, of thickness 25 mil) which provides a solenoidal path to the collected flux and prevents its diffusion on the other side of the shield. This layer also serves as the base plate for the other layers. The standard size for this base plate could be 12×12 inches which would be a convenient unit size for paneling/overlay construction such as on vaults, cabinets, room walls (partial or complete), etc.

Wrap-around shielding jacket for iron/steel or PVC pipes: It is similar to flat-type structure except that this is a wrap-around jacket structure which could be exclusively designed for pipe shielding. It is compatible with heat dissipation problems associated with iron pipes enclosing high ampacity conductors. The relevant structure is depicted in Figure 21.27. It may be noted that the high-permeability material is a single peripheral strip of definite width per unit jacket length. In practice, these units can be periodically repeated as *wrap-arounds* along the pipe length.

Both structures as described above can be prefabricated in commercial applications. The user can apply them as wrap-arounds on the surface to be shielded. They are thus useful in shielding iron pipe or PVC-pipe encased transmission lines.

Though the cross-sectional/geometrical aspects of the shields for the iron and PVC pipes are identical, the basic differences are as follows:

- For the shield intended for iron pipes, the structure can have a thin base of mild steel shim/plate, just to comply with the requirements as a base support.
- On the other hand, for the shield intended for PVC pipes, this base mild steel shim/plate should have a larger thickness so as to provide a return permeable path for the flux lines

illustrated in Figure 21.27. (In the iron pipe case, the pipe itself will facilitate this requirement.)

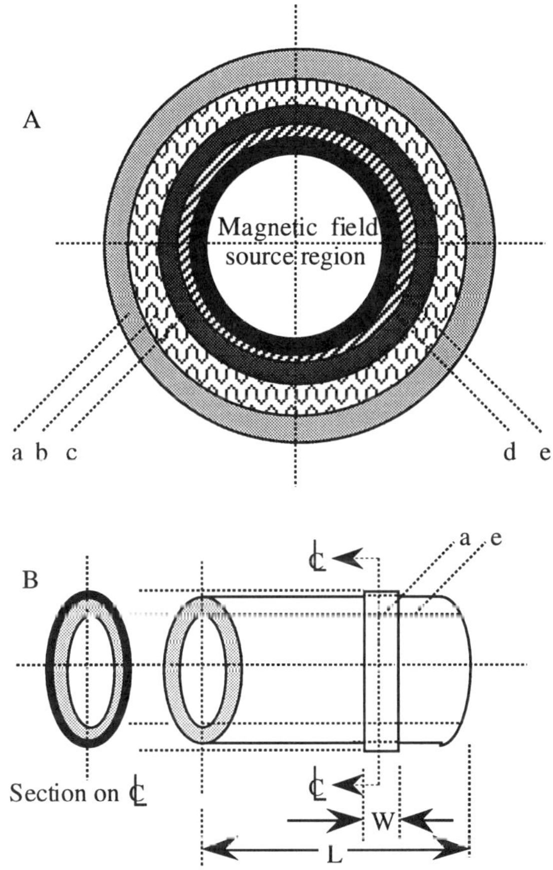

Figure 21.27 Wrap around composite magnetic shielding jacket compatible for shielding pipes.
(a) Medium permeability steel shine; (b) Steel wool plus epoxy and ferric oxide; (c) High permeability material; (d) Aluminum foil; (e) Steel or PVC pipe.

The shielding effectiveness of a simple monolithic material as described earlier is specified as the sum of absorption, reflection, and multiple reflection losses at the shield-to-air interfaces. The composite magnetic shielding structures being layers of different materials the shielding effectiveness can be predicted as follows:

(1) If the arrangement of the different layers of the composite shield is perpendicular to the direction of an unidirectional incident magnetic field, the shielding effectiveness *(SE)* can be written as :

$$SE_U = SE_1\,\theta_1 + SE_2\,\theta_2 + \ldots + SE_n\,\theta_n \qquad (21.21)$$

where SE_1, SE_2, ..., SE_n are the individual shielding effectiveness (as ratio) of the different layers and θ_1, θ_2, ..., θ_n are their respective area fractions. The method of determining the values of θ_1, θ_2, ..., θ_n depends on the specific shielding structure and is graphically represented in Figure 21.28. Since in the composite shield designated as structure I, the shielding is essentially provided by the plate of iron/steel with high-permeability strips placed on it, the *SE* is predicted by:

$$SE_U = (SE_{iron}\ \theta_1) + (SE\ \theta) \qquad (21.22)$$

where θ_1 = area fraction of iron = 1 (since it is the base plate) and θ = area fraction of high-permeability strip.

(2) Similarly, when the arrangement of the layers is parallel to the direction of the incident field, the corresponding *SE* would be given by:

$$SE_L = (1/\theta_1/SE_1 + \theta_2/SE_2 + \ldots + \theta_n/SE_n) \qquad (21.23)$$

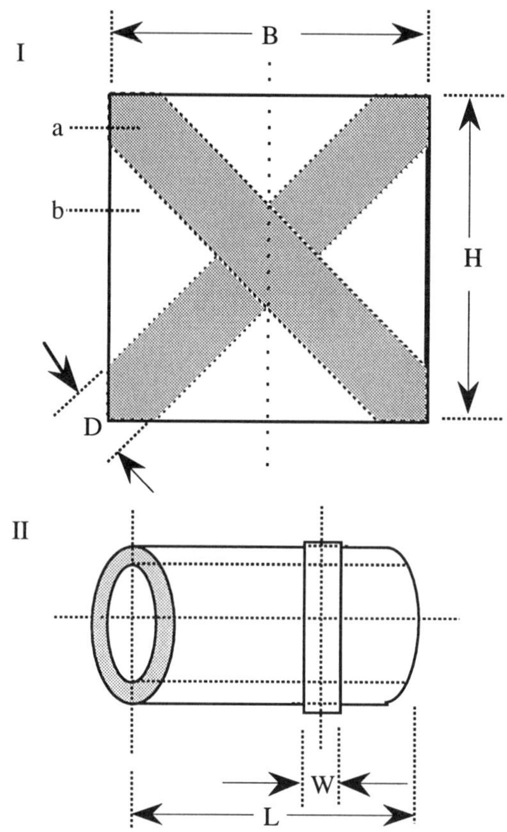

Figure 21.28 Area fraction calculations for the theoretical evaluation of magnetic shield effectiveness.
(a) High-permeability strip.
(b) Iron/steel base. If $\theta = 1$ is the area of the base plate, $\theta_1 = [2D(H^2 + B^2)^{1/2}/HB]$ is the area fraction of the high permeability material used in the flat shielding structure I. $\theta_1 = W/L$ in the wrap-around shielding structure II.

(3) In the case of an unspecified (complex) field orientations (as normally encountered in practice) incident on the shield, the *SE* could be predicted by an algorithm similar to the so-called logarithmic law of mixing (similar to that of dielectric mixture theory presented in Chapter 4), namely:

$$SE_P = \theta_1 log(SE_1) + \theta_2 log(SE_2) + ... + \theta_n log(SE_n) \quad (21.24)$$

The practical value of SE_P is bounded within the two extreme limits SE_U and SE_L corresponding to the perpendicular and parallel fields designated with respect to the shield layers.

Material selection: Essentially, the composite structure intended for magnetic shielding is composed of the following generic monolithic materials:

1. Aluminum foil
2. High permeability (high-μ) strips
3. Steel wool
4. Conducting polymer or similar epoxy adhesives
5. Ferric oxide powder
6. Mild steel shim/plates

Aluminum foil: This could be an inexpensive commercial grade/recycled aluminum foil of thickness (1 to 2 mil). No specific or special requirement is needed. Basically in large-scale manufacturing, procurement of this material should be based on cost considerations alone.

High permeability strips: This is an important constituent of the design. There are several versions available in the market. Again the selection criteria are based on the relative permeability value (on the order of 30,000 or higher) and the cost of the material. The dimensions of these strips are decided by the extent of shielding performance warranted. Examples of typical high-μ materials available are:

1. Conetic-AA™ and Netic-AA™ (Magnetic Shield Corp.), (4/10 mil)
2. Hipernorm™ (Amuneal Manufacturing Corp.)

Steel wool: This material essentially diffuses the magnetic field through the shielding structure and provides a lossy eddy current medium in conjunction with the polymeric adhesive plus ferric oxide impregnated in it. It hence provides a magnetic dissipative loss to a certain extent. A typical steel wool available commercially is International Steel Wool.

Conducting polymer/epoxy resin adhesive: This is used mainly as a bonding agent. A variety of such epoxies are commercially available. For example, Master Bond and Fiber materials Inc. are commercial sources of such materials. Again, cost considerations should be the design objective.

Ferric oxide: This medium dispersed with the bonding agent would provide partial magnetic and/or paramagnetic fluency for the magnetic flux permeation. Again, this is a low cost material available in powder form (suitable for blending with adhesives).

Mild steel shims/plates: Standard commercial grade, low-cost plates of thickness (20 to 25 mil) can be used for panel structures. For the wrap-around structures, flexible shims (4 to 10 mil) available in the market can be used.

Shielding performance: For a test structure of Figure 21.27, keeping all the dimensions as constants, the width of the high-permeability material when varied and the corresponding values of shielding effectiveness measured in each case, are presented in Figures 21.29 and 21.30 corresponding to iron pipe and PVC pipe shielding, respectively. The relative cost in each case (as a function of the strip width) is also depicted in Figures 21.29 and 21.30. In Figures 21.29 and 21.30, the test structure with a given strip width is designated as a specific grade on the basis of its relative cost factor. The theoretical estimation of the shielding effectiveness (as per Equation 21.22) is also indicated.

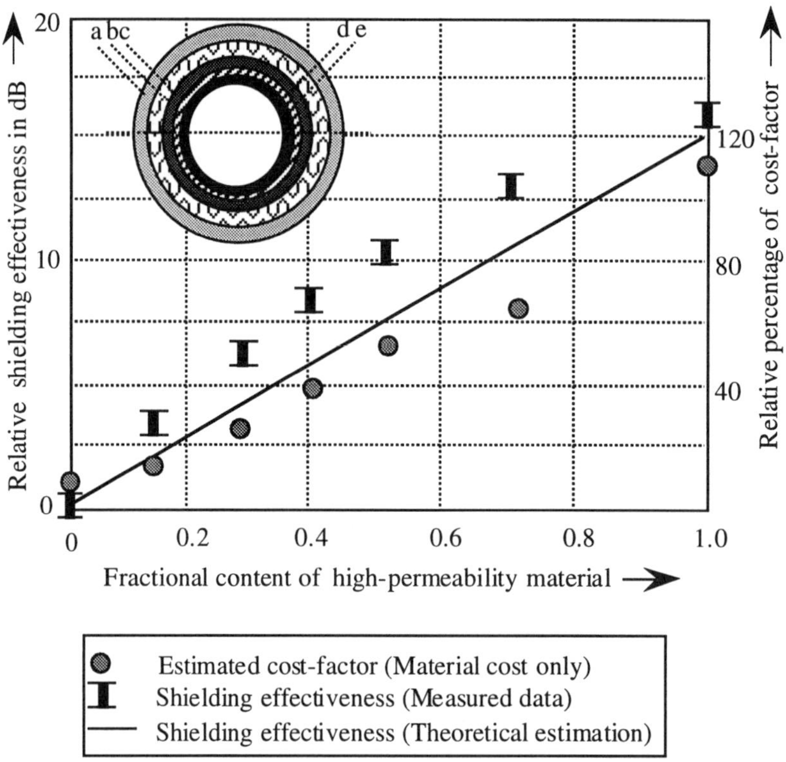

Figure 21.29 Shielding performance of a wrap-around jacket type magnetic shield on a iron pipe.
(a) Steel shim (6mm); (b) Steel wool + resin; (c) High permeability (Conetic AA™) strip (10mil); (d) Aluminum foil. (e) Iron pipe (40 mil).

These results show that a cost-effective shielding performance can be realized by controlling the width of the high-permeability material placed on a unit length of the wrap-around jacket shield. Specifically referring to Figure 21.29, it can be observed that FE composite grades 38 and 51 provide a good cost *versus* shielding compromise while in Figure 21.30 PVC composite grade 55 performs well. The unit length of the jacket shield can be fixed by considering an achievement of say, a cost factor of 55% reduction, with a width/length (W/L) ratio of 40%*.

Panel-type shields (Figure 21.25): Figure 21.31 depicts the performance test results of the panel shields with the cost factor controlled by altering the diagonal strip width of the high-permeability material used. For each cost factor, the performance grades (*SE* value) achieved are shown. Theoretical evaluation of *SE* shown corresponds to multiaxial field component formula (Equation 21.24). It can be seen that composite such as grade 45 offers the best compromise.

---

* The results presented on multiconstituent composite magnetic shields are the outcome of the research pursued by the author and have not been published in open literature yet.

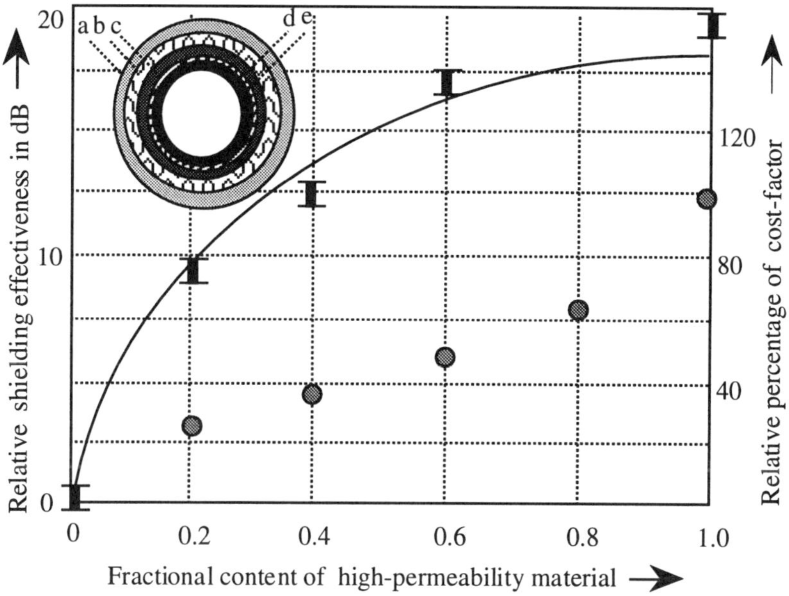

Figure 21.30 Shielding performance of a wrap-around magnetic shielding jacket for PVC pipes.
(a) Steel shim (10 mil); (b) Steel wool + resin; (c) High permeability (Conetic AA™) strip (10 mil); (d) Aluminum foil; (e) PVC tube (0.16 in. thick).

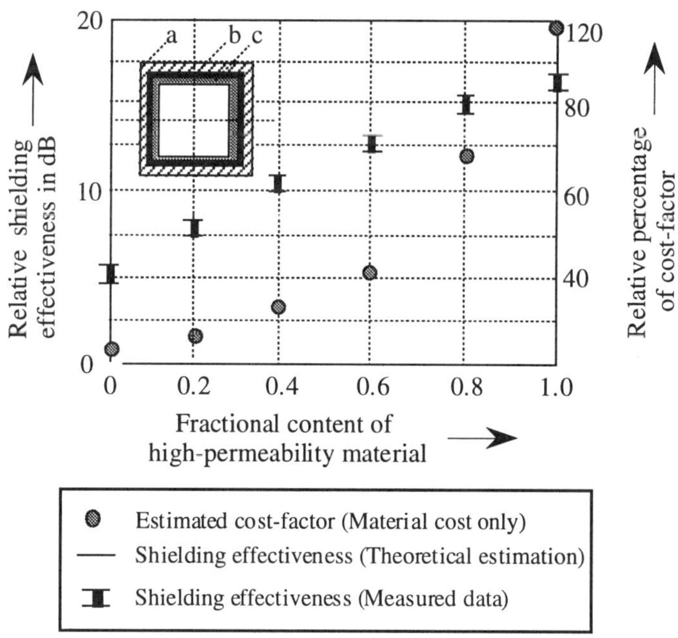

Figure 21.31 Shielding performance of panel-type shields.
(a) Steel shim 50 mil; (b) Steel wool plus resin; (c) Aluminum foil backed by high permeability (Conetic AA™) 4 mil strip.

Actively compensated magnetic shielding composites: These structures provide enhanced magnetic field shielding on the basis of *"shaking" technique* [19-21]. The technique of shaking refers to effectively increasing the permeability of a ferromagnetic material by applying a relatively strong alternating magnetic field into the material. Then the average magnetization follows an hysteresis or ideal magnetization curve, where permeability is higher. This in turn enhances shielding. The principle of shaking is illustrated in Figure 21.32. It requires shaking coils to be wound as an integral part around the shielding composite in such a manner that the generated field H_s circulates along the walls. Such active compensation against hysteresis behavior of the material would increase the effective permeability several times when biased near zero field. In case of low biasing fields, it is possible to find an optimum shaking field amplitude. The frequency of the shaking field is not, however, critical; and usually the power supply frequency (50/60 HZ) or radio frequency can be used. In designing composites which incorporate the shaking principle the following precautions should be observed:

- Shaking coil excitation should not demagnetize the shielding material
- Shaking field should not introduce interference in the region being shielded

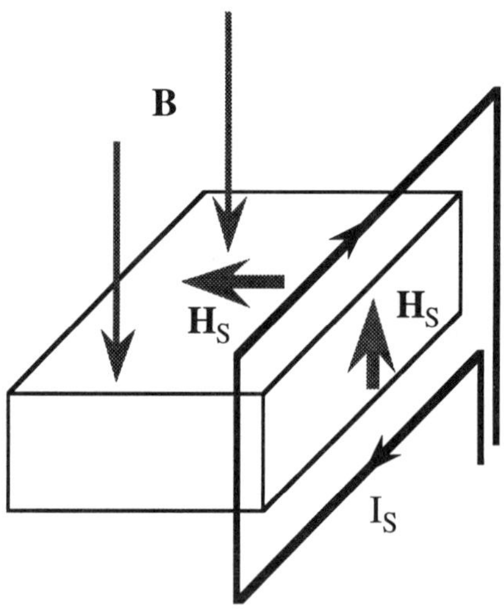

Figure 21.32 Principle of shaking.

Typical enhancement in relative permeability (μ_r) of material such as METGLAS-2705M amorphous magnetic ribbon with the application of shaking is from 3.41×10^4 to 5.27×10^5.

The principle of active shielding can be judiciously adopted to realize smart magnetic shielding inasmuch as the shaking facilitates an external mode of altering the effective permeability of the shielding composite.

21.24 Concluding Remarks

The principle of effectively using available electromagnetic materials for shielding purposes not only is governed by the electromagnetic material properties but also dictated by the geometrical outlay of the shielding media. This bifaceted requirement places stringent conditions on realizing EM shields of desirable design features, both on the choice of the materials as well as on the shielding structures. Such conditions also show new directions towards the emergence of better EM shields in the future through appropriate research.

References

[1] D. R. J. White: *A Handbook on Electromagnetic Shielding Materials and Performance*. (Don White Consultants, Inc., Gainsville, VA: 1980).

[2] B. E. Keiser: *Principles of Electromagnetic Compatibility*. (Artech House Inc., Dedham, MA: 1981).

[3] S. C. Jewell: Aluminum Foil RF Shielding Systems. Interference Technology Engineers' Master 1988, 64-71.

[4] J. Coniglio: Application considerations for conductive coatings. Interference Technology Engineers' Master, 1988, 230-238.

[5] W. Ginrup and R. R. Vinson: A logical approach to EMI shielding. Interference Technology Engineers' Master, 1986: 184-185.

[6] J. H. Ling: EMI shielding: Selection of materials for conductive coatings. Interference Technology Engineers' Master, 1986: 186-187.

[7] A. R. Henn and R. M. Cribb: Modeling the shielding effectiveness of metallized fabrics. Proc. IEEE Int. EMC Symp. (August 17-21, 1992, Anaheim, CA), pp. 283-286.

[8] J. J. Toon: Metal fibers and fabrics as shielding materials for composites, missiles and airframes. Conf. Record, IEEE Instrum. Meas. Tech. Conf., (May 14-16, 1991, Atlanta, GA), pp. 5-7.

[9] K. H. Joyner, P. R. Copeland and I. P. MacFarlane: An evaluation of a radio-frequency protective suit and electrically conductive fabrics. IEEE Trans. Elec. Mag. Compat., vol. 31(2), 1989: 129-137.

[10] A. C. Hart: Conductive fillers for EMI/RFI shielding coatings. Metal Finishing, June 1989.

[11] S. K. Bhattacharya: *Metal-filled polymers: Properties and Applications*. (Marcel Dekker, Inc., New York: 1986).

[12] J. R. Gaier: Intercalated graphite fiber composites as EMI shields in aerospace structures. IEEE Trans. Elec. Mag. Compat., vol. 34(3), 1992: 351-356.

[13] Y. Trenkler and L. E. McBride: Shielding improvement by multilayer design. 1990 IEEE EMC Symp. Records (August 21-23, 1990, Washington, D.C.), pp. 1-4.

[14] K. F. Casey: Electromagnetic shielding by advanced composite materials. Interaction Notes/Note # 341 date June 1977, Kansas State University, Manhattan, KS.

[15] M. Gunasekaren: Concrete/polymer concrete hybrid composite for EM shielding. Proc. 7th Intl. Zurich Symp. on EMC, (3-5 March, 1987, Zurich), pp. 533-535.

[16] Y. Miyazaki and K. Tanone: Electromagnetic absorption and shielding properties of lossy composite multilayers. 1990 IEEE EMC Symp. Record, (Aug. 21-23, 1991, Washington, D.C.), pp. 370-374.

[17] K. Naishadhau and P. K. Kadalea: Measurement of the microwave conductivity of a polymeric material with potential applications in absorbers and shielding. IEEE Trans. Microwave Theory and Tech. vol. 39(7), 1991: 1158-1164.

[18] V. R. Iyer, V. Ungvichian and P. S. Neelakanta: A titanium oxide based ceramic composite dispersed with conducting inclusions as an EMI shielding material. Proc. 1993 IEEE Intl. Symp. EMC (August 9-13, 1993, Dallas, TX), pp. 168-169.

[19] D. Cohen: Enhancement of ferromagnetic shielding against low frequency magnetic fields. Appl. Phys. Letts., vol. 10(3), 1967: 67-69.

[20] I. Sasada, S. Kubo, R. C. O'Handley and K. Harada: Low frequency characteristics of the enhanced incremental permeability by magnetic shaking. J. Appl. Phys., vol. 67(9), 1990: 5583-5585.

[21] V. Kelhä, J. M. Pukki, R. S. Peltonen, A. J. Pettinen, R. J. Ilmoniemi and J. J. Heino: Design, construction, and performance of a large-volume magnetic shield. IEEE Trans. Magnetics Mag., vol. 18(1), 1982: 260-268.

[22] K. Subramaniam: Application of Stochastical Mixture Theory in the Design of Electromagnetic Composite Materials. M.S.E. Thesis, Department of Electrical Engineering. Florida Atlantic University, Boca Raton, FL, April 1992.

[23] P. S. Neelakanta and D. De Groff: Smart shielding may modify performance to fit. EMC Technology, vol. 9(3), 1990: 25-29.

Defining Terms

Electromagnetic interference (EMI): Interference due to unwanted emissions of electromagnetic energy.

Electromagnetic compatibility (EMC): Refers to the extent of compatibility between an EMI source and a susceptor.

Shielding: Confining electromagnetic energy in a specified region.

Shielding effectiveness: A measure of EM shielding provided by a shielding structure expressed in dB.

CHAPTER 22

Electromagnetic Wave Absorbing Materials

22.1 Introduction

For specific applications, there is a need for materials which absorb the electromagnetic energy incident on them. These nonreflective, *electromagnetic wave* (EM) *absorbing materials* [1,2] are useful as *radar absorbing materials* (RAMs) [3,4], as surface coverings in electromagnetic *anechoic chambers,* etc. Their primary function is to absorb and dissipate the electromagnetic energy to which they are exposed so that the reflected and/or scattered electromagnetic component is significantly small. In RAM applications, when a target is surface-treated with an absorbing material, the radar echo is minimized facilitating an *electronic countermeasure* technique of low probability of detection of the target by the radar. Anechoic chambers have walls, floors, and the ceiling covered by electromagnetic wave absorbers so that any electromagnetic measurements conducted in these chambers are minimally influenced by reflected energy from the enclosures of the chambers. Electromagnetic energy absorbing materials are also candidates for the construction of *electromagnetic shields* and useful as *electromagnetic phantom* materials [5]. The general requirements and properties of electromagnetic wave absorbers are decided by the following considerations:

- Frequency of operation – whether the absorber is intended for resonance absorption (at a single or multiple, discrete frequencies) or for broadband applications
- Monolithic or composite medium – whether the absorber is constituted by a single (monolithic) material or by a set of discrete media
- Extent of absorption – as defined by the *transmission loss* through the absorbers
- Power-handling capabilities – governed by the dissipation ratings (thermal with-standability)
- Geometrical considerations – thickness and surface area to volume ratio to achieve a specified absorption level
- Mechanical stability with aging – physical and/or chemical degeneration due to continual exposure to electromagnetic radiation
- Ease of fabrication – feasibility aspects of molding or forming into required shapes and sizes
- Weight considerations such as lightweightedness in airborne/space-borne applications

22.2 Classification of EM Wave Absorbers

22.2.1. *Application-based versions*: Pertinent to various applications, the tailor-made electromagnetic wave absorbers are classified as follows:

- Radar absorbing materials (RAMs)
- Materials for anechoic chambers
- Shielding materials
- EM wave absorbers in high frequency transmission lines such as microwave plumbing, dummy loads, etc.
- EM phantom materials simulating biological tissues, bones, fluids, etc.

22.2.2. *Material-based versions*: Variability in the constituent materials of EM wave absorbers leads to the following classifications:

- Monolithic, passive, lossy materials
- Monolithic, active, lossy materials
- Composite, passive, lossy, textured media
- Composite, active, lossy, textured media
- Lossy fluids
- Lossy gel or colloidal sols or slurry
  (The lossy materials could be those which offer *ohmic loss* due to transport of particles (electrons and/or ions), or those which pose *dielectric (molecular) polarization loss* and/or those which present magnetic polarization losses.)

22.2.3. *Based on forms of the end products*:

- Thin/thick film coatings
- Impregnants for foams, etc.
- Bulk and shaped layers on conformal surfaces
- Stand-alone bulk media
- Filling media for pouches, voids, etc.

22.2.4. *Based on bandwidth of operation*:

- Single or multiple, discrete frequency resonant absorbers
- Broadband absorbers

22.2.5. *Based on frequency of operation*:

- Static (transient) state and/or quasistatic absorbers
- Radio frequency wave absorbers (HF/VHF)
- UHF wave absorbers
- Microwave/millimeter wave absorbers

22.3 Types of EM Absorbing Materials

Passive material absorbers: These are classical versions of EM wave absorbers. They are essentially high-loss materials loaded with dielectrically and/or magnetically lossy constituents and characterized by a large value of loss tangent. Polymers and epoxies with ionic impurities (being present naturally or added deliberately) pose high conductivity and hence are suitable as EM wave absorbers. For example, polymers/epoxies or paraffin wax loaded with graphite/carbon, aluminum powder, salts, etc. can be synthesized to yield a specific complex permittivity as discussed in Chapters 7 and 8. Likewise magnetically lossy ferrites which offer complex permittivity can also be used as EM absorbing materials.

One type of EM wave absorber suggested [2] is a mat of curled animal hair impregnated with a mixture of aluminum flake, graphite, or conducting carbon black in a rubber/latex solution. Examples of this type of absorbers are:

NRL Type I : It consists of two layers of curled animal hair mat each 1 inch thick, density 3.5 oz/sq. ft. The top layer is dipped in 60% graphite plus 40% neoprene made up to 20% solids in xylene. The bottom layer is dipped (3 times) in a mixture of 75% graphite plus 25% neoprene made up to 30% solids in xylene.

NRL Type II: It is made of 1 inch thick curled animal hair dipped twice in a mixture of 45% Statex A, 55% neoprene made up to 20% solids in xylene. Final density (dry): 9 oz/sq. ft.

NRL Type III : It is fabricated from 1/2 inch thick curled animal hair with dippings as in Type II. Final density (dry): 8 oz/sq. ft.

Synthetic (chemical) absorbers: Polystyrene foam or cement impregnated with fine iron, graphite, or aluminum powder can be shaped into a variety of simple EM absorbing structures. Commercial grades [4] of similar absorbers are available from different manufacturers* who use a variety of epoxies and/or resins to form magnetically and/or dielectrically loaded materials compatible for single or multiple frequency applications. Examples: Plessy series™ –(i) M-series tuned to a discrete frequency between 0.4 to 40 GHz offering 20 dB power reduction over 10% bandwidth. (ii) HP-series –These are highly magnetically loaded in the tunable range, the same as M-series, providing a peak absorption of 15 dB with higher bandwidth response. (iii) Dual-band series –These are similar to M-series except that peak absorption is exhibited at two specified frequencies.

The base elastomers in synthesizing dielectric/magnetic loaded absorbers are: Natural isoprene, neoprene, nitrile, fluoroelastomer, silicone, and urethane. Selection of these materials is based on tensile strength considerations, weathering capabilities, resistance to salt spray, exposure to oxidation and ozone resistance, tolorence to abrasions, etc. These materials are heat and pressure vulcanized to provide ultimate material and long-term stability. These base materials permit the end products to be conformed to contoured surfaces and can be bonded to metallic or nonmetallic surfaces. Properties of typical base elastomeric materials suitable for synthesizing EM absorbers are listed below [4].

Elastomers as a class in constituting an EM absorber share some basic characteristics. They are elastic, flexible, tough, and relatively impermeable to both water and air. Beyond these common characteristics each elastomer has its own properties. Properties of the finished EM absorber also depend on the magnetic and dielectric materials compounded with the rubber, since the loading of these materials varies to produce the desired electrical characteristics. Tensile strength and elongation properties of elastomers are difficult to quantify. In general, however, as loading increases the tensile strength, tear strength, and elongation decrease. Characteristics of typical elastomeric materials used in EM absorbers are as follows:

• Natural Rubber and Isoprene
Chemical name: Natural and synthetic polyisoprene.
Temperature range: –65 to 194°F.
Hardness: Shore"A" 75-90.
Advantages: Outstanding resilience; high tensile strength; superior resistance to tear and abrasion; good low temperature flexibility.
Limitations: Fair resistance to heat, ozone, and sunlight. Little resistance to oil, gasoline, and hydrocarbon solvents.
Remarks: Natural rubber has good overall physical properties. Its dielectric properties and ability to accept a high loading of magnetic material make it an excellent absorber. It is limited in its environmental resistance.

• Neoprene
Chemical Name: Polychloroprene.
Temperature range: –40 to 212°F.
Typical hardness: Shore"A" 70.
Advantages: Very good resistance to weather, ozone, and natural aging; good resistance to abrasion and flex cracking; moderate resistance to oil and gasoline.
Limitations: Fair resistance to aromatic and oxygenated solvents; limited low temperature flexibility.

---

* For example, Plessy Microwave and Emerson Cumming Inc.

Remarks: Neoprene is an excellent all-purpose elastomer. Its resistance to weathering makes it ideal for outdoor applications. Neoprene is also resistant to saltwater and stack gas, which makes it the elastomer of choice for naval applications.

• Nitrile
Chemical name: Acrylonitrile-butadiene.
Temperature range: –65 to 300°F.
Typical hardness: Shore"A" 75-80.
Advantages: Excellent resistance to oil and gasoline; superior resistance to petroleum-based hydraulic fluids; good high temperature performance; good resistance to oxidation and sunlight.
Limitations: Poor resistance to oxygenated solvents.
Remarks: Nitrile is a widely used elastomer because of its resistance to fuels and solvents. It is the elastomer of choice when used in an area subject to contact with fuels and hydraulic fluids. Nitrile also has a higher temperature capability than neoprene or natural rubber.

• Fluoroelastomer
Chemical name: Fluorinated hydrocarbon.
Temperature range: –65 to 400°F.
Typical hardness: Shore"A" 60-70.
Advantages: Excellent resistance to oil, gasoline, hydraulic fluids, and hydrocarbon solvents; very good heat resistance; very good resistance to weather, oxygen, ozone, and sunlight.
Limitations: Poor resistance to tear and cut growth.
Remarks: Plessey offers two excellent all purpose fluoroelastomers. Bondability of fluoroelastomers is more difficult than most elastomers.

• Silicone
Chemical name: Polysiloxane.
Temperature range: –80 to 700°F.
Typical hardness: Shore"A" 60-70.
Advantages: Outstanding heat resistance; excellent flexibility at low temperature; excellent resistance to weather, ozone, sunlight, and oxidation.
Limitations: Fair resistance to oil, gasoline, and solvents; poor resistance to abrasion, tear, and cut growth.
Remarks: Silicone is the elastomer of choice for both high and low temperature applications. A silicone-based adhesive is necessary to bond the material.

• Urethane
Chemical name: Polyurethane diisocyanate.
Temperature range: –90 to 290°F
Typical hardness: Shore"A" 80-85.
Advantages: Outstanding resistance to abrasion and tear; high tensile strength and elongation; good weather resistance; good resistance to oil and gasoline.
Limitations: Poor resistance to acids and alkalis; inferior resistance to hot water.
Remarks: Urethane is noted for its excellent abrasion resistance. However, it is more difficult to work with than the other elastomers and generally is not recommended, unless abrasion resistance is specifically required.

22.4 Composite EM Absorbers Structured with Passive Materials

Either for selective frequency absorption or for tailored needs of absorption characteristics, EM absorbers can be constructed as different geometries using multiple passive materials. The general principles behind such geometry-based absorbers are:

- The geometrical arrangement poses a resonant structure with selective absorption characteristics.
- The composite assembly constitutes selective transmission window(s) based on the extent of path length (or thickness of the composite) exposed to the EM wave (*Salisbury screen*).
- The absorbing composite is designed as a tapered lossy transmission line so that it offers a continuous spatial extent of absorption ("matched") path for the incident EM energy.
- The material can be designed with a balanced permeability (magnetic) and permittivity (dielectric) losses so that the structures offer effectively a matched termination to the incident radiation.

Practical aspects in implementing the aforesaid strategies to realize EM absorbers are as follows:

22.4.1 *Resonant absorbers*

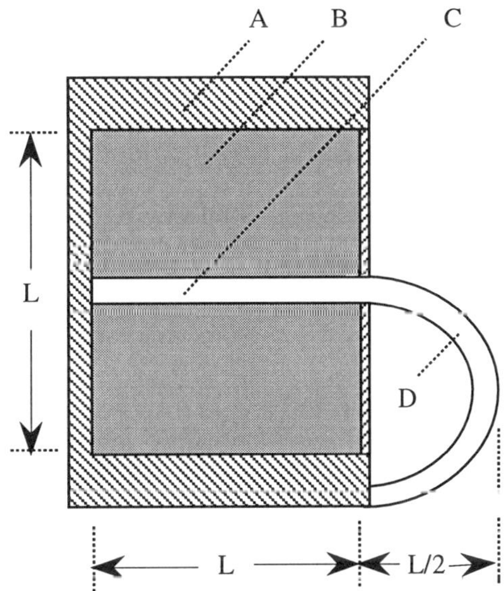

Figure 22.1 EM energy absorbing magnetic dipole.
A. Metallic cup; B. Lossy ferromagnetic material; C. Central conductor coaxial to the metallic cup; D. Conducting loop.

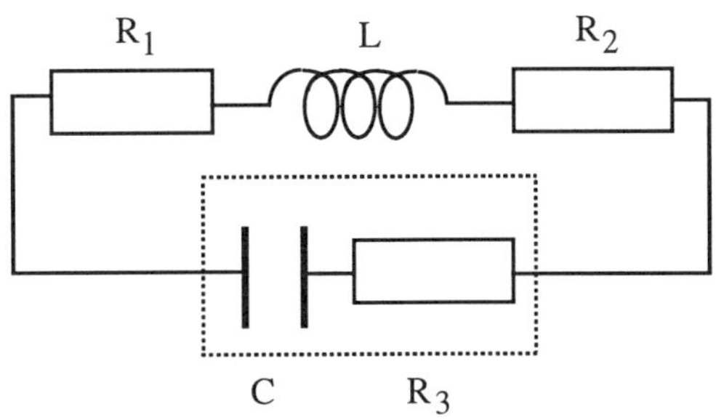

Figure 22.2 Equivalent circuit of the magnetic dipole of Figure 22.1.
R_1: Ohmic resistance; R_2: Radiation resistance; R_3: Resistance representing the power absorption by the lossy ferromagnetic material; C: Capacitance of the coaxial structure housing the lossy material.

The composite geometry can be structured as an effective resonant absorber by means of using a large number of magnetic dipoles suitably distributed on a plane metallic sheet such that the structure would exhibit pronounced reflection or absorption characteristics depending on the orientation and distribution profile of the dipoles. Meyer et al. [6] suggested such a structure with uniform distribution (two-dimensional array) of magnetic dipoles in 1954 and Chatterjee et al. [3] improvised this technique to realize a two-dimensional absorber for microwaves compatible for RCS reduction in aircraft structures. The general description of a resonant absorber structure is as follows:

Absorbing magnetic dipoles:The magnetic dipole used in the structure forms part of a coaxial guide in which the center conductor is extended outside the guide to a length less than a quarter of a wavelength and bent into the form of a loop, the free end of which is brazed to the outer conductor of the coaxial guide (Figure 22.1). The other end of the center conductor is brazed to the bottom plate of the guide. The length of the guide is made greater than a quarter wave but less than half a wavelength, so that it offers a capacitive reactance. The loop and the guide therefore form a resonant circuit. Pure grey iron powder of proper weight is placed inside the guide which is then sealed with wax to prevent moisture being absorbed by the powder. The energy picked up by the loop from the incident wave sets the circuit into resonant oscillation when the frequency of the incident wave is the same as the frequency of the equivalent circuit (Figure 22.1) formed by the loop and the guide. The resonant oscillation is damped into the dissipation of energy by the lossy iron powder. The total loss of power in the resonant circuit is due to the power dissipated in the iron powder, power lost due to radiation from the loop, and ohmic loss in the loop.

The equivalent resistance of the resonant circuit can be set approximately equal to the resistance of the iron powder only as the power lost due to the other two causes is very small compared to the amount of dissipation in the iron power.

Dipole array: A set of a large number of magnetic dipoles associated with their coaxial guides described above and illustrated in Figure 22.3 is mounted on a plane duraluminum sheet to form a two-dimensional array. The loops are backed by the metal sheet and the coaxial guides are supported at the backside of the metal plate. Suitable arrangement can be provided to orient the loops in perpendicular directions when desired. The number of dipoles mounted on the plate can be decided by the choice of spacing between the dipoles and the dimension of the individual dipole which in turn is determined by the consideration of wavelength of the incident radiation.

Absorbing material filling the dipole structure: Several materials such as grey iron powder, ferrosilicon powder, etc. are suitable as the absorbing material in the coaxial guide. For the same weight grey iron powder has been found to exhibit maximum absorption at an operating frequency of 9.4 GHz [3].

The absorption characteristics of this absorber surface have been found to depend on:

- Weight of the absorbing medium (iron powder) used in the dipole structure
- Number of dipoles per unit area
- Relative orientation of the dipoles
- Angle of incidence of the EM radiation for a given set of dipoles with a specified orientation

A minimum reflectivity of this array structure is therefore controlled by the inhomogeneity of the surface which is a function of density and distributions of the resonating elements and their associated damping resistance.

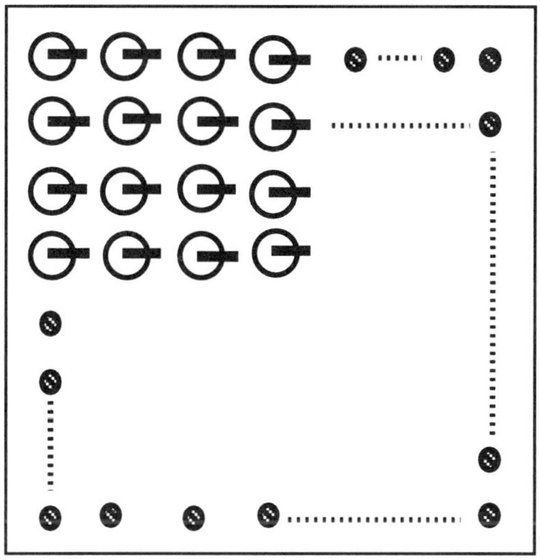

Figure 22.3 A two-dimensional matrix array of resonant, magnetic dipole EM absorbers (Figure 22.1) mounted on a metal backing.

22.4.2 *Salisbury screen*

A thin screen of EM absorbing materials (such as a glass-fiber cloth impregnated with a lossy resin plus graphite or carbon black) which provides a characteristic impedance close to that of free space (namely, $Z_o = 120\pi$ ohms) per square and placed at a distance which is an odd multiple of $\lambda_o/4$ (λ_o being the working design wavelength) from a metal surface constitutes a quarter-wave window. This quarter-wave window could provide a total absorption of the incident EM energy at λ_o for normal incidence of the EM energy (Figure 22.4).

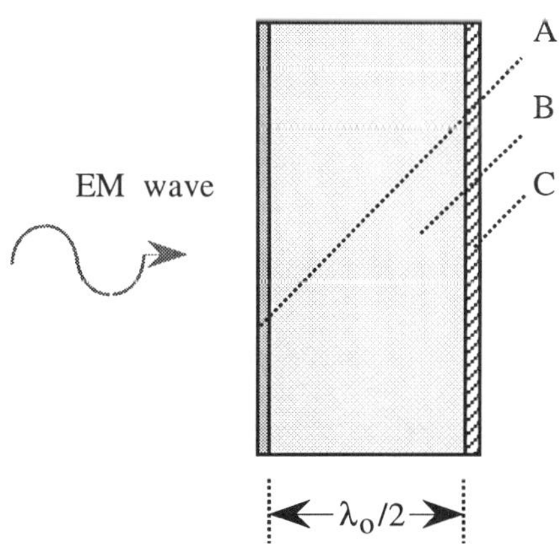

Figure 22.4 Salisbury screen.
A: Glass-fiber or cloth impregnated with a lossy material. ($Z_o \approx 120\pi$ ohm per square.)
B: Foam or spongy separator.
C: Metallic backing.

For oblique incidence at an angle θ, the corresponding power reflection coefficient $|\rho|^2$, at any wavelength λ and a screen resistance $R_s \neq Z_o$ is given by:

$$|\rho|^2 = [(1 - M)^2 + \cot^2(\pi\lambda_o\cos\theta/2\lambda)]/[(1 + M)^2 + \cot^2(\pi\lambda_o\cos\theta/2\lambda)] \quad (22.1)$$

where

$$M = 120\pi \cos\theta/R_s, \quad \text{for parallel polarization}$$
$$= 120\pi \sec\theta/R_s, \quad \text{for perpendicular polarization}$$

Equation 22.1 is valid under the following assumptions:

- Screen thickness << λ_o or λ
- Loss tangent of the screen >>1
- Interposed material between the screen and the metal (such as polyfoam, sponge etc.) has low relative permittivity

This quarter-wave structure functions on the basis of *phase cancellation* of the incident energy by out-of-phase energy that is reflected through the material by the conductive (metallic) backing. The two waves destructively interfere with each other and therefore a cancellation results. Commercial versions of this type of absorbers (such as Plessy M-series/HP series and dual-band series)™ are useful for tuned frequency absorption in the range of 0.4 to 40 GHz with typical reflectivity reductions of 25-30 dB.

These structures are loaded (in the interspace between the screen and the metal backing) with lossy dielectric and/or magnetic materials for absorption enhancement. Further, the structure can be modified for multiple frequency absorptions also.

22.4.3 *Tapered-line absorbers*

Lossy dielectric materials can be stacked up to form a thick wall wherein the characteristic impedance could vary exponentially from 120π ohms at the surface facing the free space to near-zero value at the rear end of the absorber which is normally terminated with a metallic or highly conducting surface (Figure 22.5).

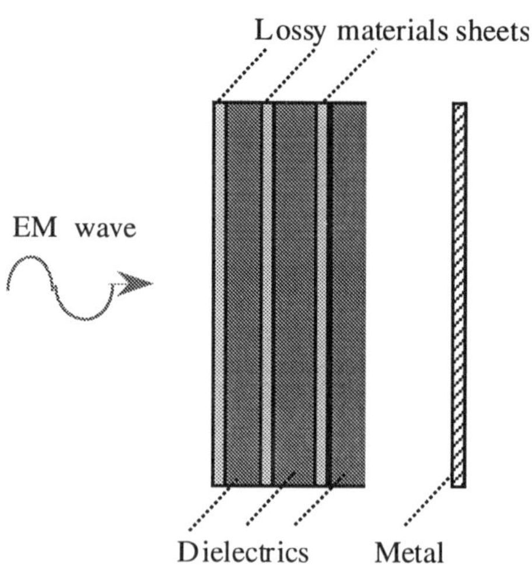

Figure 22.5 Tapered-line absorber.

The tapered-line characteristics can be emulated by a system of conical or pyramidal structures made of lossy materials forming an indented surface (Figure 22.6). The size of the cone (or the depth of indentation) is decided by the wavelength of operation.

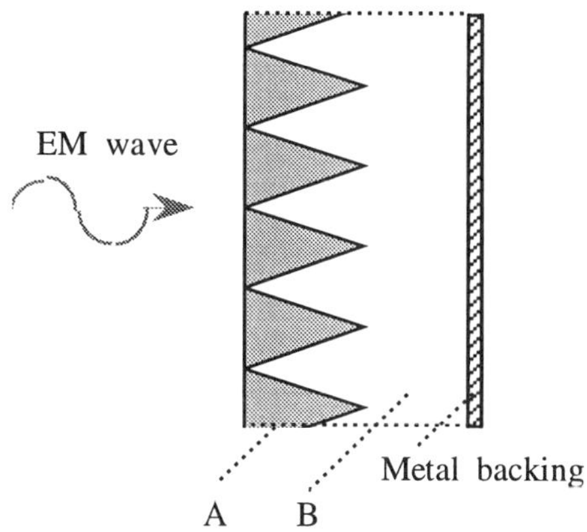

Figure 22.6: Indentated pyramidal EM absorber
A: Foam-filler mechanical rigidity and surface protection
B: Lossy dielectric (For example, polystyrene plus aluminum powder)

22.5 Applications of EM Absorbers

Radar cross-section (RCS) reduction

The concept of using radar absorbing material (RAM) to reduce the radar cross-section of vehicles was first advocated by the Germans using the quarter wavelength absorber in World War II on their U-boats. Since then, RAM technology has been a growing avenue in EM absorber evolution [4].

In modern times reduction of radar cross-section is being viewed as an integral part of the electronic *countermeasures* suite. It is important in a number of ways:

- Reducing a vehicle's "window of vulnerability"
- Decreasing the amount of *chaff* that needs to be carried and dispensed, saving total vehicle weight
- Decreasing the amount of power necessary to successfully jam an enemy's radar. This again is important in reducing total vehicle weight, as well as cost

There are two basic methods for RCS reduction [7,8]:

- Geometrical alteration of scatterers to redirect reflected energy away from the radar, that is, shaping the target surface
- Altering the materials of scatterers to make them transparent or absorptive to the incident EM energy

Though shaping is regarded generally as the first consideration, it is, however, subject to some shortcomings [7]:

- Reducing the cross-section at one aspect of the scatterers increases it at another aspect
- Shaping must be considered in the design stages and is not suitable for retrofitting existing vehicles

The other alternative is therefore the use of radar absorbing materials The type of RAM chosen is dependent on both physical and electrical constraints. Physically, consideration is given to weight, thickness, temperature, and environmental properties of the material. Electrically, considerations are given to frequency, power, and type of radiation to be absorbed.

- Specular radiation absorption is a primary consideration. Large flat areas are the usual areas of concern where strong attenuation (>20 dB) at normal angles of incidence (which constitutes specular radiation) is required.
- Corner reflectors and cavities are used in conjuction with absorption coatings. This type of situation requires an absorber that is good at off-normal angles of incidence. Because of the multibounce situation prevailing in corner reflectors and/or cavities, an absorption of 10 to 15 dB could be realized which provides, sufficient radar reduction (Figure 22.7)
- Specular radiation incident on a conducting surface may generate not only a reflection, but also a small current tangential to the conductor's surface. This condition is known as a *surface or traveling wave.* The traveling wave will continue along the conductor's surface until it strikes a discontinuity; at this point it will be reflected back. Returns from traveling waves are generally only significant at very low levels of radar signature. A thin magnetic RAM coating applied on the conducting surface is very effective in reducing returns from traveling waves.

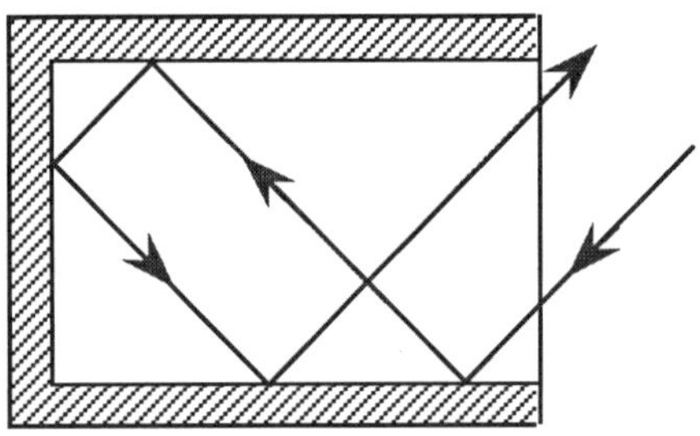

Figure 22.7 Resonant cavity type RCS reducer.

- Vehicles with very large signatures cannot use RAM to make them invisible to radar. However, the use of RAM to reduce their cross-section provides two advantages: It enhances the use of both chaff and jammers, which increases the vehicle's survivability. RAM also disguises the vehicles; that is, a very large ship may appear to be a smaller, less menacing ship to enemy radar. In this case the type of RAM desired is broadband, specular, and environmentally tough.
- Another type of radar cross-section reduction lies in the form of camouflage screens and netting. These generally provide for radar, visual, and even IR camouflage. They must be flexible, lightweight, and environmentally tough. It is also desirable for them to be as broadband as possible.

The tradeoffs considerations are the weight and thickness penalties that RAM involves. However, selective use of RAM at optimum location can pay off with RCS reduction benefits.

EM absorbers as EMI reducers

A common application of EM absorbers in naval ships is to use them at appropriate locales to prevent EMI (electromagnetic interference) emanating from the number of antennas used in the ship-borne systems. Such EMI would otherwise affect the performance of electronic systems.

EM absorbers in antenna applications

EM absorbing materials are also used to remove unwanted reflections, reduce side lobes, and enhance system performance of antennas.

EM absorbers as protective screens

Microwave absorbers can be used as shielding screens in test environments where powerful radars may cause nonionizing radiation hazards to near by personnel.

EM absorbers as bioelectromagnetic phantoms

EM absorbers can be synthesized to exhibit characteristic complex permittivity and/or permeability values which emulate biological substances such as tissue/skin/muscle, bone, and body fluids like blood and fat [5]. These materials are known as bioelectromagnetic phantoms used in evaluating the interaction between EM energy *versus* biological media. These materials are elaborated in Chapter 26.

EM absorbers as packaging materials sealable via RF heating

Some plastic materials used as thin webs in packaging media should be compatible for sealing *via* RF heating [9]. This is required when the packaged content (such as photographic material) may be damaged if sealing is done otherwise, by direct heat or infrared, etc.

Electromagnetic (RF) heating process warrants the sealable material being absorptive to EM energy. In other words, these are electromagnetic lossy materials. Such materials with enhanced electromagnetic power dissipation characteristics can be developed by "loading" plastics with conductive inclusions (see Chapters 7 and 8). Studies indicate [9] that plastics such as polyvinyl chloride or polystyrene can be mixed with particles (in the form of powders, flakes, and/or fibers) of aluminum or nickel above a certain critical volume fraction so that the end product is a conductive plastic and absorbs the EM (RF) energy to which it is exposed. The absorbed energy is dissipated as heat in a localized region facilitating sealing. The metallic loading also enables heat conduction conducive for sealing process. Normally frequencies centered on 13.56 and 27.12 MHz are used for EM excitation and these EM absorbers are useful in the thermoforming techniques.

Broadband EM absorbers

One method of realizing broadband EM absorption is to design a tuned-type absorber with low Q characteristics. Such broadbanding can be done with single peak or with dual and/or multiple resonant structures. A typical microwave absorber (such as Plessy HP series [4]™) has a specular peak absorption of 15 dB and an absorption spread of 6 to 10 dB over 6-35 GHz band. These are useful in multibounce RCS reduction techniques, traveling wave reduction efforts, and in providing a broadband insertion loss. Also such materials offer good off-normal incidence performance.

Construction-wise these materials use an elastomeric base loaded heavily with high-loss magnetic materials. Bandwidth performance is specified by ±5% around the peak (resonant) absorption frequenc(ies).

Nonresonant broadband absorbers can also be constructed with reinflated (open-cell) foam loaded with lossy (magnetic and/or dielectric) materials with conducting or metallic backing.

22.6 Design Aspects of Broadband EM Absorbers

The EM absorbers can, in general, be classified as thin, thick and band-pass types. The EM absorption characteristics of these types are as follows [10]:

22.6.1 *Electrically thin absorbers*

A thin absorber has a thickness much less than the skin depth. A lossless, conductor-backed layer of material of any thickness has the same power reflection coefficient as the conductor. A lossy thin film material supports a reflection from its front face and another from the penetrating wave that reflects off the backing conductor and returns through the front face. The two fields add as vectors; depending upon the conditions, they augment or cancel. The design task with thin absorbers is primarily *front phase cancellation*.

When a single uniform, thin layer of material (of thickness d) is placed on a conducting back plate and used as an EM absorber, its power reflection coefficient (R) is given by:

$$R = [(1 - \mu''k_od)^2 + (\mu'k_od)^2]/[(1 + \mu''k_od)^2 + (\mu'k_od)^2] \tag{22.2}$$

where $(\mu' - j\mu'')$ is the complex (relative) permeability of the absorbing material and k_o is the phase constant equal to $2\pi/\lambda_o$ with λ_o being the free-space wavelength. Minimizing R, leads to $\mu''k_0d = 1$ with $\mu'' > \mu'$. This condition, namely, $\mu'' = \lambda_o/2\pi d$ specifies that μ'' should linearly decrease with frequency over the bandwidth of interest. This can be accomplished, if the operation bandwidth lies above the permeability resonance (magnetic relaxation) frequency of the material. Further, an electrically thin absorber stipulates $(\mu\varepsilon)^{1/2} << \lambda_o/2\pi d$, that is, $\varepsilon << (\mu'')^2$. This condition is difficult to be met inasmuch as the *magnetic relaxation process* is much slower than the dielectric response of the medium.

In contrast to thin EM absorbers, the $\lambda/4$ thick absorber is largely controlled by dielectric parameters (instead of the permeability μ).

22.6.2 *Electrically thick absorber*

Thick, uniform layers operate independently of their backing; no significant amount of power returns to the front face to affect the impedance. For a normally incident plane wave, the reflection is dependent upon the permeability-to-permittivity ratio of the medium; an ideal thick absorber meets the following condition:

$$\mu = \varepsilon \tag{22.3}$$

When this condition is met, the reflection is nil with all the power absorbed in the layer itself. This condition must, however, be met over the entire frequency range of interest. Generally speaking, electric-based susceptibilities of solids are essentially independent of frequency and are small, in the range of five to fifteen. Conductive materials at optical and lower frequencies generally satisfy the following relation:

$$\varepsilon'' = \sigma/\omega\varepsilon_o >> \varepsilon' \tag{22.4}$$

where σ is the conductivity, and ω the radian frequency = $2\pi f$. Even for poor conductors, μ must generally be large to jointly meet the conditions stipulated by Equation 22.3 and Equation 22.4. Since large permeabilities decrease the skin depth, highly resistive materials such as CoFeB or CoFeZr amorphous alloys are needed. If the condition of Equation 22.3 is met and inasmuch as since ε'' generally decreases as 1/f with ε' remaining small and invariant, μ' should vary as $1/f$ with μ'' remaining constant for wideband operation.

22.6.3 *EM band-pass absorbers or windows*

EM absorbing materials can be viewed as band-pass filters. For example, copper film acts as a low pass filter; given a thickness suitable for shielding a certain frequency, it is transparent (pass) to lower frequencies, and reflective (stop) to higher frequencies. The low-pass nature of conductors can be ascertained by calculation of the transmission coefficient for a wave traveling in the x-direction incident upon a surface of any material.

With reference to Figure 22.8, the absorbing layer of thickness d is characterized by a wave impedance (normalized with respect to that of free space) given by η; and the wave number of Region 2 is taken as $k_2 = 2\pi/\lambda_2$. The corresponding transmission coefficient is given by:

$$\tau = 4\eta/[(\eta + 1)^2 e^{jk_2d} - (\eta - 1)^2 e^{-jk_2d}] \tag{22.5}$$

where τ is the transmission coefficient and the product of τ and its complex conjugate gives the power transmission coefficient. It may noted that:

$$\tau = 1 \qquad \text{for} \qquad k_2 d = 0 \tag{22.6a}$$

Also it follows that:

$$\tau = e^{-jk_2 d} \qquad \text{for} \qquad \eta = 1 \tag{22.6b}$$

For non-magnetic conductors, the relative permeability $\mu_r = \mu/\mu_o = 1$ and

$$\varepsilon_{total} = \varepsilon_o(\varepsilon' - j\sigma/\omega\varepsilon_o) \tag{22.7a}$$

where ε' is in the order of ten, $\varepsilon_o = 10^{-9}/36\pi$, and $\sigma = 6 \times 10^7$ siemen/meter for a good conductor such as copper. Therefore,

$$\varepsilon_{total} = \varepsilon_o(10 - j10^{13}/f) \tag{22.7b}$$

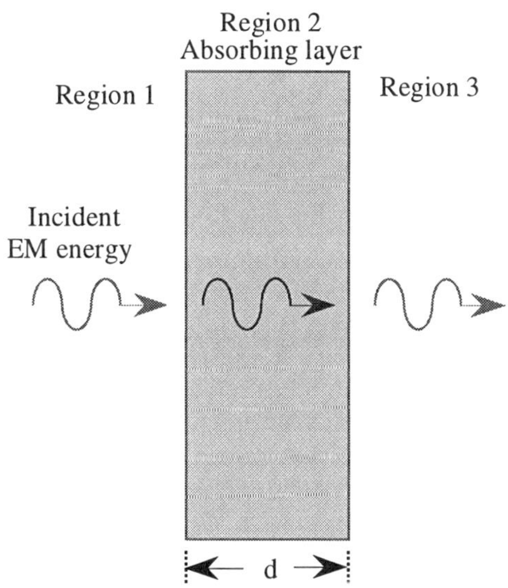

Figure 22.8 Band-pass EM absorber.

Unless f is greater than (approximately) 100 GHz, the imaginary part always dominates. Therefore,

$$\varepsilon \cong \sigma/j\omega\varepsilon_o \tag{22.8a}$$

$$\eta \cong \sqrt{j\omega\varepsilon_o/\sigma} \tag{22.8b}$$

$$k_2 d \cong \omega d/c\sqrt{\sigma/j\omega\varepsilon_o} = (1 - j)d\sqrt{\sigma\omega\mu_o/2} \tag{22.8c}$$

where c (= 3×10^8 meter/second), denote the phase velocity of the EM wave in free space. With both d and ω small and $k_2 d$ being small the above results approximate those of (Equation 22.6). If ω is large, above about 1000 GHz, $k_2 d$ is no longer small and absorption occurs. Although an exhaustive form of Equation 22.5 is needed to describe the filter action, it is nearly true that when the layer is more than a skin-depth thick, the beam is reflected and the result corresponds to a low-pass filter.

The frequency dependency of materials can also be viewed by the considerations of the skin depth. Given an electric field intensity E_s on the surface of a conductor, it travels into the conductor with a changing phase and at exponentially decreasing magnitude as:

$$E = E_S\, e^{-(1+j)x/\delta} \tag{22.9}$$

where x is the distance into the conductor and δ is the skin depth given by:

$$\delta = \sqrt{1/\pi\mu f\sigma} \tag{22.10}$$

This is, skin depth is the distancc into the conductor at which the field magnitude drops to $(1/e)$ of the surface value. Thicknesses less than the skin depth are effectively transparent and thicknesses greater than a few skin depths are effectively opaque to the penetration of EM energy.

High pass EM materials are currently receiving greater interest. These materials are optically transparent but electrically conductive (EC) enough to provide RFI shielding. Transparent EC coatings are generally semiconductive metal oxides such as the tin oxide or indium-tin oxide (ITO). The conductivity of the metal oxide films is frequency dependent undergoing a resonance at optical frequencies making them transparent. Therefore relatively thick layers can be used to provide higher levels of EM reflection characteristics. EC materials are essentially nonabsorbing. However, their controllable reflectivity can be judiciously utilized in combination with EM absorbing materials through appropriate window structures.

Very thin films of optically transparent EC metals such as copper or gold can be used to provide the control of transmission of infrared frequencies (IR). The thickness of the metal films is kept below the optical skin depth, approximately 20 nm. Although high pass metal oxide or thin metal films of equal conductivity can be used interchangeably to achieve a given level of attenuation, their optical properties are, however, quite different. For example, metal oxides are generally opaque in the ultraviolet (UV) band due to their band-gap energies. A SnO_2 crystal with a band-gap of about 3.7 eV will not transmit wavelengths below about 0.34 μm. In comparison, thin metal coatings transmit UV wavelengths down to 0.1 μm, corresponding to the plasma frequency of the charge carriers. Light transmittance is a function of the resistivity of the material and a desired electrical *versus* optical properties is a design compromise.

For IR frequencies and above, very precise high pass filters can be made using multilayer thin films. The performance of multilayer stacks are dependent upon the internal structure of the films.

An alternative to transparent EC coating for high pass EM characteristics is to use wire grids and meshes. A periodic grid of highly conductive metallic bars will, depending upon the polarization of the incident wave, transmit and reflect without diffraction wavelengths much greater than the period. Both polarizations are reflected by a mesh formed by combining two orthogonal grids. Replacing the mesh by its complementary pattern, that is, conducting squares and open wires, the transmittance and reflectance performance could be interchanged. Therefore a passband for a mesh pattern would become a stopband for the complementary pattern.

22.7 Magnetic and Dielectric Materials for EM Absorber Applications

Magnetic absorbers have losses dependent upon the magnetic field. Although magnetic metal alloys (such as Permalloy™) have very high permeabilities, since the skin depth is small the conductor attributes dominate above a few Hertz. Therefore non-conducting magnetic material is desired for EM absorption applications. Examples are ferrospinels, magnetoplumbites, ferroxplana, and ferrogarnets. Of these types, the ferrospinels and ferroxplana are particularly useful and the generic term "ferrite" is commonly applied to all of them. From 50 MHz through about 800 MHz the ferrite with ferrospinel structure is best suited; from 800 MHz through approximately 1.6 GHz a ferroxplana structure is regarded as the most suitable. In either case, the density is about 5 g/cm^3, and, depending upon specific shielding details, a thickness of 2 to 5 millimeters is needed. The result is a weight penalty of approximately 10 to 25 kg/m^2. Since maximum electric fields occur away from conductors, conductor-backing is not important for dielectric absorbers. Instead, thick layers are needed. They often contains lossy, conductor-loaded fibers enmeshed in plastic. Absorption is by I^2R power loss. In thick layers, the absorption requirement is that the internal wave impedance be nearly that of free space. Since the relative permeability of dielectric absorbers is one, it follows that the relative permittivity should be as close to one as possible. However, even low permittivity organic materials have values of ε that are too large for direct use. For example, in the MHz frequency range the permittivity of polystyrene is about 2.3. The front-face reflection from a thick layer with $\varepsilon = 2.3$ is $R = 0.16$, a number that is usually too large to be acceptable. Inserting free space into the dielectric as a dispersion reduces the effective value of ε, and hence R can be reduced. A common technique to accomplish this is to place a layer of epoxy pregnated fiberglass over a honeycomb structure. The honeycomb is made of thin pieces of an organic dielectric chosen for its structural qualities. A loss mechanism is provided by coating the hexagonal structural cells with an organic cement containing lossy conductive material such as graphite. The result is a layer whose permittivity is determined by the thickness and size of the honeycomb structure.

Nonuniform dielectric absorbers continuously reflect the incoming wave. If the material is electrically thick enough, reflections with different phases may be combined to reduce the net return. This is most conveniently done by exponentially increasing the permittivity with depth. This permits a narrow layer to be electrically long and increases the possibility of phase cancellation. Such materials are quite difficult to make and only in rare instances does the decreased thickness justify the extra cost over the use of a uniform dielectric.

In chiral media (see Chapter 25) the magnetic field generates an electric moment and *vice versa*. An example is a dielectric with embedded metallic spirals, the axes of which are parallel with the surface. A time-changing magnetic field passing through it generates a current around the spiral, and creates a magnetic dipole moment. The spiral also translates rotary currents into longitudinal ones which, in turn, create an electric dipole moment. The result is that chiral material uses the magnetic field to produce an electric loss (or *vice versa*). Therefore, chiral material is practical for use both as thin film as well as thick EM absorbers.

22.8 Ferrite Grid Absorbers

Some newer structures have been developed to widen the bandwidth of EM absorption with ferrite materials. Such structures could supress reflection ≤ -20 dB from 30 to 700 MHz, and ≤ -15 dB up to 1000 MHz [9]. Typically a ferrite fin absorber could be adopted for this purpose (Figure 22.9).

It consists of an array of sintered ferrite slabs placed on a conductor plate. For the electromagnetic wave whose electric field is polarized vertically, this structure suppresses reflection effectively.

Ferrite is typically a dispersive material. That is, its permeability is a function of frequency and can described by a relation given by:

$$\mu_r = 1 + K/(1 + j f_r/f) \tag{22.11}$$

where f is the operating frequency. For a sintered ferrite the other parameters are typically $K = 1000$ and $f_r \cong 6$ MHz. The permittivity (dielectric constant) of a ferrite remains relatively constant over the frequency. For sintered ferrites, $\varepsilon_r \cong 16$.

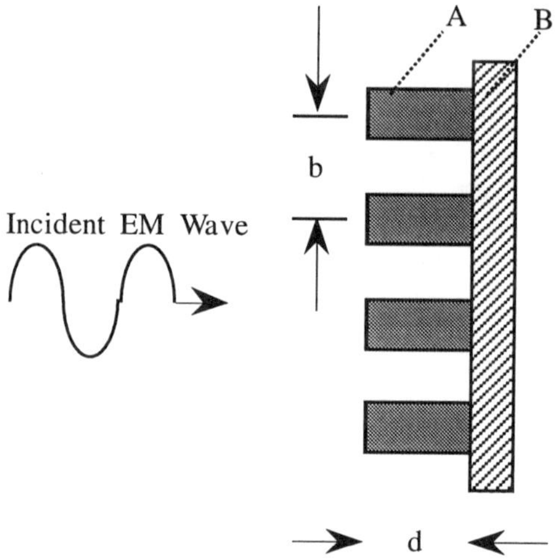

Figure 22.9 Ferrite-fin EM absorber.

In order to realize polarization independence, that is low reflection for both vertical and horizontal polarizations, a grid structure shown in Figure 22.10 can be adopted.

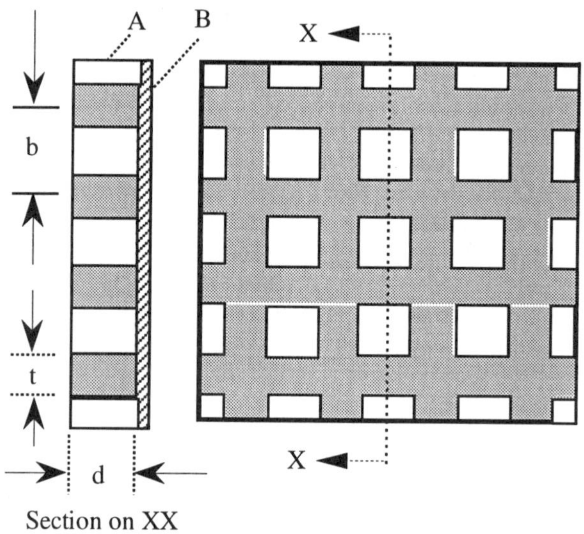

Figure 22.10 Ferrite-grid EM absorber.
A: Ferrite grid; B: Metallic backing.

It can be thought that the higher frequency limit of the EM compliance regulations will be increased in the future. To deal with this, a scheme to improve the bandwidth of grid absorber has been developed. In the case of ferrite, the single-layered absorber, its bandwidth can be increased by adopting multi-layered structures. As multi-layered structures, there are several possible candidates, such as inserting a low-permittivity layer between the ferrite and the back conductor, or attaching a pair of sintered ferrite and low-permittivity dielectric layers in front of the sintered ferrite [8]. Also, it is shown that attaching a pair of rubber ferrite and low-permittivity dielectric layers is effective to increase the bandwidth of a single-layered ferrite absorber [9,10].

Studies also have shown that inserting a low permittivity layer between the ferrite grid and the back conductor does not effectively improve the bandwidth performance. On the other hand, if a second magnetic layer is attached to the grid structure with a low-permittivity dielectric inserted in between, the wideband performance is improved even for oblique incidence of EM wave. The second magnetic layer recommended is a ferrite of appropriate thickness.

Also by attaching a single layer of lossy dielectric in front of a single-layer ferrite has been shown to offer a suppression of EM reflection at frequencies above 30 MHz. Urethane containing carbon is a compatible candidate for the lossy dielectric used for this purpose.

22.9 EM Absorption by Composites with "Active" Materials

Electromagnetic absorption by a composite material can be controlled "actively" if the material has certain "active" inclusions. Such inclusions correspond to pyrosensitive solid electrolytes [10] and *polyconductors* [11] and certain photosensitive materials.

Solid electrolytes are described in Chapter 16. These materials at a critical temperature change their phase from dielectric to conductor (β to α phase transitions). In the α-phase the conduction is essentially due to ions. It is possible to synthesize a composite material constitute by dispersing a solid electrolyte in a host medium; and, with appropriate thermal excitation, the composite material poses significant EM absorption. Chapter 16 elaborates relevant experimental studies performed by the author and reported in [10]. A variety of solid electrolytes are available which can be judiciously adopted in synthesizing EM absorbing composites. Another family of conductors known as *polyconductors* have also shown active EM absorption under thermal and/or optical excitation.

Synaptically configured arrays of photosensitive materials [12] (such as Photistor™, CMI Inc., CA) forming a matrix of optically excited nodes have been found useful as synaptic antenna for reconfigurable array applications. Use of this photosensitive material could possibly lead to a composite with photosensitive elements constituting an "active" EM absorber.

The pyro- and/or photosensitive element-based composites in general can be structured as "*smart EM absorbers*" as indicated in Chapter 23.

22.10 Concluding Remarks

Stealth technology and electronic camouflaging of modern warfare systems have given an impetus of significance to develop and comprehend a variety of EM absorbers compatible for almost the entire span of EM spectrum. Yet, the complexity of EM interaction with a wide range of natural and man-made materials is yet to be analyzed and understood before the technology EM absorbers take a leap. Hence, the relevant engineering research and development efforts would stay in the forefront in the years to come.

References

[1] A. F. Harvey: *Microwave Engineering*. (Academic Press, London, 1963), pp. 601-605.

[2] H. Jasik (Ed.): *Antenna Engineering Handbook*. (McGraw-Hill Book Co., New York, 1961), Chapter 32, pp. 35-40.

[3] S. K. Chatterjee, H. Kaushal and R. Chatterjee: A two dimensional array absorber for microwaves. J. Ind. Inst. Sci., 51(1), 1969, 103-113.

[4] Microwave Materials, Plessey Microwave, Product Catalog.

[5] G. Hartsgrove, A. Kraszenski and A. Surowiec: Simulated biological materials for electromagnetic radiation absorption studies. Bioelectromagnetics, 8(1), 1987, 29-36.

[6] E. Meyer, H. Severin and G. Umlauft: Resonazabsorber für elektromagnetische Wellen. Z. Phy., 138, 1954, 465-477.

[7] S. K. Chatterjee, P. S. Neelakantaswamy and (Mrs) R. Chatterjee: Measurement of back scattering cross-sections of metallic bodies of revolution at X-band. Instn. Engrs. (India) J., vol. ET-49, 1969: 87-93.

[8] P. S. Neelakantaswamy, D. K. Banerjee and T. Parthasarathy: Modified radar cross-section of a dielectric cylinder with conducting circumferential loop loading. Archiv. Elek. u- Übertrg., vol. 27(4), 1973: 192-193.

[9] L. R. Egan: Development of a Material with Enhanced Electromagnetic Power Dissipation Characteristics. M.S.E.E. Thesis, Rochester Institute of Technology, Rochester, NY, June 1987.

[10] C. A. Grimes: Broadband EMC absorbing materials. 1993 Int'l. Symp. EMC Record, (August 9-13, 1993, Dallas, TX), 245-249.

[11] Y. Naito, H. Anzai and T. Mizumoto: Ferrite grid electromagnetic wave absorbers. 1993 Int'l. Symp. EMC Record, (August 9-13, 1993, Dallas, TX), 254-259.

[12] P. S. Neelakanta, J. Abello and C. Gu: Microwave reflection at an active surface imbedded with fast-ion conductors. IEEE Trans. Microwave Theory Tech., 40(5), 1992, 1028-1030.

[13] M. Hamid: Polyconductor beam power sensor. Int. J. Electronics, 71(2), 1991, 363–381.

[14] R. C. Dempsey and R. M. Bevensee: The synaptic antenna for reconfigurable array applications. 1989 IEEE Antennas and Propagat. Int. Symp. Dig., 1989, pp. 760-761.

Defining Terms

Anechoic chamber: A room or a space constructed with its walls, floors, and ceilings covered with an electromagnetic absorbing material so that any incident electromagnetic energy is absorbed and very little energy is reflected.

Broadband EM absorbers: A set of EM absorbers structured to absorb EM energy over a stretch of frequency band with specified absorption bounds.

Dielectric polarization loss: Represents the loss associated with a dielectric material due to the molecular relaxation process when subjected to time-varying electromagnetic excitation.

Electromagnetic absorbing materials: A class of monolithic and/or composite materials which are designed/structured to absorb effectively the electromagnetic energy incident on them.

Electromagnetic phantom materials: Materials synthesized to emulate biological media (such as tissues, fat, bone, blood, etc.) and mimic their corresponding response to electromagnetic radiations.

Ferrite fin/grid absorbs: A fin/grid structure designed to offer EM absorption characteristics.

Magnetic polarization loss: The Power dissipation in a magnetic material due to the magnetic relaxation process at molecular level under time-varying electromagnetic excitation.

Ohmic loss: The electric power loss associated in a material due to its finite conductivity.

Radar absorbing materials (RAMs): A set of electromagnetic materials designed/structured as cladding and/or coatings on surfaces so as to absorb the electromagnetic energy emanating from a radar. It is a electronic countermeasure strategy to reduce the window of vulnerability of target (surface) being detected by a radar and forms the basis of modern stealth "technology".

Radar cross-section (RCS): The effective surface area posed by a target to a radar.

Resonant absorbers: A composite geometry which offers EM absorption at a single or multiple resonant frequencies.

Salibury screen: A structure of finite thickness and electromagnetic transmission characteristics designed to offer a characteristic window of EM absorption.

Smart EM absorbers: A class of EM absorbers facilitated with active ingredients which smartly (intelligently) alter the EM absorption characteristics depending on the input stimulus received.

CHAPTER 23

Electromagnetic Smart Materials*

23.1 Introduction

Smart materials refer to a class of materials and/or composite media having inherent intelligence together with self-adaptive capabilities to external stimuli. Also known as *intelligent materials*, they constitute a few subsets of the material family that "manifest their own functions intelligently depending on environmental changes" [1]. Electromagnetic (EM) smart materials are specific subsets of smart materials which can adaptively change their EM characteristics when an external stimulus is applied proportional to a sensed EM response.

Classically, such intelligent material systems have been conceived in the development of mechanical structures which contain their own sensors, actuators, and self-assessing computational feasibilities so as to modify their structural (elastic) behavior *via* feedback control capabilities. The relevant concepts have stemmed from an intelligent form of natural (material) systems, namely, living organisms; and hence in modern concepts smart/intelligent materials/systems are conceived as those which "mimic the life functions of sensing, actuation, control, and intelligence".

The inherency of intelligence and self-adaptable control of man-made smart materials should be "programmable" in terms of the constituent processing, microstructural characteristics, and defects so as to permit the self-conditioning to adapt in a controlled manner to various extents of stimulus.

The dividing line between *smart materials* and the so-called *intelligent structures* is not, however, distinct. In simple terms, intelligent material systems are constructed of smart materials with a dedicated, discrete set of integrated actuators, sensors, etc; and the smart materials contain largely a built-in or embedded set of distributed sensors. In general, the term "smart materials" usually connotes the structural constituent in which the discrete functions of sensing, actuation, signal processing, and control are tangibly integrated. "Intelligent structures", as an extension, are constructed with smart materials so as to respond to the environment around them in a predetermined (desired) manner.

Intelligent or smart materials which manifest their own functions intelligently *vis-a-vis* the changes in the surrounding are capable of performing, in general:

- Primary functions specifying the adaptive roles of the sensor, the effector, and processor capabilities (including the memory functions)
- Macroscopic functions which enclave the extensive or global aspects of the intelligence inherent in the materials
- Built-in social utility aspects with an instilled human-like intelligence with hyper-performance capabilities

23.2 Smart and Intelligent Structures

The framework of intelligent structures as a subset in the gamut of conventional material-based systems is illustrated in Figure 23.1. This general classification of material structures refer to [2]:

- Sensory structures "which possess sensors that enable the determination or monitoring of system states or characteristics"

---

* This chapter is largely adapted from the following author's contribution: Smart Materials, Chapter 55, *The Electrical Engineering Handbook* (R. C. Dorf, Ed.), (CRC Press, Boca Raton, FL: 1993), pp. 1173-1189.

- Adaptive structures which possess actuators to facilitate the alteration of system states or characteristics in a controlled manner
- Sensory systems which may contain sensors, but no actuators
- Adaptive systems which contain actuators, but no sensors

Referring to Figure 23.1, the intersection of sensory *versus* adaptive structures depict the controlled structures with a feedback architecture. That is, the active structure has an integrated controlled unit with sensors and/or actuators which have structural as well as control functionality. Hence, the logical subset that defines an intelligent structure is a highly integrated unit (with controlled logic, electronics, etc.) that provides the cognitive element of a distributed or a hierarchic controlled structure.

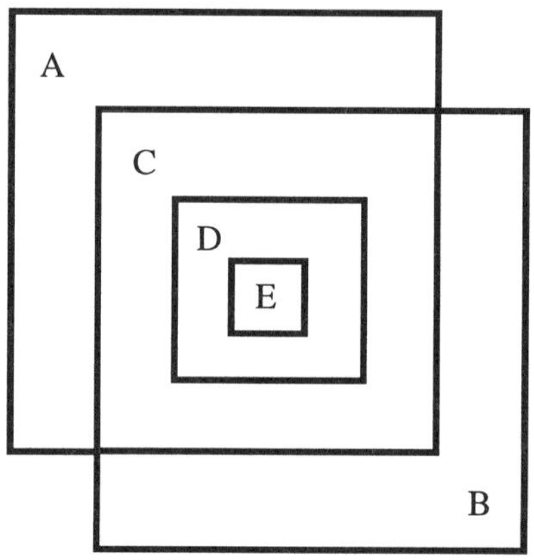

Figure 23.1 Set of structures. A. Adaptive structures; B. Sensory structures; C. Controlled structures; D. Active structures; E. Intelligent structures.

23.3 Classification of Smart/Intelligent EM Materials

Smart magnetic shielding materials: As warranted by the surroundings, the self-adaptive shielding effectiveness to magnetic fields at low frequencies (power frequencies such as 60/50 Hz) can be achieved by means of an integrated set of magnetic field sensors and actuators (magnetic biasing, current elements, etc.) plus a control system arrangement [3].

High-frequency smart shielding materials: Corresponding to radio and higher frequency environments, the shielding requirement warrants curtailing both electric and magnetic fields. Hence, the relevant self-adaptive intelligent shielding system would consist of an array of distributed electromagnetic sensors with appropriate elements (actuators) and a control system.

Smart radar absorbing materials (smart RAMs): Absorption of microwave/millimeter wave energy at radar frequencies is useful in *radar stealth* applications. Adaptively controllable smart RAMs can be synthesized with integrated distribution of electromagnetic detectors (sensors) with appropriate actuators and control system [4].

Smart optical surface materials: These can be envisioned as those in which the surface optical properties (hue, intensity, etc.) can be adaptively controlled by means of an intelligent sensor/actuator combinational control system.

Pyrosensitive smart materials: Electromagnetic active surfaces constituted by pyrosensitive inclusions have been successfully developed to manage the electromagnetic reflection and/or absorption characteristics from the active surface by means of thermal actuation of the pyrosensitive nodes embedded in the medium [4]. With the inclusion of a feedback system, "smart" operation in adaptively manipulating the active surface characteristics can be achieved.

Electroelastic smart materials: These are classical versions of smart materials using the mechanical/elastic properties of a structure which can be modified adaptively by means of an embedded distribution of such materials. *Piezoelectric materials* or their modified versions form the base materials for electroelastic smart applications.

Magnetoelastic smart materials: Applicationwise, these are smiliar in purpose to those of smart electroelastic materials. Their response, however, is magnetically wielded instead of by electric field force. *Magnetostrictive materials* are the core constituents for smart magnetoelastic applications.

23.4 Material Properties Conducive for Smart EM Applications

There are certain specific characteristics of materials which make them suitable for smart electromagnetic applications. The generic list of such properties is:

- Piezoelectric effect
- Magnetostrictive effect
- Electroplastic effect
- Electrorheological properties
- Nonlinear electrooptic properties
- Nonlinear electroacoustic properties
- Nonlinear electromagnetic properties
- Pyrosensitive properties

SMART / INTELLIGENT MATERIALS

STRUCTURAL APPLICATIONS

Electroelastic effect
(Piezoelectric effect)

Magnetoelastic effects
(Magnetostrictive effect)

Electrorheological effect

Shape-memory effect

Electroplastic effect

ACOUSTICAL APPLICATIONS

Piezoelectric effect

Magnetostrictive effect

Raman active effect

ELECTROMAGNETIC APPLICATIONS

Nonlinear feeroelectric effect

Nonlinear ferromagnetic effect

Pyrosensitive effect

OPTICAL APPLICATIONS

Nonlinear electro-optic effect
(Kerr effect & Pockel's effect)

Electrochromic effect

Figure 23.2 Application-specific classification of smart/intelligent materials.

23.4.1 Piezoelectric Effect

Piezoelectric property of a material refers to the ability to induce opposite charges at two faces (correspondingly to exhibit a voltage difference between the faces) of the material as a result of the strain due to mechanical force (either tension or compression) applied across the surfaces. This process is also reversible in the sense that a mechanical strain would be experienced in the material when subjected to opposite electric charging at the two faces by means of an applied potential.

In the event of such an applied voltage being alternating the material specimen will experience vibrations. Likewise, an applied vibration on the specimen would induce an alternating potential/charge between the two faces. The most commonly known materials which exhibit piezoelectric properties are the natural materials like quartz and a number of crystalline and polycrystalline compounds as well. More details on these materials are furnished in Chapters 13 and 14.

The strain *versus* the electric phenomenon perceived in the piezoelectric materials is dictated by a coefficient which has components referred to a set of orthogonal coordinate axes (which are correlated to standard crystallographic axes). For example, denoting the *piezoelectric coefficient* (ratio between piezoelectric strain component to applied electric field component at a constant mechanical stress or *vice versa*) as d_{mn}, the subscript n (1 to 3) refers to the three Euclidean orthogonal axes; and $m = 1$ to 6 specifies the mechanical stress-strain components. The unit for d_{mn} is meter/volt which is the same as coulomb/newton.

In the piezoelectric phenomenon, there is an electromechanical synergism expressed as a *coupling factor K* defined by K^2 which quantifies the ratio of mechanical energy converted into electric charges to the mechanical energy impressed on the material. Being a reversible process, a relevant inverse ratio is also applicable.

23.4.2 Magnetostrictive Effect

This refers to the structural strain experienced in a material subjected to a polarizing magnetic flux. A static strain of $\Delta\ell/\ell$ is produced by a d.c. polarizing magnetic flux density B_o such that $\Delta\ell/\ell = CB_o^2$ where C is a material constant expressed in (meter^4/weber^2) taking the units for B_o as weber/meter^2 (or tesla).

The magnetic stress constant (Λ) in (newton/weber) is given by $\Lambda = 2CB_oY_o$ where Y_o refers to Young's modulus of a linearly strained free bar. The coefficient (Λ) could be both positive or negative. For example, nickel contracts with increasing B whereas magnetic alloys such as 45 Permalloy™ (45% Ni + 55% Fe), and Alfer™ (13% Al, 87% Fe) exhibit positive *magnetostrictive coefficient* [5]. (For more information, see Chapter 15.)

23.4.3 Electroplastic Effect

The *electroplastic effect* (EPE) refers to the plastic deformation of metals with the application of high density electric current with an enhanced deformation rate (that persists in addition to that caused by the side effects of the current such as joule heating and the *magnetic pinch effect*). The plastic strain rate resulting from a current pulse is given by $\varepsilon_I/\varepsilon_A = \alpha J^2 \exp(\beta J)$ where ε_I is the strain rate occurring during the current pulse, ε_A is the strain rate in the absence of the current pulse, J is the current density, and α and β are material constants. Typically the EPE has been observed in zinc, neobium, titanium, etc.

23.3.4 Electrorheological Property

It is the property exhibited by certain fluids which are capable of altering their flow characteristics depending on an external applied electric field. These fluids have a fast response time, being only a few milliseconds. Once the external field is applied there is a form of progressive gelling of the fluid proportional to the applied field strength. Without the applied field the fluid flows freely. If the electrified *electrorheological (ER) fluid* is sheared by an applied force larger than a certain critical value, it flows. Below this critical value of applied shear force, the electrified fluid remains in the gel phase [6].

An electrorheological fluid requires particles (1 to 100 micro-meter in diameter) dispersed in a carrier fluid. Sometimes, a surfactant is also added to help the dispersion of particles in the fluid. The surfactant is used to prevent particle interaction which could otherwise result in a tendency for the particulates to clump together when the fluid is allowed to stand still over a stretch of time. The tendency of the particles to clump together is referred to as *settling*.

The applied electric field to perceive the electrorheological phenomenon is usually on the order of 4 kilovolt/millimeter. When the electric field is applied the positive and negative charges on the suspended particles are separated forming a dipole of charges. These dipoles then align (polarize) themselves by mutual forces of attraction and repulsion to other similar dipoles resulting in unique flow characteristics. In the absence of an electric field, there is no dipole separation of charges, and hence the fluid returns to its normal flow.

An ideal electrorheological fluid is one that has a low viscosity in the absence of an applied field and that transforms into a high viscosity gel capable of withstanding high shear stresses when the field is on. Further, it must also have a low power consumption. The first reported ER fluid consisted of finely dispersed suspensions of starch or silica gel in mineral oil nearly forty years ago. Comprehensive details on ER fluids are presented in Chapter 24.

23.4.5 Nonlinear Electrooptic Properties

In certain materials which are optically transparent when subjected to an external electric field the refractive index of the material would change. Invariably, the electric field *versus* optical effect experienced is nonlinear with the result that a time-varying electric field will modulate the refractive index and hence a phase shift is experienced by the light passing through the medium. In materials which have a central symmetry, this phenomenon is called the *Kerr effect*; in noncentrosymmetric materials, it is referred to as *Pockel's effect* [7]. See Chapter 17.

23.4.6 Nonlinear Electroacoustic Properties

Electroacoustic synergism is experienced in certain classes of materials in which the mechanical atomic vibrations are influenced by the electronic polarizability with the result the non-linear interaction between the atomic displacements *versus* the electric field would cause modulation effects resulting in the generation of new sideband frequencies. Such sidebands (labeled as *Raman frequencies*) and the response function of a *Raman active medium* has the form:

$$H(\omega) = A_1 E(\omega) + A_2\, E^2(\omega) + A_3\, E^3(\omega) + \dots \qquad (23.1)$$

23.4.7 Pyrosensitive Properties

The pyrosensitive property is governed by a class of materials known as *solid electrolytes* (Chapter 16). On thermally energizing such materials, they exhibit *superionic electric conduction* (also known as *fast-ion conduction*). With the result, the medium which is dielectric at cold conditions becomes conducting at elevated temperatures. Correspondingly, the media which are embedded with solid electrolytes show different extents of electromagnetic reflection/transmission characteristics at low and high temperatures, and hence can be manipulated thermally [4].

Typical solid electrolytes which can be adopted for such pyrosensitve applications are, for example, AgI and $RbAg_4I_5$. The materials such as β-AgI and β-alumina show increasing conductivity with increasing temperature. The compound β-AgI exhibits superionic conductivity with an abrupt transition at a temperature close to 147°C. This transition is known as the β- to α-phase transition and there are a host of other materials which exhibit this phenomenon. For example, material such as $RbAg_4I_5$ has a high electrical conductivity even at room temperature. It has also been observed that solid electrolytes provide sufficiently high electrical conductivity in the α–phase even when included in low volume fractions in a mixture with a non-solid electrolyte host [4].

23.4.8 Nonlinear Electromagnetic Properties

Basically, the nonlinear electromagnetic properties can manifest as two subsets of material characteristics, namely, nonlinear dielectric properties and nonlinear magnetic properties.

Nonlinear dielectric properties: Dielectric materials whose permittivity has a distinct dependence on the intensity of the applied electric field are referred to as active or *nonlinear dielectrics*. Such materials demonstrate very high values of permittivity (order of several thousands), pronounced dependence of dielectric parameters on the temperature and a loop of electric hysteresis under the action of an alternating voltage.

Ferroelectrics are the most typical examples of nonlinear dielectrics. Rochelle's salt (potassium sodium tartarate) was the first substance in which the nonlinearity was discovered. All ferroelectrics, however, possess the nonlinear properties only within a definite temperature range. The temperature transition points over which the ferroelectric materials gain or lose their ferroelectric properties are referred to as Curie points. The arsenates and dihydrogen phosphates of alkali metals are also examples of ferroelectric materials (Chapter 12).

Piezoelectrics also fall under the category of active dielectrics. *Electrets* which are capable of preserving an electric charge for a long period of time (hence regarded analogous to permanent magnets) exhibit highly nonlinear dielectric properties (Chapter 12 and 13).

Nonlinear magnetic properties: Ferromagnetic materials are materials in which the permanent magnetic dipoles align themselves parallel to each other. These materials have a characteristic temperature below and above which their properties differ greatly. This temperature is referred to as the Curie temperature. Above this temperature they behave as paramagnetic materials, while below it they exhibit the well-known hysteresis *B versus H* curves. Examples of such ferromagnetic materials are iron, Mu-metal™, Supermalloy™, etc. Ferrimagnetic materials are similar in their hysteresis properties to ferromagnetic materials but differ from them in that their magnetic dipoles align themselves antiparallel to each other. Ferrites are the most popular ferrimagnetic materials and they are of the greatest interest in electrical engineering applications (Chapter 15).

23.5 State-of-the-Art Smart EM Materials

23.5.1 *Piezoelectric smart materials*

These find applications primarily in intelligent structures deploying electroelastic synergism and a class of ceramics (popularly known as ferroelectric ceramics) has emerged in recent times for such applications. Typically such ceramics include the base polycrystalline piezoelectrics such as $BaTiO_3$, $CdTiO_3$, $PbZrO_3$, $PbTiO_3$, etc., formulated with various stoichiometric proportions. Another class of piezoelectric flexible composite which has the potential for smart applications is a compound consisting of $PbTiO_3$ and chloroprene rubber. A set of glass ceramic composites containing the crystalline phases of Li_2SiO_3, $Li_2Si_2O_5$, $Ba_2TiSi_2O_8$, $Ba_2TiGe_2O_8$, $Li_2B_4O_7$, etc. are also emerging materials in smart material engineering [2].

Piezoelectric smart materials can also be made from the family of polymers, namely, polyvinylidene fluoride (PVDF). The main advantage of using this polymer is that it can be formed into very thin sheets and has excellent mechanical strength combined with high sensitivity to pressure changes.

Another piezoelectric material recently developed in the NTK Research facility in Japan is a kind of rubber-based material referred to as piezoelectric rubber. This material is composed of a base material of synthetic rubber, namely, chloroben dispersed with fine particles of a popular piezoelectric ceramic, called PZT (lead zirconium titanate). Piezoelectric rubber combines the favorable properties of PZT, namely, high sensitivity, chemical inertness, linearity, and simplicity with that of the rubber base, namely, flexibility. The main drawback with piezoelectric rubber is in making an electrical contact with it. This problem has been circumvented by the development of a coaxial cable connection which is easier to use. See also Chapter 13.

23.5.2 *Magnetostrictive smart materials*

Materials with a high degree of magnetostriction are deployed in modern intelligent structures. Typically the amount of strain inducible with intelligent materials in the current state of the art is 2000 ppm. These are alloys made with iron and rare-earth materials such as terbium (Te), dysprosium (Dy), niobium (Nb), etc. A commercially known material of this category is Terfenol™ [5]. Magnetostrictive transducers for smart applications have also been developed with a certain class of metallic glass materials.

23.5.3 *Electroplastic smart materials*

Electroplastic materials are useful as smart elastic media inasmuch as the stimulus which modifies the elastic deformation is the electric current which can be controlled externally. The usefulness of these materials for smart systems under room temperature conditions is still under investigation.

23.5.4 *Electrorheological smart fluids*

Current research on electrorheological fluids is focused towards development of carrier-particle combinations which result in the desirable characteristics to achieve smart elastic behavior [6]. The earlier versions of electrorheological fluids contained adsorbed water which limited their operating temperature change (up to 80°C). Particles in the newer electrorheological fluids are, however, based on polymers, minerals, and ceramics which have a higher operating range (200°C). Also, increase in power consumption is lower with temperature increments in the recent anhydrous systems. The most commonly used carrier fluids are silicone oil, mineral oil or chlorinated paraffin which offer good insulation and compatibility for particulate dispersion.

23.5.5 *Electrooptic smart materials*

Typically, potassium dihydrogen phosphate (KDP) exhibits electrooptic behavior. Synthetic materials which have the ability to alter their refractive index (and hence the optical transmission and reflection characteristics) in the presence of an electric stimulus can be comprehended as viable smart sensor applications.

23.5.6 *Electroacoustic smart materials*

Though classically the nonlinear interaction of a vibrational (acoustic) wave and an electromagnetic wave have been studied in reference to Raman active media, relevant concepts can be exercised for smart engineering applications using those materials which exhibit a strong vibrational *versus* piezoelectric characteristics. NTK piezorubber, PZT ceramics, $LiNbO_3$, PZT with donor additives, insolvent additives, etc. are viable candidates for smart applications in addition to piezoelectric polymers.

23.5.7 *Pyrosensitive smart materials*

These are useful in realizing intelligent electromagnetic active surfaces, radar absorbing materials, electromagnetic shielding, etc. For example, it has been demonstrated in [4], that the microwave reflection characteristics at a surface of a composite medium comprised of thermally controllable, solid-electrolytic zones (made of AgI pellets) show broadband microwave absorption/reflection characteristics under elevated temperatures. This principle can be adopted in conjunction with an electromagnetic sensor to provide a controllable feedback for thermal activation of fast-ion zones reconfigurably so as to achieve smart active-surface characteristics. Exclusive for this application, depending on the temperature limited conditions, the solid electrolytes can be chosen on the basis of their α– to β–phase transition characteristics. In order to keep the cost of the system low, a mixture phase can also be adopted, in which, commensurate with the elevated temperature operation, the host medium of the mixture could be a ceramic (dielectric).

23.6 Smart Sensors

These refers to smart sensing transduction applications of electromagnetic materials. The following are relevant examples:

23.6.1 *Fiberoptic-based sensors*

The field of sensing technology has been revolutionized in the past decade by the entry of fiber optics. The properties of fiber optics which have made them suitable for communications are responsible for their being successful as sensors as well. Fiber optic sensors are of two types, namely, extrinsic and intrinsic. In the extrinsic type the fiber itself acts only as a transmitter and does no part of the sensing. In an intrinsic type, however, the fiber acts as a sensor by using one of its intrinsic properties such as induced birefringence, electrochromatism, etc. to detect a phenomenon or quantify a measurement. Relevant to smart systems, use of fiber optics in conjunction with optical (sensors) is based on changes in optical effects such as refractive index, optical absorption, luminescence, chromic properties due to alterations in the environment in which the fiber is embedded. Such alterations refer to strain or other elastic characteristics, thermal and/or electromagnetic properties as well [9]. Surfaces located with smart fiber sensors are known as smart skins.

23.6.2 *Piezoelectric-based sensors*

The most conventional form of sensing technology is that of piezoelectric materials which generate an electrical response to a stimulus. In recent times the piezoelectric materials have been improved to a large extent in mechanical strength and sensitivity. Pressure and vibration can be directly sensed as a one-to-one transduction effect resulting from elastic-to-piezoelectric effect. Bending on the other hand can be sensed *via* piezoabsorption characteristics.

23.6.3 *Magnetostriction-based sensors*

Use of metallic glass as a distributive magnetostrictive sensor has been studied. Typically, in the embedded smart sensing applications using the magnetostrictive property the magnetic field is in the submicrogauss regime and the nonlinearity associated with the hysteresis of magnetostriction provides detectable sensor signal. Pressure/force, which cause static or quasistatic magnetic fields, as well as the vibrations which induce alternating magnetic fields can be regarded as direct magnetostrictive sensor responses. In the bending mode, corresponding magnetostrictive absorption can also be sensed *via* loss in the Q-factor due to absorption losses in a magnetostrictively tunable system.

23.6.4 *Shape-memory effects-based sensors*

The latest form of sensing technology utilizes shape-memory materials, namely, Nitinol™ alloys. The Nitinol™ sensors are used for measuring strain and consist of superelastic Nitinol™ wires. The basic concept is to measure the change in resistance of a Nitinol™ wire used as the unbalanced arm of a Wheatstone bridge as a function of the strain. The desirable properties of Nitinol™ in such a sensing application are its high sensitivity and superelastic nature (which permits strains up to 6% to be accurately and repeatedly measured). The piezoelectric and Nitinol™ sensing materials can also be used for actuation applications [10].

23.6.5 *Electromagnetic-based sensors*

Smart electromagnetic sensors are simple deviations of classic electric/magnetic probes, more properly known as antennas or pickups. Depending on changes in the surroundings *vis-*

a-vis the electromagnetic characteristics, these sensors respond and yield a corresponding signal. Again, the environmental changes refer to possible alterations caused by elastic, thermal, optical, magnetic, electric, and/or chemical influences.

23.6.6 *Electroacoustic smart sensors*

These are embedded acoustic (vibration) sensors (similiar to a microphone) which adaptively yield a signal proportional to the acoustic input. Such inputs could result from changes in the alterations in the surroundings caused by elastic, thermal effects, etc.

As far as smart sensor technology is concerned, in fact all the synergistic responses and effects between the electric and nonelectric phenomena discussed before can be judiciously adopted. However, considering state-of-the-art technology and practical considerations the existing smart sensors are limited to the aforesaid versions. Future trends could, however, include other possible electric to nonelectric synergistic responses.

23.7 Examples of Intelligent/Smart Systems

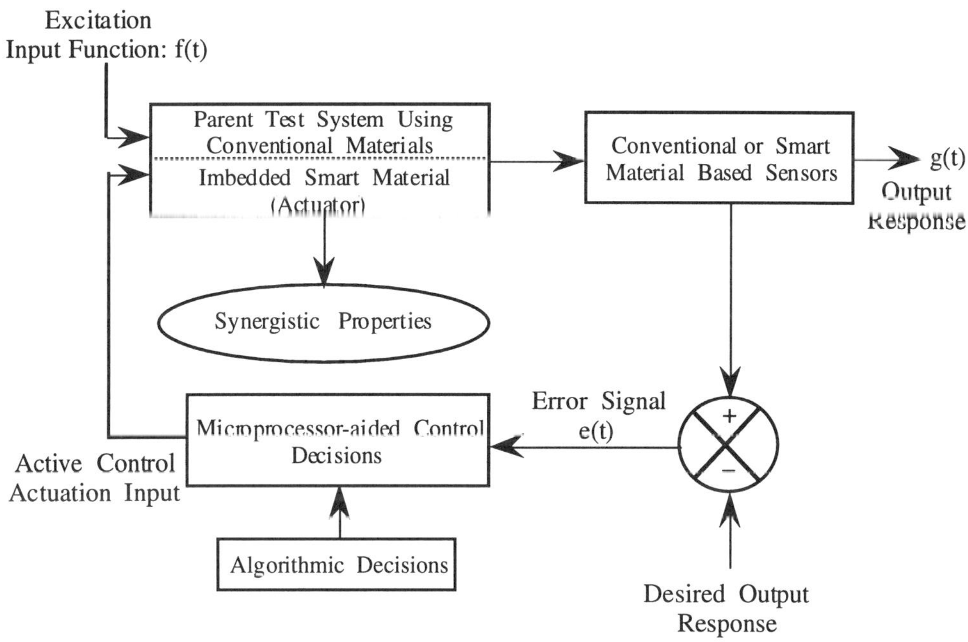

Figure 23.3 Schematic of a smart system.

The method of synthesizing a smart/intelligent system is illustrated in Figure 23.3. The output response under a given set of input condition(s) of a parent test system is normally decided by the properties of the constituent (conventional) materials. However, if the system states (changes) under the influence of external inputs are sensed, an appropriate feedback control can be used to "actuate" an embedded "smart" material in the parent unit so that output will track adaptively a desired response. The feedback path may include relevant electronic hardware (such as microprocessors) for on-line processing of the feedback signal to optimize the system performance.

Essentially, the smart materials can be adopted in two regimes of the system shown in Figure 23.3. The "sensing unit" can be zones of an integrated set of smart material which senses the response of the parent system on real time basis. (Sometimes, conventional sensors/tranducers can serve this purpose, as well.)

The "actuating unit", built-in as a part of the parent structure consists of a smart material, which upon receiving the electric signal from the feedback loop modifies the response of the parent system as dictated by the input signal. Thus, the actuation is based on the synergism between the electric input to the corresponding material property of the parent structure being altered.

The feedback control unit may consist of decision logics which can relatively modify the error-signal being fed to the actuator. The decision logic(s) refer to, for example, response linearization, time-averaged smoothing, amplitude-limiting, bandwidth control, etc. On the basis of the general schematic depicted in Figure 23.3, a few examples of application-specific intelligent systems using smart materials follow:

23.7.1. *Structural engineering applications*

Active control of vibrating beams: Illustrated in Figure 23.4, is a smart vibration control strategy in structural beams. Normally, the parent beam is made of conventional material(s) and its vibrational characteristics are decided by the elastic behavior of the constituent materials. Suppose a smart material is embedded in the test beam. This material could be one of the types indicated in Figure 23.2. A vibration sensor yields an electric output proportional to the vibration. Suppose the dynamic response of the beam (as observed at the output of the sensor) deviates from the desired characteristics. Then an error signal will be generated which in turn can be used to develop an optimal control signal; and this control signal can be fed back to the smart material whose elastic behavior is then altered as a function of the control input. As a result, the vibration characteristics of the entire (parent) structure are modified; or the system is dynamically tuned in an adaptive manner.

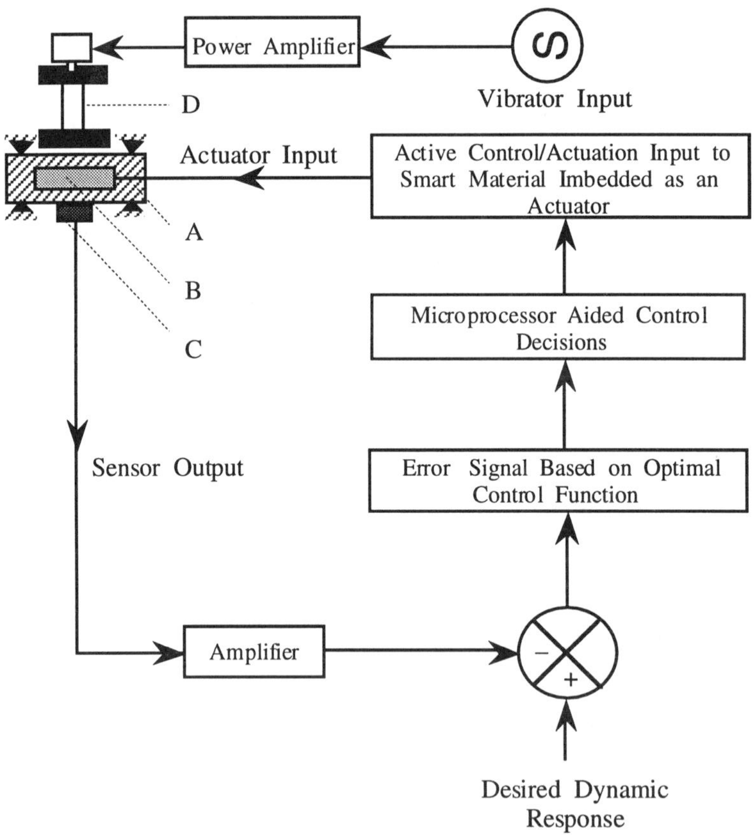

Figure 23.4 Active control of vibrating beams.

The vibration sensor used either can be a conventional transducer (such as resistive, capacitive, inductive or optical displacement versions) or it can be a smart sensor by itself.

For example, an optical fiber with a leaky sheath (which permits the light energy to leak from the core to the outside surface) can be embedded in the parent structure. When the structure is deformed, the extent of light leakage from the fiber to the surrounding will modify proportionately. Hence, the detected light signal from the fiber optics when detected delivers information on the deformation or the dynamic structural characteristics of the test beam. This sensor can be made "smart" by integrating a distributed set of fibers which can sense the strain, vibration, temperature (if needed), etc. so that the network implemented with appropriate algorithms will provide exhaustive data for a comprehensive adaptive feedback control strategy.

Though the scheme illustrated in Figure 23.4 refers to vibration control (or damping) in structures, judicious choice of subsystems, materials, etc. would also permit adaptive control of other structural aspects such as the strain, bending moment, and redistribution of load path in response to failures, etc.

23.7.2. *Electromagnetic applications*

The smart material/structural techniques can be adopted in electromagnetic systems. The following are possible applications:

- Smart low frequency magnetic shields
- Smart high frequency electromagnetic shields
- Smart electrostatic dissipative/conductive surfaces
- Smart radar absorbing materials (smart RAMs)
- Smart linear and aperture antennas

In all the above applications, the basic consideration is that the relevant structure can smartly and adaptively change its electromagnetic properties (normally specified *via* dielectric permittivity, magnetic permeability and electrical conductivity parameters) so that the desired electromagnetic performance is achieved. Two typical systems are detailed below.

Electromagnetic active surface embedded with ferroelectric inclusions

Figure 23.5 illustrates the concept. The surface is made of a mixture of polyacrylamide, ferrite and barium titanate on a ceramic substrate. This skin material which represents a lossy, nonlinear electromagnetic medium with anisotropic ferroelectric and ferromagnetic properties offers different extents of surface impedance in the presence and absence of an electric voltage stimulus applied to it. Hence, the reflection coefficient of this material to electromagnetic energy can be altered *via* electric stimulus. Relevant feedback can facilitate adaptive "smart" responsiveness of the system as illustated [4].

Smart electromagnetic aperture

The aperture radiation of microwaves can be "smartly" controlled by using a pyrosensitive material as illustrated in Figure 23.6. A set of solid-electrolyte (AgI) pellets interconnected *via* nichrome heating elements is placed at the aperture of a microwave horn. At room temperature, the pellets behave as dielectrics (β-phase AgI). However, when heated, the β-phase AgI changes to a highly conducting medium (α-phase) which would "mask" a part of the aperture, thus modifying the radiation pattern of the horn antenna. Again, an appropriate feedback loop would render the functioning of this system intelligent [4].

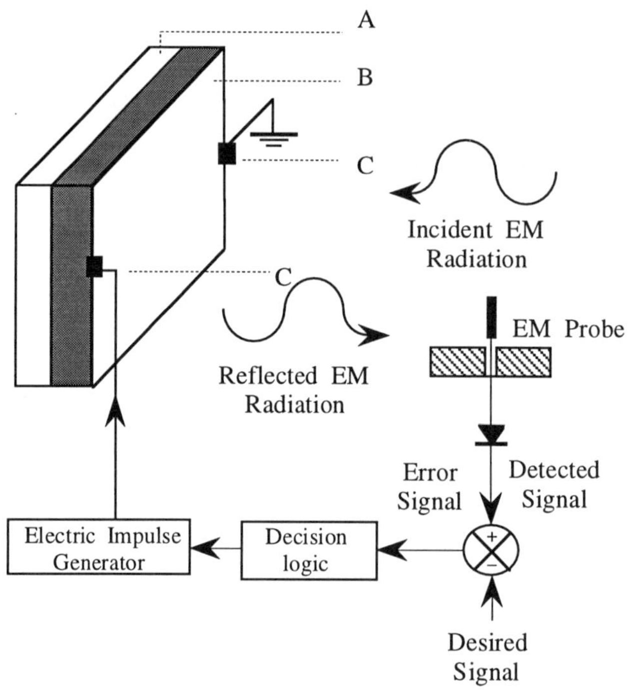

Figure 23.5 Smart EM active surface.

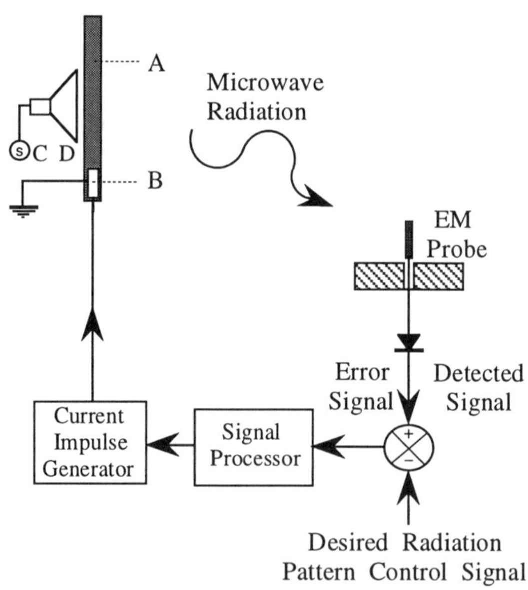

Figure 23.6 Smart electromagnetic aperture radiation control.

23.8 High-Tech Application Potentials

Though smart material technology is in its infancy pending significant efforts to make them usable on a wide-scale basis, the existing results and ongoing research have confirmed the usability of these materials in several avenues of modern high technology systems.

The use of intelligent materials currently imaginable enclaves not only structural engineering but also other areas such as electromagnetics, biomedical, optical, and biological techniques. Relevant research has also been focused heavily on aerospace, aeronautics, marine vessel, and robotic applications.

Adaptive, self-monitoring of well-being by a system which has an integrated set of smart devices to self-assess its performance, diagnosing any malfunctions/failures, and ability to change the system characteristics *vis-a-vis* the environment have been the objectives of the relevant seed-research pursued until now. For example, self-checks health by aircraft *via* a network of smart-skin sensors offer real-time monitoring of the structural well-being of tomorrow's aircraft [9]. The protocol in such efforts includes self-diagnosis, prediction/notification, and self-repair strategies relevant to mechanical structures (such as aircraft bodies).

Another domain of smart material application refers to self-induced morphologies in the infrastructure of the material with "self-adaptive" adjustments to the surroundings. Examples of this category include: Materials usable over a wide range of temperatures (as in space shuttles, etc.), with a smart adaptability to transform according to the environment. Similarly, in radar stealth applications, the target skin could offer variable electromagnetic absorption over a broadband of radar frequencies.

Extensions of smart material concepts can cover selective acoustical absorptions, adaptive chromic controls in glasses, mirrors, etc. In short, viable smart systems can be conceived with various combinations of material characteristics discussed earlier together with the advent of new conventional materials, innovative sensors, advances in microcomputers, artificial intelligence, neural networking, and other upcoming technologies. Currently imaginable "outlets" for smart materials are summarized below:

23.8.1 *Structural/mechanical engineering*

- Airborne/space-borne systems with "smart skins" for adaptive self-health check feasibilities
- Earthquake resistant intelligent buildings
- Large deployable space structures
- Nondestructive evaluation of large structures

23.8.2 *Thermal engineering*

- Adaptive heat transfers and heat-resistant structures (space shuttles, etc.)

23.8.3 *Optical engineering*

- Adaptive hue, optical transparency, reflection, opaqueness control in glasses and mirrors

23.8.4 *Electromagnetic engineering*

- Magnetic and electrostatic shielding
- High frequency shielding
- Radar absorbing materials
- Active surfaces
- Adaptive scattering/radiation control

23.8.5 *Acoustical engineering*

- Active absorption/reflection of sonar radiations
- Adaptive anechoic chambers

23.8.6 *Chemical engineering*

- Materials with adaptive adsorption characteristics
- Adaptive corrosion-resistant materials

23.8.7 *Biomedical engineering*

- Materials with "smart" structural properties usable as artificial limbs
- Materials with adaptive biochemical properties

23.8.8 *Warfare systems*

- Smart shelters
- Shock-resistant structures

23.9 Conclusions

The quest for new materials in scientific endeavors and engineering applications is everlasting. The emergence of the smart material concept has set a trend that science and technology in the coming years will rely on to a large extent in the development of exotic materials, with "intelligent materials" being the leading candidates. Such materials will be "hyperfunctional" with "unstereotyped purposeful response to novel and changing situations".

References

[1] C. A. Rogers and R. C. Rogers: *Recent Advances in Adaptive and Sensory Materials and Their Applications.* (Technomic Publishing Co., Inc., Lancaster, PA: 1992).

[2] K. P. Chong, S. C. Liu and J. C. Li (Eds): *Intelligent Structures.* (Elsevier Publishing Co., London: 1990).

[3] P. S. Neelakanta and K. Subramaniam: Controlling the properties of electromagnetic composites. Adv. Materials and Process, vol.141(3), 1992: 20-25.

[4] P. S. Neelakanta, J. Abello and C. Gu: Microwave reflection at an active surface imbedded with fast-ion conductors. IEEE Trans. Microwave Theory Tech., Vol. MTT-40(5), 1992: 28-30.

[5] R. S. Reed: Shock Isolation Using an Active Magnetostrictive Element, Proc.59th Shock and Vibration Symp. Vol. IV, (Albuquerque, NM, Oct.18-20, 1988).

[6] M. V. Gandhi and B.S.Thompson: A new generation of revolutionary ultra-advanced intelligent materials featuring electrorhealogical fluids, in *Smart Materials, Structures, and Mathematical Issues.* (Technomic Publishing Co., Inc., Lancaster, PA:1989), 63-68.

[7] I. P. Kaminow: Parametric principles in optics. IEEE Spectrum, vol. 2(4), 1965: 35-43

[8] R. Ting: The hydroacoustic behavior of piezoelectric composite materials. ferroelectrics. vol. 102, 1990: 215-224.

[9] R. O. Claus: Fiber sensors as nerves for smart materials. Photonics Spectra, Vol.25, (4), 1991: 75.

[10] C. M. Jackson, H. J. Wagner and R. J. Wasilewski: 55-Nitinol- The Alloy with a Memory: Its Physical Metallurgy, Properties and Application, NASA-SP-5110, 1972.

[11] K. Talat: Fiber sensors take wings in smart-skin applications. Photonics Spectra, Vol. 25(4), 1991: 85-88.

Defining Terms

Electroacoustic smart materials: Materials which have self-adaptive characteristics on their acoustical behavior (such as transmission, reflection and absorption of acoustical energy) in response to an external stimuli applied as a function of the sensed acoustical response.

Electromagnetic smart materials: Materials such as shielding materials, radar absorbing materials (RAMs) and electromagnetic surface materials in all of which some electromagnetic properties can be adaptively controlled by means of an external stimuli dictated by the sensed electromagnetic response.

Electrooptic smart materials: Materials in which their optical properties are changed self-adaptively with an external electric stimulus proportional to the sensed optical characteristics.

Electroplastic effect: Plastic deformation of metals with the application of high density electric current.

Electroplastic smart materials: Materials with smart properties of elastic deformation changes proportional to a controlled electric current applied in proportion to the sensed deformation.

Electrorheological property: Property exhibited by some fluids which are capable of altering their flow characteristics depending on an externally applied electric field.

Electrorheological smart fluids: Fluids with smart flow characteristics dictated to change self-adaptively by means of an electric field applied in proportion to the sensed flow parameters.

Intelligent structures: Structures constructed of smart materials with a dedicated, discrete set of integrated actuators, sensors, etc., so as to respond to the environment around them in a predetermined (desired) manner.

Magnetostrictive effect: Structural strain experienced in a material subjected to a polarizing magnetic flux, or reversibly experiencing magnetic property changes to external mechanical stresses.

Magnetostrictive smart materials: A class of materials with elastic properties self-adaptively modifiable in response to a magnetic field applied in proportion to a sensed and fed-back stress-strain information.

Nonlinear dielectric property: The distinct dependence of the electric permittivity of certain dielectric materials on the intensity of an applied electric field.

Nonlinear electrooptic property: Nonlinear changes in the refractive index of certain optically transparent materials with change(s) in the externally applied electric field.

Nonlinear magnetic property: Nonlinear dependence of the magnetic susceptibility of certain materials on the intensity of an applied magnetic field.

Piezoelectric property: Ability of a material to induce opposite charges at two faces (correspondingly to exhibit a voltage difference between the faces) of the material as a result of the strain due to a mechanical force applied across the faces; reversibly, application of a potential across the faces would induce a mechanical strain.

Piezoelectric smart materials: Materials capable of changing their elastic characteristics (by virtue of their piezoelectric property), self-adaptively in response to an externally applied electric potential proportional to the observed elastic behavior.

Pyrosensitive properties: Exhibited by materials known as solid electrolytes whose electromagnetic properties could be altered by temperature.

Pyrosensitive smart materials: Materials which manage the electromagnetic surface characteristics of active surfaces constituted by pyrosensitive inclusions self-adaptively ("smartly") in response to an external temperature-inducing stimulus applied as per the fed-back information on electromagnetic characteristics.

Shape-memory effects: Mechanism by which a plastically deformed object in the low-temperature martensitic condition regains its original shape when the external stress is removed and heat is applied.

Shape-memory smart materials: Materials which smartly change their elastic characteristics by virtue of their shape-restoration characteristics achieved by means of an external stimulus in proportion to the magnitude of sensed shape changes.

Smart (or) intelligent materials: A class of materials and/or composite media having inherent intelligence together with self-adaptive capabilities to external stimuli applied in proportion to a sensed material response.

Smart sensors: Sensors with inherent intelligence *via* bulit-in electronics.

Smart structural materials: Materials in which the mechanical (elastic) properties can be modified adaptively through the application of external stimulus.

Smart thermal materials: Materials which can influence their thermal states (temperature or thermal properties such as conductivity) self-adaptively by means of an external control in response to environmental demands.

CHAPTER 24

Electrorheological Materials

24.1 Introduction

The *electrorheological effect* refers to the abrupt change in viscosity in certain *colloidal sols* when subjected to an electric field. The phenomenon of electrorheology (ER) was first reported by W. M. Winslow in 1947 [1] and therefore, is also known as the *Winslow effect.*

An ER fluid changes its flow characteristics (due to the viscosity changes) in the presence of a high voltage, low current electric field and, if the strength of the electric field is sufficient, the fluid behaves much like a solid. The adaptive response of an ER fluid (which takes only milliseconds) is in the form of progressive gelling which is proportional to the field strength. In the absence of electric field, the fluid flows freely like water. By electrification, if the ER gel is sheared (with a sufficient force), it flows. However, if the applied shear force is below a critical value, the gel reacts as a solid with a measurable stiffness.

The ER phenomenon is conducive to many engineering applications such as nonslip fluid clutches, valves without moving parts, tunable dampers, vibration isolators, controlling the flow of liquids through narrow channels, friction devices, switching components in fluidic control devices, controlled leak seals, *smart* actuators, etc.

The direct use of electricity without intermediate transformations to control the property of the material makes the ER fluid-based devices more energy efficient than their mechanical counterparts.

Essentially, all devices employing ER effect are assemblies of electrodes of different geometry with the ER suspension being either stationary or pumped through them; and the ER fluid for engineering applications refers to a colloidal sol which consumes low electric power (upon electric energization) and withstands high shear stress. Also, it should pose a low viscosity when the electric field force is removed. To meet these requirements, the materials with ER characteristics which have emerged as successful candidates are discussed in the following section.

24.2 ER Fluids: State-of-the-Art Materials

The classical versions of ER fluids are based on mineral oils (such as kerosene) and used silica gel particles or starch with absorbed water to provide the charge separations. However, the addition of water increases the charge transfer from particle-to-particle, especially at elevated temperatures. Some studies have indicated the possibility of developing particles with no water content for ER applications and to operate at higher temperatures.

The state-of-the-art ER fluids are based on polymers, minerals, and ceramics as suspensions. Specific materials used in practice are: Silica, gypsum, clay minerals, metal oxides, cellulose, starch, Dextran™, alginic acid, ion-exchange resins and polymers (such as lithium salts of polymethacrylic acid, polyvinyl alcohol, and polystyrene containing lithium chloride). There are also a few proprietary materials which are used in the anhydrous dispersing phase and are semiconductors in nature (for example, Ge, Fe_3O_4, Fe_2O_3, ZnS, SnO, and pyrolyzed polyvinyl chloride). These materials are again only substantially anhydrous.

A comprehensive listing of ER active fluid dispersions has been compiled by Block and Kelly [2] which is reproduced in Table 24.1.

The carrier fluids must be good insulators and should be compatible with the materials they contact. The popular versions of carrier fluids are: Silicone oil, mineral oil, and chlorinated paraffin. (The halogenation makes the carrier fluid heavy and density matches with the solid phase.)

Table 24.1 Composition of ER-Active Fluid Dispersions

Dispersed Phase	Dispersant	Additive
Alginic acid	Polychlorinated biphenyls, poly(tri-fluorovinyl chloride), o-dichlorobenzene, p-chlorotoluene, xylene, plus mixtures of above	Water
Aluminum dihydrogen tripoly-phosphate	Mineral oil	Water
Aluminum oxide (as coating on aluminium)	Spindle oil or water	-
Calcium titanate	Naphthenic oil	Non-ionic sufactant
Carbon (coal)	Transformer oil, olive oil or mineral oil ("heavy oil")	Water and surfactant
Carboxymethyl dextran	Polychlorinated biphenyls, poly(trifluorovinyl chloride), o-dichlorobenzene, xylene, or mixture of above	Water and sorbitan mono-oleate or sorbitan mono-sesquioleate
Cellulose	Chlorinated insulator oil	Water
	Liquid paraffin or hydraulic oil or dibutyl sebacate or oleic acid or chlorotoluene or silicone oil	Aqueous ammonium chloride and other electrolytes
Clays-diatomite, kaolinite, montmorillonite, vermiculite, polygorskite	Kerosene, plus 1% poly(isobutene)	Water
	Vaseline oil	Water
	Hydrocarbons	Water
	Transformer oil	Diethylamine
	Lubricating oil	Water
Copper phthalo cyanine	Paraffin grease or silicone oil	None
Gelatin	Transformer oil or olive oil or mineral oil	
Gypsum	Transformer oil or olive oil or mineral oil	

(continued...)

Dispersed Phase	Dispersant	Additive
Ion exchange resins-strong and weak, acid and base, but otherwise unspecified	Di-n-butylphthalate, di-n-octylphthalate, di-n-decylphthalate, di-isodecylphthalate, tri-n-octyl trimellitate, tri-2-ethylhexyl trimellitate, tri-isodecyl trimellitate, or tri-cresyl phosphate	Water
Iron(II) oxide	Petroleum fractions or dibutyl sebacate or di-2-ethylhexyl adipate	Water and surfactant
Iron(III) oxide	Petroleum fractions or dibutyl sebacate or di-2-ethylhexyl adipate	Water and surfactant
Lime	Transformer oil or olive oil or mineral oil	
Pentaerythritol	Mineral oil or p xylene or poly(p-phenylmethyl siloxane)	Water and glycerol oleates
Phenol-formaldehyde-type ion exchange resins based on resorcinol or a-resorcylic acid or 1,5-dihydroxy naphthalene or 2,2', 4,4'-tetrahydroxy benzophenone as lithium, sodium, potassium or guanidium salts	Brominated diphenyl methanes	Water
Phthalocyanine	Silicone oil	None
Piezo-ceramic powder (unspecified)	Mineral oil or p-xylene	Water and gylcerol oleates
Poly(acene-quinone radicals) based on anthracene, naphthalene, terphenyl, ferrocene, pyrene, or phenanthren	Chlorinated hydrocarbons, liquid paraffin/paraffin grease, or silicone oils	None
Poly(acrylic acid) cross-linked with divinyl benzene as lithium salt	Chlorinated hydrocarbons or fluorolube FS-5	Water
Poly(methacrylic acid) as lithium salt	Chlorinated hydrocarbons or fluorolube FS-5	Water
	Dipolar halogenated aromatic or penta-chlorophenyl phenyl ether	Water

(continued...)

Dispersed Phase	Dispersant	Additive
Poly(methacrylic acid) cross-linked with divinyl benzene-lithium or guanidium or mixed lithium/chromium salt	Chlorinated hydrocarbons or fluorolube FS-5	Water
Poly(vinyl alcohol)	Hydrocarbons	Water
Silica	Kerosene or dibutyl sebacate	Water and soaps, sorbitol or fatty acid esters
	Naphthenic oils	Non-ionic surfactant
	Kerosene plus 1% poly(iso butene)	Water
	Hydrocarbons	Water
	Cetane	Diethylamine
	Mineral oil, xylene, or silicone oil	Water and glycerol oleates
	Petroleum distillate, transformer oil, or silicone oil	Water or water/glycerol and surfactant
Sodium carboxymethyl cellulose	Paraffin or silicone oils	Water
Sodium carboxymethyl dextran	Polychlorinated biphenyls, poly(tri-fluorovinylchloride), o-dichlorobenzene, p-chlorotoluene, xylene, mixtures, or the above	Water and sorbitan mono-oleate or sorbitan mono-sesquioleate
Starch (flour)	Mineral, transformer oil, or olive oil Petroleum spirit, or transformer oil	Water and sorbitan mono-oleate or sorbitan mono-sesquioleate
	Hydrocarbons	Water
	Vaseline oil	Water
Stone	Transformer oil, olive oil, or mineral oil	Water

(continued...)

Dispersed Phase	Dispersant	Additive
Sulfopropyl dextran	Polychlorinated biphenyls, o-dichlorobenzene, xylene, or mixtures of the above	Water and sorbitan mono-oleate or sorbitan mono-sesquioleate
Tin(II) oxide	Petroleum fractions, dibutyl sebacate, or di-2-ethylhexyl adipate White spirit-vaseline mixture	Water and surfactant
Titanium dioxide		Glycerol oleate plus a "low molecular weight polyamide" or triethanol-amine
		Water and glycerol oleate
	Mineral oils, p-xylene, or poly-(phenylmethylsiloxane) White spirit-alkyd resin mixture	

Note: Relevant references on these materials are indicated in [2].
Adapted from H. Block and J.P. Kelly, Electrorheology, *J. Phys. D* (Appl. Phys.) 21, 1663, 1988. With permission of the authors.

24.3 Application-Based Characteristics of ER Materials

In engineering applications of ER fluids used in practical devices, the following characteristics are essential:

- Overcoming the tendency of the particles to settle out if the fluid is static (stands undisturbed over a period of time)
- Less temperature-dependent characteristics of ER effect and electrical conductivity
- Low power consumption
- Adaptive response time
- Effective ER effect at low electric fields
- Critical shear force at which the ER effect is enunciated
- Noncorrosive/electrolytic action on the electrode assembly
- Adequate charge separation
- Stability *vis-a-vis* shelf-life and while in use
- Low volatility

24.4 Synthesizing ER Fluids

The basic considerations involved in synthesizing ER fluids refer to meeting the requirements as above and the following techniques are adopted in the synthesizing procedure(s):

- The particulate settling is controlled by adjusting the particle size and by using surfactants that inhibit particle interaction and the tendency to clump together.
- Another stabilization technique is to match the base-liquid density to that of the particles. However, this matching technique will limit the choice of particles and carrier fluids. Further, even if matching is achieved, it holds only for a limited temperature range

inasmuch as the thermal expansion coefficients of the particles and the fluid could differ significantly.

- Since water-activated systems are limited by the loss of ER effect and extremely high electrical conductivity at high temperatures, anhydrous systems are more appropriate in ER fluid synthesis. Studies indicate that hydrous systems are limited up to 80°C whereas anhydrous systems could operate only −10 to +120°C. Further, anhydrous systems show lower power consumption *versus* temperature characteristics.

24.5 ER Parameters

Yield stress: ER fluids are regarded a Bingham plastic in the sense that the associated flow is observed only after exceeding a minimum yield stress. The relevant equation for a *Bingham body* is:

$$\sigma = \sigma_S + \eta_{pl}\,(d\gamma/dt) \tag{24.1}$$

where σ_o is the static yield stress, η_{pl} is the plastic viscosity, and $d\gamma/dt$ is the shear rate. With the application of electric field, the yield stress increases significantly while the plastic viscosity almost remains invariant.

The flow curves are obtained with viscosimeters (such as the rotational type with concentric cylinder cone and plate, parallel plate, cone, and cone or double cone/plate arrangements). Typical flow curves of an ER fluid are shown in Figure 24.1.

The occurrence of a field stress is a distinct indicator of the presence of ER effect in the fluid.

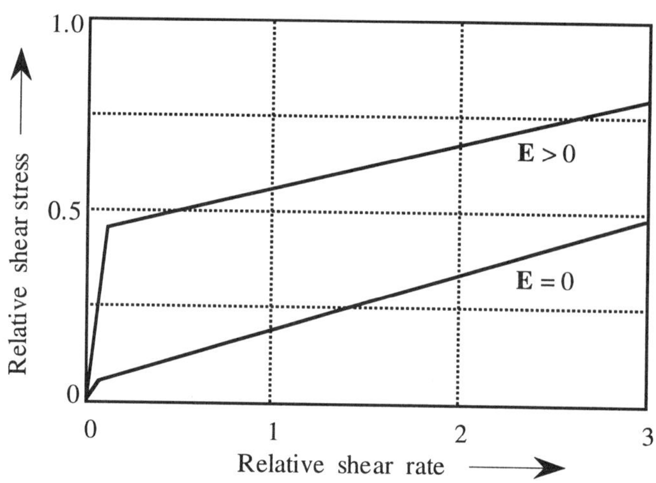

Figure 24.1 Flow characteristics of an ER fluid (**E**: Applied electric field).

Steady shear viscosity: Displacement of the fluid layer(s) at a constant velocity refers to a steady simple shear and the corresponding viscosity is defined as:

$$\eta = \sigma/\dot{\gamma} \tag{24.2}$$

where $\dot{\gamma} = d\gamma/dt$. ER fluids are *non-Newtonian* and tend to be *pseudoplastic* (shear thinning) or *dialant* (shear thickening) characterized by the apparent viscosity decreasing with an increase of the shear rate. At low shear rates, η (apparent) is proportional to $\dot{\gamma}$. Consequently, in the region of sufficiently low shear rates, ER fluids have a constant shear stress, reflecting solid-like properties in the presence of an electric field. The difference between the apparent viscosity with and without the electric field is known as the *electroviscosity* ($\Delta\eta$).

24.6 Electroviscous Property

Electroviscous effects describe the influence of charged colloidal particles on the rheology of the sols. Though sometimes synonymously used to refer to ER effect, it should be noted that ER effect is distinctly attributed to nonaqueous suspensions subjected to external electric fields. On the contrary, electroviscous effect may refer to any charged particles in a colloidal solution without external electric field.

The viscosity of the suspension η (with no electric charge) can be specified by

$$\eta = \eta_o \exp\{5\phi/2[1 - (\phi/\phi_{max})]\} \tag{24.3}$$

where η_o is the viscosity of the dispersing medium and ϕ is the volume fraction of the dispersed particles; and ϕ_{max} refers to the maximum volume fraction of the dispersed particles. (For example, for monodisperse spheres $\phi_{max} = 0.74$.)

The above expression (Equation 24.3) also accounts for particular crowding in concentrated suspensions. With the effect of electric charge on the particles, there is an increase in the viscosity of the primary electroviscous effect quantified by:

$$\eta = \eta_0\{1 + 5\phi/2[1 + (\zeta\varepsilon/2\pi)^2/\sigma\eta_0 a^2]\} \tag{24.4}$$

where σ is the particulate radius, ζ is the *zeta-potential*, and ε is the permittivity of the dispersing medium. The *zeta-potential* is the potential at the surface of shear of separation between two layers existing on the charged particulates. The first layer is a distinct monolayer about an ion thick closely adhering to the solid surface. The other one is a diffused thick layer extending some distance into the dispersing phase.

Pertinent to concentrated suspensions of charged particles, a *secondary electroviscous effect* is normally observed which refers to the enhancement of viscosity. Due to such excessive concentration, interparticulate interaction and the greater dissipation of shear energy due to repulsion and overlapping of double layers cannot be ignored.

The *tertiary electroviscous effect* describes the behavior of polyelectrolytes. That is, if ions are introduced into the polymer chains, the electrical repulsion energy increases the total size of the random coil. This would enhance the hydrodynamic resistance and hence, the viscosity.

24.7 Viscoelastic Properties

Viscoelastic materials in general produce a stress with components both in phase and out of phase when subjected to deformation. Therefore, they can be described by a complex dynamic modulus given by:

$$G^* = (G' + jG'') \tag{24.5}$$

where G' is the storage (in-phase) modulus and G'' is the loss (out-of-phase) modulus.

Relevant to ER fluids, linear viscoelastic properties have been recognized.

24.8 Characteristics of the Constituent Media of the ER Fluid

As indicated before, ER fluids with very few exceptions are invariably constituted by moist particulates dispersed in a nonconducting fluid. The characteristics of these two phases are as follows:

24.8.1 *Dispersed particulate phase*

An appropriate dispersed phase for an ER fluid should have:

- Hydrophilicity (water loving).
- Porosity (for aqueous retention).
- Size: 0.04 to 50 micron-meter (Smaller particles have forces due to Brownian movement that competes with the electrical forces upsetting the ER effect; and larger particles would respond slowly in an electric field and sediment due to gravitational or centrifugal forces.) Oversized particles (200 to 5000 μm) forming an *electrofluidized* bed may exhibit shear-stress *versus* electric field strength characteristics similar to ER effect. The particle size in general has influence on ER effect only when thermal forces dominate the polarization forces.
- Particulate shape: Existing studies are sparse and incomplete. Particulates of large aspect ratio which are highly anisotropic when used may tend to align along the electric field. Present studies, however, are insufficient to indicate whether such polarization would slow down the response of the ER fluid. A recent study due to Park [3] indicates that the spheroidal particles with significant eccentricity produce electric field concentrations an order of magnitude greater than that of spherical particles along the axis of the particles. This large field intensity along the interparticle axis could enhance the occurrence of dielectric breakdown in the suspending fluid, inasmuch as the applied field intensities to achieve ER effects are already of significant magnitude. Such breakdown would normally lead to polymerization of the dielectric fluid, which can then form a rigid mechanical bond between particles. This results in increased electrical conductance of the composite (non-ohmic behavior) as well as establishing an irreversible mechanical process that would render the ER composite useless as a cyclic engineering material. Further, the nonspherical particles may produce inferior interparticle attraction forces as compared to the spherical particles as a result of the anisotropic nature of the electric field distribution coupled with the particle geometry over which the field is integrated in producing the uniaxial force. Therefore, it is concluded in [3] that the nonspherical particles in the shape of prolate spheroidal are not suitable for applications as electrorheological dispersants.
- Chemistry: The chemical characteristics of the dispersed phase is largely linked to the behavior of the additive to the particles. For example, in water-activated particulates, the mobile ions are spared by the included water.
- Other properties: Ions and electrons polarizability, dielectric anisotropy, and piezoelectric characteristics.

24.8.2 *Dispersing medium*

The characteristics of the dispersant in an ER fluid are:

- Continuous, homogeneous fluid
- Low dielectric permittivity
- Good electric insulation properties
- Low conductivity or high electrical resistivity (with low power consumption)
- Hydrophobicity (water hating)
- Low viscosity
- High boiling point and low volatility
- Low freezing point
- High dielectric breakdown strength
- Density matching the dispersed solid phase

24.8.3 Additives to solid phase

Invariably, ER fluids are constituted with a dispersion of a solid phase which is aqueous activated. Presence of water in the ER fluid plagues the engineering propects of ER fluids in practical devices in terms of poor range of operational temperature, loss of water, low dielectric breakdown, corrosion, and high power consumption characteristics.

Presence of water presumably increases the particulate permittivity, strengthening particle interactions, thereby increasing the ER effect. Porosity of the particle facilitates water adsorption; and interfacial effects may also play a role in the polarization associated with ER effects [6].

Another theory suggests [6] that water contributes high adhesion between particles through surface tension effects permitting the enunciation of ER effect. Further, in water-activated ER fluids, *electroosmotic movement* would cause water to be repelled from the pores of the particles under electric field. By containing these particles within the high field region(s) between particles in sufficient quantity would result in setting up a bridge between the particles enhancing their interactions.

Other theories consider the presence of water provides ionic polarization mechanism; that is, water acts as a vehicle for mobile ions causing interfacial polarization and electric double layer.

Proton mobility has also been considered as a plausible cause for ER effect in water-activated ER fluids.

Maximum ER effect has been reported [6] to occur in the concentration range of inclusions wherein water is still being strongly absorbed with a corresponding increase in the conductivity of the suspension. However, excessive water content would form free gaskets across the continuous medium of the suspension reducing the effective electrorheological process. Further, such high water content would reduce the dielectric strength of the medium.

Fundamental understanding of the role of water in ER fluids is still in pursuit so that water could be substituted by suitable anhydrous fluids emulating the same phenomenological influence as water on ER effect. Polar liquids (such as alcohol, ethylene glycol, dimethylamine and formamide) *in lieu* of water provide ionic conductance. However, high surface tension of these liquids may cause particulate adhesion, hampering the ER effect.

24.8.4 *Solid additives*

ER effect has also been studied with the addition of:

- Surfactants which presumably improve the stability of the suspension and could possibly create *mesomorphic micellar bridges* between particles thereby improving the ER effect.
- Salts to improve the ER activity *via ionic polarization* mechanism.

24.9 Parameters Influencing ER Effect

24.9.1 *Electrical Parameters*

The rheological properties of ER fluids are significantly affected by the characteristics of the electric field applied. The controlling electrical parameters are:

Electrical field strength: This refers to the electric field intensity or the electric coulombic force exerted on unit charge. This force field could affect (1) the static yield stress of the medium, and (2) the electroviscosity.

Studies reveal that a minimum (threshold) level of electric field intensity is required to set on the ER effect. Beyond this threshold value, the static yield stress is linearly proportional to the electric field strength until a saturation level is reached. Algorithmically, the yield stress (σ_y) can be specified by:

$$\sigma_y = AE^B \tag{24.6}$$

where A and B are constants and E represents the applied field.

Like the yield stress response, the apparent viscosity *($\Delta\eta$) versus* electric field strength follows a square-law (parabolic) dependency reaching a saturation like a sigmoidal curve. The slope of $\Delta\eta$ *versus* E also depends on the volume fraction of the dispersed phase.

Mason number *(M_n)*: This is a dimensionless parameter which cohesively includes the material variables as follows:

$$M_n = \eta_d \dot{\gamma}/(2\varepsilon_o\varepsilon_d\beta^2E^2) \tag{24.7}$$

where (η_d, ε_d) are the viscosity and dielectric constants of the dispersant, respectively, and $\beta = (\varepsilon_p - \varepsilon_d)/(\varepsilon_p + 2\varepsilon_d)$, where ε_p is the dielectric constant of the particulate phase. Further, ε_o is the absolute permittivity of the free space, and $\dot{\gamma}$ is the shear rate. In terms of this *Mason parameter*, the steady shear response of an ER fluid can be described comprehensively by:

$$\eta/\eta_\infty = (M_n^*/M_n) + 1 \tag{24.8}$$

where η_∞ is the limiting high value of the apparent viscosity; and M_n^* is a critical Mason number at which a transition from shear-dependent to Newtonian viscosity takes place at a specific shear rate or over a range of electric field strength. M_n^* has been correlated with the volume fraction (ϕ) of the particulate inclusions by a relation $M_n^* = 0.45\phi\ (0.09 \leq \phi \leq 0.34)$.

Dependence of the dynamic modulus $(G' + jG'')$ of the ER fluid on E has been sparsely addressed in the literature. In general, G' monotonically increases with E and the shear modulus exhibits a characteristic peak value at a specific electric field strength. Again in both cases it has been observed that the concentration of the particles (ϕ) would affect the dependency characteristics.

Frequency of the applied field: With the application of an alternating electric field ER fluids exhibit decreasing apparent viscosity characteristics with increasing frequency of the applied field. This is related to the dielectric dispersion effects which affect implicitly the overall ER characteristics.

Dielectric parameters: The complex permittivity $(\varepsilon' - j\varepsilon'')$ of the ER fluid represents the effective dielectric response of a multiphase system (or at least by biphasic constituents, namely, the particulates and the dispersant). The rheological response of an ER fluid has been observed to mirror the dielectric response. The underlying mechanism of the dielectric and/or rheological relaxation effects (over the frequency of the applied field) has been attributed to the polarization and interfacial effects.

Electrical conductivity: The electrical conduction process normally refers to electronic conduction in pure dielectrics or charge conduction in multiphase (or biphase) mixtures. Electrical conduction in ER fluids is not, however, well understood. It has been attributed to the double-layer polarization and the relevant conductance is invariably nonohmic.

24.9.2 *Nonelectrical Parameters*

Temperature: Pronounced ER effect has been observed at a characteristic temperature for a given electric field applied. This "resonance"-like ER effect *versus* temperature (as depicted for a typical sample in Figure 24.3) refers to enhancement of apparent viscosity and dielectric constants with increasing temperature.

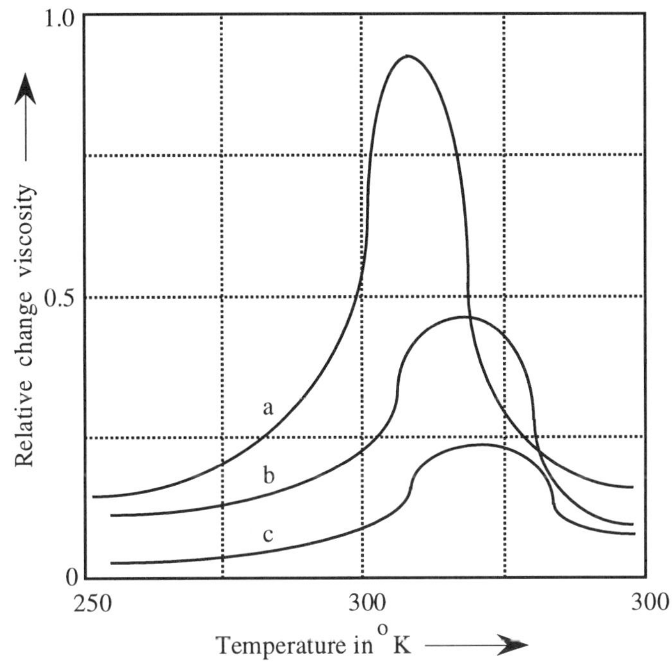

Figure 24.2 Change in apparent viscosity of a typical ER fluid with temperature. a, b, c refer to the curve with different levels of electric field (E) excitation ($\mathbf{E}_a > \mathbf{E}_b > \mathbf{E}_c$).

The reason for this has been surmised as that with increasing temperature, the viscosity of the dispersant is lowered thereby increasing the mobility of ions which facilitates more polarization of the double layer.

The drop in ER activity at even higher temperatures has been attributed to reflect the role of water additive. Water has high dielectric sensitivity *versus* temperature characteristics which renders insufficient polarization of the dispersed phase. Also, loss of water (due to evaporation) at high temperatures (>70°C) would inhibit the ER activity.

Further enhanced Brownian movement activity at high temperatures could disrupt the suspension structure. The effect of thermal forces has been specified by a parameter $\lambda = \pi\varepsilon_o\varepsilon_d a^3(\beta E)^2/(k_B T)$, where $k_B T$ is the Boltzmann thermal energy at the temperature T. However, such thermal forces are not significant unless the dispersed phase is small (particle diameter < 0.1 micron-meter) and the temperature approaches 550°C. Otherwise, the thermal force comes into play only at impractical temperatures on the order of 12000°C.

Concentration of the particulate phase: The ER activity quantified in terms of yield stress (σ_y) increases with increasing volume fraction (ϕ) of the particulate phase at a given temperature and electric field intensity. The relation is approximately specified by $\sigma_y \propto \phi^{1.5}$ and more elaborately by the following algorithm based on fibrillation theory:

$$\sigma_y = 180F_{max}[(1 - \sqrt{\pi/6}/(L\ \tan(\theta_{max})\ \phi^{1/2}/a)] \qquad (24.9)$$

where L is the electrode spacing, a is the particle diameter, F_{max} is the restoring force acting to realign a doublet (dipole) along electric flux lines after being perturbed to the maximum allowable strain, θ_{max}.

The monotonicity of Equation 24.9 does not, however, explain the critical volume fraction at which ER effect is perceived. Microstructured aspects of the ER fluid have been considered to find a plausible explanation for the critical behavior. Details concerning fibrillation and microstructural theory of ER fluids are presented in the next section.

The concentration of solid phase in the ER fluid also affects the apparent viscosity. In general, the apparent viscosity (η_A) of the suspension increases with increasing volume fraction (ϕ) of the particulates. Using the model of closely packed spheres, the following relation has been developed:

$$\eta_A = K_m E^2/\dot{\gamma}\,[(\phi_m/\phi)^{1/3} - 1]^2 \tag{24.10}$$

where K_m is a material constant and f_m is the maximum packing fraction. K_m represents the *double-layer polarizability*; or, as observed by others, it refers to bulk particle polarizability.

Considering the dynamic modulus of the ER fluid studies indicates that increasing the concentration of particles strengthens the ER effect by way of increasing the elastic modulus G'.

24.10 Theory of Electrorheology

There are at least three theories in vogue to explain the electrorheological properties. They are based on:

- Interparticulate interaction and fibrillation
- Deformation of double layer and overlap
- Anisotropic alignment of discrete particles

All these theories explain the observed enhancement of stresses in an ER fluid when subjected to an electric field. In general, they predict the E^2 dependency of the ER effect differing only in the scale factors.

24.10.1 *Fibrillation structure formation*

As originally observed by Winslow [1], ER fluids fibrillate upon the application of E. That is, at microscopic level, the particles of the suspension form chains which bridge the gap between the electrodes. The fibrillated structure may sometimes assume three-dimensional dendrite skeleton-like structure. The strength of the fibrillated structure indicates directly the magnitude/extent of the ER effect.

The particle chains under electric field have been observed to increase in length with time. Also, their length increases with increased field strength and volume fraction of the solid phase. However, the chain length decreases with the increase in frequency of the applied field. These trends have been attributed to the characteristics of the polarization force between the interacting particles. Such a force for dilute suspension (with large distance between the particles) has been expressed by the relation:

$$F_p = 6\varepsilon_p(a/2)^6 E^2/d_p^4 \tag{24.11}$$

where d_p depicts the average distance between the particles. The corresponding aggregate size of the conglomerated particles (in the *pearl-chain* configuration) is therefore proportional to E^2/d_p^4. The frequency dependency of the chain length arises from the decreased polarization as a result of associated dielectric dispersion in the fluid.

When a distance mismatch between ε_p and ε_d exists, external dielectric field may deform the shape of the particles causing nonuniform field distribution in the suspension. This would result in a translational motion (*dielectrophoresis*) of the particles culminating as clusters or fibrillated structures termed as *"bunching" or "chaining"*. The chain-forming mechanism has been modeled as a field-induced selective polarization or alignment of particles *vis-a-vis* the anisotropic structure of the particulates. Microscopically, considering the Brownian motion and the associated hydrodynamic field forces, Jordan and Shaw [6] have predicted the viscoelastic storage modules of the ER suspension as a function of oscillating flow frequency. Such a model is free from adjustable (empirical) factors in the algorithms explaining the ER mechanism.

Once the chains are formed, a minimum force is required to rupture the structure. This breaking stress corresponds to the *Bingham model* attributed to ER fluids. Again, the breaking stress is dependent on the aspect ratio (length/radius) of the fibrils, the concentration of the particles, and a dipole strength factor.

With the application of electric field, the particle with its double layer undergoes deformation (elongation) stretched along the E field. This stretching increases the mutual influence of one particle on the other. As a result, the ER effect enhances with increased viscosity in shear and other related issues discussed earlier.

24.11 Applications of ER Fluids

The technological utility and engineering applications of ER fluids have grown in recent times thanks to observed results from the scientific studies which have indicated the unique characteristics of these materials, namely, the ER fluid is comprehendible as a solid or as a fluid by simple application or removal of a voltage. As mentioned earlier, the liquid-to-solid/semisolid-phase transition controlled by electrical means permits the utility of ER fluid in mechanical systems such as ER clutches, brakes, hydraulic valves, fluidic controls, etc. The important requirements of ER fluids for electromechanical applications stem from various intrinsic properties.

Consistent in meeting the relevant requirements, a variety of devices/systems have been conceived which are summarized as follows [5]:

24.11.1 *Hydraulic coupling*

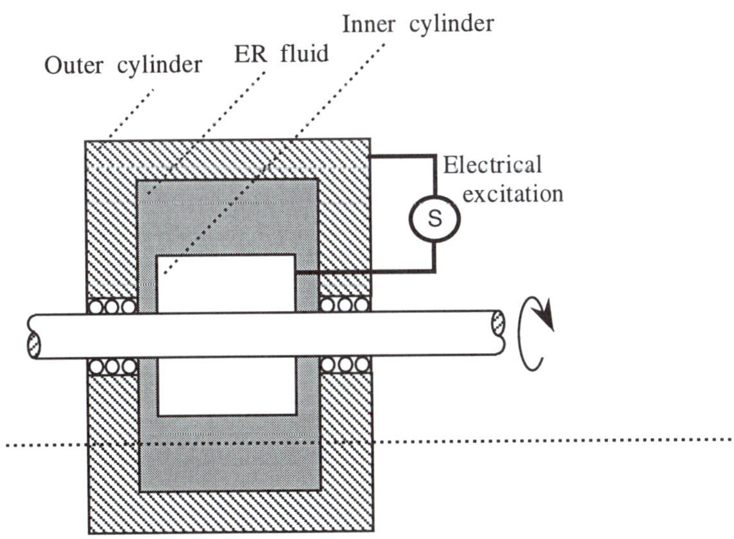

Figure 24.3 Controllable hydraulic coupling *via* electrically excited ER fluid.

Figure 24.3 illustrates a simple hydraulic coupling mechanism using ER fluid. The components of the system are: A cylindrical rotor mounted on a driving shaft, an outer cylinder coaxial to the rotor and mounted on bearings, and an ER fluid filling the interspace between the rotor and the cylinder. An electric field is applied to the ER fluid *via* appropriate electrode systems. The rotational coupling between the rotor and the cylinder is achieved through the ER fluid. This hydraulic coupling could be manipulated by changing the electric field applied to the fluid. Hence, the torque perceived through the coupling can be altered.

24.11.2 *Hydraulic damper*

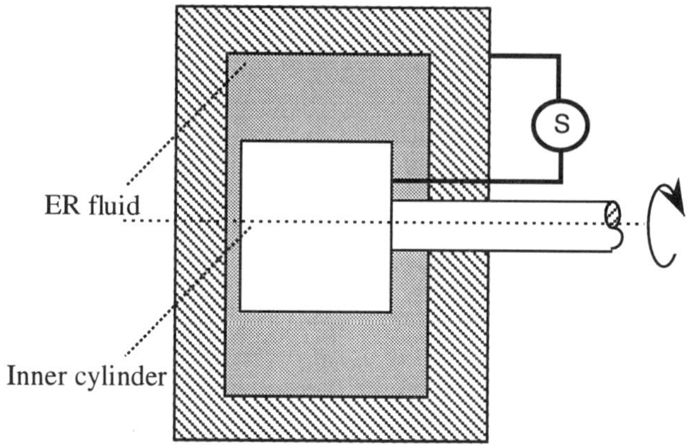

Figure 24.4 Controllable torsional damping *via* electrically excited ER fluid.

A rotor-piston assembly as illustrated in Figure 24.4 is housed in a cylinder with an ER fluid filling the interspace. The ER fluid receives electric field excitation *via* appropriate electrode arrangement. Any torsional vibration on the piston/rotor assembly is damped out due to the viscosity changes exerted on the ER fluid by the applied electric field.

24.11.3 *ER valve in a reciprocating piston system*

A two-valve, two-pump system shown in Figure 24.5 is a valve mechanism that permits a reciprocating movement of a piston. The electrorheological suspension is pumped through a cylinder and two ER valves. Alternate application of electric pulses to the valves induces them to operate as pistons which set the connecting rod, rigidly fastened to them, into a reciprocating motion with the corresponding frequency of the alternating electric field.

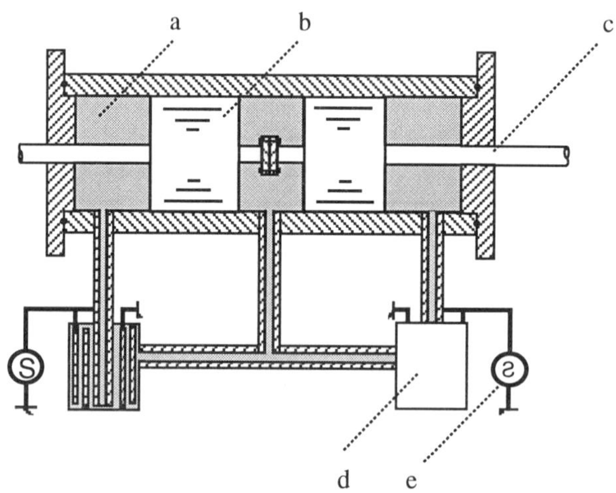

Figure 24.5 ER fluid-based reciprocating system. a: ER fluid; b: Reciprocating piston; c: Reciprocating shaft accessible outside; d: ER value; e: Electrical excitation.

24.11.6 *Dielectric suspension pump*

Shown in Figure 24.6 is a system intended to pump dielectric suspensions. It has a cylinder equipped with two valves, one of which (the intake valve) is capable of reciprocating motion when moved by the drive and the other (the discharge valve) is fastened rigidly to the cylinder and maintains the pressure in the pressure system. When the intake valve moves towards the discharge valve, sufficient voltage is applied to the intake valve to obtain a maximum increase in the effective viscosity of the ER suspension, that is to block the flow. Then, serving as a piston, the intake valve pushes on the liquid contained in the volume between the valves and expels it through the discharge valve into the flow system. When the intake valve reaches the dead point in travel, a potential difference is applied to the discharge valve so as to prevent the pressure in the system from dropping; at the same time, the intake valve is disenergized. On the return stroke, the ER fluid flows through the intake valve into the space between the valves. Repetition of the above cycles produces the pumping action to move the liquid. The flow rate and pressure can be adjusted by varying the operating voltage.

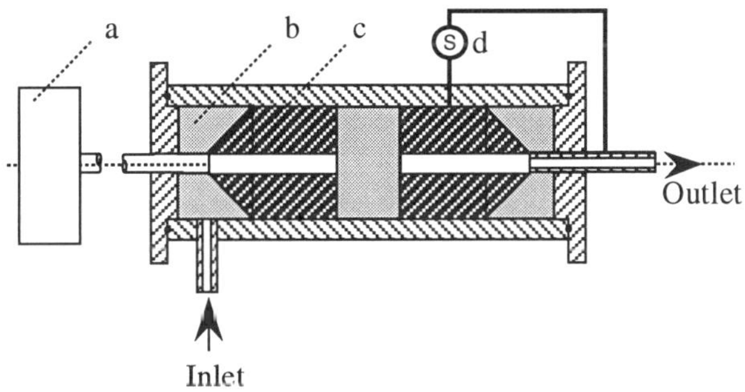

Figure 24.6 Pumping of a dielectric fluid using ER principle. a: Pump drive; b: ER fluid; c: ER intake valve; d: ER discharge valve; e: Electrical excitation.

24.11.5 *Hydraulic brake*

An electrohydraulic brake actuator system corresponds to an electric motor (Figure 24.7) which rotates a rotor wheel forcing the ER suspension to flow though a cylinder. An ER valve and a duct are provided to empty the fluid at the inlet to the cylinder beneath the rotor wheel. When the valve is energized, it begins to operate as a piston moving together with the connecting rod rigidly fastened to it. The rod transmits the force to the lever system of the brake. The brake is easily controlled by energizing and deenergizing the valve; the response time (after application of the electric field) is less than 0.05 sec.

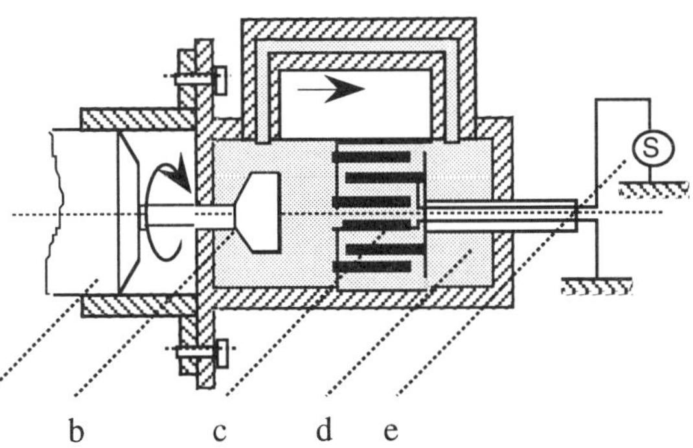

Figure 24.7 Hydraulic brake system using ER fluid. a: Motor; b: Brake drum; c: ER valve; d: ER fluid; e: Electrical excitation.

24.11.6 *Safety valve*

The electrorheological suspension is made to behave as a quasi-solid by application of an electric field so that depending on the voltage applied, it can resist a certain pressure of liquid or air in the line. When a critical pressure is reached, a diaphragm made of any low-strength material (paper or plastic film) ruptures. Its remnants flow to the drain together with the suspension, allowing the liquid or air to vent freely into the atmosphere or into a surge vessel (Figure 24.8). The critical pressure for which the valve is preset is adjusted by varying the effective viscosity of the electrorheological suspension which depends on the voltage applied to the valve.

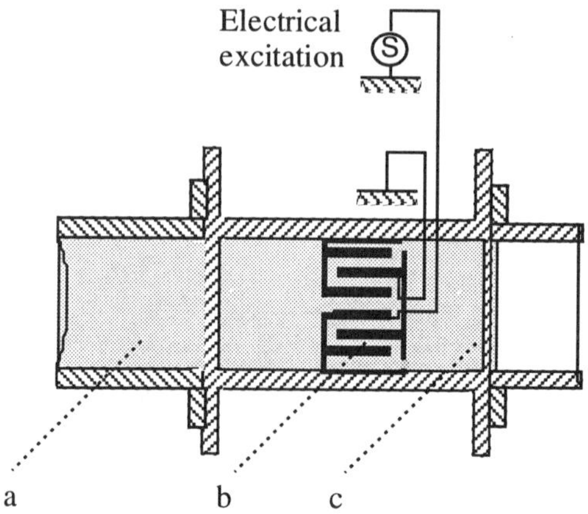

Figure 24.8 ER principle-based safety. a: ER fluid; b: ER valve; c: Diaphragm.

24.11.7 *Pulsating pressure generator*

Generation of pressure waves in a liquid can be achieved through the use of ER fluid. Electrorheological suspension contained in a cylinder flows through the clearances of the electrorheological valve to the bypass (Figure 24.9). When a high-voltage electric pulse is supplied from a generator, the rising pressure from the pump resulting from the increasing hydraulic resistance of the fluid forces the valve to move (downward in Figure 24.9) together with the connecting rod and the slid-valve. As pressure equilibrium is approached, the side valve gradually throttles down the flow to the bypass (with its shoulder). This happens because, as the valve moves, the slide valve also moves down until the forces acting on it from top and bottom equalize it. Because of this, the rate of flow to the bypass decreases whereas the output pressure to the test element rises. When the ER valve is deenergized, the spring pushes it up together with the slide valve and the pressure of the liquid at the output decreases.

When the flow rate of the electrorheological suspension is changed by an undesired external force, this flow change is picked up by the flow meter and is transmitted as an (electric) error signal to the generator which delivers a compensating high voltage signal of a proper amplitude. This signal then corrects the hydraulic resistance of the electrorheological valve and stabilizes the output oscillations.

The amplitude and frequency of forced pressure fluctuations in the liquid that leaves the device replicate these changes in the pressure of the flow of ER suspension as well as in the motion of the valve with the rod and slide. The nature of the fluctuations can be adjusted to be sinusoidal, impulse, or square wave.

It should be possible to use this effect in chemical processes and instrumentation by modifying the existing devices as well as in new designs.

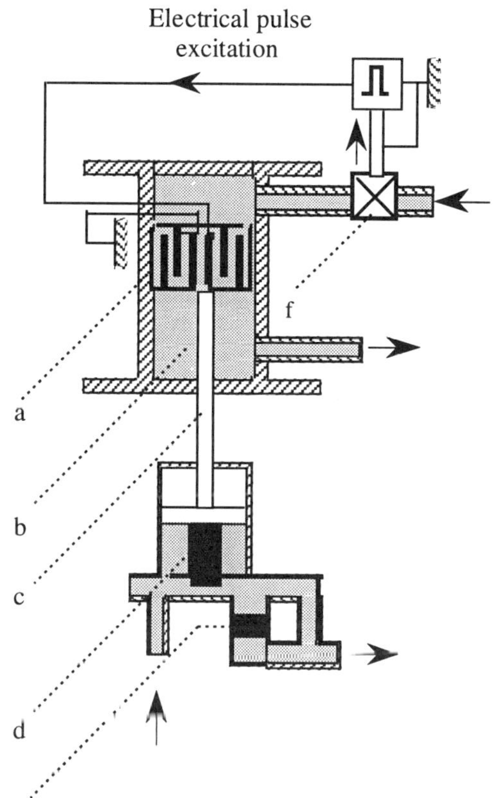

Figure 24.9 Pulsating pressure generator in fluids. a: ER valve; b: ER fluid; c: Piston; d and e: Sliding valves.

24.11.8 *Pulverization*

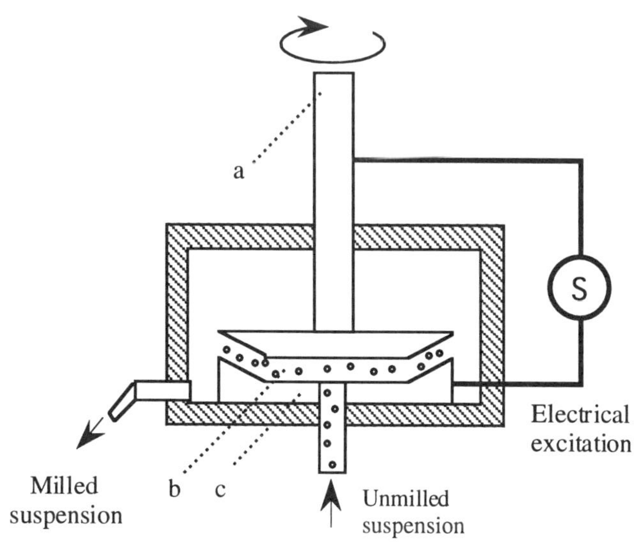

Figure 24.10 Colloidal mill for wet pulverization. a: Rotating shaft; b: Particulate suspension of an ER fluid under pulverization; c: Stator electrode.

In wet pulverization colloid mills (Figure 24.10) a stator and a rotor are electrically insulated from the housing and are connected to a d.c. or a.c. source. The suspension to be milled (in which the dispersing liquid is of low electrical conductivity and the solid particles are activated, that is, retained mildly wet by 6 to 7 percent by weight) is subjected to the action of an electric field in the working gap. The material being milled is decelerated by the field, so that the dwell time of particles in the gap is increased. This time can be controlled by changing the supply voltage.

24.11.9 *Filtering*

One of the characteristic examples of using electrorheological effect in a process refers to precoalescing of particles prior to filtration suspension (where the suspending liquid is of low electrical conductivity).

Prior to being fed into the filter, the suspension is fed to a coagulating vessel, the geometry of which can be varied over a wide range depending on specific conditions (Figure 24.11). When an electric field is applied to the liquid, there is cross-linking or structure formation to produce aggregates of the solid particles. For a given combination of electric field strength and velocity of the supplied suspension, the so-produced aggregates of appropriate particle size will not pass through pores in the filter so that filtration is improved. This is particularly important during the formation of the first layer of the filter cake.

This device can simultaneously be used as a remotely controlled variable-flow valve.

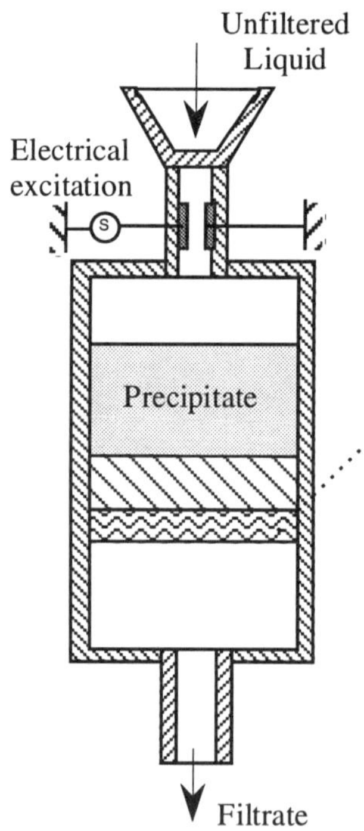

Figure 24.11 ER-based filtering.

24.11.10 *Sedimentation or settling*

Coalescence of solid particles of suspensions under an electric field enables in certain cases the speed up of precipitation significantly. For this purpose, the suspension as in the case of filtration is passed through a precoagulation vessel. This method is best used in two- and multistage settlers since the existing designs are easily adapted to it (Figure 24.12).

The advantages of this method for certain classes of suspensions are obvious, since the coventional procedure of adding a flocculant may result in contamination of the final product or its chemical modification.

In addition to the above applications, the electrorheological effect can be useful in mixing, separation, control of catalysis, and specific kinds of chemical reactions; and also in purification units.

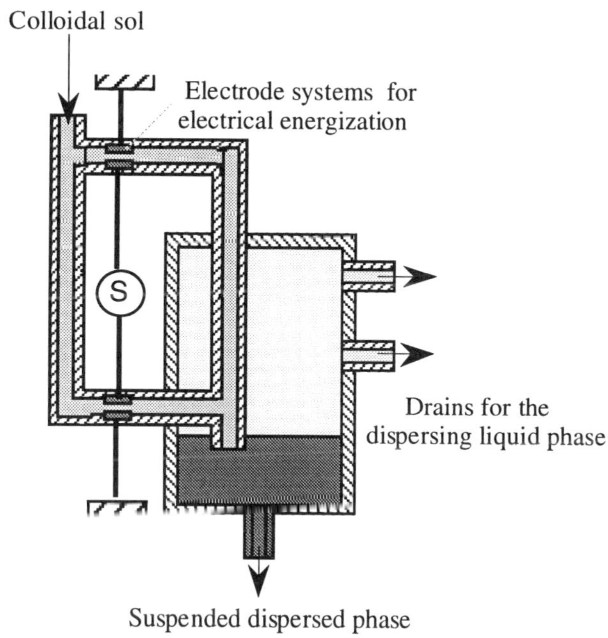

Figure 24.12 ER-based coagulation of colloidal sols.

24.11.11 *Electrodilatancy*

Certain engineering developments employ the effect of *electrodilatancy* of electrorheological suspensions. The physical phenomenon of spontaneous uncontrolled swelling (volumetric expansion) of certain ER systems made of fairly large solid particles is known as dilatancy, as described in Reiner's classical studies [11].

Electrorheological suspensions which are systems formed by finely sized particles, have very special properties: When no electric field is applied, their volume remains constant. When a field is applied their volume increases, not randomly as in *Reynolds dilatancy,* but to some specific value proportional to the electric field strength. This process is reversible and the volume returns to its initial value when the field is removed.

The devices/systems based on electrodilatancy are:

- Membrane transducer

A membrane transducer, consists of a cavity enclosed between two electrodes with dielectric gaskets between them. One of the electrodes is a membrane of an elastic dielectric material, the inner surface of which is coated with an electrically conducting material. The electrodes are connected to a high-voltage source. When an electric field is applied to the suspension filling the space between the electrodes, its volume increases and this causes the membrane to deflect outward. When the field is removed, the membrane is restored to its original position. The magnitude of the deflection of the membrane depends on many factors, namely, cell geometry, elasticity of the membrane, composition of the suspension, and applied voltage.

- Grinding and polishing

When a voltage is applied to the membrane transducer, the membrane deflects outward causing the grinding tool to move in the radial direction towards the machine surface (Figure 24.13). The pressure of the tool on the abrasive fluid increases locally which in turn increases the rate of machining at a given spot. The instrument can be equipped with a system of membrane transducers with tools covering the entire workpiece with independent control for each tool by a programming device.

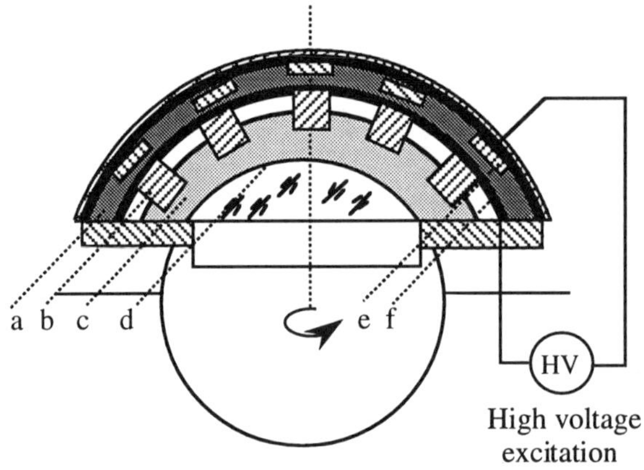

Figure 24.13 Polishing of surfaces with ER-based pressure control on polishing tools. a: ER fluid; b: Polishing discrete tools; c: Abrasive fluid; d: Surface under polish; e: Discrete rigid electrodes; f: Flexible membrane with metallized surface to act as a continuoues electrode.

- Peristaltic pump

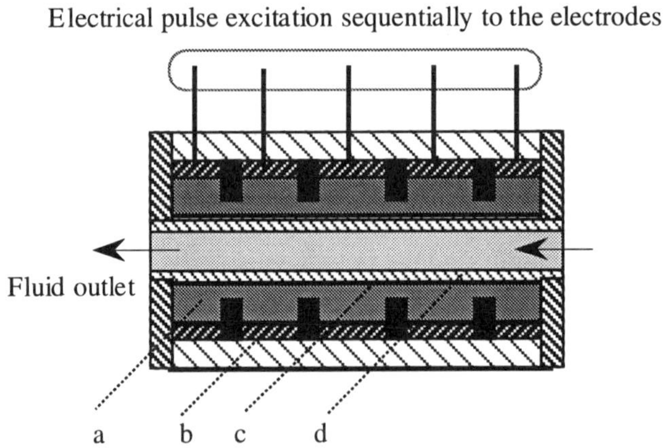

Figure 24.14 ER-based peristaltic pump. a: ER fluid; b: Assembly of discrete electrodes; c: Continuous electrode coating; d: Flexible pipeline to carry the fluid being pumped.

When suitable voltage is applied to a pair of electrodes of a membrane transducer (Figure 24.14), the wall of an elastic hose deflects inward squeezing out the liquid and thus pumping it. If a traveling electric field is applied in sequence to several successive, baffle-separated individual areas of the electrodes, the liquid is propelled in a specific direction. The frequency of the electric pulses applied to the electrodes controls, the "compression-extension" periods of the membrane.

24.11.12 *Rotary viscosimeter*

The ability of electrorheological suspensions to smoothly and reversibly change their effective viscosity in an electric field has been incorporated in the design of rotary viscosimeters operating on the comparison principle.

A longitudinal section through such a viscosimeter is given in Figure 24.15. The test liquid is placed in the space between the inner cup (which acts as the rotor) and an outer cup. The electrorheological suspension fills the gap between inner cylindrical electrode and the outer cylindrical electrode. The cups are made to rotate by a common drive. The resistance forces arising in the liquids will force the inner cylinder and the electrode to turn about the axis through an angle proportional to the viscosity of the liquids. A pointer is made to move to the zero mark on a scale by gradually increasing the voltage. The viscosity of the electrorheological suspension corresponding to the given voltage is found from a calibration table.

This viscosimeter measures in a short time interval the viscosity of liquids over a wide range (from 5 to 500 centipoise) under laboratory as well as industrial conditions.

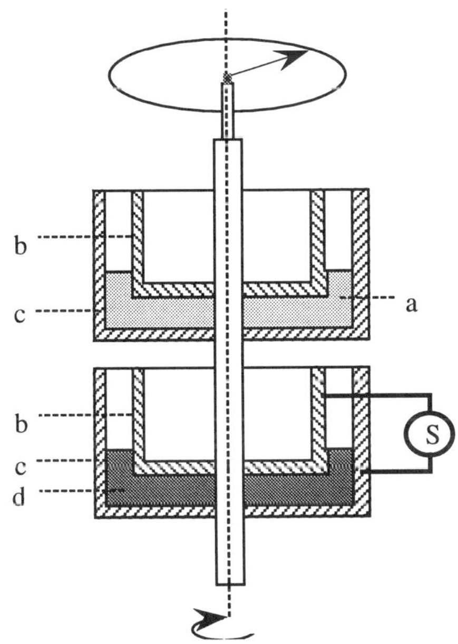

Figure 24.15 ER-based rotary-type viscosimeter. a: Test fluid; b: Inner cup(s); c: Outer cup(s).

24.11.13 *Electroelastic "smart" applications*

ER fluids find applications as "smart" materials in controlling the elastic behavior structures. Specific details on this one are furnished in Chapter 23.

24.12 Concluding Remarks

The resurgence of interest in ER fluids in recent years as an electromagnetic material has stemmed from the unique properties of these fluids which permit the use of such materials in "electrically controllable" devices. These are several other possible applications of ER fluids which have not been adequately investigated. For example, ER fluid as a propagating medium for electromagnetic waves, with controllable lossy dielectric (propagation) characteristics, may add new dimensions to EM transmission systems. Further, ER fluids as electroelastic materials in "smart" implementations have taken a leap in the research towards synthesizing intelligent structures. Composities with ER fluid ingredients need to be studied in depth so as to avail of their unique properties in technological applications. Thus ER fluids are important EM materials in the present and for the future.

References

[1] W. M. Winslow: Induced fibration of suspensions. J. Appl. Phys., vol. 20, 1949: 1137-1140.

[2] H. Block and J. P. Kelly: Electro-rheology. J. Phys. D: Appl. Phys., vol. 21. 1988: 1661-1677.

[3] J. C. Park: Stochastical and Neuromimetic Aspects of Modeling Electromagnetic Composite Materials. Ph.D. Dissertation, Florida Atlantic University, Boca Raton, FL. April 1994.

[4] J. P. Coulter, K. D. Weiss and J. D. Carlson: Engineering applications of electrorheological materials. J. Intell. Mater. Syst. Structures, vol. 4, 1993: 248-259.

[5] R. G. Gorodkin, Y. V. Korobko, G. M. Blokh, V. K. Gleb, G. I. Sidorova and M. M. Ragotner: Applications of the electrorheological effect in engineering practice. Fluid Mechanics-Soviet Research, vol. 8(4), 1979: 48-61.

[6] T. C. Jordan and M. T. Shaw: Electrorheology. IEEE Trans. Elec. Insulation, vol. 24(5), 1989: 849-880.

[7] P. M. Adriani and A. P. Gast: A microscopic model of electrorheology. Phys. Fluids, vol. 31(10), 1988: 2757-2768.

[8] T. C. Halsey: Electrorheological fluids. Science, vol. 258, 1992: 761-766.

[9] H. Block: Electrorheological fluids. Chemtech, June 1992: 368-373.

[10] Y. Choi, A. F. Sprecher and H. Courod: Response of electrorheological fluid-filled laminate composites to forced vibration. J. Intell. Mater. Syst. Structures, vol. 3, 1992: 17-28.

[11] M. A. Reiner: *Lectures in Theoretical Rheology.* (North Holland Publishers, Amsterdam: 1965).

Defining Terms

Electrorheology: Study of fluid flow subjected to an electric field.

Electrorheological (ER) fluids: Colloidal sols which exhibit abrupt change in viscosity when subjected to an electric field.

Electrorheological effect: Viscosity change effect observed in ER fluids under electric field force.

Winslow effect: Same as electrorheological effect.

CHAPTER 25

Electromagnetic Chiral Materials

25.1 Introduction

A special class of electromagnetic (EM) materials referred to as *chiral materials* are emerging in engineering applications. A chiral medium is one whose electric and magnetic fields are *cross-coupled.* The characteristic aspect of such materials is the intrinsic *handedness* (right or left) present in their physical structure.

Optically active, natural materials exhibit mirror-asymmetric molecular structure(s) and have been originally known as chiral materials. Natural chiral structures include a diverse array of sugars, amino acids, DNA and certain mollusks as well as winding vegetations while the man-made versions encompass such objects as a helix, a Möbius strip, or an irregular tetrahedron. For example, a random suspension of metallic helical springs in a dielectric host constitutes a typical *electromagnetic chiral* medium.

As stated earlier, inherently, a chiral medium has left- or right-handedness in its microstructure with the result that a circularly polarized electromagnetic wave propagating through it would experience different phase velocities and/or absorption depending on it being left or right circularly polarized; and a rotation of the plane of polarization will be caused in a plane wave transmission through such a medium.

The concept of chiralic behavior of materials at suboptical (such as microwave, millimeter) wavelengths is of interest due to the feasibility of synthesizing such media as new types of electromagnetic composite materials. Considering a simple, isotropic, two-phase *achiralic medium*, the constitutive relations refer to $\boldsymbol{D} = \varepsilon_{eff}\boldsymbol{E}$ and $\boldsymbol{B} = \mu_{eff}\boldsymbol{H}$ where $\boldsymbol{D}$ and $\boldsymbol{B}$ are the electric and magnetic flux densities, respectively, and $\boldsymbol{E}$ and $\boldsymbol{H}$ depict the corresponding electric and magnetic field intensities. The macroscopic EM properties of materials, in general, are quantified by the effective permittivity (ε_{eff}) and the effective permeability (μ_{eff}) parameters. However, in the case of a chiralic mixture, the electric and magnetic fields are cross-coupled with the result the effective medium is modeled through the *cross-coupled constitutive relation(s)* written in the matrix form as:

$$\begin{pmatrix} D \\ B \end{pmatrix} = \begin{pmatrix} \varepsilon & \alpha \\ \alpha^* & \mu \end{pmatrix} \begin{pmatrix} E \\ H \end{pmatrix} \quad (25.1)$$

where α is a complex number depicting the dimensionless *cross-coupling coefficient* and α^* is its conjugate. Explicitly $\alpha = (\chi - j\kappa)$ where χ is called the *Tellegen parameter* [1] measuring the nonreciprocal property of the medium. When $\chi = 0$ the medium is designated as a nonreciprocal chiral or *Pasteur medium* [2]; and the parameter κ decides the *degree of chirality.* When the chirality vanishes (with $\kappa = 0$), the medium represents a simple nonreciprocal achiralic material called the *Tellegen medium.* There are three well known versions of the cross-coupled constitutive relation given by Equation (25.1). These are known as *Post relation(s)*, *Condon-Tellegen relation(s)* and *Drude-Born-Fedorov relation(s)* [3]. Essentially, they all represent the electromagnetic constitutive relations of a chiralic medium but manifest in different algorithmic formats.

The chirality (right- or left-handedness) is a geometry-induced property of the medium which renders the medium to rotate the plane of polarization of a transmitted plane wave with respect to that of the incident plane wave. Classically, this property is referred to as the *optical rotary dispersion* (ORD). The cross-coupling between the field components in a chiral medium refers to the feasibility of an electric field *(**E**)* force inducing not only the electric displacement *(**D**)* or dielectric polarization but also a magnetic flux *(**B**)* or the magnetic polarization. Likewise, the magnetic field *(**H**)* impressed on a chiral medium would produce both magnetic and dielectric polarizations. The extent of such cross-coupled magnetoelectric effect is quantified by the chirality parameter κ.

Further, the handedness of the medium is represented by the quantity κ. When $\kappa > 0$ the medium is right-handed; and when $\kappa < 0$ the medium is left handed and the magnitude of κ *(0 to ±1)* decides the amount of angular rotation that an incident plane wave would suffer in traversing such a medium. Also, the amount of rotation depends on the distance traveled in the medium; and this implies that the optical activity occurs not only at the surface but throughout the chiral medium. The constitutive parameters, namely, ε_{eff} and μ_{eff} are dependent only on the magnitude of the chirality factor (κ); that is, they remain the same for both right- or left-handedness of the medium.

25.2 State-of-the-Art Models of Chiralic Mixtures/Composite Materials

Although chiral materials have received attention only recently in electromagnetic applications, the concept of chirality and its role in a variety of fields like mathematics, chemistry, optics, and life sciences date back to the early 19th century. Electromagnetic chirality embraces both optical activity and *circular dichroism.* Optical activity refers to the rotation of the plane of polarization of optical waves by a medium while circular dichroism indicates a change in the polarization ellipticity of optical waves by a medium. The phenomenon of *optical activity* was first discovered by Arago in 1811 who found that a crystal of quartz rotates the plane of polarization of linearly polarized light. Pasteur [2] and Fresnel [4] also studied the phenomenon of optical activity.

Chirality and its effects attracted the attention of the electromagnetic community with the simple but illuminating microwave experiments of Lindman [5]. Regarding the analysis of wave interaction with the chiral media the work of Bassiri [6], Jaggard et al. [7], Silverman [8], and Lakhatia et al. [11,12] are more recent to note. Concerning the modeling of the effective parameters (permittivity and permeability) of a chiralic mixture the works of Sihvola and Lindell [9,10], Lakhatia et al. [11,12] are well known. A few well-researched applications of such chiral composites have been elaborated in [13-15]. Yet another possible approach in modeling the effective parameters of different types of chiralic mixtures usable at suboptical frequencies has been proposed by the author [16] and Subramaniam [17] as described below.

25.3 EM Chiralic Mixtures with Spherical Inclusions

The *logarithmic law* of mixtures (see Chapter 4) can be extended to a chiralic medium constituted by spherical chiralic inclusions dispersed in an achiralic host. Considering a two-phase, isotropic mixture formed by spherical chiralic inclusions of volume fraction θ and achiralic properties (ε_r, μ_r), embedded in an achiralic host medium of volume fraction equal to $(1 - \theta)$ with electromagnetic property specified by (ε_2, μ_2), the corresponding cross-coupled values of the effective permittivity (ε_{eff}) and the effective permeability (μ_{eff}) of the mixture can be written as:

$$\varepsilon_{eff} = (C_1{}^{\theta} C_2{}^{(1-\theta)})\, \{\varepsilon_{log} \mp \gamma\, (1/\eta_{log})\} \tag{25.2a}$$

$$\mu_{eff} = (C_3{}^{\theta} C_4{}^{(1-\theta)})\, \{\mu_{log} \mp (1/\gamma)\, \eta_{log}\} \tag{25.2b}$$

where, $\varepsilon_{log} = \varepsilon_1{}^{\theta}\varepsilon_2{}^{(1-\theta)}$, $\mu_{log} = \mu_1{}^{\theta}\mu_2{}^{(1-\theta)}$, $\eta_{log} = (\varepsilon_{log}/\mu_{log})^{1/2}$ and γ is a cross-coupling coefficient. C_1, C_2, C_3, and C_4 are the weighting coefficients. These weighting coefficients are implicit parameters chosen to offer the attributes of logarithmic law of mixing to the effective permittivity and permeability properties. It may be noted that these coefficients weighted by the volume fraction exponents, namely, θ and $(1 - \theta)$, are in the same analytical (logarithmic) form as described in Chapter 4. Further, the terms involving γ in Equation 25.2 represent the magnetoelectric crosscoupling due to the presence of chiralic inclusions.

Though, in general, ε_{eff} and μ_{eff} can be related (*via* a set of weighting functions) to the constituent values (namely, ε_r, μ_r, ε_2, μ_2, and γ) by any arbitrary function F_1 of Equation 4.8, the logarithmic format is currently chosen in conformity with the statistical/probabilistic attributes of the mixture as conceived by Lichtenecker and Rother [18].

The dimensional consistency in the above expressions is maintained *via* γ and $(1/\gamma)$ being used as appropriate. Further, (ε_1, μ_1) are the chiralic material parameters of the inclusions decided by their achiralic counterparts (ε_r, μ_r) and by the chirality factor ζ_1 of the inclusions. That is,

$$\varepsilon_1 = \varepsilon_r - (\eta_c\, \zeta_1)^{\eta\zeta_1 + 1} \tag{25.3a}$$

and

$$\mu_1 = \mu_r - (\eta_c\, \zeta_1)^{\eta\zeta_1 - 1} \tag{25.3b}$$

where η is the intrinsic impedance due to the achiralic parameters, namely, $(\mu_r/\varepsilon_r)^{1/2}$ and $\eta_c = \eta/[1 + (\eta\, \zeta_1)^2]^{1/2}$. In the above expressions it may be noted that in the case of the inclusions being achiralic (that is, $\zeta_1 \to 0$), $\varepsilon_1 = \varepsilon_r$ and $\mu_1 = \mu_r$.

The weighting coefficients of Equation 25.2, namely, C_1, C_2, C_3, and C_4, can be evaluated explicitly with the *geometrical mean constraint* that the effective characteristic impedance of the mixture, namely, $\eta_{eff} = (\mu_{eff}/\varepsilon_{eff})^{1/2}$, at the equivolume condition $(0 - 0.5)$ would tend to the geometrical mean of its limiting values at $\theta = 0$ and 1; that is, at $\theta = 0.5$,

$$(\mu_{eff}/\varepsilon_{eff}) = \{(\mu_2/\varepsilon_2)^{1/2}\, (\mu_1/\varepsilon_1)^{1/2}\} \tag{25.4}$$

The foregoing geometrical mean constraint is the basis and the underlying principle of the logarithmic law expressed in the most general form by Equation 25.4. That is, in a truly stochastic mixture Lichtenecker [19] contended that the geometrical mean of the properties at the extremities of the volume fraction should correspond to the property at the mid-value of the volume fraction (that is, at $\theta = 0.5$). Hence, using this geometrical mean constraint specified above, the unknown coupling cofficient γ can be determined as 0, +1 or –1. The zero value applies when the inclusions are achiralic; and the +1 and –1 values refer to the right- or left-handedness of the inclusions and hence the mixture, respectively. Also the prefixed signs ($\mp$) or ($\pm$) for γ in Equation 25.2 account for the invariance of the effective parameters (ε_{eff} and μ_{eff}) with the handedness of the mixture.

Accordingly, the weighting coefficients C_1, C_2, C_3, and C_4 can be explicitly specified in terms of the material parameters as:

$$\begin{aligned} C_1 &= \varepsilon_1/\{\varepsilon_1 \mp \gamma(\varepsilon_1/\mu_1)^{1/2}\} \\ C_2 &= \varepsilon_2/\{\varepsilon_2 \mp \gamma(\varepsilon_2/\mu_2)^{1/2}\} \end{aligned} \tag{25.5a}$$

and

$$\begin{aligned} C_3 &= \mu_1/\{\mu_1 \pm (1/\gamma)(\mu_1/\varepsilon_1)^{1/2}\} \\ C_4 &= \mu_2/\{\mu_2 \pm (1/\gamma)(\mu_2/\varepsilon_2)^{1/2}\} \end{aligned} \tag{25.5b}$$

The effective chirality (ζ_{eff}) of the mixture is given by:

$$\zeta_{eff} = [(1+\zeta_1)^{\theta} - 1]\, \{(1/\eta_{eff})^2 - (1/\eta_{log})^2\}^{(1-\theta)/2} \tag{25.6}$$

The above expression for the effective chirality of the mixture is derived on the basis of the characteristic impedance relation:

$$\eta_{eff} = \eta_{log}/[1+(\eta_{log}\ \zeta_{eff})^2]^{1/2} \quad (25.7)$$

where η_{eff} is the chiralic, and η_{log} the achiralic effective characteristic impedance of the mixture.

Inasmuch as the test medium represents a random mixture, the values of ε_{eff} and μ_{eff} should also be constrained by their corresponding upper and lower limits specified by the Wiener limits mentioned in Chapter 4. That is,

$$1/[\theta/\varepsilon_1 + (1-\theta)/\varepsilon_2] \leq \varepsilon_{eff} \leq \theta\varepsilon_1 + (1-\theta)/\varepsilon_2 \quad (25.8a)$$

$$1/[\theta/\mu_1 + (1-\theta)/\mu_2] \leq \mu_{eff} \leq \theta\mu_1 + (1-\theta)/\mu_2 \quad (25.8b)$$

The weighting coefficients as expressed in Equation 25.5 are the optimal values for a truly stochastic mixture inasmuch as they are derived on the basis of the geometrical mean constraint. If they are derived in any other possible way (such as on the basis of arithmetic mean constraint), the resulting values would not apply to a truely stochastic system.

The logarithmic law formulations of Equation 25.2 which refer to spherical chiralic inclusions can be extended to shaped chiralic inclusions as well *via* Fricke's formula by incorporating the explicit dependency of the results on the aspect ratio or eccentricity of the inclusions.

25.4 Chiralic Composites with Shaped Inclusions

In the previous section, analytical descriptions for the effective parameters of a simple chiralic mixture randomly dispersed with spherical chiralic inclusions were indicated. The problem of shaped chiralic inclusions randomly dispersed in an achiralic host could also be addressed on the basis of mixture theory. The effective values of the dielectric permittivity and magnetic permeability of such a composite medium are derived by modifying Fricke's formula on the basis of the logarithmic law of mixing. The resulting expressions are thus *ad hoc* extensions of the approaches due to the well-known Fricke's and logarithmic law formulations.

In general, as discussed in Chapter 4, the particulate inclusions are referred to as shaped if two or more of the lateral dimensions are significantly different as in the case of ellipsoids, prolate/oblate spheroids, needles, and disks. For a spheroidal geometry with semi-axes *a*, *b*, and *c* and taking $b = c$, the aspect ratio is equal to (a/b). When this aspect ratio is of significant value (either large or small compared to unity) the corresponding eccentricity (e) would play a significant role in the polarization of the particles when the mixture is submitted to an external field; and the depolarization arising from the relative disposition of the particles due to the random nature of the particle dispersion (and/or orientation) in the mixture would become another effective stochastic parameter to be duly considered. The works of Wiener, Fricke, Sillars, Lewin, Hamon, and Boned and Peyrelasse are the well-known endeavors directed towards the elucidation of the dielectric properties of simple achiralic mixtures with shaped inclusions as discussed in Chapter 4.

Considering a chiralic mixture composed of an achiralic host dispersed randomly with shaped chiralic inclusions (such as helices) which have an inherent shape factor associated with them, no well-known formulations are presently available. Hence, in the following section analytical descriptions for such a mixture are developed and some theoretical results and experimental data are presented.

25.5 Effective Parameters of Chiralic Mixtures with Shaped Inclusions

For a spheroidal geometry of the inclusions with semi-axes *a*, *b* and *c* and taking $b = c$, the aspect ratio is equal to (a/b); and in Fricke's formulation as well as in [20], a shape factor denoted by x was chosen to represent the dependency of the permittivity on the aspect ratio. The corresponding factor for the permeability is taken as y currently. Using the method of [20], the modified shape factor(s) for a chiral mixture constituted by an achiralic host and shaped chiralic inclusions can be deduced as follows.

In terms of Fricke's formulation [20], the effective permittivity (ε_{eff}) is given by:

$$\varepsilon_{eff} = \{\varepsilon_1\varepsilon_2 (1 + x\theta) + \varepsilon_2^2 x(1 - \theta)\}/\{\varepsilon_1(1 - \theta) + \varepsilon_2(x + \theta)\} \quad (25.9)$$

Analogously the effective permeability (μ_{eff}) of the mixture is given by:

$$\mu_{eff} = \{\mu_1\mu_2 (1 + y\theta) + \mu_2^2 y(1 - \theta)\}/\{\mu_1(1 - \theta) + \mu_2(y + \theta)\} \quad (25.10)$$

Setting these expressions identically equal to the respective relations of Equation 25.10, the shape factors x and y are obtained as:

$$x = M N_x/D_x \quad (25.11a)$$

where $N_x = \{C_1^{\theta} C_2^{(1-\theta)}[\varepsilon_1(1 - \theta) + \varepsilon_2\theta](\varepsilon_{log} \mp \gamma(1/\eta_{log})) - \varepsilon_1\varepsilon_2\}$

and $D_x = \{\varepsilon_2[\varepsilon_2(1 - \theta) + \varepsilon_1\theta - C_1^{\theta} C_2^{(1-\theta)}\{\varepsilon_{log} \mp \gamma(1/\eta_{log})\}]]$

$$y = N N_y/D_y \quad (25.11b)$$

where $N_y = \{C_3^{\theta} C_4^{(1-\theta)}[\mu_1(1 - \theta) + \mu_2\theta](\mu_{log} \pm \gamma(1/\eta_{log})) - \mu_1\mu_2\}$

and $D_y = \{\mu_2[\mu_2(1 - \theta) + \mu_1\theta - C_3^{\theta} C_4^{(1-\theta)}\{\mu_{log} \pm \gamma(1/\eta_{log})\}]]$

The parameters M and N in Equation (25.11) are factors dependent on the (a/b) ratio of the spheroidal inclusions. For an oblate spheroid $(a > b)$ or disk-like $(a >> b)$ inclusions, $M = 2/(m - 1)$ if $\varepsilon_1 \geq \varepsilon_2$ or $(m - 1)/2$ if $\varepsilon_1 \leq \varepsilon_2$. Likewise $N = 2/(m - 1)$ if $\mu_1 \geq \mu_2$ or $(m - 1)/2$ if $\mu_1 \leq \mu_2$. Here m refers to the depolarization factor given by [23]:

$$m = 1/\{1/(1 - f^2) - [f \arccos(f)/(1 - f^2)^{3/2}]\} \quad (25.12)$$

where $f = (b/a) < 1$ and the eccentricity e (of the oblate spheroid) is equal to $1 - (b/a) = (1 - f)$.

For prolate spheroidal $(a < b)$ or needle-like $(a << b)$ inclusions, $M = 2/(m - 1)$ if $\varepsilon_1 \leq \varepsilon_2$; or $(m - 1)/2$ if $\varepsilon_1 \geq \varepsilon_2$. Likewise $N = 2/(m - 1)$ if $\mu_1 \leq \mu_2$; or $(m - 1)/2$ if $\mu_1 \geq \mu_2$. In this case m is given by [23]:

$$m = 1/\{1/(g^2 - 1) - [g \ln(g + (g^2 - 1)^{1/2})/(g^2 - 1)^{3/2}]\} \quad (25.13)$$

where $g = b/a > 1$; and the eccentricity e (of the prolate spheroid) is equal to $1 - (a/b) = 1 - (1/g)$.

Hence, the effective permittivity (ε_{eff}) and effective permeability (μ_{eff}) of the test mixture can be calculated from Equations 25.9 and 25.10 with the values of x and y of Equation 25.11. The effective chirality (ζ_{eff}) of the mixture can then be evaluated from Equation 25.6.

Sample computations on the above algorithms were performed with the following data. A test mixture is presumed to consist of an achiralic background with shaped chiralic inclusions of volume fraction $\theta = 0.4$. Two sets of hypothetical ingredients were considered, namely, ($\varepsilon_r = 78.3$, $\mu_r = 1000$, $\varepsilon_2 = 2$, and $\mu_2 = 55$) and ($\varepsilon_2 = 78.3$, $\mu_2 = 1000$, $\varepsilon_r = 2$, and $\mu_r = 55$). In each case, an arbitrary chirality factor of $\zeta_1 = 0.0001$ was presumed. That is, the chosen value of ζ_1 represents the degree of chirality decided by the distinct shape of the inclusions (such as helical); and the magnitude of ζ_1 (taken here as 0.0001) would alter the achiralic parameters (ε_r, μ_r) of the inclusions to the corresponding chiralic values, namely, (ε_1, μ_1) *via* the relations given by Equation 25.11. Thus ζ_1 controls the degree of chirality and can be designed by appropriate geometry of the inclusions. Inasmuch as the inclusions are chiralic, the resulting mixture would also exhibit chiralic characteristics with an effective chirality factor given by Equation 25.6.

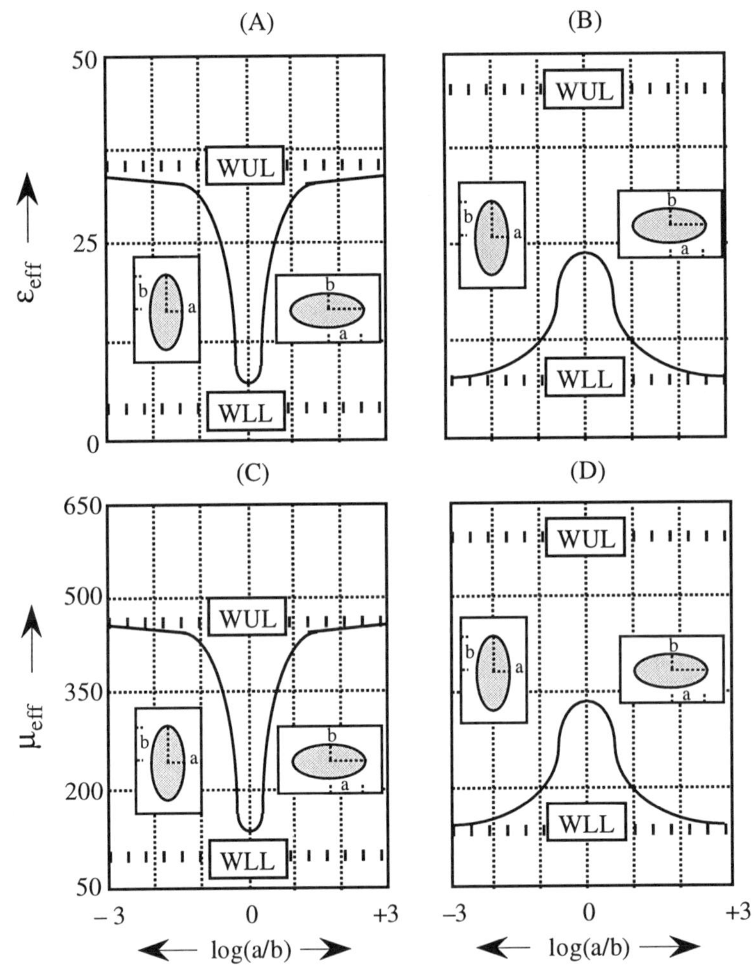

Figure 25.1 Effective values of permittivity (ε_{eff}) and permeability (μ_{eff}) of a chiralic mixture *versus* aspect ratio (a/b) of the inclusions of volume fraction θ. ($\theta = 0.4$; WUL: Wiener's upper limit; WLL: Wiener's lower limit. For A&C: $\zeta = +1$, $\zeta = 0.0001$, $\varepsilon_r = 78.3$, $\mu_r = 1000$, $\varepsilon_2 = 2$, $\mu_2 = 55$; For B&D: $\zeta = +1$, $\zeta = 0.0001$, $\varepsilon_2 = 78.3$, $\mu_2 = 1000$, $\varepsilon_r = 2$, $\mu_r = 55$.)

In the foregoing theoretical considerations, the chiralic particulates dispersed in the host medium can be "stretched" or "compressed" so that each particle would assume an axially asymmetric or "shaped" chiralic geometry. In this case the particulate eccentricity *(e)* or the aspect ratio *(a/b)* should also be considered. Hence, for different values of *(a/b)*, the computed results on the two hypothetical samples considered are presented in Figure 25.1.

The inferences from the results pertinent to Figure 25.1 are:

- The effective dielectric permittivity and magnetic permeability of a random chiralic mixture are functions of the shape factor of the inclusions as in the case of achiralic mixtures.
- The material parameters given by Equations 25.9 and 25.10 reduce to that of a mixture with spherical inclusions as in Equation 25.10 when a/b = 1.
- Also, the formulations of Equations 25.9 and 25.10 are in a closed form.
- Use of the logarithmic law of mixing confirms the proportionality postulation applicable to a statistical mixture.
- The expressions for ε_{eff} and μ_{eff} satisfy conditions at the extreme limits of $\theta = 0$ or 1.
- The results indicated are bounded by the Wiener limits (see Chapter 4).
- Last, these formulations based on the logarithmic law of mixing refer only to randomly dispersed spheroidal inclusions (disordered systems) in an achiralic host and do not apply when the shaped inclusions are aligned/oriented specific to the electric field direction. The algorithms, however, can be modified to suit such orderly disposed inclusions and are detailed in Section 25.8.

From the results presented, it could be evinced that, for a given set of constituent parameters, ε_{eff} and μ_{eff} vary significantly with respect to the particulate eccentricity and eventually approach their Wiener bounds at the limiting values of the aspect ratio corresponding to the particulate shape becoming disk-like or needle-like as depicted in Figure 25.1.

25.6 Practical Considerations: An Experimental Study

To illustrate the practical aspects of using the algorithms of the previous section in synthesizing a chiral electromagnetic composite for applications at quasioptical frequencies, the following experimental study as reported in [17] can be considered.

A square slab of 11.85 x 12.5 x 2.72 cm of a test composite was fabricated with a host medium of paraffin wax ($\varepsilon_2 = 2.35$, $\mu_2 = 1$) embedded with a large number (approximately 120) of miniature right-handed helical springs made of a high permeability metal alloy with $\mu_r = 30{,}000$. The radius of each helix is 0.15 cm and the pitch is 0.25 cm and it contains 3.5 turns. The metallic volume is about 2% of the cylinder of radius 0.15 cm and height 0.9 cm. Considering the total volume of the slab, this cylindrical volume constitutes a volume fraction of 0.2. At the test wavelength of about 3 cm, the chiral mixture so fabricated is fairly homogeneous by virtue of the size of the helices being small in comparision to the operating wavelengths. However, the chiral inclusions should not be too small lest they become "invisible" to the propagating wave. Further, the helices were disposed randomly in the wax medium so as to emulate a truly statistical mixture and ensuring isotropic performance.

The metal concentration of the right-handed helices, namely, 2% refers to the corresponding cylindrical volume fraction occupied by the helices of the order of 20%. Hence, notwithstanding the actual volume fraction of the metal appearing low, the apparent cylindrical volume enclosed by the springs is quite high. Considering the dimensions of the springs, the ratio of the length of the cylindrical volume enclosed to the radius of the helix specifies the aspect ratio *(b/a)* of the inclusions. In the present case, it is equal to about 6.

The influence of the effective electromagnetic properties of the test sample on microwave transmission was studied at X-band frequencies (8 to 10 GHz). For this purpose, a microwave transmitter-receiver arrangement was used with the test slab irradiated by a focused beam of microwaves emerging from a microwave horn.

For a given frequency setting with the transmitter horn launching a vertically polarized transmitted (E_{yI} component only) beam, the receiver (also set to receive the vertically polarized wave) was calibrated and normalized so that the detected output corresponds to 0 dB. This refers to a total free-space transmission with a coefficient of 1. As a next step, the test chiralic slab was introduced between the transmitter and the receiver horns and the following resulting effects at the receiver were measured:

(1) The power transmission coefficient due to vertically polarized transmitted and received E_{yT} waves (that is, $|E_{yT}/E_{yI}|^2$)

(2) The power transmission coefficient due to vertically polarized transmitted wave and the cross-polarized (orthogonal) received E_{zT} wave (that is, $|E_{zT}/E_{yI}|^2$)

These measurements were performed at different spot frequencies over 8 to 10 GHz range. Relevant results are presented in Figures (25.2 and 25.3). To compare the experimental data with the theoretical calculations, the values of ε_{eff} and μ_{eff} as given by Equations 25.9 and 25.10 were computed with the following data relevant to the test slab: $\varepsilon_2 = 2.35$, $\mu_2 = 1$, $\mu_r = 30000$, $\theta = 0.2$ and $a/b = 1/6$. Inasmuch as the inclusion is made of metal and its corresponding value of ε_r is not definable, the use of the logarithmic law of mixing (when the inclusions are metallic) is usually questioned. However, the author [21] had developed an exclusive method to obviate this difficulty by extending the logarithmic law to complex dielectric susceptibility which accounts for such metallic inclusions. Hence, following the method given in [21], for the metallic inclusions of conductivity σ_I, the expression for ε_{log}, ε_r and ε_I of Equation 25.9 can be written as:

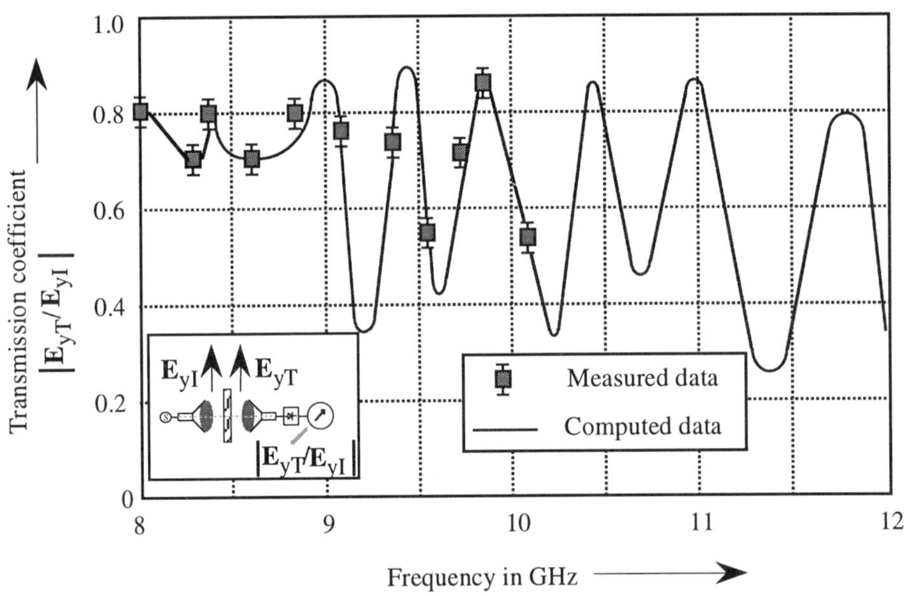

Figure 25.2 Computed and measured data on transmission coefficient *versus* frequency for a normally incident E-polarized beam wave on a chiralic composite slab.

$$\varepsilon_r \rightarrow (\sigma_I/\omega\varepsilon_0)$$
$$\varepsilon_I \rightarrow \Omega = [(\sigma_I/\omega\varepsilon_0) - (\eta_c\zeta_I)^{\eta\zeta_I} + 1] \quad (25.14)$$

and

$$\varepsilon_{log} \rightarrow \Omega^{\theta}\{(\varepsilon_2 - 1)^{1-\theta}\cos(\pi\theta/2)\} + 1$$

where $\omega = 2\pi x$ frequency and ε_0 = free-space permittivity.

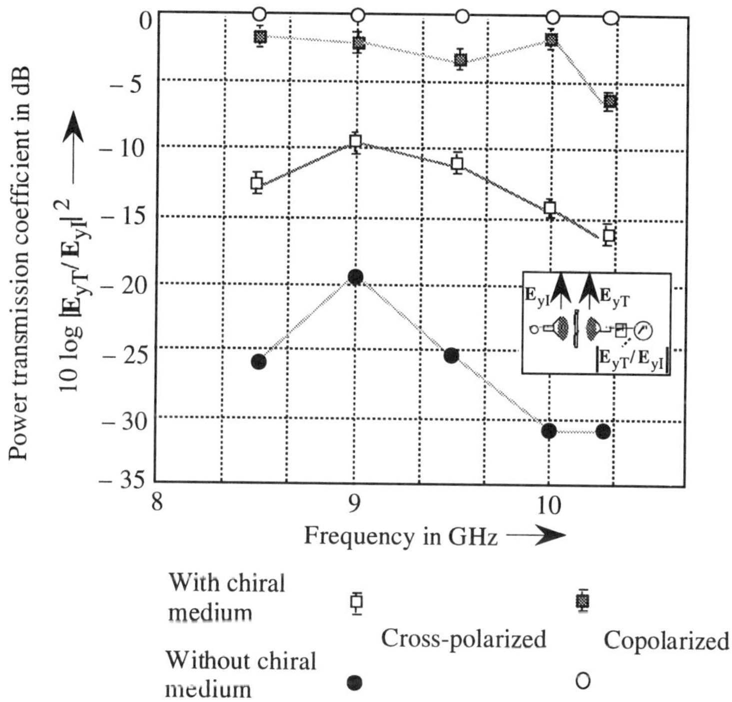

Figure 25.3 Data on co- and cross-polarized power transmission coefficients *versus* frequency measured with a E-polarized beam wave normally incident on a chiralic composite slab.

It may be noted that the above expressions are frequency dependent. Therefore, the corresponding results of Equations 25.9 and 25.10 would also be frequency dependent. For the test sample under discussion the metallic inclusions have $\sigma_l = 1.8 \times 10^6$ S/m. Hence, to pursue the calculations with Equations 25.9 and 25.10, the only unknown left is the chirality factor ζ_l of the inclusions.

For a given set of values, namely, (ε_{eff}, μ_{eff} and η_{eff}), the corresponding transmission coefficient (for normal incidence) at the metal-backed test composite is given by:

$$|T| = \{|\Gamma - \Gamma exp[-j4\pi(\varepsilon_{eff}\mu_{eff})^{1/2}d/\lambda_0]|\}/\{|1 - \Gamma^2 exp[-j4\pi(\varepsilon_{eff}\mu_{eff})^{1/2}d/\lambda_0]|\} \tag{25.15}$$

where $j = \sqrt{-1}$, d is the thickness of the slab, and λ_0 is the free-space wavelength.

Considering a test frequency (in the range 8-12 GHz), the computed data on ε_{eff}, μ_{eff}, and η_{eff} of the test sample are presented in Figures 25.4–25.6. These results show that regardless of the value of ζ_l (in the range $0 < \zeta_l < 1$), the calculated values of ε_{eff}, μ_{eff}, and η_{eff} remain almost unchanged.

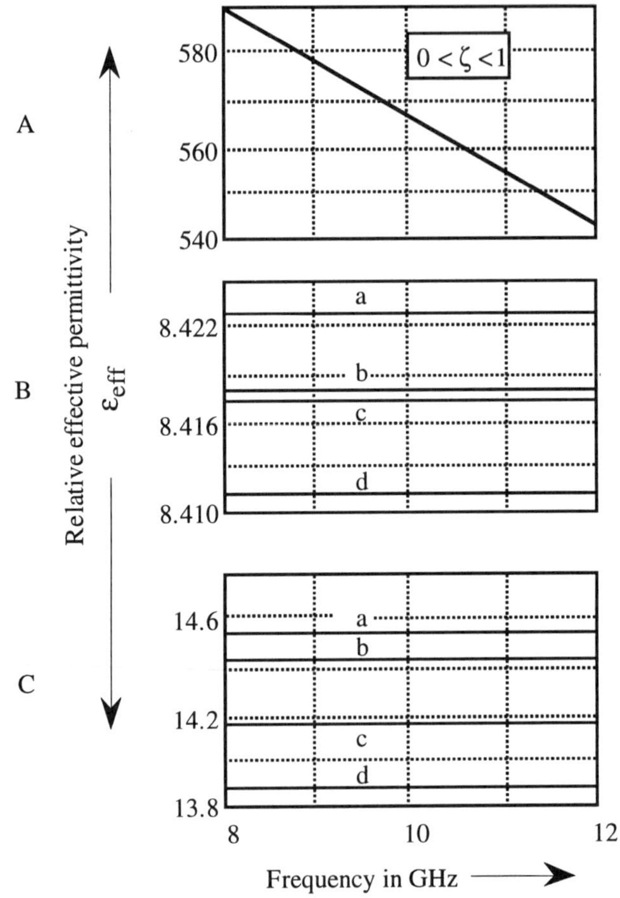

Figure 25.4 Computed data on the relative effective permittivity *versus* frequency pertinent to chiralic composite media with $0 < \zeta_1 < 1$.
Data on the test samples: A: $\varepsilon_2 = 2.35$, $\mu_2 = 1$, $\sigma_1 = 1.8 \times 10^6$ siemen/meter, $\mu_r = 30000$.
B: Hypothetical sample 1: $\varepsilon_2 = 2$, $\mu_2 = 55$, $\varepsilon_2 = 78.3$, $\mu_r = 1000$.
C: Hypothetical sample 2: $\varepsilon_r = 2$, $\mu_r = 55$, $\varepsilon_2 = 78.3$, $\mu_2 = 1000$.

For the hypothetical samples on the other hand with nonmetallic composition ($\varepsilon_r = 78.3$, $\mu_r = 1000$, $\varepsilon_2 = 2$, and $\mu_2 = 55$) and ($\varepsilon_2 = 78.3$, $\mu_2 = 1000$, $\varepsilon_r = 2$, and $\mu_r = 55$) as considered in the previous section, the corresponding values of ε_{eff}, μ_{eff}, and η_{eff} (as presented in Figures 25.4 to 25.6 for a volume fraction $\theta = 0.2$ and $a/b = 1/6$ over the frequency range of 8-12 GHz) show that for different values of ζ_1 (in the range of 0 to 1) the effective parameters of these hypothetical samples may vary significantly.

As regards the hypothetical samples and the test sample, the major difference in their electromagnetic constitutive characteristics is that in the test sample the fabricated inclusions are conductors contributing a susceptance term ($\sigma_1/\omega\varepsilon_0$) of excessive magnitude whereas this is replaced by the ε_1 term in the hypothetical samples which is relatively of very small value. Further, unlike in the test sample, ε_1 and μ_1 are assumed to be frequency independent in the hypothetical samples.

Hence, it can be surmised that the overwhelming dielectric susceptance contribution due to the conductor inclusions make the effective parameters of the mixture dependent on the frequency but insignificantly dependent on the ζ_1 values. In contrast, the low dielectric susceptance due to the small values of ε_I (such as those for dielectric inclusions considered in the hypothetical samples) render the effective parameters dependent on ζ_I, but independent of the frequency.

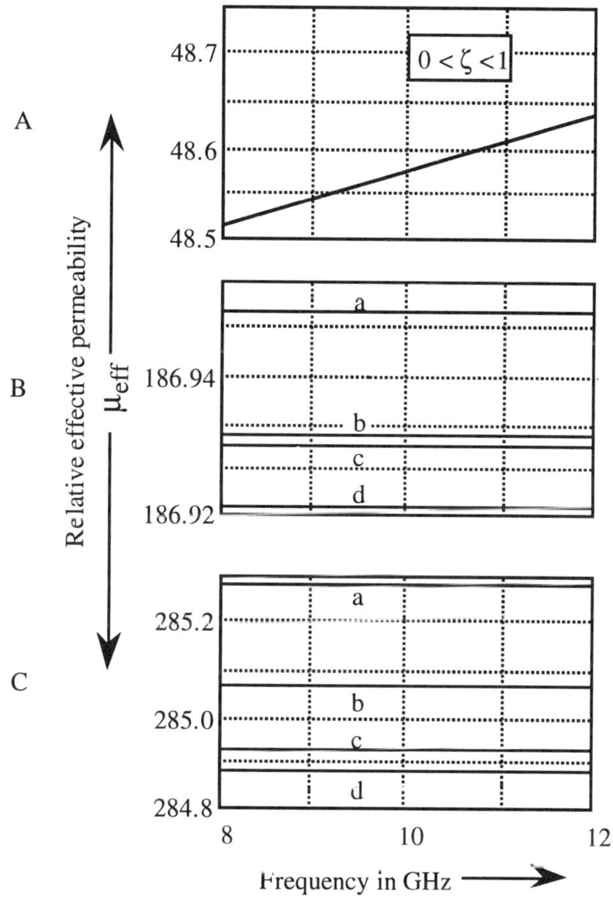

Figure 25.5 Computed data on the relative effective permeability *versus* frequency pertinent to chiralic composite media with $0 < \zeta_1 < 1$.
Data on the test samples: A: $\varepsilon_2 = 2.35$, $\mu_2 = 1$, $\sigma_1 = 1.8 \times 10^6$ siemen/meter, $\mu_r = 30000$.
B: Hypothetical sample 1: $\varepsilon_2 = 2$, $\mu_2 = 55$, $\varepsilon_2 = 78.3$, $\mu_r = 1000$.
C: Hypothetical sample 2: $\varepsilon_r = 2$, $\mu_r = 55$, $\varepsilon_2 = 78.3$, $\mu_2 = 1000$.

Now referring to Figure 25.2, the calculated values as per Equation (25.15) of the transmission coefficient corresponding to the co-polarized components of the incident (vertically polarized) beam wave, vary with frequency exhibiting resonant windows. This is confirmed by the measured values.

Further, the presence of the test slab in the transmission path causes a rotation of the plane of polarization of the incident beam wave. This is evinced from the measured cross-polarized (orthogonal) field component at the receiver. The ratio of the co-polarized and cross-polarized field components at the receiver therefore refers to the ellipticity of polarization caused by the chiralic property of the test sample. Hence, it should implicitly depict the effective chirality of the test medium, namely, ζ_{eff}. From ζ_{eff} using Equation 25.14 the corresponding values of ζ_I can be determined. Relevant results are also presented in Table 25.1. It may be noted that both ζ_{eff} as well as ζ_I are almost invariant with frequency suggesting that they are more dependent on the geometry of the included chirals rather than on the external electromagnetic field forces.

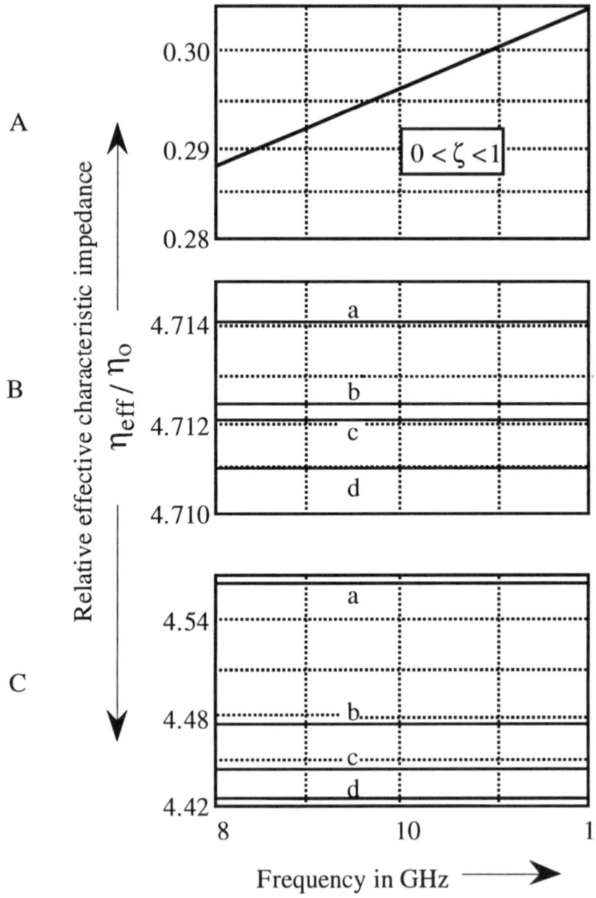

Figure 25.6 Computed data on the relative effective characteristic impedance *versus* frequency pertinent to chiralic composite media with $0 < \zeta_1 < 1$ ($\eta_o = 120\pi$ ohms).
Data on the test samples: A: $\varepsilon_2 = 2.35$, $\mu_2 = 1$, $\sigma_1 = 1.8 \times 10^6$ siemen/meter, $\mu_r = 30000$.
B: Hypothetical sample 1: $\varepsilon_2 = 2$, $\mu_2 = 55$, $\varepsilon_2 = 78.3$, $\mu_r = 1000$.
C: Hypothetical sample 2: $\varepsilon_r = 2$, $\mu_r = 55$, $\varepsilon_2 = 78.3$, $\mu_2 = 1000$.

25.7 Discussions on the Theoretical Considerations

On the basis of the theoretical aspects presented and the experimental results furnished in the previous sections the following can be inferred:

1. Although the formulations indicated are extrapolations of the logarithmic law of mixing as applied to dielectric permittivity, its applicability to practical systems is evident from the theoretical and experimental results furnished. Close correlation between the theoretically evaluated transmission coefficients (in terms of ε_{eff} and μ_{eff}) and the experimentally determined values at different frequencies (Figure 25.2) validates the algorithms of Equations 25.9 and 25.10.
2. A chiralic mixture constituted by a random dispersion of conducting chirals in an achiralic host offers electromagnetic absorption characteristics as is evident from the measured transmission coefficient results depicted in Figure (25.2).

3. Such electromagnetic absorption is frequency dependent as could be seen in Figure (25.2). This frequency dependency arises from:
   (a) The loss tangent of the host dielectric. (This, however, being very small for paraffin wax, could be neglected).
   (b) The dissipative loss is due to conducting inclusions as dictated by Equation 25.14. This is rather a more predominant factor than the loss-tangent effects of the host medium.
4. Both theoretical calculations and experimental data (Figures 25.2 and 25.3) indicate that the absorption characteristics may exhibit resonances. Although this corresponds to limited bandwidth of operation, these resonances can be quenched with high concentration of inclusions as indicated in [22].
5. Further, the dispositions of inclusions may also affect the bandwidth performance. Especially, the inclusion-to-inclusion contiguity will decide the relaxation and hence the effective absorption process.
6. Such contacts between the inclusions would also affect the chirality of the composite due to induced surface currents on the helices [22].
7. The indicated study has addressed implicit definitions for the effective chirality (ζ_{eff}) of the mixture and for the chirality of the inclusions (ζ_I) in terms of experimental parameters as depicted in Table 25.1. In the relevant work as well as in the other existing studies [9,10], there is no method of knowing the value of the intrinsic chirality parameter (ζ_I) of the inclusions on *a priori* basis, except that it is controlled by the size and geometry of the inclusions. However, as proposed above a strategy to measure the effective chirality of the mixture-medium (namely, ζ_{eff}) is feasible, and thereby the value of ζ_I on *a posteriori* basis can be deduced (Table 25.1).

Table 25.1 Estimation of the Intrinsic Chirality of the Inclusions

Frequency in GHz	Measured Cross-polarization power C_p in dB relative to free-space transmission	Proposed measure of effective chirality $\varepsilon_{eff} = \{\text{antilog}[-C_p/10]\}^{-1/2}$	ε_1 as determined from ε_{eff} *via* Equation (25.6)
8.5	10	0.3162	0.9860
9.0	10	0.3162	0.9930
9.5	12	0.2512	0.7500
10.0	11	0.2818	0.8700
10.2	10	0.3162	0.9990

8. The results further indicate that ε_{eff} and μ_{eff} are frequency dependent (Figures 25.4 and 25.5) mainly due to the complex permittivity of the inclusions. The effective chirality of the overall mixture as given by Equation 25.14 is, however, only slightly frequency dependent as decided by the transmission coefficient(s) involving the electrical thickness (d/λ_0) of the test sample. This is confirmed by a minor variation in the measured chirality-dependent cross-polarized component *versus* frequency in Figure 25.3.

To conclude, the approach pursued in [17] refers to the extension of the well-known logarithmic law of mixing on an *ad hoc* basis to determine the effective electromagnetic constitutive parameters of chiralic mixtures.

The theoretical results so obtained are supplemented by some experimental results to portray the design feasibilities of realizing chiral composites at microwave frequencies. A variety of applications of such composites have been considered in practice [13–15]. To name a few, EMI shields, radar absorbing materials, polarizing lenses, etc. can be designed to match certain specific characteristics with the type of chiralic mixtures discussed.

The formulations indicated in this chapter are useful in synthesizing chiralic composites with chiralic inclusions of size comparable to the wavelengths of operation. If the size of the inclusions is too small (in comparison to the wavelength), it will render the medium achiralic. Then the present formulations will refer to an achiralic mixture medium with $\zeta_I = 0$.

Last, the following should be noted concerning the theoretical considerations of this chapter which are based on combining Fricke's formula and the logarithmic law of mixing as applied to a chiralic composite. Unlike Maxwell-Garnett's formulation, Fricke's formula (see Chapters 4–7) is devoid of dilute-inclusion approximation; and the shape factor is controlled in the analysis presented here by the logarithmic law, permitting the mixture to be viewed as a stochastic entity with macroscopic random attributes to its constitutive parameters. Thus, although not derived on the basis of the first principles, the algorithms of this chapter are in line with the electrostatic aspects of Fricke's formula and the probabilistic characteristics of a statistical mixture.

25.8 Orderly-Textured Chiralic Mixture Media

In Chapter 4 (Section 4.5), the case of an orderly-textured simple (achiralic) dielectric mixture was discussed and relevant formulations were presented. Now, relevant to the discussion on chiralic mixtures a new class of chiralic mixtures called *orderly-textured chiralic mixtures* are considered. Such mixtures can be constructed by an ordered arrangement of shaped, chiralic inclusions in an achiralic host. The effective electromagnetic properties of such a mixture would be considerably different from those of both simple orderly-textured media as well as random chiralic media discussed earlier.

In order to describe the effective electromagnetic properties (permittivity and permeability) of an orderly-textured chiralic mixture, the formulations indicated in Chapter 4 can be utilized and a strategy similar to that of the foregoing sections can be adopted. In the previous sections the logarithmic law was extended in a modified form to describe the effective properties of a chiralic mixture comprised of an achiralic host and randomly dispersed with shaped chiralic inclusions. Accordingly, the formulations for the effective permittivity and the effective permeability of such a mixture are as given by Equations (25.9) and 25.10.

Now utilizing the above formulations and following a weighting strategy similar to that adopted for the orderly-textured simple (achiralic) mixture, analytical formulations for the effective properties of an orderly-textured chiralic mixture can be developed. That is, corresponding to the Equations of 4.23 and 4.24, the formulations for the effective permittivity (ε_{eff}') and the effective permeability (μ_{eff}') for an orderly-textured chiralic mixture can be written as follows:

$$\varepsilon_{eff}' = \varepsilon_{eff}(\varepsilon_U/\varepsilon_{eff})^r \tag{25.16a}$$

$$\mu_{eff}' = \mu_{eff}(\mu_L/\mu_{eff})^r \tag{25.16b}$$

and

$$\varepsilon_{eff}' = \varepsilon_{eff}(\varepsilon_L/\varepsilon_{eff})^n \tag{25.17a}$$

$$\mu_{eff}' = \mu_{eff}(\mu_U/\mu_{eff})^n \tag{25.17b}$$

All the notations and other conditions or statements as regards to the orientation specified with respect to the electric field direction remain the same as those for a simple orderly-textured dielectric mixture discussed in Chapter 4.

The above formulations (Equations 25.16 and 25.17) are concerned only with the interaction of an orderly-textured chiralic mixture with the direction of an external applied electric field. However, it should be noted that inasmuch as an externally applied magnetic field can also evoke a response in a chiralic medium, the corresponding formulations for ε_{eff}' and μ_{eff}' would be different from Equations 25.16 and 25.17.

It should be observed that even though the effective permittivity (ε_{eff}) alone was discussed for the case of orderly-textured achiralic mixtures presented in Chapter 4, formulations for both ε_{eff}' and μ_{eff}' are to be considered in the case of such chiralic mixtures. This is due to the fact that in a chiralic mixture (unlike an achiralic mixture) the electric and magnetic fields are cross-coupled.

In summary, proceeding from the simple logarithmic law of mixing weighted on the basis of Langevin's function, the effective permittivity of a simple, orderly-textured dielectric mixture can be elucidated. Hence, the effective permittivity and permeability of an orderly-textured, chiralic mixture can also evaluated with the aid of the results presented in the previous sections.

25.9 Sample Results on Orderly-Textured Chiralic Mixtures

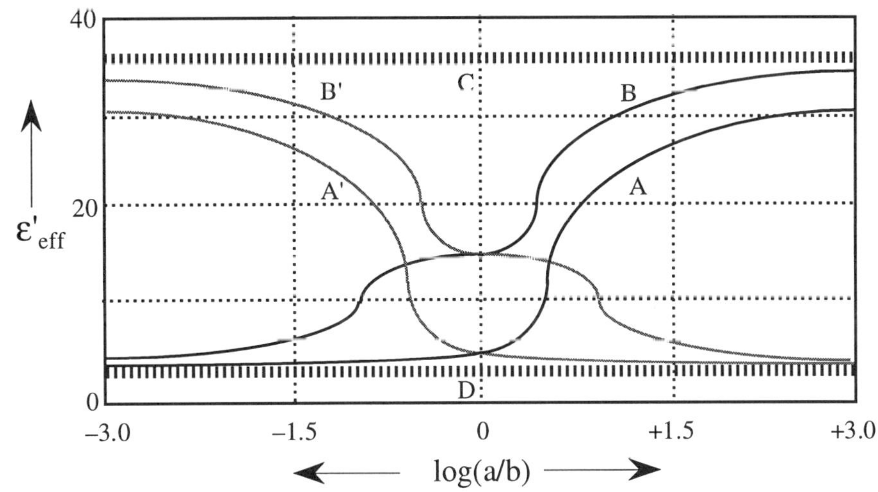

Figure 25.7 Effective permittivity (ε_{eff}') of an orderly-textured chiralic mixture *versus* the (a/b) ratio of the inclusions of volume fraction, $\theta = 0.4$.
(Data: $\varepsilon_r = 78.3$; $\mu_r = 1000$; $\varepsilon_2 = 2$; $\mu_2 = 55$; $\gamma = +1$; $\zeta = 0.0001$.)
AA': Sihvola and Lindell's [10] formulations for perpendicular orientation (A) or parallel (A') to the applied electric field (**E**); BB': Corresponding results with the formulations of Equations 25.16 and 25.17; C: Wiener's upper limit; D: Wiener's lower limit.

The results corresponding to an orderly-textured chiralic mixture as per Equations 25.16 and 25.17 are compared with those due to Sihvola and Lindell [10] in Figure 25.7.

- It is observed from Figure 25.8 that for prolate spheroidal inclusions the value of ε_{eff}' tends towards Wiener's upper limit and μ_{eff}' to Wiener's lower limit in the limiting case of all inclusions being aligned parallel to the applied electric field. Likewise, when all the inclusions are antiparallel to the applied electric field, the value of ε_{eff}' tends towards Wiener's lower limit while μ_{eff}' tends towards the upper limit.

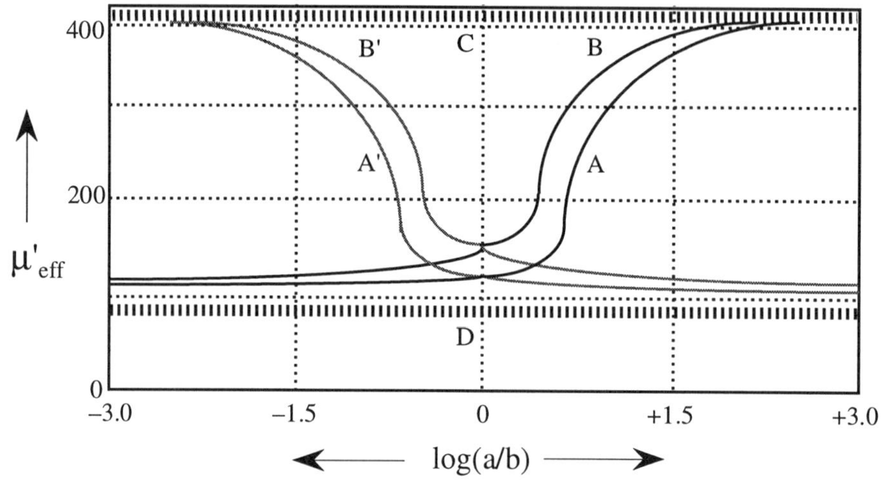

Figure 25.8 Effective permeability (μ_{eff}') of an orderly-textured chiralic mixture *versus* the (a/b) ratio of the inclusions of volume fraction, $\theta = 0.4$.
(Data: $\varepsilon_r = 78.3$; $\mu_r = 1000$; $\varepsilon_2 = 2$; $\mu_2 = 55$; $\gamma = +1$; $\zeta = 0.0001$.)
AA': Sihvola and Lindell's [10] formulations for perpendicular orientation (A) or parallel (A') to the applied electric field (**E**); BB': Corresponding results with the formulations of Equations 25.16 and 25.17; C: Wiener's upper limit; D: Wiener's lower limit.

- Again as before, the above are reversed for oblate spheroidal inclusions.
- Unlike Taylor's formulations discussed in Chapter 4 of ε_{eff} at $e = 0$, Sihvola and Lindell's formulation underestimates the values of both ε_{eff}' and μ_{eff}' at $e = 0$ when compared to the logarithmic law. This is could again be reasoned as due to the dilute phase approximation of Maxwell-Garnett theory on the basis of which Sihvola and Lindell's formulations were derived.
- Use of Langevin's theory enables the construction of the ordered texture from the disordered dispersion regardless of particulate concentration and it also implicitly accounts for the interparticulate interaction within the macroscopic test mixture.

25.10 Applications of Electromagnetic Chiralic Materials

Chirosorb™: A novel material which is invisible to incident to EM waves (by virtue of its zero reflectance characteristics) has been synthesized [14] by embedding randomly oriented identical microstructures (such as microhelices) in an isotropic host medium. This material could be used for RCS reduction purposes.

Chirowaveguide materials: When a waveguide is filled with an appropriate electromagnetic chiralic material, the propagating TE and TM modes become coupled with a coefficient proportional to the chiralic admittance parameter. Possible use of this principle in polymer waveguides and integrated optical devices is indicted in [13].

Chiralic substrate materials: For microstrip antennas use of a chiralic material as a substrate has been suggested. Since chiral materials are polarization sensitive, use of these materials in such structures would enable controlling the antenna beam characteristics, bandwidth, and radiation efficiency.

Chiral-coated EM shieldings: Chiral-coated surfaces are feasible as EM shields compatible for common EM ambients and pulsed electromagnetic (EMP) environments. Such surfaces can also be synthesized for RCS control [24]. *Salisbury shield* realization with chiral substances is also a conceivable product [25].

Chiralic material for EM focusing: Named as *"chirolens"* of spherical geometry, Enghetta and Kowartz have indicated [24] a novel method of realizing two focal points corresponding to two eigen modes of the EM field components present in the chiralic lens material. It is indicated that such lens can focus one of the modes and defocus the other. Potential use of a chirolens as couplers for waveguides, polarization filters, etc. has been suggested.

References

[1] Tellegen, B. D. F.: The gyrator, a new electric network element, Phillips Res. Repts., vol. 3(2), 1984: 81-101.

[2] Pasteur, L.: Sur les relations qui peuvent exister entre la forme cristalline, la composition chimique st le sens de la polarisation rotatorie, Ann. Chim. Phys., vol. 24, 1848: 442-459.

[3] Lakhatia, A. et. al.: Time-harmonic electromagnetic fields in chiral media, in *Lecture Notes in Physics*. (Springer-Verlag, Berlin: 1989), pp. 335.

[4] Fresnel, A.: Memoire sur la double refraction que les rayons lumineux eprouvent en traversant les aiguilles de cristal de roche suivant des directions paralleles a laxe, Oeuvres, vol. 1, 1822: 731-751.

[5] Lindman, K.F.: Über die durch ein aktives Raumgitter erzeugte Rotationspolarization der elecktromagnetischen Wellen, Ann. Phy., vol. 69, 1922: 270-284.

[6] Bassiri, S.: Electromagnetic Wave Propagation and Radiation in Chiral Media, Doctoral Dissertation, Division of Engineering and Applied Science, California Institute of Technology, Pasadena, CA: 1987.

[7] Jaggard, D.L., et. al.: On electromagnetic waves in chiral media, Appl. Phys., vol. 18, 1979: 211-216.

[8] Silverman, M.P.: Specular light scattering from a chiral medium: Unambiguous test of gyrotropic constitutive relations, Lett. Nuova Cimento., vol. 43, 1985: 378-382.

[9] Sihvola, A.H. and Lindell, I.V.: Chiral Maxwell-Garnett mixing formula, Electron. Letts., vol. 26(2), 1990: 118-119.

[10] Sihvola, A.H. and Lindell, I.V.: Polarisability and mixing formula for chiral ellipsoids, Electron. Letts., vol. 26(14), 1990: 1007-1009.

[11] Lakhatia, A. et al.: Effective properties of a sparse random distribution of non-interacting small chiral spheres in a chiral host medium, J. Phys. D., vol. 24, 1991: 1-6.

[12] Lakhatia, A., et al.: Scattering and absorption characteristics of lossy dielectric, chiral, non-spherical objects, Appl. Opt., vol. 24, 1985: 4146-4154.

[13] Pelet, P. and Enghetta, N.: Theory of chirowaveguides, IEEE Trans. Antennas Propagat, vol. AP-38(1), 1990: 90-97.

[14] Jaggard, D.L. and Enghetta, N.: Chirosorb™ as an invisible medium, Electron. Letts., vol. 25(3), 1989: 173-174.

[15] Jaggard, D.L. et al.: Chiroshield: A Salisbury/Dallenbach shield alternative, Electron. Letts., vol. 26(17), 1990: 1332-1333.

[16] Neelakanta, P.S., Subramaniam, K. and Chaoli, Gu.: Permittivity and permeability of chiralic mixture: Application of logarithmic law of mixing. Electron. Letts., vol. 27(6), 1991: 496-497.

[17] Subramaniam, K.: Application of Stochastical Mixture Theory in the Design of Electromagnetic Composites. M.S.E. Thesis, Department of Electrical Engineering, Florida Atlantic University, Boca Raton, FL, April 1992.

[18] Lichtenecker, K. and Rother, K.: Die Herleitung des logarithmischen Mischungsgesetzes aus allegemeinen Principien der stationaören Strömung. Physik. Zeitschr, vol. 32, 1931: 255-260.

[19] Lichtencker, K.: Mischkorpertheori als lichkeitswahrscheis Problem. Physik. Zeitschr., vol. 30, 1929: 805-809.

[20] Kisdnasamy. S. and Neelakantaswamy. P.S.: Complex permittivity of a dielectric mixture: Modified Fricke's formula based on logarithmic law of mixing. Electron. Letts. vol. 20(7), 1984: 291-292.

[21] Neelakanta, P.S.: Complex permittivity of a conductor loaded dielectric, J. Phys. (Condensed Matter), vol. 2, 1990: 4935-4947.

[22] Guire, T. et al.: Influence of chirality on the reflection of EM waves by planar dilectric studies. IEEE Trans. Electromag. Compat, vol. EMC-32(4), 1990: 300-303.

[23] S. R. Wallin: Dielectric Properties of Heterogeneous Media, Ph. D. Thesis, University of Wyoming, 1985.

[24] N. Enghetta and M. W. Kowarz: Chiropens as a bifocal lens, Abstract of 1990 URSI Radio Science Mtg. (May 7-11, Dallas, TX), pp. 134.

[25] N. Enghetta: Chiral materials and chiral electrodynamics: Basic physical principles and background. Presented at the Workshop of Chiral and Complex Materials, Progress in Electromagnetic Research Symp. (PIERS), Cambridge, MA, July 1-5, 1991.

[26] A. K. Bhattacharya: Control of radar cross-section and cross-polarization characteristics of an isotropic chiral sphere. Electron. Letts., vol. 26(14), 1990: 1066-1077.

Defining Terms

Chirality: (Right or left)-handedness.

Chiralic composite: A composite medium in which an achiralic host medium forms a receptable for the dispersion of chiralic inclusions.

Chiralic materials: Medium wherein electric and magnetic fields are cross-coupled.

Optical rotary dispersion (ORD): Geometry-induced rotation of plane of polarization of a plane wave propagating across a medium.

CHAPTER 26

Electromagnetic Phantom Materials

26.1 Introduction

A state-of-the-art interest in electromagnetic research is to elucidate and characterize the interactions between *nonionizing* radiations *versus* living systems. The purpose is three-fold: (1) To understand the mechanism(s) of interactions involved; (2) to investigate any deleterious effects which may result from such interactions; and (3) to consider the possibilities of such interactions inducing beneficial effects in the living systems.

A subset of relevant studies refers to electromagnetic dosimetry and hyperthermia considerations. In both cases, it is of interest to know the extent of absorption of nonionizing radiation by living systems.

Nonionizing radiation in general refers to electromagnetic (EM) waves from the quasistatic region to the end of visible radiation. Characteristically, unlike ionizing radiation such as X-rays and/or radioactive emissions, nonionizing radiation is not of sufficient energy to ionize the molecules of the medium (such as air) through which it proliferates.

Simulating materials to depict the dielectric and thermal characteristics of various biological structures over different frequency ranges refer to synthesizing the EM phantom materials. The biological materials are in general not monolithic and are invariably constituted by two are more substances. Development of materials which mimic the behavior of biological substances in their response to EM interactions, is therefore directed at synthesizing composite media as appropriate. Such endeavors warrant clear specifications on the complex dielectric parameters of various biological materials.

Such characteristics are, however, largely dependent on data measured under *in vivo* or *in vitro* conditions, temperature, and living systems from which the test specimens were gathered.

Therefore unique specifications for the complex dielectric parameters of biological materials are not feasible. However, on the basis of various experimental works, the data gathered are summarized in the next section which represents the average values of the relevant parameters, as reported in the available literature.

26.2 Complex Dielectric Properties of Biological Materials

Essentially, biological substances can be classified into the following categories:

- Biological fluids
- Water-borne tissues, muscles, skin etc. (with large water content)
- Bones, fat, and tissues (with low water content)

Depending on the frequency band of interest, the measured data on specific resistance and the complex permittivity parameters (namely, $\varepsilon' - j\varepsilon''$) have been widely reported in the literature, a compendium of which is presented in the following tables (Tables 26.1 to 26.6):

26.2.1 Biological Fluids

Table 26.1 The Complex Dielectric Constant of Human Blood at Room Temperature

Relative Permittivity of Blood: $\varepsilon_B = (\varepsilon_B' - j\varepsilon_B'')$

Frequency in MHz (f)	Volume Fraction of RBC* (Hematocrit Value) (c)	Measured Values due to Bianco et al. [1]	
		ε_B'	ε_B''
100	0.14	68.7466	–196.5908
	0.28	67.6619	–181.7745
	0.41	66.9199	–164.5834
	0.53	65.9390	–155.2289
	0.84	63.6034	–123.1350
500	0.14	65.5428	–42.4751
	0.28	62.7100	–39.6517
	0.41	61.0684	–36.4393
	0.53	59.0783	–34.4925
	0.84	54.4693	–28.4197
1000	0.14	63.8943	–23.9896
	0.28	60.9095	–22.2137
	0.41	59.1531	–20.6747
	0.53	57.1505	–19.6454
	0.84	52.4498	–16.3249
1500	0.14	63.0756	–18.3516
	0.28	59.9311	–17.0036
	0.41	57.9560	–15.9502
	0.53	55.8107	15.3838
	0.84	50.9991	12.9609
2000	0.14	63.0435	–15.8471
	0.28	59.5073	–14.8409
	0.41	57.5331	–14.0258
	0.53	55.2680	–13.5188
	0.84	50.2958	–11.4703

Note: RBC*: Red blood corpuscles.

Table 26.2 Conductivity of Human Blood

Frequency in MHz	Hematocrit Value c	Conductivity of Blood σ_B siemen/meter
		Measured Values Bianco et al. [1]
100	0.14	1.0937
	0.28	1.0112
	0.41	0.9156
	0.53	0.8636
	0.84	0.6850
500	0.14	1.1815
	0.28	1.1029
	0.41	1.0136
	0.53	0.9594
	0.84	0.7905
1000	0.14	1.3300
	0.28	1.2358
	0.41	1.1502
	0.53	1.0929
	0.84	0.9082
1500	0.14	1.5314
	0.28	1.4190
	0.41	1.3310
	0.53	1.2807
	0.84	1.0816
2000	0.14	1.7632
	0.28	1.6512
	0.41	1.5606
	0.53	1.5041
	0.84	1.2762

Note: $\sigma_B = (55.631452 \times 10^{-12}) \times \varepsilon''_B \times$ frequency (in Hz) siemen/meter.

Table 26.3 Resistivity of Blood [5]

Substance	Resistivity (ohm-cm)	Frequency	Temp. (°C)	Remarks
Human blood	150	d.c.	40	
	155	20–5 KHz	40	
	165 aver. (148-176)	1 KHz	37	Normal subject
	137.8	1 KHz	37	34.4% hematocrit
	170.3	1 KHz	37	37.2% hematocrit
	169.1	1 KHz	37	40.2% hematocrit
	176.0	1 KHz	37	42.5% hematocrit
	131.2	1 KHz	37	43.9% hematocrit
	230.9	1 KHz	37	50.9% hematocrit
	199.8	1 KHz	37	55.6% hematocrit
	180.0	1 KHz	37	56.4% hematocrit (all calculated values)
	160.0	d.c.	37	
	154 aver. (of two methods)	120 KHz	36.3 aver.	Flowing venous blood
	200.0	d.c.	20	
	195.0	20–5 KHz	20	
	230.0	d.c.	18	
	363.0	120 KHz	1.3	Flowing blood
Human blood	63 aver. (61-67)	1 KHz	37	Normal subjects
	100	d.c.	18	
	70	d.c.	Body	
Dog blood	108	100 KHz	Body	29% hematocrit
	118	100 KHz	Body	33% hematocrit
	120	100 KHz	Body	36% hematocrit
	129 and 158	100 KHz	Body	40% hematocrit
	155	100 KHz	Body	41% hematocrit
	153	100 KHz	Body	47% hematocrit
	156-243	Inductorium	38	Measured within 1 min
	207 aver. (185-230)	1 KHz	Body	Approx. 50% hematocrit
Dog serum	138 aver. (98-178)	1 KHz	Body	

(continued...)

Substance	Resistivity (ohm-cm)	Frequency	Temp. (°C)	Remarks
Cow blood	145	20–5 KHz	38	
	131	d.c.	37	
Cow-pig blood	137 aver. (119-152)	Audio	37	
	192 aver.	Audio	20	
Cow blood	190	d.c.	20	
	180	20– 5 KHz	20	
	169.6 aver. (164.5–174)	1 KHz	Room	Stationary 70% cells
	164 aver. (159–168)	1 KHz	Room	Flowing at 15 cm/sec–70%cells
	271.75	1 KHz	Room	100% cells stationary
	249.75	1 KHz	Room	100% cells flowing–15 cm/sec
	116.15	1 KHz	Room	40% cells stationary
	114.65	1 KHz	Room	40% cells flowing–15 cm/sec
	80.75	1 KHz	Room	5% cells stationary
	80.75	1 KHz	Room	5% cells flowing–15 cm/sec
	196.5 aver.	1 KHz	Room	80% hematocrit–stationary
	189.25 aver.	1 KHz	Room	80% hematocrit flowing–15 cm/sec
	145.4 aver.	1 KHz	Room	60% hematocrit–stationary
	142.25 aver.	1 KHz	Room	60% hematocrit flowing–15 cm/sec
	91.5 aver.	1 KHz	Room	20% hematocrit–stationary
	91.35 aver.	1 KHz	Room	20% hematocrit flowing–15 cm/sec
Cow-pig blood	91 aver.	25–100 MHz	37	
	133 aver.	25–100 MHz	20	
	99 aver.	25–100 MHz	37	
Cow blood	89 aver. (89–96)	200–900 MHz	27	

(continued...)

Substance	Resistivity (ohm-cm)	Frequency	Temp. (^{o}C)	Remarks
Cow plasma	65	20 Hz – 5 KHz	40	
Cow-pig serum	62.5	Audio	37	
	83	Audio	20	
Cow plasma	90	20 Hz–5 KHz	20	
Calf serum	89.4	87 KHz–4.52MHz	21.6	
Cow-pig serum	62.5	25–100 MHz	37	
	83	25–100 MHz	20	
Horse red cells	3890	1 KHz	25	Centrifuged and packed
	840	400 KHz	25	Centrifuged and packed
	410	1 MHz	25	Centrifuged and packed
	285	2 MHz	25	Centrifuged and packed
	232	3.5 MHz	25	Centrifuged and packed
Rabbit blood	128 (117–136)	1 KHz	39	
	129–176	Inductorium	39	
Turtle serum	120	Inductorium	16	
	110	Inductorium	20	
	103 aver.	Inductorium	24	
	91 aver.	Inductorium	30	
	78 aver.	Inductorium	38	

Table 26.4 Resistivity of Other Body Fluids

Substance	Resistivity (ohm-cm)	Frequency	Temp. (^{o}C)	Remarks
C.S.F.				
Human	64.6 (64.0–65.2)	1–30 KHz	24.5	
Cat	65.7 (65.5–66.1)	1–30 KHz	24.5	
Rabbit	55.9 aver. (51–62)	1 KHz	39	
Bile				
	60	Audio	37	
Cow-pig	78	Audio	20	
	59	50 MHz	37	
	76	50 MHz	20	
Rabbit	66.2 aver. (61–72)	1 KHz	39	
Amniotic fluid				
Sheep	65	1 KHz	25	Measured after 2 days of refrigerated storage
	39	1 KHz	37.5	
Urine				
Cow-pig	30	Audio	37	
	39	Audio	20	
Physiological solutions				
Saline	72	d.c.	18	
Saline 0.9%	57 aver. (56–58)	200–900 MHz	27	
Tyrode	52	Not given	37	
Saline 0.9%	70	1 KHz	20	
	50.5	1 KHz	37.5	
3 M KCL	4.28	–	20	
	3.26	–	37.5	
	3.85	–	23.5	
	3.70	–	38	
Saline 1%	55.5	–	23.5	
	50.0	–	38	
Saline 2M	7.14	–	23.5	
	5.88	–	38	

26.2.2 Biological Solid Substances

Table 26.5 Dielectric Properties of Fat, Bone, and Tissues with Low Water Content

Frequency (MHz)	Dielectric Constant ε_L	Conductivity σ_L (millisiemen/meter)
1		
10		
27.12	20.0	10.9–43.2
40.68	14.6	12.6–52.8
100	7.45	19.1–75.9
200	5.95	25.8–94.2
300	5.70	31.6–107
433	5.60	37.9–118
750	5.60	49.8–138
915	5.60	55.6–147
1500	5.60	70.8–171
2450	5.50	96.4–213
3000	5.50	110–234
5000	5.50	162–309
5900	5.05	186–338
8000	4.70	255–431
10000	4.50	324–549

Table 26.6 Dielectric Properties of Muscle, Skin, and Tissues with High Water Content

Frequency (MHz)	Dielectric Constant (ε_H)	Conductivity σ_H (mho/m)
1	2000	0.400
10	160	0.625
27.12	113	0.612
40.68	97.3	0.693
100	71.7	0.889
200	56.5	1.28
300	54	1.37
433	53	1.43
750	52	1.54
915	51	1.60
1500	49	1.77
2450	47	2.21
3000	46	2.26
5000	44	3.92
5900	43.3	4.73
8000	40	7.65
10000	39.9	10.3

In addition to the data presented in the above tables, more details can be seen in [2-5].

26.3 Electromagnetic Phantom Materials: Synthesizing Concepts

Basically an EM phantom material is a lossy dielectric material constituted by appropriate lossy/loss-less substances with the end result of emulating the complex dielectric characteristics of a given biological medium at a specified temperature and frequency. The complex permittivity of a medium is specified by the Debye expression, namely:

$$\begin{aligned} \varepsilon &= \varepsilon_\infty + [(\varepsilon_s - \varepsilon_\infty)/(1 + j\omega\tau)^{1-\alpha}] - (\sigma/\omega\varepsilon_o) \\ &= \varepsilon_b' - j\varepsilon_b'' \end{aligned} \tag{26.1}$$

where

ε_∞	Dielectric constant of the material at infinity frequency
ε_s	Static dielectric constant of the material
τ	Relaxation time of the material in second
ω	$2\pi f$, with f frequency in Hertz
α	Empirical parameter
σ	Ionic conductivity of the material (siemen/meter)
ε_o	8.854×10^{-12} farads/meter (free-space permittivity)

A phantom is designed to exhibit the similar dielectric properties as dictated by Equation 26.1. That is, if the complex permittivity of a biological medium, namely, $(\varepsilon_b' - j\varepsilon_b'')$, is known, the corresponding phantom material should have the permittivity values $\varepsilon_p' = \varepsilon_b'$ and $\varepsilon_p'' = \varepsilon_b''$. In the following sections, the methods of emulating such dielectric properties in phantoms are addressed.

26.4 Saline Solution as a Phantom Material

The simplest way to realize the lossy biological characteristics (such as that of a tissue material) is to use saline solution as the phantom material. Water with the addition of NaCl becomes a lossy dielectric whose dielectric permittivity can then be expressed by the following Debye formula [6]:

$$\varepsilon_W = (\varepsilon_\infty)_W + [(\varepsilon_s)_W - (\varepsilon_\infty)_W]/[1 + j\omega\tau_W]^{1-\beta} - j[\sigma_W/\omega\varepsilon_o] \tag{26.2}$$

where

$(\varepsilon_\infty)_W$	Dielectric constant of the material at infinity frequency = 4.9
$(\varepsilon_s)_W$	Static dielectric constant of the water
τ_W	Relaxation time of the material in second
ω	$2\pi f$, f frequency in hertz
β	Empirical parameter $\approx (0.02 \pm 0.007)$
σ_W	Ionic conductivity of the material (siemen/meter)
ε_o	8.854×10^{-12} farads/meter (free-space permittivity)

If T indicates the temperature in °C and S is the salinity (in parts per thousand) of the salt water, then,

$$(\varepsilon_s)_W(T) = 88.045 - 0.4147T + 6.295 \times 10^{-4}\, T^2 + 1.075 \times 10^{-5}\, T^3 \tag{26.3}$$

$$\sigma_W(T,S) = \sigma_W(25,S)\, \exp(-\Delta\gamma) \tag{26.4}$$

with

$$\Delta = (25 - T),$$
$$\gamma = 2.033 \times 10^{-2} + 1.266 \times 10^{-4}\Delta + 2.464 \times 10^{-6}\Delta^2$$
$$- S(1.849 \times 10^{-5} - 2.551 \times 10^{-7}\Delta + 2.551 \times 10^{-8}\Delta^2) \quad (26.5)$$

and

$$\sigma_W(25,S) = S(0.182521 - 1.46192 \times 10^{-3}S + 2.09324 \times 10^{-5}S^2$$
$$- 1.28205 \times 10^{-7}S^3) \quad (26.6)$$

Further,

$$\tau_W(T,S) = \tau_W(T,O)\, b(S,T) \quad (26.7)$$

where

$$\tau_W(T,O) = 1.768 \times 10^{-11} - 6.086 \times 10^{-13}T + 1.104 \times 10^{-14}T^2 - 8.11 \times 10^{-17}T^3$$

and

$$b(S,T) = 1.000 + 2.282 \times 10^{-5}ST - 7.638 \times 10^{-4}S - 7.760 \times 10^{-6}S^2$$
$$+ 1.105 \times 10^{-8}S^3$$

The *salinity factor (S)* is related to the percentage concentration of the salt content by the following relation:

$$S = (1.805 \times chlorinity + 0.030) \quad (26.8)$$

and

$$\text{Chlorinity} = \frac{(6.0657)}{(\text{Specific gravity of the solution})}\,\frac{(\text{Weight of NaCl} \times 100)}{(\text{Volume of the solvent})} \quad (26.9)$$

The specific gravity of NaCl solution for a given percentage concentration (that is, weight/volume %) can be extrapolated from the following data:

NaCl Concentration (Weight/vol)%	3.30	3.28	3.21	3.20	3.15	3.14	3.12	3.07	2.94	2.78	1.48
Specific gravity	1.0250	1.0249	1.0243	1.0242	1.0238	1.0238	1.0236	1.0232	1.0216	1.0209	1.0104

Though a saline solution offers a simple phantom synthesization, it is useful only for the representation of a simple biological medium (such as a simple structure to emulate the specific absorption rate (SAR) pertinent to human adult, etc.). The dielectric behavior of saline solution is highly temperature sensitive and also it is sensitive to the concentration of NaCl content. Further, being a fluid, its use in a compact geometry is rather futile. Nevertheless, columns of saline solution are being used as human phantoms to study their proximity effect in the performance of devices like pagers. Nonliquid (semisolid) phantoms *in lieu* of water are preferred in practice because of their moldability to required shapes. Various such semisolid phantoms developed and reported in [7-11] are briefly described in the following sections.

26.5 Polyacrylamide Gels as Phantom Materials

A convenient way of emulating biological materials such as tissues is to use the polyacrylamide gel which offers the following advantages as phantom materials: (i) Easily shaped into complex forms, (ii) an elasticsolid, (iii) readily prepared with a complete range of highly reproducible values of electrical parameters, (iv) optically transparent facilitating the insertion probes, and etc. (v) low cost material.

The base material to constitute a polyacrylamide phantom is acrylamide (C_3H_5NO) which with the addition of water becomes gel-like. NaCl is used to dope the gel to realize desired lossy characteristics. As reported in [9], the following are the techniques associated in realizing viable phantom materials using polyacrylamide:

Constitutents of the gel:

1. Base material: Polymeric acrylamide (C_3H_5NO)
2. Water
3. Polymerization catalysts:
   MBA (N, N'-methylene-bis-acrylamide, $C_7H_{10}N_2O_2$)
   TEMEDA (N, N, N', N'-tetramethyl-ethylene-diamine, $C_6H_{16}N_2$)
4. Primer: Ammonium persulfate (AP) $(NH_4)_2S_2O_8$

Recipè:

Phase 1: 15 grams of acrylamide
+0.1 gram of MBA
+0.5 gram of TEMEDA
plus water to make a total volume of 90 cm^3 + adequate amount of NaCl to acquire a specified electrical conductivity.

Phase 2: AP in water (1.3% by weight) to form 10 cm^3

Phase 3: Solutions of phase 1 and phase 2 mixed to set the polymerization

Conductivity control:

Electrical conductivity of the above material is primarily dictated by NaCl content. With no salt added the lower limit of conductivity is 0.15 siemen/meter.

Permittivity control:

This is accomplished by varying acrylamide content (maintaining the stoichiometric proportions of MBA and TEMEDA). Nominal dielectric constant $\approx$60. This value can be lowered by substituting water with low permittivity liquids such as dioxane, pyridine, or ethylene glycol. Stability of the composition is dependent on loss of water content *via* evaporation.

Thermal properties of these materials have also been assayed as reported in [9].

26.6 Other Semisolid Gels to Simulate Soft-Tissue Materials

Soft-tissue phantoms can be fabricated from the composition of a saline solution, polyethylene powder, and gelling agent. Resulting phantom materials are very viscous, pliable and putty-like and are suitable to simulate bisected phantoms for thermographic studies.

Another gel-like phantom material is constituted from hydroxethyl cellulose (HEC) and saline solution. For frequencies above 100 MHz, polyethylene or sugar is also added.

A moist gellied plastic to simulate muscle is formed as described below.

This material is constituted from 76.5% (by weight) of saline solution (equivalent to 12 grams of salt/liter), 15.2% (by weight) of powdered polyethylene, and 8.4% (by weight) of Super Stuff™ which is a gelling agent supplied by Whamo Manufacturing Company, San Gabriel, CA. The dielectric properties of this phantom material are: Dielectric constant ≈ 49-58 and loss tangent 0.33-1.7. The other physical properties are: Density ≈ 1.0 and specific heat 0.24-0.30.

26.7 Simulation of Bone and Fat Phantoms

Bone and adipose tissue (fat) have low water content. They can be simulated in dry plastic-like form using: 84.81% (by weight) of laminac polyester resin + 0.45% (by weight) of a catalyst (for example, methyl ethyl ketone peroxide 60%) + 0.24% (by weight) of acetylene black + 14.5% (by weight) of aluminum powder. The dielectric constant of this phantom is 4.6-6.2 and the loss tangent is 0.70-0.55. The density of the material is 1.3 and the specific heat is 0.24-0.30.

Other possible compositions of biological phantom materials are presented in Table 26.7.

Table 26.7 Composition of Typical Phantom Materials

Biological Media	Compositions in % by Weight				
	Aluminum Powder	Polyethylene Powder	Gel Agent	Water	NaCl
Brain (2450 MHz)		29.8	7.01	62.6	0.582
Muscle (2450 MHz)		15.2	8.45	75.4	0.907
Muscle (27 MHz)	13.7		9.49	76.6	0.153
	Aluminum Powder	XC–72*	Acetylene Black	Polyester Resin**	
Adipose (27 MHz)	29.46	0.94	0.30	69.3	

*XC–72 ("Fluffy") carbon powder, Cabot Corporation.
**Laminac 4110.

Table 26.8 Synthesis of Typical Biological Equivalent Phantoms

1. Muscle and brain equivalent phantom

Material	Percentage by Weight	
	Muscle	Brain
Water	52.4	40.4
Salt (NaCl)	1.4	2.5
Sugar	45.0	56.0
HEC	1.0	1.0
Bactericide	0.1	0.1

2. Lung material

Material	Percentage by Volume
Muscle material (above)	47
Microspheres	53

3. Bone material (castable)

Material	Percentage by Weight
Two-ton epoxy	35.0
Epoxy	35.0
Hardener	28.0
KCl solution	–

4. Bone material (liquid)

Material	Percentage by Weight
TWEEN	57.0
n-Amyl alcohol	28.5
Paraffin oil	9.5
Water	4.5
Salt (NaCl)	0.5

26.8 Thermal Properties of Phantom Materials

In dosimetry applications, knowing the precise values of physical and thermal parameters of phantoms are rather essential. Typical values of these parameters *vis-a-vis* a variety of phantom materials as reported in [7-11] listed in the following table.

Table 26.9 Thermal Properties of Phantom Materials

Simulated Phantom	Thermal Conductivity (W/m-K)	Specific Heat ($\times 10^3$ J/kg-K)	Density ($\times 10^3$ kg/m^3)	Thermal Diffusivity ($\times 10^{-6}$ m^2/s)
Ethylene glycol	0.259 + 0.005	–	–	–
Brain (2450 MHz)	0.478 + 0.015	3.41	0.98 0.03	0.143
Muscle (2450 MHz)	0.535 0.003	3.70	1.00 0.02	0.145
Muscle (27 MHz)	0.657 0.028	3.58	1.00 0.01	0.167
Adipose tissue (27 MHz)	0.352 0.004	1.07	1.43 0.01	0.230

Actual biological substances have the following physical and thermal charcteristics as estimated on an average basis.

Table 26.10 Thermal Properties of Actual Biological Media

Material	Thermal Conductivity ([W/(m – °C)] $\times 10^3$)	Specific Heat [(J/kg – °C)] $\times 10^3$)	Thermal Diffusivity ([m^2/s] $\times 10^{-6}$)	Mass Density ([kg/m^3] $\times 10^3$)
Muscle *in vitro*	0.197–0.545	3.51	0.183	1.07
Brain *in vitro*	0.497–0.566	3.47	0.134–0.143	1.07
Fat tissue i*n vitro*	0.132–0.371	1.21–1.55	0.230	1.06
Bone	–	1.26–2.97	–	1.25–1.79

It should be noted that the thermal properties of the phantom materials are also largely influenced by the conducting constitutents of the composite such as aluminum powder and/or NaCl. For example, if the weight fraction of aluminum is changed by a 1:4 ratio, thermal conductivity is increased more than twice and the specific heat is altered by about 22%.

26.9 Concluding Remarks

Tissue-equivalent electromagnetic phantoms offer a means to measure the extent of EM power deposition which manifests as the thermal power in the medium assessable through thermographic/thermometric methods. A variety of phantom materials have been designed for use, especially at radio and microwave frequencies. Electrical as well as physical/thermal properties of these materials have been evaluated.

One of the greatest problems associated in the practical use of these materials is maintaining the stable characteristics, both electrical and thermal. Presence of materials like water poses evaporation problems and growth of parasitic fungi is encouraged. Further, synthesizing phantoms to match the exact properties of biological substances is based more on trial and error basis rather than by algorithmic recipes.

In effect, phantoms are composite dielectrics. However, the heterogeneous nature of such materials is so complex, no simple analytic modeling is feasible. Development of appropriate algorithms using computer-aided design strategies, would lead to comprehensive and user friendly, ready-made mixture formulations. Also, conceiving new compositions with a variety of possible ingredients offers a wide scope for further research in EM phantom technology.

References

[1] B. Bianco et al.: Measurements of complex dielectric constant of human sera and erythrocytes. IEEE Trans. Instrum. Meas., vol. IM-28, 1979: 290-295.

[2] K. R. Foster and H. P. Schwan: Dielectric properties of tissue – A review, in *Handbook of Biological Effects of Electromagnetic Radiation.* (CRC Press, Cleveland, OH: 1986).

[3] M. A. Stuchly and S. S. Stuchly: Dielectric properties of biological substances – tabulated. J. Microwave Power, vol. 15, 1980: 19-26.

[4] R. Pethig: *Bioelectric and Electronic Properties of Biological Materials.* (John Wiley and Sons, New York: 1979).

[5] L. A. Geddes and L. E. Baker: The specific resistance of biological material – A compendium of data for the biomedical engineering and physiologist. Med. & Biol. Eng., vol. 5, 1967: 271-293.

[6] L. A. Klein and C. T. Swiff: An improved model for the dielectric constant of sea water at microwave frequencies. IEEE Trans. Antennas Propagat., vol. AP-25, 1977: 104-111.

[7] G. Hartsgrove, A. Kraszewski and A. Surowiec: Simulated biological materials for electromagnetic radiation absorption studies. Bioelectro Magnetics, vol. 8, 1987: 29-36.

[8] C. K. Chou, G. W. Chen, A. W. Guy and K. H. Luk: Formulas for preparing phantom muscle tissue at various radio frequencies. Bioelectromagnetics, vol. 5, 1984: 435-441.

[9] M. G. Bini, A. Ignesti, L. Millanta, R. Olmi, N. Rubino and R. Vanni: The polyacrylamide as a phantom material for electromagnetic hypertherma studies. IEEE Trans. Biomed. Eng., vol. BME- 31(3), 1984: 317-322.

[10] J. Leonand, K. R. Foster and T. W. Athey: Thermal properties of tissue equivalent phantom materials. IEEE Trans. Biomed. Eng., vol. BME-31(7), 1984: 533-536.

[11] O. P. Gandhi (Ed.): *Biological Effects and Medical Applications of Electromagnetic Energy*. (Prentice-Hall, Englewood Cliffs, NJ: 1990), pp. 165-169.

Defining Terms

EM phantoms: Phantoms fabricated with materials that simulate the dielectric properties of various biological tissues (such as fat, muscle, brain, bone, or a body fluid).

Nonionizing radiations: Electromagnetic radiations in the radio/microwave frequency ranges which do not cause ionization of molecules in their transit across a medium like air, unlike ionizing radiation like X-ray, etc.

Subject Index